The Bimolecular Lipid Membrane: A System

MAHENDRA K. JAIN
Department of Chemistry
Indiana University

 VAN NOSTRAND REINHOLD COMPANY
New York   Cincinnati   Toronto   London   Melbourne

Van Nostrand Reinhold Company Regional Offices:
New York   Cincinnati   Chicago   Millbrae   Dallas

Van Nostrand Reinhold Company International Offices:
London Toronto Melbourne

Library of Congress Catalog Card Number: 72-075099
ISBN: 0-442-24086-4

Manufactured in the United States of America

Published by Van Nostrand Reinhold Company
450 West 33rd Street, New York, N. Y. 10001

Published simultaneously in Canada by Van Nostrand
Reinhold Ltd.

15 14 13 12 11 10 9 8 7 6 5 4 3 2 1

Library of Congress Cataloging in Publication Data

Jain, Mahendra K.   1938–
    The bimolecular lipid membranes.

    Includes bibliographical references.
    1.   Membranes (Biology)   2.   Lipids.   I.   Title.
QH601.J35   574.8'75   72-075099

To
D I P T I
I
still remember
when she first walked

Preface

In the fashionable field of molecular biology, the first model at a molecular level to describe the *system properties* of a living system was probably the bimolecular lamellar membrane model. In some ways it has been disappointing in its limited scope. However, this simple model suggested by Gorter and Grendel, and, so well articulated by Davson and Danielli, has certainly been effectual in forming a tenuous bridge between the physical world and the biological world. Biomembranes are characterized by a multiplicity of phenomena which have been described at the macroscopic as well as the microscopic level. Nevertheless, there is in this breadth and multiplicity of phenomena an underlying set of physical principles.

An attempt is made in this book to present an exposition of these principles sufficient to provide a basis for understanding the specialized literature and/or more complex phenomena. Even in this effort, limitations imposed by space and the need for conciseness preclude the inclusion of several broad areas. Thus, the special situations considered herein are chosen on the basis of their frequent occurrence and particular interest. It is, therefore, necessary when reading this book to remember that it is written in an era in which the molecular aspects of biological organization are barely established, and that, in time, the detailed interpretive positions described in the following pages will probably be modified and improved.

A complete resumé of the literature on the function and structure of biological membranes would be monumental, confusing, contradictory, and of little aid at this stage. No attempt has been made to this end. However, for students of the biomembrane and its cognate fields, the treatment may be helpful not only to provide a general picture of the membrane and its complexities, but also for the information on membrane properties which is collected under one cover. A modest attempt has been made to go through the breadth of the subject, the border lines of which are, however, diffuse.

It is a great pleasure to acknowledge my indebtedness to Professor E. H. Cordes for his help and encouragement in all the phases of this work. I am particularly thankful to Dr. P. Shakespeare for critically reading and editing the manuscript from cover to cover. My thanks are also due to Drs. A. Strickholm, P. Mueller, H. Singh, A. Koch, B. Mehrotra, R. P. Rastogi, R. Apitz-Castro, I. Parikh, H. Mahler, K. Koehler, F. P. White, C. Zervos, B. Sears, P. Hamill, and E. Williams for pointing out quite a few errors and mistakes in various phases of this work. I also avail myself of this opportunity to thank various authors and publishers who extended their kind permission for reproducing various figures and tables. Last but not least, I should express my indebtedness to various women who kept me possessed and obsessed with various facets of life in all its shades.

Considerable help has been received from several friends and the editorial staff of Van Nostrand Reinhold Company, particularly Mr. George Narita; however, the residual errors of fact or fancy are assignable to the perversity and ignorance of the author. If this book turns out to be of any use, especially to the beginners and graduate students sequestered in various disciplines, the labor would be considered a success. Finally, any comments on any aspect of this book are welcome.

MAHENDRA KUMAR JAIN

Contents

1 / Introduction and Perspectives

The living cell is a dynamic system concerned with the processes of energy capture, transfer, and conversion. Each activity is complex, consisting of myriads of individual integrated biochemical reactions that collectively constitute the processes characterizing life, such as growth, maintenance, response, differentiation, and reproduction. It is the integration and regulation of these processes, according to conditions in the external environment and within the organism itself, which is essential for the functioning and survival of complex organisms. On purely theoretical grounds, it is evident that by its very nature, life in all its manifestations is a reflection of continuous exchange of matter between an organism and its surroundings. Indeed, each cell type is unique in terms of its structure and function, and its own characteristics are established by the architecture of its own components.

All living cells in nature are surrounded by one or another kind of extraneous membrane or jellylike coating. These coatings may be delicate or tough, thin or thick, soft or hard, elastic or plastic, rigid or viscous. These extraneous coats may be removed mechanically or by physicochemical treatments so that finally only a limiting surface remains. This surface is commonly termed the protoplasmic surface film (or membrane), and it cannot be removed without destroying the protoplasmic unit (the cell) or be torn without damaging the protoplasm.

In cells which have been denuded of this extraneous structure, various phenomena, such as shrinking of the cell in hypotonic solution and hindrance to movement of ions and molecules between cytoplasm and surroundings, suggest the presense of a boundary membrane which acts as a barrier to free diffusion.

1

This *protoplasmic membrane* is in fact ubiquitous in all living organisms and its universality attests to its early origin in evolution and implies an essential role as a functional structure in the survival and the development of the living systems.[1,2] It is no longer the ubiquity of membranes that is most interesting to biologists, but the unsolved problems of membrane structure and function.

The point of view employed here is that the functional interface between the intracellular and the extracellular compartments is a domain that we shall call the greater membrane;[3] it consists of a thin (~100 Å thick) inner zone which shall be hereafter referred to as biological, natural, or plasma membrane, and an outer zone, sometimes called the cell coat. The molecules in these zones dynamically interact with each other and with those of the intracellular and extracellular compartments. Morphologically, membranes separate the nucleus from the cytoplasm, the cellular constituents from the surrounding medium, and they also subdivide the cytoplasm into several different kinds of vesicles, such as mitochondria, lysosomes, endoplasmic reticulum etc. To facilitate the readers' grasp of terminology used in the text, simplified diagrams of a typical animal and plant cell are presented in Fig. 1-1.

When isolated and examined morphologically, each structure enclosed within a membrane is found to contain its characteristic portion of metabolic activities. It is generally believed that membranes play a fundamental role in processes associated with life. The most notable among these are: to provide physical and biochemical compartmentalization, to provide regions for organization of energy transduction mechanisms such as photoreceptors, and to act as sites for enzymatic reactions. Furthermore it appears that membranes play roles in regulatory functions such as protein synthesis, active and passive transport and consequent electrolyte metabolism, volume regulation, excitation, neural communication and probably information transduction, photosynthesis, oxidative metabolism, genetic regulation, cell division and differentiation, tissue adaptation, embryogenesis, morphogenesis, platelet reactions, coagulation, protoplasmic movements, pinocytosis, and phagocytosis. To focus attention on membrane-associated functions and the disturbances which attest to their existence, these functions and processes are collected in Tables 1-1 and 1-2. Some of the membrane functions and processes are linked with other vital processes, for example transport with metabolism. These functions sustain even under extremely adverse conditions in which the most primitive forms of life survive. The participation of membrane and associated structures in many diverse functions must have led to the establishment of a satisfactory set of mechanisms during the course of evolution. Indeed, the phenomenological description of various membrane-governed processes can be given in terms of relatively simple system properties of membranes; thus permeability may be associated with metabolism or energy transduction. As we shall see later on, such a conviction of unity in the diversity of membrane functions is not altogether unreasonable. The basic problem is however to correlate this diversity at the molecular level of structure and organization.

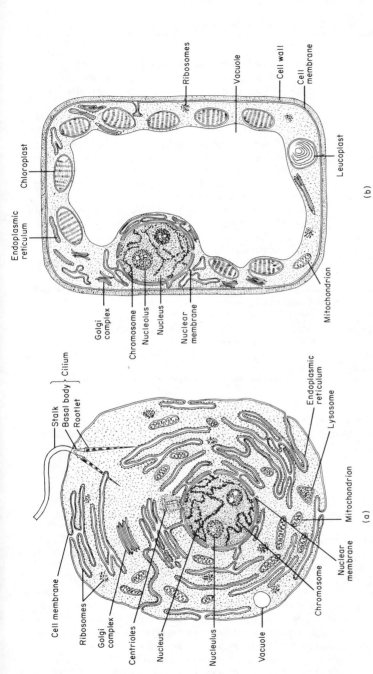

Fig. 1-1. Diagram of a 'typical' animal (a) and of a plant (b) cell illustrating relationships of the cell membrane and various organelles. The cell membrane is shown as a pair of lines separated by a light interzone. The invaginations of the cell surface (cf. animal cell) are indicated in several areas. Some of these endoplasmic reticular structures extend for a considerable distance into the cell. The nuclear membrane is composed of flattened sacs of the endoplasmic reticulum, and may be in continuity with cytoplasm. Golgi complex is shown here as modified ER but separated from cellular membrane. Mitochondria is shown with its cristae formed by invagination of its inner membrane. Chloroplasts have similar but not identical structure. Lysosomes represent a class of cellular organelles whose condition within a tissue may vary under different stimuli. More commonly found morphic forms of the lysosomes include: autophagic vacuoles or cytolysome, autolytic vacuoles, cytosome, residual bodies, multivasicular bodies, microbodies, storage granules or digestive vacuoles (phagolysome). Various organelles vary in form, size and in location within the cell. From (4).

TABLE 1-1 SOME FUNCTIONS OF BIOLOGICAL MEMBRANES

Membrane Type	Function
All plasma membranes	Compartmentalization, maintenance of osmotic balance, general and selective diffusion of metabolites and ions and their physiological regulation; active transport, phagocytosis, pinocytosis, and secretion
	Cell adhesion, aggregation; specificity, and plasticity of membrane contact
	Motility, cytoplasmic streaming, coalescence, contraction and expansion; translational and rotational movements of membrane-bound organelles
	Receptor for surface antigens, hormones, transmitters, and other biodynamic substances
	Limitation of organ growth; impaling and conveyance of cytoplasmic and luminal matrix; reversible geometric transformations between various forms, such as spherical, tubular; transformations from elastic to rigid forms, and *vice versa*. Formation of tissue—specific patterns with specialized functions (brain)
	Differentiation and dedifferentiation; DNA replication and RNA stabilization
Myelin	Electrical insulation
Nerve plasma membrane	Generation and conduction of nervous impulse
Sensory neurons	Sensory transduction
Chloroplast and mitochondrial membranes	Photo- and oxidative phosphorylation—energy transduction in general

Thus far we have not defined the term membrane, since the definition of the cell membrane is a matter of contention.[5] It is usually convenient to treat membranes as permeability barriers* in a functional sense.

---

*Some attention has been given to postulation of a membrane as permeability barrier; it has been suggested that cell membranes may not be universal rate-limiting barriers to the transport of water and other solutes between the cell interior and exterior; rather it is cytoplasm which offers the effective resistance to the movements of solutes.[6–9] Ling[6,7] has been particularly articulate in proposing the so-called association-induction hypothesis which emphasizes both the state of water in the living cells and the nature of adsorption of solutes, particularly electrolytes, in the cytoplasm. According to these views, solute movement, ion selectivity, volume regulation, active transport, etc., are considered to be integrative functions of protoplasm. This line of argument has not been given serious scrutiny by other investigators. However, in some tissues at least, there is little reason to undermine these alternative viewpoints. As we shall see later in our discussion of various membrane-associated phenomena, replacement of cytoplasm has very little effect on them. Thus, it has been assumed, and is further supported by other evidence, that the nature of protoplasm is of little importance for the situations with which we are concerned. This may not, however, be generally true.

TABLE 1-2 MALFUNCTIONS OF MEMBRANE[a]

Membrane Type	Malfunction and Consequence
Plasma membrane	Aberrations: lack of contact assembly appears to be the common feature of tumor cells as inferred from (a) asocial tissue organization, (b) abnormal host-parasite integration and altered cellular adhesiveness, (c) defective contact inhibition of movement and growth, abnormal electrophoretic mobility of the cell; abnormal fusion capacity, permeability; loss of intercellular communication
	Metabolic aberrations and disorders due to defective transport systems may characterize all membrane type and may sometimes be genetically determined, for example in hyperprolinemia, iminoglycinuria, cystinuria, Hartnup disease (tryptophan transport defect) result from one or more transport defects of kidney
	Increased permeability or active transport of Na results in spherocytosis of red cells
Nerve cell membrane	Defects in transport functions and ion metabolism result in many forms of psychoses and in cystic fibrosis
	Defects in membrane biosynthesis result in neurohypophysis and in multiple sclerosis (demyelinating disease)
	Accumulation of lipids and lipoproteins result in Rufsums disease
Mitochondria	Abnormal (generally reduced) oxidative phosphorylation results in defects in basal metabolism; abnormal fatty acid synthesis is encountered in all hepatomas
Endoplasmic reticulum	Morphologic aberration, abnormal behavior of membrane-bound polysomes, and lack of regulation of cholesterol synthesis all result in hepatic carcinogenesis
	Defects in hormonal functions
	Disturbance in transport mechanism lead to nephrogenic diabetes insipidus (water), renal glycosuria (glucose), resistant rickets (phosphorus), glycinuria (glycine), Fanconi syndrome (a general breakdown of transport mechanism)
Jejunal epithelial cell membranes	Abetalipoproteinemia—inborn error of lipid metabolism
Basement membranes	Morphologic alterations have been observed in a number of diseases including poststreptococca glomerulonephritis, nephrotic syndrome, diabetes mellitus, lupus erythemetosus and various others

[a]See also Refs. 1194 and 1195.

PERSPECTIVE

The molecular structure of biological membranes has thus far been a refractory problem. The very definition of membrane implies, to a first approximation, a spatial and temporal continuity. The major difficulty in approaching any direct study arises from the fact that the plasma membrane is too thin to be observed directly with the light microscope. Some visual evidence for the existence of plasma membranes is obtained in the elegant experiments in which naked starfish eggs stained with natural red are placed in an isotonic ammonium chloride solution.[10] The cytoplasm rapidly gives an alkaline reaction due to the selective penetration of un-ionized ammonia formed by the hydrolysis of part of ammonium chloride. If ammonium chloride is placed in the cytoplasm the whole cell becomes acidic due to selective diffusion of ammonia out of cell. As implied by this simple experiment, ions do not move in and out of the cell freely. It was the recognition of this phenomenon at a fairly early stage by Overton[11] that gave rise to identification and the operational definition of the plasma membrane as the component of the "greater membrane" (total structure of the boundary of the cell) which is mainly responsible for limitation and control of the free diffusion of solutes. The concept implies that the permeability process is the basic characteristic of the plasma membrane. In fact one of the first successful attempts to evaluate the chemical nature of the plasma membrane took this function into serious consideration (Table 1-3). Thus the invisibility of the membrane under the light microscope led to the adaptation of indirect methods for deciphering the structure of the plasma membrane and its functional significance. Our present concepts of the structure of the plasma membrane and other related membranes seem to have developed from the application of indirect techniques. The course of development as outlined in Table 1-3 may be considered in the following three stages: (a) postulation of lipidic nature of plasma membrane; (b) postulation of "paucimolecular" lamellar structure; and (c) postulation of "unit membrane" concept as a unifying hypothesis for membrane structure in general. In the following paragraphs we shall describe some of the historical observations which aided the development of these concepts.

The plasma membrane shows low permeability to ions and relatively high permeability to lipid-soluble substances. By the end of the last century this observation had been interpreted to suggest that the permeability barrier is mainly lipoidal in character, since nonpolar molecules have greater solubility in lipid solvents.[11] The first detailed ideas about the molecular organization of the cell membrane were based on an experiment of most elegant simplicity carried out by Gorter and Grendel.[12] They extracted lipids from a sample of a known number of erythrocytes. The benzene solution of the total lipids when dispersed over a trough partly filled with water, left a film of polar lipid molecules on the surface of the water. If an appropriate quantity of the lipid is dispersed, the film thus formed is only one molecule thick. When a thread was drawn across the surface from the edge of the trough, this surface layer was compressed, until at a certain quite sharply defined area the surface layer began to offer resistance for further compression. At this point the polar lipids presumably formed a close-packed monomolecular layer with their hydrocarbon chains sticking up in the air and their polar groups in the water. Resistance to further compression

TABLE 1-3 MEMBRANEOLOGY: A PERSPECTIVE

Date	Major Concepts	Main Events
1855	Nägeli suggests presence of membrane	Nägeli observed that the surface of unicellular plants, algae, fungi and mosses were impermeable to pigments present in exo- or endoplasmic solution
1895–1900	E. Overton suggests lipid nature of membrane	It is found that the rate of penetration of various compounds into plant cells parallel the partition coefficients of these compounds between oil and water
1917	I. Langmuir stressed propensity of lipids to form monolayers	Fats and fatty acids were shown to stack in parallel array by Van der Waals forces between hydrocarbon chains and by repulsion between the water and hydrocarbon chains
1925	Gorter and Grendel suggest that the plasma membrane is two molecules thick	Lipids were extracted from erythrocytes of several species of mammals; ratio of area of monolayers formed from this lipid to the surface area of the erythrocytes was $1 \cdot 8$ to 2
1935	Danielli and Davson suggest idealized plasma membrane as bimolecular lamellar lipid layer	According to this model, surface of living cell consists of a thin lipid layer with protein films adsorbed on the polar surfaces
1936–60	Many experimental advances essentially support the DD model	X-ray diffraction, optical polarization, electrical, surface chemical, permeability measurements seem to support DD-model; however, only generally speaking
1940–60	Evidence accumulates which appears to be inexplicable by DD-model	Development of sophisticated experimental techniques such as voltage clamp, microelectrodes, fast recording techniques, and isotope flux studies leading to characterization of transient and steady-state phenomena associated with membranes: action potential, active transport, etc.
1960	Electron microscopic studies by Robertson support and extend DD model; concept of unit membrane	Physicochemical considerations of electron micrographs suggest that tripartite structure arises from apolar structure separating polar ones
1961–62	Mueller, Rudin, Tien, and Wescott succeed in making black lipid layer of DD-type	Even though BLM shows only static properties of biological membranes, it almost conclusively suggests need for modification of unit membrane model of Gorter-Grendel-Danielli-Davson-Robertson.
1963–	BLM proves to be an excellent tool	Several of the functions of biological membranes could be induced in modified BLM; these include action potential, interionic permselectivity, photovoltaic effect, active transport, etc.

TABLE 1-3 cont.

Date	Major Concepts	Main Events
1965–	Concept of mosaic membrane evolves	Freeze etching technique in electron microscopy shows presence of globular structures interspersed in bimolecular membrane; proteins in lipid matrix seem to be generally involved in membrane architecture

arises from the fact that such compression must either push some hydrocarbon chains down into the water, or bring some polar groups out of the aqueous phase (to be elaborated further in Chapter 2).

A comparison of the area of the compressed monolayer and the estimated area of erythrocytes from rabbit, pig, man, guinea pig, sheep, and goat showed that the ratio was almost 2 : 1 in all of these cases. It implies that the lipid is arranged at the cell surface in the form of a bimolecular leaflet. On the basis of these observations and in compliance with the physical behavior of aqueous lipid solutions, it was suggested that the hydrated polar groups are at the interfaces and the hydrocarbon portions of the molecules make up the interior of the membrane. Later studies have supported these results and conclusions, especially if the area of the lipids spread at the air/water interface is measured at low surface pressure (less than 3 dyne-cm^{-1}) to prevent excessive packing of the lipids.[13] Similarly, subsequent studies on birefringence of erythrocyte ghosts, nerve myelin sheath, and retinal rods have suggested that their membranes contain radially arranged lipid molecules and the layer is not more than a few molecules thick.[14,15] Many other experimental observations such as facile passage of erythrocytes into the interior of an oil layer[16-18] and low electrical conductivity of the cell membrane to direct and alternating currents[19-23] are also in accord with the proposed two-layer lipid structure of the plasma membrane. The permeability characteristics can also be satisfactorily explained by assuming that the primary permeability barrier to the passage of ions and molecules is lipid and that only the specialized regions permit the passage of specific ions and molecules.[24,25] Thus the solubility-diffusion mechanism for transport implied in the lipid bilayer model of the plasma membrane came into general usage. Whether or not Gorter and Grendel's estimate of a lipid layer two molecules thick around the cell is absolutely correct in the light of present day refinements in technique seems of less importance than the fact that the concept acted as a direct stimulus to much of the later work. Still more important, however, is the concept that the main underlying structural element of membranes is a bimolecular lipid leaflet.

One of the main objections to this concept was its inability to account for extremely small interfacial tension, about 0.2 dyne-cm^{-1}, measured at the lipid-water interface of the membrane.[26-28] In contrast, the interfacial tension at the lipid-water interface of a lipid droplet in salt solution is considerably larger—about 20–40 dyne-cm^{-1}, depending upon the type of lipid. These observations were reconciled at that time by assuming the presence of substances other than lipids, presumably proteins or polysaccharides, at the surface of the plasma membranes.[29-31] Thus a lamellar bimolecular lipid membrane was considered to be sandwiched between two layers of protein and/or polysaccharides (Fig. 1-2); this model has been known ever since as "pauci-molecular" or Davson-Danielli model. It was further postulated by these authors on the basis of the properties of proteins that the proteins associated with the membrane are spread as a fully opened macromolecular layer. Moreover, it was suggested that protein might also be incorporated in such a fashion as to interrupt the lipid layer in places, forming protein-lined aqueous pores across the lipid layer. Such a suggestion gave sufficient latitude to accommodate the observed data on hemolysis, effects of anesthetics and drugs, electrical and elastic properties of the unspecialized membranes, and with certain additional assumptions it enabled the explanation of some of the specialized electrical and transport properties.[31]

The paucimolecular model for the structure of the plasma membrane has been quite successful in the sense that it has withstood the test of time and no significant rival theory or hypothesis has gained momentum during the subsequent years. Needless to say, the model originally proposed by Davson and Danielli[29-31] has undergone some major modifications. With the advent of electron microscopy of thin sections of tissues, the problem of extent and nature of membrane structure became subject to direct examination. Through a

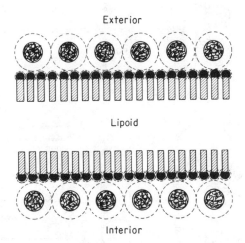

Fig. 1-2. Schematic drawing of 'molecular conditions' of the cell membrane as proposed by Danielli and Davson. From (31).

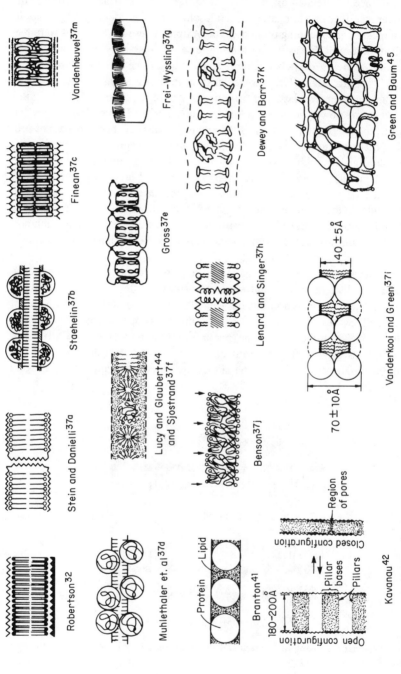

Fig. 1-3. Schematic representation of various models for biomembranes. It may be noted that these proposals were made with several different but somewhat specific aspects in perspective.

skillful interpretation of electron microscope and X-ray diffraction data, the paucimolecular model has been elaborated to a general concept of a 'unit membrane' structure.[32-34] In its original form it suggests that there is one basic structure to which all membranes or most portions of all membranes of all cells and of all species conform. The basic structure is assumed to correspond to that postulated by the paucimolecular model. However, the different chemical reactivities of the outside and inside strata of the membrane to fixing agents used in electron microscopy has led to the postulate that the membrane is chemically asymmetric (Fig. 1-3). Electron micrographs of red cell membrane is shown in Fig. 1-4. In simplified form, the interpretation of dense lines separated by less dense lines rests on the assumption that only the polar parts of membranes are accessible to polar 'staining' reagents, which show up as dark lines. Unstained portions are assumed to be due to hydrophobic portions of lipid molecules. Generally speaking the unit-membrane concept has gathered considerable support not only from electron microscopy but also from X-ray diffraction studies on biological as well as artificial model lipid systems. Although direct evidence in support of this model is still lacking, and specific details of molecular architecture need to be elaborated, it may be pertinent to discuss the role played by X-ray diffraction and electron-microscopic techniques in elaborating the presence of membranes at the boundary of cells and subcellular organelles. We shall discuss the structure of myelin as an illustration.

Myelin is a multilayered structure surrounding single axons of peripheral nerves in a variety of organisms. The multilayered structures seen in electron micrographs (Fig. 1-5) are formed by spiraled infolding of the surrounding Schwann cell membranes. In unmyelinated nerves only one layer of the Schwann cell plasma membrane covers the axon while the myelinated axons are sur-

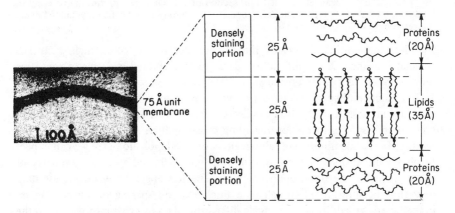

Fig. 1-4. Electron micrograph of the plasma membrane from erythrocyte. The schematic diagram shows the interpretation as generally developed by Gorter-Grendel-Davson-Danielli-Robertson (see text). Other interpretations are given in Fig. 1-3.

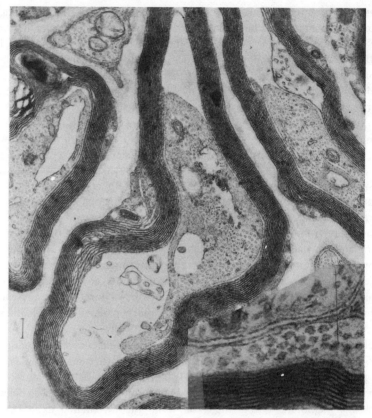

Fig. 1-5. EM of a negatively stained thin section of cat sciatic nerve, showing an array of "railroad tracks" of "unit" plasma membranes surrounding the so-called myelinated axon. Bar represents 1500 Å. *Courtesy H. Cohen.*

rounded by multiple layers of the plasma membranes of a single cell. The original low-angle X-ray diffraction data was interpreted in terms of repeating units containing two bimolecular leaflets of aggregated lipids approximately 140 Å thick interspersed with a protein layer of approximately 25 Å.[35] However, recent X-ray diffraction studies on fresh, wet, and unfixed nerves have given more useful information regarding broad features of their molecular architecture. If the membrane contains regions of fixed stoichiometry we can expect to obtain dimensions of the repeating cell (unit cell) and also the symmetry or space group of the lipoprotein complex. The low-angle X-ray diffraction data may thus be interpreted to calculate thickness and spacing of adjacent membrane layers. In addition, large-angle diffraction data can give information in the range of 1 to 10 Å, mostly about the packing of lipid molecules in the membranes. Thus X-ray diffraction data and their interpretation can be compared with the electron-microscopic studies on the same preparation.

The high-resolution electron micrographs of myelin fixed in osmium tetroxide show a series of dense lines about 25 Å thick, separated by about 125 Å. Some of these tracks wiggle and weave, indicating dimples or wrinkles in the surface. Furthermore, significant differences may be noted from species to species and even from tissue to tissue. Occasionally an intraperiod line is seen which is much thinner, less dense, and more irregular than the major dense lines. When potassium permanganate is used as electron stain, an intraperiod line of thickness similar to that of the major dense lines is always seen.[34,36] The dimensions of the bilayer roughly correspond to that of two phospholipid molecules joined tail-to-tail. This may be interpreted as experimental evidence for the presence of a bimolecular lipid layer as the basic unit for membrane architecture. The exact dimension of such structures remains uncertain due to a variety of factors to be discussed later. Nevertheless, some recent studies do suggest relationships among various features of the membrane substructure as observed in electron micrographs and their functional variation, such as the formation of "pits" or "pores" in digitonin-treated membranes. Thus, the results obtained from electron microscopic studies are in *general accord* with the conclusions arrived at from X-ray diffraction data.[37]

As has been pointed out earlier the weaknesses inherent in the methods of electron microscopy and the involved chemistry of staining make it difficult to come to conclusions concerning the exact molecular structures of biological membranes. Thus the method as such can not rule out possible variants in the unit membrane structure, for example, small areas of pure protein interrupting an otherwise continuous lipid sheet, or a netlike mosaic-type membrane of protein with lipid plaques enmeshed. The artifacts introduced by the preparative techniques employed in electron microscopy are not restricted to fine details; they also include artifacts of distortion arising from structural shrinkage. In both electron microscopy and X-ray diffraction studies, information is obtained about the membrane components which are organized in a more regular array, and no information is obtained about any irregularly arranged material which might compose as much as 20–30% of the membrane structure. In electron microscopy the specimen is fixed and stained with organic agents and heavy metals. The chemistry involved in these processes is not very well understood and some of the "stains" are not specific.[38] Osmium tetroxide, for example, interacts with a variety of proteins, with phospholipid polar groups, and with the double bonds of unsaturated hydrocarbon chains. Potassium permanganate, though it seems to interact less strongly with protein, is still a relatively non-specific electron stain. These observations imply that it is almost impossible to decide which component of the cell will be more "osmophilic" or otherwise. It is also possible that as a result of chemical change the hydrocarbon chains may become reoriented to some extent during fixation to bring their unsaturated groups to the surface of the lamella, with the osmium forming crosslinks between them, and then reorientation of the newly formed polar groups toward the aqueous phase.[39] If the reaction with the electron stain takes place at the aqueous interface, a layer of lipoprotein may produce a similar profile no matter what the arrangement of lipid and protein may be within the layer. Thus the constituent molecules and their relative positions may undergo considerable rearrangement, and the electron micrographs would thus fail to give precise information about their actual orientation. Indeed, evidence to this effect can be inferred from the X-ray diffraction pattern of a specimen fixed with osmium tetroxide.[40] It was shown that after a lamellar phase system reacted with osmium tetroxide the repeat pattern of the lamellar structure was preserved, although the 4.5 Å band due to orientation of the chains disappears completely. It is assumed therefore that after fixation the lamellae are held together by relatively few crosslinks, which are formed between fatty acid chains on the opposite sides of the bimolecular layers, and that the packing of the remaining fatty acid chains must be severely disturbed. It is also conceivable that, in certain circumstances, the layer-forming property of lipids may be subordinated to a structural pattern in protein or polysaccharide, etc., without being noticed by any of these techniques. These various artifacts can be minimized by better fixers, improved embedding techniques such as use of epoxy resins, or sublimation of the embedding medium by use of *Vestopal*. However, there are other limitations and possible sources of error. Alterations may occur in thin sections during bombardment with electrons which can cause loss of lighter atoms. Another technical difficulty which arises in comparing sets of electron micrographs is due to inexactitude in sectioning, focussing and magnification.

Yet another technique in electron microscopy is the freeze etching technique. The specimen is frozen in *Freon* cooled by liquid nitrogen and fractured at that temperature. For electron microscopy the fractured surfaces are etched by vacuum sublimation at low temperatures. The results obtained from freeze-etched samples (Fig. 1-6) from a variety of plant and animal materials suggest that the plane of fracture is usually between the two halves of the unit membrane.[41] The faces usually show globular structures measuring up to 300 Å in diameter, interspersed on the hydrophobic surfaces. This has been interpreted to suggest that the globular sub-unit structures may be embedded in the membrane phase. It is however possible, though less likely, that these are surface structures. Further studies have shown that various lamellar lipid membranes, both natural and artificial, undergo similar mechanical rupture while in the vitrified frozen state.

It may be reasonably concluded at this stage of the discussion that the various measurements of membrane thickness do not detract from the concept of the plasma membrane as a bimolecular lipid layer. It may be emphasized that the detailed structure and the dynamic relationships can only be hinted at by electron microscopy. Even though the evidence for the unit membrane concept may be convincing, and the bimolecular lipid layer structure may be acceptable in relation to the plasma membranes, it may not be nearly so suited for some of the cytoplasmic membranes for which the concept is to be treated in a generalized form. Quantitatively, small deviations from the bimolecular lamellar structure may produce disproportionately large effects in terms of cell physiology, and it is questionable whether such deviations can be detected by the existing techniques. See Refs. 1053–1056 for further discussion.

The immediate question posed even by the limited acceptance of Gorter-Grendel-Danielli-Robertson unit-membrane concept concerns the nature of interactions between proteins and polar lipids. It was originally suggested that the extended polypetide chains of the protein lie along the surface of the lipid leaflet in such a way as to intersperse its short nonpolar side chains between the chains of the lipid. This can be accomplished in a variety of ways. For example proteins may be held by electrostatic interactions or hydrogen bonding to suitable membrane components, or their nonpolar chains may extend into the lipid to make hydrophobic interactions. Also the protein may assume a conformation in which the nonpolar groups are at the surface and the polar groups are buried inside the structure, thus enabling the macromolecules to "dissolve" in the nonpolar phase of the membrane. It is becoming increasingly evident that the catalytic properties of the membrane-bound proteins may be manifested in a number of membrane phenomena such as active transport, sensory transduction, excitability, etc., wherein the proteins appear to be coupled to permeability characteristics of the membrane. Thus protein may be incorporated in the membrane either as such, without any change in the structure (even the quaternary), or with complete loss of secondary and higher order structures depending upon its function. However, little is known of membrane-associated protein structure except for the inferential and indirect conclusions based on the solution behavior of proteins and macromolecules in general. Thus our concept of membrane-associated proteins and their function is at present in a state of flux, particularly since most of the "extrapolations" cannot account for the multiplicity of functions associated with biological membranes. In fact, the structure-function interrelation of the biological membrane suggests that it is a far more complex system (rather than an entity) than is suggested by and embodied in various primarily static structures and models. The departure from the static models and concepts results while considering the entire spectrum of known and probable functions and actions of membranes together with the physical and colloid chemistry of their constituent molecules. Indeed, a dynamic picture of membrane structure and function has been proposed.[42,43] Similar views are germane in micelle ⇌ bilayer transitions advocated by Lucy.[44] According to these views, the biological membrane is an exceedingly complex (essentially) lipid structure, which interconverts between several different configurations in the course of different membrane functions. The structure showing two dense lines enclosing a less dense line in electron micrographs is proposed to be but

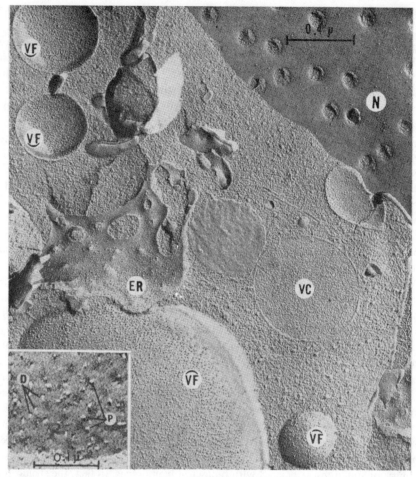

Fig. 1.6(a)

Fig. 1.6. (a) Freeze-etch views of various cell membranes in an onion root-tip cell. Both concave faces (VF) and convex faces (V̂F) of vacuolar membranes have been exposed, as well as the nuclear membrane (N) and endoplasmic reticulum (ER). Inset shows membrane particles (P) and depressions (D) from which some particles have been fractured.[41] (b) Part of exposed surface of an axon membrane surrounded by a sheath. Bands of protusions extend across the axon surface arranged in a regular lattice with unit cell dimensions of approximately 115–150 Å. Magnification marker is 1000 Å This electron micrograph was obtained from a freeze-etching replica of a fracture surface in the region just below the basal lamina of the photoreceptor epithelium in *Deilephila elpenor*, and the lattice seems to be specific to the electrically excitable part of the photoreceptor cell.[1047] (c) Molecular configurations which could best account for the freeze-etch results. Particles seen in the fracture faces of frozen membranes are interpreted as (*1–3*) lipid micelles in equilibrium with a lipid bilayer, or (*4*) protein elements in the hydrophobic interior of the membrane. See Refs. 1033 and 1034 for further discussion.

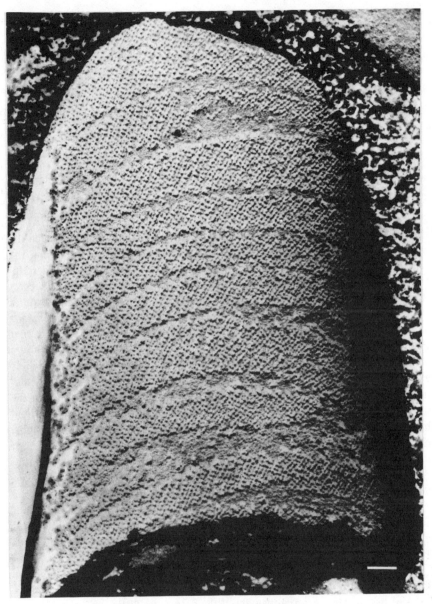

Fig. 1.6(b)

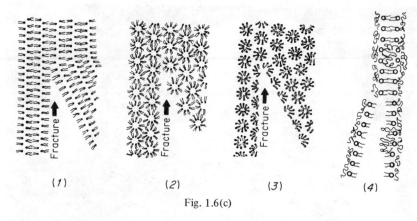

(1)   (2)   (3)   (4)

Fig. 1.6(c)

the desiccated remnant of the most stable of these configurations to which the membrane transforms automatically in the course of dehydration during the initial preparative treatment. However, these qualitative and essentially descriptive models relating membrane structure with functions have had little success thus far. It is in its general form that the unit-membrane concept constitutes a broad unifying theme in membrane morphology and is thus the common starting point for most descriptions of the plasma membrane. It is in this respect that the model derived from it is fairly successful as a concept designed to give a static profile of membrane structure. Inherent in the above concept is the whole experience of macromolecular organization for which the tacit working hypothesis is that such systems are built up from repeating units. Although the original unit-membrane hypothesis was formulated and formalized with a lipid bilayer structure as the core, the new chemical, structural, and functional evidence requires molecules other than lipids, and, consequently, other forms of molecular aggregation must also be considered. In this wider context, the concept of "elementary particles" as applied to the repeating unit in the inner-membrane of mitochondria,[45] and of quantasomes in chloroplasts can be incorporated.* The success of this *generalized concept* lies in the fact that it allows for the activity, specificity, and adaptability attributed to the biological membrane in both its static structural aspects and dynamic functional aspects. The details, however, remain rather obscure.

It follows from the foregoing discussion that in spite of the growth in our knowledge regarding function and composition of biological membranes and in spite of having some general idea about their structure, little can be said by way of correlation of structure with function at the molecular level. There are cer-

---

*A detailed discussion of this theme of membrane structure is presented in Chapters 7 and 9.

tainly a number of reasons for this ignorance. It is obvious from the foregoing discussion that the direct study of structure and function in living systems is complicated due to the nature of the system: the complexity of the molecular organization and the inapplicability of conventional methods of structure determination. The membrane components tend to be insoluble and therefore difficult to characterize. More generally, membrane phenomena occur in or on a highly structured matrix and are therefore not amenable to study by standard techniques of solution chemistry. Based on such observations some workers have looked upon the membrane as a highly organized inactive structure. In these cases, functionally important centers may be too few or too transient to induce any structural reorganization or periodicity detectable by physicochemical measurements which depend upon averaging techniques. Thus it is not surprising that the physicochemical techniques and the analytical studies have hardly given any information on the functional characteristics and the structural correlates of biological membranes at the molecular level. However, see Chapter 9 for further discussion.

In view of the shortcomings of the methods and techniques described thus far, there have been extensive but not always definitive attempts to find suitable models to describe the behavior of the membrane. The words "models" and "analogs" have been used often and in many connotations, and generally they are more illustrative than descriptive. From this viewpoint both analog and molecular models have been brought forward to describe membrane phenomena dealing with motility, excitability, ion transport, adsorption, adhesion, and several other biochemical, pharmacological, and physiological processes. Analog models, for example, electrical circuitry, mechanical dashpots and springs, ion-exchange resins, etc., have been useful in describing certain aspects of the biologically functioning membrane. In this respect artificial membranes showing any one of the properties of biological systems have been particularly useful. The molecular models in general have dealt with some specific property of the membrane components or systems as a whole, for example, the hydrophobic character of lipid chains or the core, charge transfer complexing capability, electrical double layer, etc. However, models in this latter class are based upon rather incomplete analogy, and thus generalization of results and conclusions is not always profitable. Most of these models[46,47] designed to elaborate the functional characteristics of biological membranes take into consideration interfaces between bulk lipid and aqueous solution[48] as in micellar and colloidal suspensions,[49] insoluble monomolecular layers of lipids at aqueous surfaces,[50-54] proteins and synthetic polymers at the air-water interface,[49,55] filter-supported liquid films in aqueous media,[50] and aqueous soap films.[56,57] In contrast, various other models emphasize bulk properties of the phospholipid-water system, such as those observed in supported lipid films,[55,58-60] polymerized lipid membranes,[61] and ion-exchange columns.[62] The more realistic of these approaches involve studies on isolated cell surface membranes,[63] reconstituted natural membranes,[64] and the theoretical models such as the dipole-aligned surface model[65] and the general equations for diffusion potential.[66] Besides all this, numerous studies on desorption, adsorption, coacervation, flocculation, dispersion, mobility, and surface area also pertain to membrane phenomenology. For a review of certain aspects of these see Ref. 67.

The experimental methods used to study these model systems vary from simple considerations such as temperature, salt, pH, and metal ion effects, to physical characterization by electrical measurements involving capacitance, resistance, and diffusion potentials. The more elaborate studies involve measurement of electroosmosis and permeability of water, ions, and other solutes. The theory applied for the interpretation of these experimental findings often involves thermodynamic and kinetic treatments extrapolated from bulk systems. Thus the conclusions derived from some of these studies have been useful in relating membrane function to macroscopic structure. For example, the ionic permselectivity in transport processes can be related to ionic radii, degree and energy of hydration of ions and the fixed counter charge, and to the diameter of the pore through which ions may permeate.[68] Also see Chapter 6.

The complexities of biomembranes lead one to think that the models proposed for these membranes (Fig. 1-3) would have a very open-ended character. In fact these various models cannot be regarded as mutually exclusive*. Furthermore, these models are proposed to account for the specific properties of specific systems, for example nerve cell or mitochondria. Similarly, the main advantage of studying experimental model systems is that experiments with various simple systems are easily interpreted and often give an insight into factors that are important in more complex systems. However a note of caution is warranted in extrapolating the results of such simple systems to predict the behavior of biological membranes. The main role of models is not so much to explain and to predict—though ultimately these are the functions of science—as to polarize thinking, to establish dialectics, to pose sharp questions, and above all, lead to some radical, undreamed of, unifying concept. Thus various model studies have provided an additional inferential support to the experimental data obtained from biological membranes and have substantiated the "unit membrane" concept as the unifying theme in membraneology. However, these models and model studies have failed, in a rather strict sense, to meet the criteria of analogy to biological membranes in function and structure, if not in form.

It must be realized that the lack of understanding of the membrane results in part from some unique aspects of membrane physics, chemistry, and dynamics. Thus, it is worth noting that whereas model systems for membranes always ap-

---

*It may be reminiscent of the fable in which several blind persons tried to describe an elephant as they perceived through their remaining senses, but from one side only!

pear to be symmetrical, the plasma membrane of various cell types is commonly asymmetrical, not only with respect to its environment but also its structure and function. This important feature of the membrane, which is commonly ignored, concerns the symmetry element in the orientation of membrane constituents in the plane of the membrane. For example, the membrane is a highly structured environment and some reactions which occur within it may be like those in the solid state rather than in solution. Thus yet another variable, directionality or anisotropy of orientation of membrane constituents, and vectorial character of fluxes and coupled chemical reactions, must be considered.

It is evident from this discussion that there is no doubt that these models and analogs have been intellectually stimulating, but most of them lack the characteristics of a realistic model system. Obviously, analogies are fun and useful; however, one should be careful enough not to carry a wrong impression. It is also clear from the foregoing discussion that no single approach is suitable for correlating the structure of biological membranes with their functions. The following experimental approaches are now being tried to this end:

(1) Attempts to ascertain the molecular architecture of the membrane by the physicochemical means which have been used for the elucidation of the structure of macromolecules.
(2) Attempts to isolate various membrane components in order to determine their chemical nature and biochemical function.
(3) Kinetic studies of various processes on intact systems.
(4) Study of the reconstituted membrane.
(5) Search for realistic models, both theoretical and experimental.
(6) Study of genetic control of membrane components and function.

The diversity of these methods undoubtedly reflects the growing evidence that membrane structure, function, and composition vary considerably. They also emphasize the possibility of dynamic interrelations between the several membrane components, and explicitly deny the notion of a biological membrane which is spatially and temporally uniform. We cannot but hope for a single membrane model that will have both power and flexibility to contain most of the important facts. However, a generally acceptable explanation of biological membrane structure may be expected to provide a fundamental understanding of how different membranes carry out their respective functions.

Each of the methods listed above is capable of supplying information which is unique in itself, and each one of them may be developed as far as possible, both, for use within its particular scope and as a likely component of broader generalizations. Consequently, none of the approaches by itself is able to provide a total picture of membrane dynamics. Thus the analytical approaches lead to a dissection of the system into its component parts and to the mode of their operation individually, a method well tested in biochemistry. The synthetic approaches, such as reconstitution studies, emphasize the interaction of various individual components and their specific functions. Use of genetic mutants, of immunological methods to identify components in the functional system, and the reconstitution experiments are some of the rather unconventional methods of great promise. In gene mutation for example, removal of a given protein leads to alteration of tissue function, and the properties of the system can be studied in its absence. Such studies have opened a way to a deeper understanding of transport mechanisms, their specificity, their interdependence, and their isolation and characterization.

SCOPE OF THE TEXT

Much has been learned about what the membranes are and what they are not, and much remains unknown. This book is an attempt to consolidate the properties of the lipid lamellar membranes in terms of the known laws of physics and chemistry. Each of the following chapters is designed to summarize the present state of our knowledge in various sections of this field. The concepts developed from the study of various association colloids indicate that the bimolecular lamellar membrane structure is a consequence of the peculiar chemical structure of the constituent lipids (Chapter 2). This conclusion strongly suggests that under suitable conditions the aqueous lipid solutions or sols can undergo phase transitions to give the bimolecular lamellae or some such related phase which can be shown to have the lowest free energy, and is thus thermodynamically the most favored state of the lipids in aqueous solution (Chapter 3). Indeed, such an assertion has opened the way to one of the most significant experimental advances in this field. The bimolecular lamellar lipid membrane can be readily prepared as a diffusion barrier separating two aqueous compartments, each of which is open and accessible for sampling and medium control. Within certain limits the composition of the membrane may also be controlled. The experimental evidence suggests that the structure of this membrane corresponds to the Davson-Danielli model (Fig. 1-2). The properties of this system are discussed in Chapters 4 and 5. The properties of these black lipid membranes (BLM) can be modified by incorporation of various agents. The modified BLM bear strong resemblance to biological membranes; yet they are amenable to more direct experimental study, and rather well defined as far as the nature of the constituents is concerned (Chapter 6). Such studies give way to a clearer understanding of some of the functional aspects of biological membranes, although the correspondence is presumptive. In Chapters 7 and 8 two of the most fundamental functional characteristics of biological membranes, that is, energy transduction and excitation, are discussed. The phenomenology and functional aspects of BLM and biological membranes described in Chapters 3 through 8 give strong indications for the presence of hyperstructure arising from the inter-actions of the component molecules in the plane of the membrane. The specula-tions regarding the possibility and nature of such interactions and the resulting fine structure in the lipid-protein "unit membrane" are summarized in Chapter 9.

Obviously each of these topics could form the subject matter of a separate monograph, and in fact several of these are referred to in the text. However, the main theme throughout this text is to elaborate a reasonably coherent picture of membrane function and its structural correlates. There are numerous difficulties in undertaking such an endeavor. The information regarding struc-ture and function of membranes is scattered and the studies usually reported in the literature are done with some specific purpose, such as to find the perme-ability coefficient, etc., of the membrane for a particular group of solutes. Sometimes the experimental data is presented with regard to some theoretical model. Complex as the results of such experiments are, simplifying assumptions may often be made as to which specific variables may be neglected. Thus to re-trieve such information and to obtain insight into the significant features of membrane structure in general, is possible only to a limited extent. There are only a few theoretical models solely based on functional aspects and indepen-dent of most structural features. Such a theoretical treatment is most desirable if the origin of the variables involved can ultimately fit into the framework of the known molecular configurations in the biological membranes. Such theoretical treatments have their origin in the concepts established in physical chemistry, such as the adsorption isotherm, double layer theory, electrostatic and van der Waals interactions, diffusion equations, net flux equations, conservation equations, statistical mechanical expressions for mean force and frictional co-efficient of a molecule of a given species, the concept of electroneutrality, vanishing net electric current density, and so on. The concepts are derived from macroscopic variables. Although extrapolation to the molecular level is possible through statistical mechanics, the usefulness and relevance of these concepts in analyzing the functioning of the specific structural components at the molecular level, where individual, nonstatistical interactions are of importance, has not been demonstrated. In a similar manner, an electrostatic treatment based upon statistics of time-independent functions is also of little help in certain cases. Most statistical analyses of time-independent functions are based on the Boltz-man distribution. In cases where the number of identifiable and distinct sub-groups is markedly small, Boltzman statistical theory may break down. These situations are conceivable in the consideration of such membrane functions as those involving replication of DNA, phenomena exhibited by substances having activity in parts per million in small samples, receptor response, cases where the surface charge (dipole structure) is fixed and aligned to the membrane plane, or ordering and selection of a few macromolecules as probably occurs in certain permeability characteristics of biomembranes. Situations like these are not uncommon; hence, one must be discrete in extrapolation of even well-established physical concepts. We shall refrain from going into details of such considerations, and only the broad dynamic aspects which touch upon the membrane phenomenology will be mentioned in nonmathematical terms.

2/ The Surface Chemistry of Lipids

Before we proceed to discuss lamellar lipid membranes in detail, it is necessary to consider briefly the fundamental properties of lipids, both, in the *neat* phase and in aqueous solutions or suspensions (or sols). The present treatment is not exhaustive. However, it is intended that this discussion serve the purposes of those less sophisticated readers who wish to become familiar with the physico-chemical properties of lipids. For further details the reader should consult one of the standard works on physics and chemistry of interfaces.[68-73]

The plasma membrane is known to contain appreciable quantities of lipid. Indeed, lipids appear to be ubiquitous components of biological membranes (Table 2-1). By the end of the last century a substantial amount of experimental evidence had accumulated that suggested diversity of chemical structure of lipids, and their chemical constitution has now been established beyond reasonable doubt. Lipids are a heterogeneous collection of molecules. They are united by only one common characteristic: low solubility in water. The structures of some representative common lipids are given in Fig. 2-1. Chemically speaking, the term lipid is used in stricter sense; the term refers to an ester of a long chain fatty acid with simple alcohols, or their more complex relatives such as glycerol, choline, ethanolamine, saccharides etc. The more complex lipids are usually phosphoric acid esters of glycerides or mixed esters. Chemically, sterols, carotenoids, chlorophylls, etc., are not lipids (Fig. 2-1), but by virtue of their solubility properties they are frequently treated as lipids. Nevertheless, the lipids in this monograph are considered to be derivatives of fatty acids and related compounds.

The fatty acids, the basic units of lipid molecules, largely determine their

TABLE 2-1 ESTIMATES OF PERCENT COMPOSITIONS OF SOME MEMBRANE PREPARATIONS[a]

Preparation	Bovine Myelin	Retinal ROS	Human RBC	Liver Plasma Membrane	Guinea Pig Brain Synaptic Vesicles	Rat Liver Microsomes	Rabbit Muscle Microsomes	Bovine Heart Mitochondria	Bovine Kidney Mitochondria	Guinea Pig Kidney Mitochondria
P	22	59	60	60	66	62	54	76	76	86
TL	78	41	40	40	34	32	22	24	24	–
PL	33	27	24	26	28	25	11	22.5	22	14
NL	26	2	9.2	13	5.6	7.5	–	1.5	1.9	–
PC	7.5	13	6.7	8	11.5	12	8.3	9.3	8.8	–
PE	11.7	6.5	3.4	–	4.2	4.8	1.4	8.4	8.4	3.9
PS	7.1	2.5	2.4	–	3.3	2.1	Trace	Trace	Trace	Trace
PI	0.6	0.4	Trace	–	1.3	2.5	Trace	0.75	0.75	1.0
CL	–	0.4	–	–	Trace	–	–	4.3	4.2	3.1
SM	6.4	0.5	3.6	4.0	3.0	1.5	0.8	–	–	–
AP	10.3	Trace	2.4	–	–	–	0.5	–	–	–
PA	–	0.8	–	–	Trace	–	–	–	–	–
TG	–	–	Trace	Trace	Trace	3.5	–	–	–	–
GL	22.0	9.5	Trace	–	Trace	–	–	Trace	Trace	Trace
CS	17.2	–	–	–	Trace	–	–	–	–	–
S	3.5	–	–	–	Trace	–	–	–	–	–
C	17.0	2.0	9.2	13.0	5.6	4.0	–	0.24	1.2	–

[a]From Ref. 37k. Key: P, protein; TL, total lipid; PL, phospholipid; NL, neutral lipid; PC, phosphatidylcholine; PE, phosphotidylethanolamine; PS, phosphatidylserine; PI, phosphotidylinositol; CL, cardiolipid; SM, sphingomyelin; AP, acidic phosphotides; PA, phosphotidic acid; TG, triglyceride; GL, glycolipid; CS, cerebroside; S, sulfatide; C, cholesterol.

properties. Most lipids isolated from natural sources are derived from mono-carboxylic fatty acids possessing an unbranched chain of an even number of carbon atoms. Although present in large amounts as constituents of more complex lipids, fatty acids also occur free to a very limited extent. Minor quantities of fatty acids possessing branched chains or chains containing an odd number of carbon atoms do occur naturally. Unsaturated fatty acids with one or more double bonds, both conjugated and isolated, are the most common constituents of natural lipids. The neutral (uncharged) lipids derived from glycerol and fatty acids (commonly termed glycerides) are the most abundant fatty acid derivatives in animals and plants. However, glyceryl ethers are also known to occur naturally. The charged lipids are abundantly distributed as essential components of all natural membranes. Among these, the phosphorylated glycerides are most common, and are generally known as phospholipids. The ionic groups of other naturally occurring lipids are given in Table 2-2; usually more than one polar group (charged) is present in these lipids.

There are a minimum of seven classes of phospholipids. Each class contains members with hydrocarbon chains of different length, containing a variable number and position of double bonds. Some phospholipids can occur as more complex derivatives, such as plasmogens. It is evident that the total number of molecular species occurring in biological membranes is potentially large.

The amount and composition of the lipid mixture in biological membranes varies considerably from species to species, from tissue to tissue, and even from

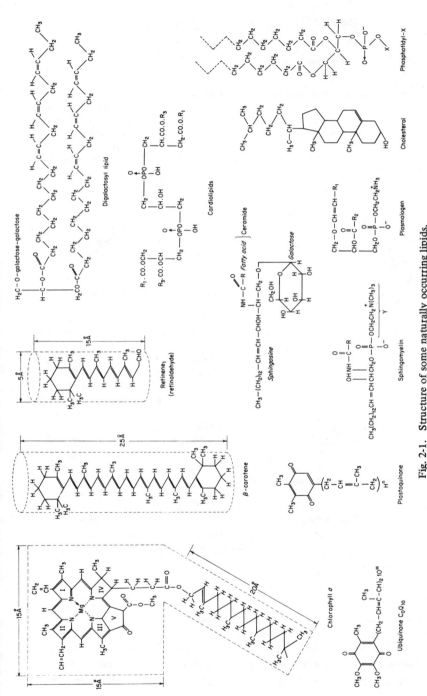

Fig. 2-1. Structure of some naturally occurring lipids.

TABLE 2-2 A CLASSIFICATION SCHEME FOR LIPIDS[a]

| | Substituents | | |
Lipid Type	Hydrophobic (Chain Length)	Hydrophilic (pK)	Composite Lipid
Fatty acid	Alkyl-(8–22 even)	COO⁻(4–5)	Fatty acids with various chain lengths, degree of unsaturation, branching and substituents.
Glyceride	Acyl-(1–3 alkyl chains)	Glycerol (12)	Mono-, di-, and triglycerides.
Glycerophospholipids	1,2-Diacyl-L-glycerol	PO_4 (1–2, 6–8)	Phosphatidic acid
	1,2-Diacyl-L-glycerol	$PO_4 + X$	Various phospholipids
		X = choline	Phosphatidyl choline (isoelectric 3–10)
		= ethanolamine (7.5)	Phosphatidyl ethanolamine
		= serine (2.2, 9.1)	Phosphatidyl serine
		= threonine (2.6, 10.4)	Phosphatidyl threonine
		= glycerol	Phosphatidyl glycerol
		= glycerophosphate	Phosphatidyl glycerophosphate
		= o-amino acid derivative of glycerol	O-amino acid ester of glycerol
		= inositol (myo-)	Monophosphoinositide
		= 4-phospho-inositol	Diphosphoinositide
		= 4,5-diphospho-inositol	Triphosphoinositide
		= inositoldimannose	Phosphatidyl (myo) inositoldimannoside
		= sulfosugar	
Sphingolipids	N-acylsphingosine	$PO_4 + Y$	
		Y = choline	Sphingomyelin
		= inositolglycoside	Phytosphingolipid
	N-acylsphingosine	Glucose	Cerebroside
		Oligosaccharide containing neuraminic acid	Gangliosides

[a]See also Ref. 76.

a membrane of one subcellular organelle to another within the same cell (Fig. 2-2, Table 2-1). Significant differences in lipid composition have been noted when the organism is reared under different conditions of diet and temperature. The characteristic patterns of lipid composition in different tissues and organelles have revealed many of the fundamental properties of the aggregated system which they form, and thus introduce some order into the con-

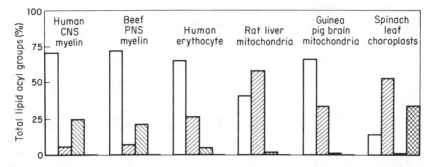

Fig. 2-2. Lipid acyl moieties in membranes. The proportions of lipid acyl moieties in each membrane are given in four bars under each membrane heading. The bars represent: ☐ medium chain fatty acids; ▨ polyunsaturated fatty acids; ◧ long-chain fatty acids; ▓ branched-chain moieties.[77]

sideration of the role of lipids in membrane structure and function. For example, the characteristic fatty acid patterns are sometimes altered owing to the occurrence of certain diseases, or owing to variation of growth environment such as electrolyte concentration, dietary conditions, growth temperature, or by genetic mutation. This is not surprising in view of the dynamic role of membranes in metabolism and other cellular functions. The biogenetic reasons and functional implications of this behavior are not completely understood. However, some correlation of the permeability, metabolic, and excitability characteristics of the membrane with their lipid composition have been made (see Refs. 74 and 75; several other specific examples are quoted in subsequent chapters). It appears that, among other factors, the nature of the polar group and the variation in the structure of the hydrocarbon chains are important determinants for molecular packing, permeability properties, lipid-protein interaction, and membrane stability and integrity of the unit membrane.

The most common structural feature of lipids is the presence of a large hydrophobic hydrocarbon chain containing ten to twenty-four carbon atoms in a straight chain (usually referred to as the *tail*), and a relatively polar hydrophilic group which can be nonionic, zwitterionic, or ionic (usually called the *head*). Physicochemical studies suggest that "staggered" bond configurations would be the most stable for the carbon-carbon bonds of the hydrocarbon chain; these being the ones in which the bonds radiating from adjacent atoms achieve the maximum interbond angles. Such an arrangement of carbon atoms in hydrocarbon chains results in a zig-zag configuration, and, as we shall discuss later, contact between such adjacent chains in close-packed structures leads to the most stable configuration for molecular aggregates.

The hydrocarbon chain, which forms the major portion of the lipid molecule, determines the solubility in nonpolar solvents, whereas the relatively polar group has the dual (*amphi-*) sympathy (*-pathi*) or liking (*-phile*) from which are derived the terms amphipath or amphiphile which are frequently employed in referring to these compounds. The amphipathic character of lipid molecules is determined both by the nature of the polar head as well as by the length, unsaturation, and stereochemistry of the hydrocarbon chain; such a duality in behavior is sometimes referred to as a hydrophil-lipophil balance (HLB).

GENERAL CHARACTERISTICS OF LIPID DISPERSIONS IN AQUEOUS SOLUTION

In aqueous solution amphipaths show some interesting properties. Liquid water has a markedly organized structure resulting from the strong mutual attraction of water molecules. Hydrophobic species do not interact favorably with water molecules and have the capacity to disrupt normal water structure and induce the formation of ice-like structure in their vicinity. This immobilization of water molecules is not favored entropy-wise and, hence, such species tend to be squeezed out of aqueous solution. This tendency will be opposed by the favorable interactions between the polar moiety of the lipid and the aqueous solvent. Thus, the balance of the hydrophilic and hydrophobic portions of the lipid molecules determines the propensity of the lipid molecules to stay in aqueous solution or to precipitate out. Simple triglycerides have a relatively small polar group and one which does not interact strongly with water; therefore, they show little solubility in aqueous solution. On the other hand, compounds such as alkali metal salts of carboxylic acids and lecithins which are capable of having fully ionized groups are more *soluble* in water when they exist in the ionized form. The amphipathic behavior of these various lipids is exhibited in their solutions, in which the individual molecules tend to aggregate. If two phases of contrasting dielectric constant are available, such as in water-air and water-oil biphases, these amphipaths tend to accumulate at the interface and form an adsorbed layer oriented so that the hydrophilic polar group stays in water and the hydrocarbon chains extend away from it occupying the phase of lower dielectric constant. Such organization results in "surface activity," from which is derived the term "surface-active agent" or simply "surfactant." The tendency of amphipaths to accumulate on the surface of the aqueous solution changes the surface contracting capacity of water,* i.e., the surface tension. Thus, depending upon the HLB

---

*The attractive and repulsive forces between the molecules constituting the bulk phase of the liquid are equal, that is the net force on any one molecule is zero in the interior of the phase. The atoms and molecules in the interfacial region, however, experience less complete interactions in the direction normal to the interface than they would have in the bulk phase. Thus the attraction away from the interface tends to reduce the number of molecules of the liquid in the interfacial region, and consequently the intermolecular distance increases. This pulling apart of molecules at the liquids requires work. Consequently, whatever the nature and strength of the intermolecular attractive forces, the energy of the molecules at the

of the amphipath, upon the differences in the dielectric constants of the media forming the interface, and upon the concentration of amphipaths in the aqueous medium, the surface tension of water may drop from about 70 dyne-cm^{-1} to almost zero. Thus interfacial characteristics of water are significantly altered by surfactants by altering molecular arrangement in the interfacial region. The overall process may be represented as:

$$\text{Surfactant in neat phase} \rightleftharpoons \text{Surfactant in bulk solution} \rightleftharpoons \text{Surfactant at the interface}$$

The equilibrium concentration of surfactant in these three states depends on surfactant concentration, its composition and structure, and on physical variables such as temperature, electrolyte concentration, etc.

Surfactants do not form solutions in the molecular sense; they show a strong tendency to aggregate whenever enough interfacial area is not available. Such aggregation results in orderly structures. In all these cases, the hydrocarbon chains orient such that they interact minimally with water; that is, they tend to "solvate" each other. Such association tends to damp thermal agitation of lipid and also tends to maximize the entropy of the surrounding water by liberating water molecules otherwise tightly organized about the hydrocarbon chains. This provides the major thermodynamic force stabilizing the aggregated structures. Various lines of experimental evidence have accumulated to support this argument.

It is implied in the preceding discussion that the orientation of amphipathic molecules at the interfaces between dissimilar dielectric media is a built-in property of their structure. Amphipaths realize their free energy minimum by ordering their chains away from the aqueous phase. It can be shown that the opposite arrangement in which the polar groups are oriented toward the less polar phase will be highly unfavorable. Consider a system in which the lipid molecules in the two states of orientation A and B are in equilibrium at a water-oil interface as shown below:

liquid interface is greater than the energy of molecules in bulk phase. Since, the free energy of a system tends to a minimum, the interface of two immiscible liquids will always tend to contract. It is usual to consider that free energy resides in the surface monolayer, although in some systems contributions from molecules of the second and third layers below the surface have been demonstrated. It may also be noted that at any ordinary temperature molecules are in rapid motion and the interfacial region is actually undergoing constant depletion and replenishment, as molecules move out of and into it.

State A has molecules with polar groups facing the aqueous phase while state B has the molecules with polar groups facing the oil phase. If a and b are the number of molecules in states A and B, respectively, the condition for equilibrium is:

$$K = b/a$$

Thus b/a is the ratio of the concentration of the molecules in states B and A, respectively. According to the Boltzmann distribution law:

$$a = A_1 \exp(-F_a/RT)$$
$$b = B_1 \exp(-F_b/RT)$$

where F_a and F_b are the free energies of the molecules in states A and B; A_1 and B_1 are constants for the system, and T and R have their usual meaning. Therefore,

$$K = b/a = B_1/A_1 \cdot \exp[(F_b - F_a)/RT]$$
$$= B_1/A_1 \cdot \exp(\Delta F/RT)$$

ΔF is the standard free energy change resulting when a given mole of amphipath changes its orientation from one state to the other. This quantity may be approximately computed. Assuming that in the reorientation of one fatty acid molecule, eight methylene residues are transferred from a predominantly non-aqueous phase to a predominantly aqueous region around the polar head groups of adjacent phospholipid molecules and that the head groups are transferred in the reverse direction. ΔF is roughly equal to the total work done in transferring such a hydrocarbon chain from oil to water (about 900 cal/mole per methyl group) and the polar group from water to oil (about 800 cal/mole for HO— group and about 5000 cal/mole for the carboxyl group). Based on the foregoing considerations and other simplifying assumptions, the energy for desorption of glycerol distearate has been calculated to be 17,000 cal/mole and the corresponding value of K is computed as about 10^{-13}. (See Ref. 78.) Whatever the exact value of K, the above approximation points out that the process of partition favors state A. Consequently, amphipaths in aqueous solution exist as aggregates which conform to the basic features of state A. The nature of the polar group and the hydrophobic chain determine the precise nature and shape of the aggregated structures. As described below, the structure and properties of such aggregates can be examined by several methods.

The physical properties of aqueous solutions of amphipaths show considerable deviation from ideal-solution behavior: some of these properties are osmotic pressure, vapor pressure, surface tension, refractive index, density, turbidity, contact angle, spreading coefficient, viscosity, and equivalent electrical conductivity. The deviation from ideal-solution behavior has been interpreted to suggest a considerable degree of association of amphipaths in aqueous solution. These observations coupled with various other direct studies (such as light scattering, birefringence, X-ray diffraction, electron microscopy, and electron diffraction) indicate that, depending upon the nature of the polar group, the medium, and the length and degree of unsaturation of the hydrocarbon chain, the amphipath can assume (Table 2-3) one or the other of the following forms through association: (*1*) emulsions and foam; (*2*) micelles; (*3*) insoluble monolayers at the air/water interface; (*4*) liquid crystals; (*5*) phospholipid dispersions in water; and (*6*) bimolecular lipid (lamellar) layers.

This classification points out that highly developed polymorphism is a common feature of lipid-water systems. A small variation in lipid concentration, temperature, and other external conditions may induce dramatic changes in the proportion of various forms of aggregates (phases) at equilibrium. Below a certain temperature, for example, micelles will not form; the hydrocarbon chains of amphipaths are not dispersed by thermal agitation and thus remain crystalline (neat phase). On increasing the temperature of lipid-water mixtures the hydrocarbon chains melt over a narrow temperature range (known as Kraft temperature), allowing the water to enter the crystalline lattice and disperse the lipid into some thermodynamically stable form(s) such as micellar solution. On re-

TABLE 2-3 BULK AND INTERFACIAL BEHAVIOR OF LIPIDS

Amphipath	Bulk Behavior	Surface Behavior
Nonpolar lipids: phytol and cholesterol esters	Insoluble in water; show polymorphism to a limited extent only	Negative spreading pressure
Polar lipids: cholesterol, fatty acids, triglycerides, carotenoids, mono acyl (C_{16-18}) and diacyl (C_{9-11}) phosphoglycerides, synthetic surfactants having one polar group attached to a single hydrocarbon chain	Slightly water soluble (?); in anyhydrous state show numerous phase transitions between crystalline and liquid phase	Form monolayers and emulsions; disrupt lamellar structures when used as mixed lipid
Monoglycerides, diglycerides, chlorophyll	Slightly water soluble (?); in anyhydrous state show numerous phase transitions between crystalline and liquid phase	Form lamellae, monolayers, and liquid crystals
Sulfatides, lysolecithin with medium-length chain, surfactants having two chains attached to one polar group	Fairly soluble; may form isotropic solutions in water	Form extended micelles and lamellar structures
Phospholipids with medium to long chain length, cerebrosides, gangliosides and other lipids having large hydrophobic groups and large hydrophilic groups	Fairly soluble; may form isotropic solution in water	Show very high degree of polymorphism; most of the phases consist of lamellar structures with large surface-to-volume ratio
Saponin, bile acids and salts, synthetic detergents with large polar groups	Fairly soluble; may form isotropic solutions; show little polymorphism in solid state	Mostly form micellas or emulsions and foams

cooling, the micellar solution may or may not revert back to the crystalline phase; there may be hysteresis. The major factors affecting the Kraft temperature are (a) the energy required for water to penetrate the ionic crystal lattice, and (b) the energy required to melt the hydrocarbon chains. The Kraft point is thus determined by the type of polar groups, the counterions in the aqueous medium, and the nature of the paraffin chain. The Kraft temperature rises with increasing chain length in a homologous series. The Kraft point for a cis unsaturated lipid is always below that of its trans isomer, and both are below that of a fully saturated homolog. Polar substitution or branching of the paraffin chain usually lowers the Kraft temperature. Finally, a mixture of two amphipaths may give a Kraft point which is intermediate between their respective temperatures. Kraft temperature of certain lipids are given in Table 2-4.

TABLE 2-4 PHASE TRANSITION TEMPERATURES (KRAFT POINTS) FOR LIPIDS

Lipid	Transition Temperature ($^{\circ}C$)	Phase	Method
Fatty acids			
C_{13}	−4	Monolayers	Molecular area
C_{14}	8	Monolayers	Molecular area
C_{15}	20	Monolayers	Molecular area
C_{16}	28	Monolayers	Molecular area
Alkylbenzene sulfonates (Na salt)			
methyl decyl-	19	Aqueous	Solubility
methyl dodecyl-	28	Aqueous	Solubility
methyl tetradecyl-	32–33	Aqueous	Solubility
methyl hexadecyl	45–46	Aqueous	Solubility
Fatty acids (Na soaps)			
C_{12}	26–28	Aqueous	Solubility
C_{14}	38–39	Aqueous	Solubility
C_{16}	50–51	Aqueous	Solubility
C_{18}	63–65	Aqueous	Solubility
Alkyl sulfates (Na salt)			
C_6	−16	Aqueous	Spectroscopic
C_8	−2	Aqueous	Spectroscopic
C_{10}	8–9	Aqueous	Spectroscopic
C_{12}	20	Aqueous	Spectroscopic
C_{14}	35–36	Aqueous	Spectroscopic
C_{16}	41–42	Aqueous	Spectroscopic
C_{18}	56–7	Aqueous	Spectroscopic
Phosphatidyl ethanolamines			
L- di C_{14} (saturated)	86	Neat phase	Sintering temperature and DTA
L- di C_{16} (saturated)	105	Neat phase	Sintering temperature and DTA
L- di C_{18} (saturated)	130–5	Neat phase	Sintering temperature and DTA
DL- di C_{16} (oleoyl)	about 42	Neat phase	Differential thermal analysis
Phosphatidyl cholines			
L- di C_{14} (saturated)	60–70	Neat phase	Conductivity, DTA, sintering, NMR
L di C_{16} (saturated)	75–80	Neat phase	Conductivity, DTA, sintering, NMR
L di C_{18} (saturated)	85–90	Neat phase	Conductivity, DTA, sintering, NMR
DL or L- di C_{16} (oleoyl)	Less than 0	Neat phase	Conductivity, DTA, sintering, NMR

Key. DTA, differential thermal analysis; NMR, nuclear magnetic resonance.

The properties of various phases present in lipid-water mixtures indicate that the built-in asymmetry of the amphipaths is manifested in the formation of various geometrically distinguishable structures. This is probably best illustrated by the relationship between molecular structure and the structure of the derived aggregates from those natural lipids containing two hydrophobic chains, such as lecithins, and those containing one such chain, such as lysolecithins. The former structures tend to form planar aggregates, such as structures (f) and (g) in Fig. 2-3, while the latter tend to form spherical micelles, structures (d) and (e) in Fig. 2-3. Nature makes good use of these tendencies; lecithins and related structures are very common membrane components, whereas in contrast, lysolecithins are not and they, in fact, *lyse* the membrane to micelles. This aggregation behavior and the shape of the aggregates is probably related to the shape of the monomer molecules itself. Lecithin is more or less cylindrical whereas lysolecithin is more or less conical.

The orientation of amphipaths in various aggregated systems is shown in Fig. 2-3. Most of these structures are anisotropic.* Because of a high degree of orientation of tightly packed molecules, the partial molar volume of the lipid molecules may sometimes exceed 90% of the total volume. The state of the molecules or substituents is probably best reflected in interfacial tension measurements. Interfacial tension is also a measure of free energy of formation of various interfaces. They are equal by definition. The interfacial free energies (or surface tensions) of some of the interfaces with which we shall be concerned are as follows:

Water/air interface	$72-73$ erg-cm^{-2}
Liquid decane/aqueous electrolyte	50
Cholesterol in decane/aqueous electrolyte	$40-45$
Triglyceride in decane/aqueous electrolyte	$20-30$
Phospholipid in decane/aqueous electrolyte	$7-15$
Phospholipid bilayer/aqueous electrolyte	$0.1-2$
Plasma membrane/aqueous electrolyte	about 0.1

These values provide good examples of the alteration of surface tension of a liquid by the addition of a second component. At the same time they are a measure of the packing of molecules in the surface layer. This phenomenon arises due to the tendency of the additive to adsorb onto the surface, the biphasic characteristics of which are suited for orientation. If γ is the surface tension of the adsorbed layer and γ_0 is that of the pure liquid:

$$\gamma_0 - \gamma = \pi = b - oa$$

*Periodically ordered structures may be grouped into three classes depending on whether the periodic repeat is in one, two or three dimensions. The first two classes are mesomorphic or liquid crystalline, as in monolayers, bilayers and dispersions of phospholipids in water. The third is the crystalline phase with three dimensional repeat units.

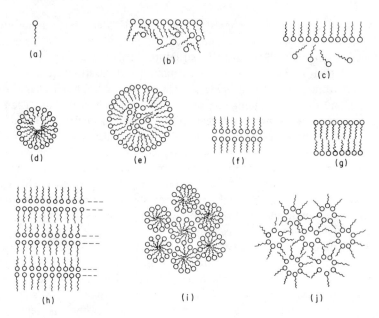

Fig. 2-3. Various aggregated forms of amphipaths: (a) lipid molecule; (b) bulk lipid; (c) lipid solution in water; (d) micelle; (e) emulsion; (f) lipid bilayer at air interfaces; (g) lipid bilayer at water interfaces; (h) myelinic, (i) hexagonal I, (j) hexagonal II phases of phospholipid dispersion.

The characteristics of these various systems may be summarized as follows.

Structural arrangement (type)	b	f and g	h and j	i	d and e	c
% Water* (approximate range)	0	5, 20-50	23-40	34-80	30-99.9	Greater than 99.9
Physical state	Crystalline	Liquid crystalline, lamellar	Liquid crystalline, face-centered cubic	Liquid crystalline, hexagonal compact	Micellar solution	Solution
Gross character	Opaque solid	Clear, fluid, moderately viscous	Clear, brittle, very viscous	Clear, viscous	Clear, fluid	Clear, fluid
Freedom of movement	None	2 directions	Possibly none	1 direction	No restrictions	No restrictions
Microscopic properties (crossed nicols)	Birefringent	Neat soap texture	Isotropic with angular bubbles	Middle soap texture	Isotropic with round bubbles	Isotropic
X-ray data	Ring pattern 3-6 Å	Diffuse halo at about 4.5 Å	Diffuse halo at about 4.5 Å	Diffuse halo at about 4.5 Å		
Structural order	3 dimensions	1 dimension	3 dimensions	2 dimensions	None	None

where π is called the *surface pressure*. A condensed film on a liquid or solid substrate exhibits a linear relationship between surface pressure and the mean area (σ) occupied per molecule in the adsorbed layer; a and b are constants. The relative magnitudes of γ and γ_0 have significant bearings on the behavior of a lipid in aqueous solution. Thus, if $\gamma_0 > \gamma$, the system forms aggregated structures which increase the area in which lipid molecules are oriented, such as micelles, bilayers, and phospholipid dispersions. If $\gamma < 0$, that is, when the surface free energy of the lipid-water interface is positive, the net effect on the interface would be such that expansion of the surface would lead to buckling or to spontaneous emulsification of the two separate phases.

Yet another aspect of the aggregate systems with which we are dealing lies in the orientation of polar groups in the aqueous solution. Since this region consists of net electrical charges, it would act as a shield for like charges, and would allow the opposite charges. This leads to a characteristic charge distribution profile, known as an electrical double layer. Its characteristics are described in the following section.

Electrical Double Layer

A difference in electrical potential usually forms at the boundary between two phases. The origin of such a potential difference lies in the transfer of electrical charge from one phase to another, that is, the transfer of ions and/or electrons on severing their bonds in one phase and reacting to form new bonds with the substance comprising the other phase. Similarly, fixed ionic groups at the interface would alter the relative mobilities of various ions across the interface thus giving rise to separation of charges. At temperatures close to absolute zero, the ionized groups of the fixed polar heads and the associated macromolecules (if any) would have their counterions associated with them at the interface. With increase of temperature the small ions would be shaken loose by thermal motion and would diffuse away from the interface. A double layer would thus arise: relatively fixed ions of one charge at the interface and the mobile ions of the opposite charge in the aqueous phase. Such an asymmetry of charge distribution could also arise at a neutral interface (such as that of alkane/water) provided one of the ions has significantly different mobility from the other ions in either of the phases. The presence of unequal amounts of mobile ions at the interface gives rise to an asymmetry of charge distribution and hence to a potential difference across the interface. At various points away from the interface, the electrical double layer gives rise to characteristic electrical potentials (Fig. 2-4).

The electrical double layer does not have sharply defined borders owing to a variety of factors, including the thermal motion of ions. Similarly, a concentration gradient exists in the vicinity of the double layer and the osmotic pressure of these ions would be the moving force for the destruction of a compact layer,

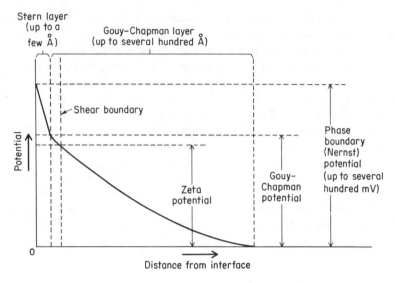

Fig. 2-4. A graphic representation of an electrical double layer. The fixed charges and counterions are not shown. The phase boundary potential is also known as the total interfacial potential. The shear boundary is shown arbitrarily in the Gouy-Chapman layer but could be at any distance beyond the Stern layer.

the osmotic pressure being higher in the double layer than in the bulk solution. Thus the concentration of counterions decreases as the distance from the interface increases. Due to ion-ion and ion-dipole interactions, the innermost counterions of the electrical double layer form a compact inner region known as the Stern layer. Generally, even at low ionic strengths, a considerable fraction of the counterions (roughly 60–85%) is found in this compact region. Due to high electrical field strength in the inner region of the double layer, the neighboring water molecules are polarized and immobilized (see Chapter 9 for further discussion). This reduces the dielectric constant in the immediate neighborhood of fixed ionic groups, and consequently the net charge of the interface region beyond the Stern layer is considerably smaller. In systems of high charge density and high ionic strength, or in which the interfacial pressure is not high, the counterions may penetrate into and beyond the surface defined by the positions of the fixed ionic groups.*

Beyond the Stern layer the distribution of the counterions obeys the Poisson-Boltzman distribution relationship. Consequently, the potential in such a diffuse double layer decreases from the Stern layer to infinity. The conglom-

---

*Distances from the interface are difficult to define exactly. Normally such a distance is referred to a tangential plane covering the charged polar groups. However, the effective radius of the polar groups, their degree of hydration, and their orientation at the interface may not be the same as in bulk solution.

eration of ions in this layer is known as the *ionic atmosphere*. It has the net charge of the counterions, which is equal in magnitude but opposite in sign to the net charge at the Stern layer.* The principle significance of double layers is that they produce potential barriers which may strongly assist or inhibit the motion of charged species across the interface. Consequently, double layers are often the most important factors determining the rate of an interfacial reaction, position of adsorption equilibria, or establishment of the electrical nature of a contact, thereby influencing the mass or charge transfer across such interfaces.

An electrostatic potential at a point is measured by the work (against electrostatic forces) which is required to bring a unit charge from infinity to that point. In the case of a diffuse double layer the electric field intensity falls rapidly to zero as distance from the interface increases. By a variety of methods, at several interfaces it is possible to discern substructure in the diffuse layer; several types of electrical potentials and consequent interfacial phenomena have thus been characterized. Referring to Fig. 2-5, the potentials at three points in the double layer are of special interest. The potential at the interface itself is the measure of total potential of the double layer; this phase boundary potential is called the Nernst potential. Similarly, the potential between the Stern and diffuse layers is known as the Gouy-Chapman potential. The third interesting point is in the diffuse layer at the boundary between the solvent adhering to the interface and that which can move with respect to the interface. This plane of shear or slipping plane essentially separates the water of hydration from free water. The potential at this point is called the *zeta* potential. In most cases, the plane of shear is only a few angstroms thick and, therefore, the zeta potential and Gouy-Chapman potential are almost identical.

In an external electric field, such charged interfaces show a variety of phenomena, collectively known as *electrokinetic phenomena*, due to interaction of an externally applied potential with the interfacial potentials. The term electrokinetic clearly connotes electricity with motion; in principle, it covers all phenomena involving both motion and electricity and could be synonymous with electric transport phenomena, although this term is misleading. In their simplest form electrokinetic phenomena can be demonstrated as follows. When an electric field is applied along the axis of a charged capillary, a force will arise which

---

*A necessary aspect of ionic double layer and other models relating charge profile to distance from the interface lies in the equilibrium which exists between the charged species. This does not necessarily imply that the electrical behavior of the double layer of charge is purely capacitative and has no ohmic leakage resistance component in parallel with it corresponding to a movement of ions. This requirement is an idealization since under most practical conditions, any charged interface will have at least a very small current passing across it corresponding to the passage of some net charge across a membrane. This would imply that the ions in the compact Stern layer are in dynamic equilibrium with those in the diffuse double layer, and the ions in these two layers therefore exchange with those in the bulk medium.

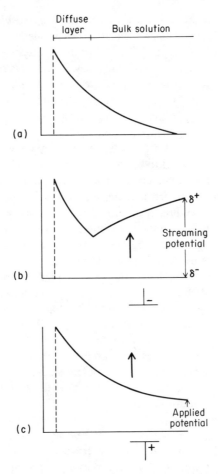

Fig. 2-5. Schematic distribution of charges and potential at a fixed negative charge-electro-lyte interface under three sets of conditions: (a) when no external perturbation is present—a simple double layer profile; (b) when liquid is set in motion with respect to a fixed charged surface, potential difference (streaming potential) sets in; (c) when a potential is applied perpendicular to the double layer, a flow of liquid sets in—electroosmosis. The direction of the arrow points in the direction of liquid flow. If the interface is not fixed an electric field would cause its migration toward the electrode, giving rise to electrophoresis or cataphoresis (not shown here). In the reverse effect, an electrophoretic potential is observed when suspended particles are set in motion in a liquid under the influence, say of gravity or of ultrasonic oscillations. Thus, the interrelationship of these various phenomena is discernible in terms of the properties of the diffuse layer and the mobility of the interface.

will tend to displace the charge of the liquid (and consequently the liquid itself) within the capillary with respect to its walls, thus bringing about electro-osmosis as shown in Figure 2-5.

Other related phenomena such as streaming potential (generated by a flow of liquid in a charged capillary), sedimentation potential (associated with a charge displacement on a particle along the direction of sedimentation), and electrophoresis (migration of charged particles under the influence of an electrical field) are known. The electrokinetic phenomena originate from the fact that, although the diffuse double layer is bound to the charged surface by electrostatic forces, the main part of this layer exists within the bulk solvent and is capable of moving along the interface. This motion along the interface may be caused either by electrical forces acting on the charges or by mechanical forces (such as drag) acting on the liquid phase and the charged interface. We have noted earlier that the plane of shear delimits this mobile part, hence the potential at this plane (zeta potential) determines the electrokinetic phenomena. Thus electrokinetic phenomena may be used for determination of zeta potentials. For instance, small charged particles suspended in an electrolytic solution move under the influence of an applied potential gradient. From the measurement of the resulting velocity of motion, the zeta potential may be computed. Similarly, if the charged interface can be held stationary, the liquid becomes the easily observed mobile phase and from its flow-rate the zeta potential can be computed. If both liquid medium and particulate material are held fixed, the pressure developed in the system may be used.

SPECIFIC CHARACTERISTICS OF LIPID AGGREGATES

Thus far we have considered some of the general characteristics of interfaces elaborated by amphipath-water aggregate systems. In fact these general characteristics of the interfaces derived from diverse theoretical and experimental approaches have been particularly useful in elaborating the nature of various aggregated phase (for reviews see Refs. 56, 57, 68–73, 79, 80). A brief description of the salient features of various aggregated amphipath-water phases (or more appropriately forms) is presented in the remainder of this chapter.

Emulsions and Foams

Foams are dispersions of a relatively large volume of gas in a relatively small volume of liquid or solid. In a true foam, the bubbles are so crowded that their shape is polyhedral; liquid or solid is reduced to thin films separating the polyhedra. When the volume of liquid is considerably greater than that of gas, the gas bubbles are, as a rule, spherical and their mutual interaction is weak. Foam may also be treated as a gas emulsion; however, the term emulsion is usually reserved for temporary or permanent dispersions of oil or other hydrophobic substances (termed the discrete or dispersed phase) in water (termed the continuous phase) or vice versa. The shape of the dispersed globules or droplets at low concentrations of the dispersed phase is generally spherical (since the surface area tends to be a minimum), but when the concentration of these droplets is so high that they influence each other, they may cease to be spherical. The diameter of droplets of the dispersed phase depends mainly upon the interfacial tension between the medium and the dispersion and the presence of electrolyte. It is usually in the range of 0.2 to 5 microns, although in theory these diameters may have almost any magnitude. These limits are not completely due to chance; droplets considerably smaller than 0.2 micron (μ) must have greater solubility (due to larger area) in the continuous phase than the larger drops. Thus the former tend to disappear and their substance enlarges the latter. Drops considerably greater than 5 μ sediment or cream too rapidly to form a significant fraction of the total population.

A closely packed assembly of spheres of equal sizes has a maximum volume $\pi/\sqrt[3]{2}$ ($\simeq 3/4$) of the total volume, which therefore sets the upper limit on the partial volume of a dispersed phase that can be accommodated in this manner. The size of the globules formed by the addition of the emulsifier is sometimes so small (less than 0.1 μ) that the emulsions are optically transparent and in this respect they resemble isotropic micellar solutions of pure lipids. The fundamental difference between the micro- and macroemulsions is that the latter have small but finite positive interfacial tensions leading ultimately to coalescence, whereas the former have negative interfacial tensions and tend toward greater dispersion and permanent separation.

Structurally, emulsions are characterized by a liquid lipid phase bound together by an organized layer of amphiphilic molecules (Fig. 2-3). Their structure resembles the structures of micelles in several important respects. A well-known example of a foam is soap suds and of an emulsion is milk.

Micelles

In a solution (approximately 10^{-5} M or more) ionic amphipaths undergo association-dissociation equilibria. The associated polymeric species are generally termed micelles*. The average time which a monomer spends in a micellar phase is of the order of 10^{-10} sec or less.[1197] Two important parameters which characterize ionic micelles in dilute solutions of surfactants are the aggregation number (the number of surfactant molecules which make up the micelle) and the degree of dissociation (the extent to which the surfactant molecules in the micelle are ionized). The formation of micelles in a solution of surfactant is characterized by sharply defined changes in several properties of the solution as illustrated in Fig. 2-6. The concentration of the surfactant at which limit of molecular solubility is reached is termed the *critical micelle concentration* (c.m.c.). Above this concentration, aggregation of the molecules into spherical or rod-shaped particles

---

*In the recent literature the term micelle is usually employed to refer to most of the associated structures of amphipathic molecules. However, for this discussion we shall adhere to the more conventional definition implied in the structure drawn for a micelle in Fig. 2-3.

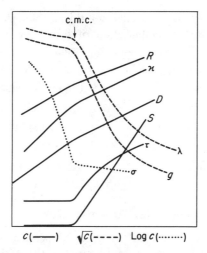

Fig. 2-6. Typical variations in the physical properties of an ionic association colloid. Characteristic plots shown are R, refractive index; D, density; κ, specific conductance; τ, turbidity; S, solubility of a water-insoluble dye such as Orange OT (all against the concentration c of the association colloid); g, osmotic coefficient; λ, equivalent conductance (against $\sqrt{c}$); σ, surface tension (against $\log c$).[79]

occurs; the hydrophobic tails aggregate and form a particle with hydrocarbon chains in the center and the ionic groups of the molecules projecting into the aqueous phase. Both the hydration and charge of the ionic heads of the molecule prevent coalescence of micelles and the separation into two phases.

The micelles are of colloidal dimensions, formed by association of usually 2 to 150 monomers and a smaller number of counterions. The micelles of nonionic amphipaths do not have any counterions. Generally speaking, micelles are spherical and physical studies suggest that the hydrocarbon chains in the core have a relatively large degree of freedom of motion. This is consistent with the observation that the monomers in the micelle are in rapid equilibrium with the molecules in the free solution. Micelles are capable of solubilizing other amphipathic and hydrophobic substances to form co- and mixed micelles. Nuclear magnetic resonance and electron spin resonance studies suggest the dynamic nature of this solubilization in which the probe (or the substrate) preserves a random spatial orientation and experiences a relatively polar time-averaged environment. This may not necessarily be true for the substrates which form micelles of their own.

In ionic micelles, the surface and electrokinetic potentials are substantial. These systems seem to be excellent models for the study of various aspects of interfacial phenomena primarily because of their thermodynamic stability and reproducibility. They have been used as models for the study of electrokinetic, electrostatic, and bulk interface properties.

The most commonly used micellar solution in biological studies is that of sodium dodecylsulfate for solubilization of proteins and lipids. Some of the physical characteristics of this system may be summarized as follows:

Critical micelle concentration	2.3 mM at 25°C in 0.05M NaCl
Aggregation number	84 (under the same conditions)
c.m.c.	0.51 mM at 25°C in 0.5M NaCl
Aggregation number	126 (under the same conditions)
Free energy of micellization	11.9 kcal/mole (0.5 NaCl solution)
Heat of micellization	-1.72 kcal/mole at 50°C
	$+0.4$ kcal/mole at 20°C

Insoluble Monolayers at Water/Air Interfaces

Most lipids, when placed on a water surface, disperse into a thin film. Under suitable conditions such films can be dispersed to the thickness of a molecule. Such a monomolecular film, when compressed by a movable barrier, resists compression beyond a certain film area. At this point where the film itself is exerting a lateral pressure, the molecules composing it are coherent and area per molecule is constant (Fig. 2-7). In this way a measured pressure may be applied to a film to produce a measured decrease in the film area, and a pressure area curve may be obtained. Such relationships suggest that the monolayers fall into two main categories: (1) The condensed film has the properties of a two-dimensional solid for which beyond the pressure of solidification a further increase in pressure produces little change in surface area. In such a situation the molecules of the amphipath are preserved in fixed positions. Most of the lipids with which we are concerned form condensed films. (2) The expanded film has the properties of a two-dimensional liquid in which the constituent molecules do not have fixed positions.

The experimental evidence further suggests that the molecules forming the monolayer are so organized that the hydrophilic part is buried in the aqueous phase, whereas the remainder of the molecule tends to extend out of the water. The electron diffraction pattern of multiple monolayers, formed by collecting individual monolayers on a glass slide,* indicate that the mean positions of the molecules of the liquid condensed films correspond at short range to a crystalline structure, the molecules being locally in a two-dimensional central hexagonal lattice.[81] The orientation of amphipaths in monolayers is further supported by the observations that the monolayers of fatty acids of various chain lengths can be compressed to equal areas; it implies that fatty acids must all form films in which the molecules are oriented identically with respect to the interface.

---

*These monolayers would be reorganized to form multibilayer structures (cf. Fig. 2-3).

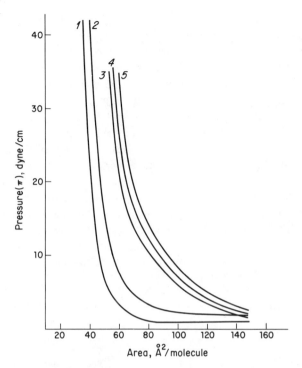

Fig. 2-7. Curves for typical synthetic phospholipids on phosphate buffer (pH 7.4) at 21–24°C. (*1*) Distearoyl-L-α-lecithin; (*2*) distearoyl-DL-α-phosphatidyl ethanolamine; (*3*) Dimyristyl-DL-α-lecithin; (*4*) γ-stearoyl-β-oleyl-L-α-phosphatidylethanolamine; (*5*) γ-stearoyl-β-oleyl-L-α-lecithin (Figure from Ref. 50 and data from Ref. 94).

The lipid monolayers reduce the surface tension of water. This lowering of the surface tension is defined as surface pressure. The variation of surface pressure (π) with area (A) available to the surfactant is represented by a π–A curve, sometimes referred to as pressure-area curve. The value of the surface pressure of monolayers always remains positive; that is, the phase tends to contract into as little an area (and consequently volume) as possible.

Monolayers are permeable to various small molecules such as water and gases. For efficient retardation of penetration of foreign molecules the hydrocarbon chains in the monolayer must be eighteen or more carbon atoms long, and the compression must be above 15 dyne-cm^{-1}. Under these conditions molecular cohesion is high and the films composed of long-chain alcohols solidify; it implies nearly complete adlineation of the chains. However, the area per molecule in such compressed films (21.6 Å^2) is greater than the cross-sectional area of the hydrocarbon chain (18.4 Å^2) in the crystalline state. Implications of these observations on the permeability of biological membranes is discussed in Chapter 5.

The effects of polar groups of the amphipaths on the properties of mono-layers is generally not specific but depends to a great degree upon the extent to which the head group affects the critical factors of orientation, packing, and ad-lineation of the chains. Film solidification can also be induced by an increase in the lateral attractive interactions between the head groups which contribute sub-stantially to the lateral cohesion between the molecules.[82] Strong ion-ion and ion-dipole interactions and divalent cation bridging can solidify the monolayer readily. Thus the properties of the condensed monolayers seem to be de-termined both by polar groups and the hydrocarbon chains of the amphipath.

In mixed monolayers, sometimes, the area per molecule is smaller than that of the pure system. For example, if a mixture of cholesterol and lecithin is spread at the air/water interface, in the compressed monolayer the area occupied by a lecithin molecule is a function of the amount of cholesterol present in the lipid mixture. At a ratio of three cholesterol molecules to one lecithin molecule, the effective area occupied by each of the lecithin molecules is reduced to 50 $Å^2$, rather than the value of 96 $Å^2$ found for pure lecithin (Fig. 2-8). At the inverse ratio of three lecithin to one cholesterol, the area occupied by each lecithin

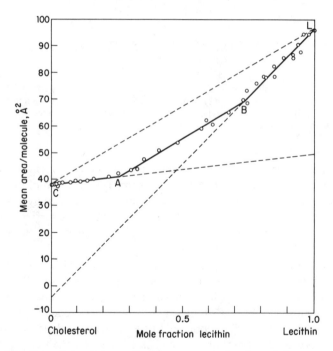

Fig. 2-8. Variation of the mean molecular area, as a function of the composition, for monolayers spread from mixtures of cholesterol and egg lecithin. The dashed straight line CL represents the variation which would correspond to a simple additivity rule of the molecular areas of cholesterol (C) and lecithin (L).[80]

molecule is 82 Å². Thus the results suggest that the critical points shown in Fig. 2-7 define the ratios for the formation of a particular form of the cholesterol-lecithin aggregates, which may have a packing order different than that of any of the components in their pure states.[68,71,80,83,84,1057,1058] Such studies suggest the possibility of hyperstructuring in the aggregate forms of lipid mixtures. These aspects are further considered in Chapter 9.

Liquid Crystals

The phase which shows the high degree of order usually associated with crystals but a mobility like that of a liquid is termed *liquid crystal phase.*[1196] This intermediate phase is stable over a characteristic temperature range. For pure substances (not solutions) more than one type of intermolecular arrangement can sometimes be obtained, and this type of polymorphism is not characteristic of lipids only. The *smectic, nematic* and *cholesteric* phases have been obtained so

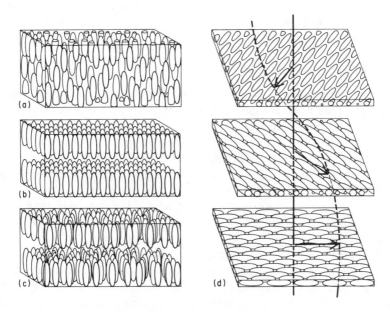

Fig. 2-9. Three kinds of liquid crystal are distinguished from one another by the arrangement of their molecules. Nematic liquid crystals (a) consist of molecules that are parallel, resembling matches in a box. Each molecule can, however, rotate around its axis and move from side to side or up and down. Smectic materials (b), (c) have a layered arrangement. The layers can slide over one another because the molecules in each move from side to side or forward and backward but not up and down. Within each layer molecules may be ordered in ranks or randomly distributed. Cholesteric molecules (d) consist of layers, as smectic crystals do. Within each layer, however, molecules are parallel, as nematic molecules are. Molecules in one layer displace those above, so that a helical pattern (*vertical line and broken helical line*) forms from layer to layer.[88a]

far in a variety of compounds including a large number of nonlipid molecules. In the smectic type (Greek for "soap like") crystals, strong attraction exists between the sides and the planes of the planar elongated molecules that generally make up such a phase. These attractions appear in adjacent molecular layers, each having a thickness of one molecule, and each containing monomers whose long axes are nearly normal to the planes existing between the layers. Strong attractions exist between molecules in each strata but little attraction is exerted between adjacent strata themselves, thus allowing them to slide relative to one another with little resistance (Fig. 2-9). The smectic crystals thus behave like positive uniaxial crystals with the optical axis normal to the strata. In the nematic (Greek for "thread") phase the molecules lie parallel but may slide relative to one another, no regular arrangement of adjacent molecules being necessary. The swarm (aggregate) of such structures move independently of each other in the surrounding isotropic liquid. In the cholesteric phase, the molecules are arranged with a parallel alignment. Since the liquid crystals are formed in a neat phase, we shall not be concerned with their properties directly; however, the properties of these aggregates resemble very closely those of phospholipid dispersions in aqueous solution as described below.

Phospholipid Dispersions in Water

Molecular aggregates resembling those in the smectic phase of liquid crystals can also be produced when phospholipids are equilibrated (or swollen) with excess water or aqueous salt solution.[80] In fact, these swollen paracrystalline preparations as ordinarily obtained contain several related phases depending upon ionic concentration, nature of ions and lipids, and temperature as characterized by X-ray diffraction studies. Three structurally distinct phases are probably best characterized: myelinic smectic, hexagonal I, and hexagonal II (Fig. 2-3). The myelinic smectic phase consists of a large number of bimolecular lamellar layers, each separated by a fixed layer of aqueous solution. The hexagonal I phase roughly corresponds to the micellar dispersion of phospholipids in water. This phase predominates when the relative amount of water in the lipid/water mixture is high (about 96-99%). The arrangement of lipid molecules is analogous to that of spherical micelles. The hexagonal II phase consists of hexagonally packed cylindrical micelles with ionic groups directed toward the core of water. This phase predominates when the water content is very low (about 5%). In the presence of appreciable proportions of water (15-60% at 22°C) the arrangement is myelinic smectic or lamellar.[83,84]

X-ray diffraction photographs of these various phases show a broad band similar to that found in liquid paraffins, and corresponding to a short spacing of about 4.5 Å. This has been interpreted to suggest that in the liquid crystalline phase the hydrocarbon chains are liquid-like. Such an interpretation has been challenged on the grounds that similar bands can be observed in the fiber diffraction pattern of sodium myristate. The band can therefore be due to rather regular packing of the chains in an extended form.[85] In plasma membranes, the nature of this band seems to depend upon the nature of the lipids present. In membranes containing high levels of cholesterol the band occurs at 4.75 Å, whereas the membranes containing lipids with high unsaturation show the band at 4.52 Å. Thus the band could be due to packing characteristics of lipids (see Chapter 9 for further discussion on this point).

Besides the three major types of aggregated structures just described, various phases have been identified in other lipid-water systems. These phases include rectangular, cubical, lamellar, rod shaped, hexagonal, and a variety of other associated and mixed structures which show long-range order and short-range disorder. In this class of structures the *elements* are generally lamellar; the lamellae may be in the form of infinite planar sheets packed in a one-dimensional lattice (see discussion of smectic myelinic phase above), infinitely long ribbons organized in two-dimensional centered rectangular, or oblique lattices, or disks organized in three-dimensional lattices. Besides these lamellar arrangements, some novel phases may result if the polar groups form planar two-dimensional networks, either square or hexagonal, embedded in a continuous matrix formed by the disordered paraffin chains. In this class the basic structural elements are identical rods, either infinitely long or of finite length and branched to form three-dimensional networks (Ref. 86 and Refs. therein).

The origin of these various aggregated forms and the reason for their stability are not known at this time. However, it is not unreasonable to speculate that many of the properties of such phases and dispersions could very well arise from order-disorder transitions. It is particularly interesting to note that highly developed organization of all these phases is achieved in spite of a great disorder of the paraffin chains (cf X-ray diffraction data). It is quite likely therefore, that the polymorphism exhibited by various lipids is a sum of factors arising from close packing of polar groups, relatively high degree of packing of hydrocarbon chains which are liquid-like at least to the extent of chain rotation (*gauche-trans*), chemical disorder (various molecular species arising from difference in unsaturation, chain length) and other disorders. Finally, the nature and the sequence of the phases are related to the structure of the lipid in at least one obvious way: phases with a high surface to volume ratio are promoted by bulky hydrophilic groups, with respect to the paraffin moiety, and vice versa.

Bimolecular Lamellar Lipid Membranes

In principle two types of bimolecular lamellar lipid membranes are possible: one with the polar groups at the aqueous interface and the apolar groups aggregated to form the so-called oil phase, the other has the polar groups joined together and sandwiched between hydrocarbon residues which are oriented toward the air interface (Fig. 2-3). The first type corresponds to the generally accepted model for the plasma membrane (as discussed in Chapter 1) and also for the black lipid membrane (BLM), and shall be considered in detail in the subsequent chapters.

The second type of bimolecular lamellar membrane is formed at the air/water/ air interfaces and is obtained in the most common form from soap bubbles under suitable conditions.[56] The soap films when first formed are usually too thick. They rapidly thin out, some of the liquid draining away either to form drops adhering to the lower edge or to accumulate in a thick ridge at the junction of the film with its support. The progress of thinning is shown by the development of interference colors and in the final stages the film looks "black," since the reflections from the front and the back are very nearly in counterphase. The black regions look like holes as reported by Hooke in 1672.[87] Newton determined the thickness of the black film to be about 50 Å, a value which is still in good agreement with the currently accepted value, which, as expected, depends upon the nature of the lipid used for making the film. This thickness is approximately equal to twice the hydrocarbon chain length of the amphipath normally used for the formation of the film; however, the thickness of the film varies as a function of the ionic strength of the soap solution.[88] It has been interpreted as being due to the presence of a varying thickness of water located between the centrally oriented polar groups, and to varying ionic and hydration forces. The limiting thickness at appreciable salt concentrations approaches 50 Å as determined by optical methods. The amount of water present in the film cannot be determined easily; however its amount must be of importance since evaporation must be prevented if the films are to be maintained for long times. The composition of these films has been measured both by radiotracer (direct analysis) and by absorption spectroscopic technique. The interfacial tension of these films has been found to be less than 1 dyne-cm^{-1}.

Factors Affecting Stability of Aggregates

At first, it would appear as though the entropy decrease associated with the existence of order in various aggregates described above would work against their stability. Yet the fact that these structures are formed spontaneously shows that the thermodynamics of their formation is favorable. It appears therefore that the stability of aggregates depends on factors other than overall entropy considerations. It may be due to the integrity of the boundary layer of hydrophilic and hydrophobic moieties. The polar group and its volume relative to that of the hydrocarbon chain will determine the packing and orientation of amphipaths, thus bringing Van der Waals and London dispersion forces into the picture. Since the dielectric constants of liquids that are immiscible with water are usually much less than that of water, it follows that unless the liquid-liquid interface is highly charged, the repulsive forces between the interacting surfaces would be very low. Thus strong double-layer repulsion prevents a close approach and greatly reduces the probability of coalescence. Since the double layer is built up of the adsorbed potential-determining charged particles, the factors such as nature and concentration of electrolytes in the medium and the magnitude of the charge due to ionic groups of amphipaths determines the magnitude of various interfacial potentials. These interfacial characteristics are further modified by various additives, which can also affect the viscosity and fluidity of surfaces.

It was also shown earlier in this chapter that the free energy which the hydrocarbon chains gain on transfer from the hydrophobic to aqueous phase is positive. It has also been suggested that water tends to aggregate around hydrophobic molecules, and when a hydrocarbon chain goes into an adsorbed layer to form an interface, this structured water is disrupted. Such disruption of water structure could also contribute to the stability of new interfaces by an increase in entropy. The freedom of motion of the hydrocarbon chain is not much affected when one end of it is attached at the interface or when it exists in the bulk phase, except when the monolayer is tightly packed. However, any decrease in the entropy by such a process will be more than compensated by a decrease in enthalpy due to cohesion, i.e., the decrease in the susceptibility to stray field transitions.

Thus it may be recapitulated that the stability of various organized structures (or phases) formed in lipid/water mixtures is a property of the amphipathic molecules. The overall negative value of the free energy change accounts for the stabilization of the molecule at the interface. The possibility of order-disorder transitions, both in organized water structure and in hydrocarbon chains is very real (however, little understood) and has far-reaching implications. The fact that small changes in temperature and concentration may induce drastic changes in the relative proportion of various phases may cast doubt upon the concept of the continuous lamellar lipid layer as a diffusion barrier in biological membranes and for that matter on the individuality of various phases. Such aspects are considered further in Chapters 3 and 9 with special reference to the structure of the bimolecular lamellar lipid membrane. However it should suffice to state at this stage that the concept of a lipidic mesophase provides a verifiable hypothesis for the unique character and origin of supramolecular aggregates such as lamellar lipid membranes whose properties are entirely consistent with the molecular order and energetics of experimental lyotropic systems.

3/ Black Lipid Membranes: Formation and Stability

Ever since the postulation of the lamellar bimolecular structure for the plasma membrane (see Chapter 1) various unsuccessful attempts have been made to make model bimolecular lipid membranes.[89-93] Stimulated by the success of lipid-monolayer studies, Langmuir and Waugh[93] in 1938 attempted to make films of lecithin and protein at the interface between bulk benzene and an aqueous solution of egg albumin. Although they had little success in preparing membranes, they clearly emphasized the potentialities of such model systems:

> We believe that the technique we have described for building protein membranes across holes in plates may be useful to the biologists in the study of the properties of these membranes .... these membranes could form partitions between two separate aqueous solutions so that permeabilities and conductivities, etc., could be studied.

These lipoprotein films, however, were unstable and no further progress was reported in this general direction. In the years to follow various lines of evidence for the bimolecular lamellar structure of the plasma membrane accumulated but the concept remained essentially theoretical even though much inferential support for this idea has been provided by studies on bulk properties of phospholipid water systems (see Chapter 2 and Refs. 80, 84, 94, 95). The first successful attempt to make membranes with a bimolecular lamellar arrangement of lipids was only recently reported by Mueller, Rudin, Tien, and Wescott in 1962.[96-99] Since then numerous studies have appeared on such membranes formed from a variety of lipids. In this chapter we shall describe the methods of formation of lipid bilayers at water interfaces and the factors governing their stability.

52

The method used by Mueller and co-workers is essentially an extension of the method used to form bimolecular lamellar layers at the air-air interface (soap bubbles). A solution of total lipids from beef brain in a chloroform-methanol mixture was painted over a small hole (1–100 mm^2) in the partition of a chamber or on a loop of suitable diameter, which was then immersed in a salt solution. The lipid solution contained some additives which prevented solidification of the film. Dispersion of the solvent into the aqueous medium eventually leaves a thin film of lipid on the hole. These membranes show properties which strongly suggest a bimolecular lamellar structure (evidence to be described in Chapter 4). It should, however, suffice to state that the morphology of black lipid membranes (BLM)* and some of their properties attest to their close similarity to plasma membranes. Furthermore, these BLM may be modified by addition of a variety of agents, and the modified BLM are able to mimic the behavior of plasma membranes in still other important respects.

PREPARATION OF BLM

The formation of BLM separating aqueous phases is conceptually simple: two compartments are separated by a thin partition and communicate through an aperture in this partition. The compartments are filled with an aqueous medium and a (thick) membrane of amphipathic lipid is formed in the aperture separating the aqueous phases. The thick membrane spontaneously thins by draining to form a ring or annulus of bulk phase lipid around the margin of the aperture (Fig. 3-1). At equilibrium, a metastable BLM continues to separate the aqueous phases. Although many practical, operational design parameters have been recognized theoretically, the following have been found to be useful experimentally by empirical, trial-and-error testing of theoretical considerations:

1. Provision for optical and electrical measurements separately and simultaneously should be provided. Optical measurements require that one of the walls of the chamber be transparent and inclined at an angle to the vertical axis in order to avoid spurious reflections. Visual inspection is by far the best method to follow the progress of BLM formation (thinning).

2. The whole assembly should be free from vibrations, and the aperture must be on a stable support. The mechanical stability of the membranes is improved markedly in chambers with one closed (sealed) constant-volume compartment.

3. The temperature in the compartment and the pressure on the membrane must be adjustable since these factors govern the solubility, viscosity, diffusion, thinning, and stability of the membrane.

---

*Incidentally, the abbreviation BLM may mean black lipid membrane or bimolecular lipid membrane or bilayer lipid membrane or bimolecular lamellar lipid membrane or even "bare lipid membrane." We shall use this abbreviation very often hereafter to refer to these artificial membranes.

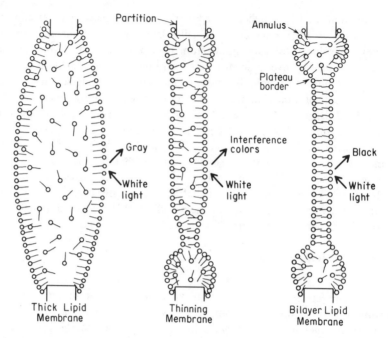

Fig. 3-1. Diagram illustrating the three stages observed during thinning of a lipid membrane in aqueous media, and indicating the patterns of reflected light.[101]

4. The low tension of the films allows practically no hydrostatic pressure difference across the plane of the membrane. The magnitude of the hydrostatic pressure difference (Δp) that is involved can be estimated from the relationship

$$\Delta p = 2\gamma_i/r$$

where γ_i is the interfacial tension of the BLM and r is the radius of curvature. When the radius of a flat circular membrane is 0.1 cm and the interfacial tension is 5 dyne/cm, the pressure difference of 10^{-4} atm is sufficient to cause the film to adopt the shape of a hemisphere; a little more pressure will rupture the film. Therefore, the chambers should be perfused without exerting any hydrostatic pressure on the membrane. This is important since bulging of the membrane changes its surface area and, consequently its properties.

5. The material used in the construction of the chamber should be stable toward water and organic solvents such as methanol. It should also have a low dielectric constant, and good machine workability.

6. Shielding must be provided to eliminate extraneous light, atmospheric static charges, and ac 60 cycle interference. The electrical measurements also require that no ionic or electronic conductive pathway exists between the two

aqueous media except via the membrane. Sometimes it is advantageous to coat the inner walls of the chamber with silicone oil.

7. The borders of the film must not become detached from the support. The Gibbs free energy connected with the surfaces of a BLM is less than twice the normal surface free energy. This causes the black film to be attached to the support and to the bulk of the membrane forming lipid with a finite contact angle.[100] The situation can be improved by providing a very narrow groove at the point where the film is attached. This would make it more difficult for the lipid to dewet the support, and would also provide a reservoir of lipid solution equilibrating with the film. This can also be achieved by using two very thin sleeves of the support material stacked together with a hole punched through both.

8. The edge of the hole must be smooth—usually made by piercing with a hot needle in a heat-thinned area of the septum. This procedure gives a smooth, rounded torus surrounding the aperture. The aperture depths are also important. The rounded edges of these apertures tend to make the membrane center itself in the middle of the edge of the aperture, rather than at one end of the edge of the aperture.

These demands on the experimental set-up pose considerable design and engineering problems. The choice of the material which can be used for the septum is limited by the requirement of its low dielectric constant, high electrical resistance, a reasonably low contact angle at the lipid/water interface, chemical inertness, reasonable mechanical strength and machine workability. Teflon, polyethylene, Delrin (polyformaldehyde), polypropylene, polycarbonate, and cast Leucite (Perspex glass) are used singly or in combination for the construction of various components. Nylon, Dacron, and Bakelite are generally less satisfactory. As the edges of the hole should be smooth, special care is necessary while drilling holes. Holes drilled in a Delrin septum may be smoothed out by soaking it in concentrated hydrochloric acid solution for a few hours. Since Delrin is slightly soluble in concentrated hydrochloric acid, the edge of the hole becomes smooth. In other cases the aperture may be polished by friction with a smooth cotton string.

Howard and Burton[101] have considered the various mechanical and engineering features in detail and designed a cell which meets most of the experimental and theoretical requirements. In actual practice however, depending upon the nature of the experiments, the gross construction can be simplified. For simple studies most of the elaborations are not required. For optical studies of the membrane a loop of metallic wire or a Teflon O-ring can replace the partition in the chamber. A detailed account of apparatuses for the study of the mode of formation of BLM under controlled conditions is given elsewhere.[78,1059] This method is claimed to be highly reproducible for the formation of interference fringes and black membrane formation under a given set of conditions. The method appears to be well suited to the characterization of physical and chemical factors governing BLM formation. The lipid solution is applied to the rim of the ring through a syringe which regulates the volume of the lipid solution delivered, and also the duration for the application of the solution.

Apparatus suitable for the measurement of water and solute permeability under osmotic gradients and for isotopic flux studies has been described in the literature.[102,104] A typical chamber consists of two compartments connected only through the BLM across which an osmotic or isotopic gradient may be maintained. When osmotic flux ensues, the hydrostatic pressure changes the area of the membrane by causing it to distend. Counterbalancing of this pressure head is achieved by manipulation of the hydrostatic pressure with a micrometer screw-operated syringe.[102] The difference in the micrometer readings thus indicate the volume flow over the recorded period of time. Distension can be followed either visually or electrically by measurement of the membrane capacitance.[105] Two types of cells described for isotopic diffusion exchange studies can also be used for the measurement of the permeability coefficient of other labelled solutes.

A very simple experimental set-up has been described which is well suited for demonstration and simple experiments.[133]

MEASUREMENT OF CHARACTERISTICS OF BLM

The measurement of the electrical properties of BLM requires insertion of suitable nonpolarizing electrodes into the compartments, and protection from atmospheric static charges. Various electrodes such as silver–silver chloride or a calomel half-cell with an agar-KCl bridge are usually quite stable. Various cells and electrical circuitary have been described in the literature for the measurement of conductivity,[106,108] capacitance,[106,107,109,110] dielectric breakdown voltage,[107-108] diffusion potential, transient electrical characteristics, and current-voltage relationships.[99,109-115] The principle involved in these measurements is essentially the same as generally employed for measurement of small currents and high resistances. Usually, an operational amplifier with suitable input devices and variable resistors is most convenient to use. The circuit shown in Fig. 3-2 has been used successfully in our laboratory.

There are numerous sources of experimental difficulties which may be encountered in various measurements. Difficulties due to instability of the membrane can generally be traced to impurities in the plastic material used for the construction of chamber, or to the oxidation of lipids, or to solidification of the film due to low temperature or high viscosity. New chambers should be freed from detergents, plasticizers, curing agents, catalysts, and release agents used in the casting process. It is usually advantageous to extract new Teflon or polyethylene pieces with chloroform-methanol (2 : 1 v/v) for several hours. After use, parts should be washed free from any lipids and/or proteins with strong sodium

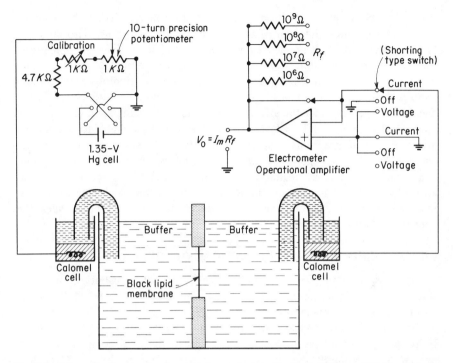

Fig. 3-2. Schematic diagram for the experimental set-up for the formation of BLM and electrical measurements.

hydroxide solution. Further complications in the formation of BLM may also arise from aging lipids which tend to oxidize, polymerize, and develop color. Samples which have undergone these changes usually give BLM of lower resistance or do not thin at all. They can also affect the medium by increasing fluorescence and absorbance. It is not unusual to observe increased time for the formation of BLM with even slightly deteriorated lipid solutions. This difficulty can be avoided by preserving lipid samples at low temperature under nitrogen, or by using completely saturated lipids. Persistent instability of the membrane can usually be traced to low viscosity of the lipid solution, or to the presence of minute bubbles of air dissolved in the lipid solution or the aqueous phase, or to the presence of traces of detergents or glass particles[1198] in water or containers.

Improved membrane stability may be obtained by preconditioning the aperture. This can be done by drying a small annular ring of the BLM-forming solution (preferably the lipid solution without the additives) around the aperture on both sides of the partition. The solution should be dried in a vacuum, especially if crude natural lipids are used.

Sometimes temperature changes of $\pm 1°C$ in the medium have been found crucial for the formation of BLM. This is not altogether unexpected in view of the liquid crystalline character of these membranes which show sharp temperature transitions at Kraft point. At lower temperature, one of the components of the BLM-forming solution may crystallize out and thus BLM formation is prevented. In such cases it is necessary to change the proportion of various components in the lipid solution. One of the serious experimental limitations is the small area of the aperture since the effect of fluctuations of pressure and vibrations on large circular membranes may be serious. In principle it is possible to increase the area of the membrane by using apertures of different shapes such as stars, rectangles, etc., which have larger rim-to-area ratios. However, border leakage and other complications may arise. In addition to all these difficulties, other complications arising from the withdrawal of solution from compartments, or due to changes in ionic concentrations, and the effects of unstirred layers are to be considered in appropriate cases.

LIPIDS USED FOR PREPARATION OF BLM

A large variety of purified and crude natural lipids, and a variety of synthetic amphipaths have been used for the preparation of BLM (Table 3-1). The composition of the membrane-forming lipid solution depends upon the nature of the lipid used. A lecithin+n-alkane mixture has been used for a variety of studies. However, as we shall see in subsequent chapters, the lipid composition is important in determining the characteristics of modified membranes. Thus addition of various lipids and their relative degree of purity is an important factor in determining the lipid composition to be used for membrane formation. Synthetic amphipaths used with cholesterol or oxidized cholesterol form stable BLM. In contrast, pure cholesterol does not form stable membranes. Oxidized cholesterol, a mixture of sterols formed by boiling cholesterol in n-octane in the presence of air for several hours, forms stable BLM and this lipid solution does not deteriorate for months at room temperature.

Crude natural lipids have been used in varying degrees of purity. In general, the chloroform-methanol (2:1 v/v) extract prepared from homogenized tissue is freed from nonlipid substances by being placed in contact with fivefold its volume of water or 0.2-fold its volume of salt solution.[114] The lower phase contains essentially all the tissue lipids and proteolipids. Usually, mixed lipid extract, dried or concentrated under reduced pressure and a stream of nitrogen and redissolved in about 5 volumes (relative to the original tissue fraction weight) of chloroform-methanol (2:1 v/v), can be used as stock solution. It contains about 2-3% proteolipids. This lipid solution mixed with α-tocopherol (about 150-250 mg for each ml of lipid solution) or n-alkane (about 10-15%), and sometimes cholesterol (about 0.5-5% w/v) have been found to be suitable for BLM formation. Several methods have been described for the purification of crude lipid mixtures and for preparation of solutions suitable for BLM

TABLE 3-1 COMPOSITION OF LIPID SOLUTIONS FOR THE FORMATION OF BLM[a]

Lipid	Additives	Aqueous Phase	Ref.
Glycerol distearate	n-Hexane	NaCl	119
	n-Octane	NaCl	120
Sorbitan tristearate	n-Hexane	NaCl	119
Dodecyl acid phosphate 0.3% + cholesterol (0.93%)	n-Decane	Various salt solutions	120, 121
Oxidized cholesterol	n-Octane	Various salt solutions	120, 122
	n-Decane	Various salt solutions	123
Cholesterol + dioctadecylphosphate	Dodecane	NaCl	119
Oxidized cholesterol (1%), decyl acid phosphate (0.25%), cholesterol (0.42%)	n-Octane + n-decane	Various salt solutions, histidine buffer	123
Cholesterol (1%) + hexadecyltrimethyl ammonium bromide	n-Dodecane	NaCl	120
Cholesterol + cetyltrimethyl-ammonium bromide	n-Decane	Various salt solutions	121, 124
Synthetic phospholipids	n-Alkane (C_{10}-C_{16})	Various salt solutions	125
Phosphatidyl ethanolamine + cholesterol	Decane		126
Distearoyl cephalin		Various salt and buffers	127
Egg lecithin	n-Alkanes	Various salt solutions	105, 108, 110, 117, 120, 128, 130
Egg lecithin	Decane + squalene (2:1)	NaCl + histidine	123
Completely hydogenated lecithin	Decane	NaCl + $MgSO_4$ + histidine	131
Beef brain lipids with or without cholesterol	α-Tocopherol	Various salt solutions	99, 132 etc.
Beef brain lipids with or without cholesterol	Dodecane and/or tetradecane	Various salt solutions	99, 101, 106, 117 etc.
Chlorophylls and xanthophylls		Various salt solutions	132
Retinal + oxidized cholesterol and/or lecithin	Decane	Various salt solutions	1290
Rhodopsin			118

[a]Various other synthetic and natural lipids have been used; this list is by no means complete.

formation (see Refs. 99, 101, and 111 and Table 3-1). Use of tocopherol retards lipid oxidation; cholesterol confers electrical stability and higher viscosity to the membrane interior; lower hydrocarbons modify drainage characteristics. Other additives such as silicone oil, fluorinated hydrocarbons, and mixtures of natural lipids may also be used.*

A large variety of salts (alkali metal salts, low concentrations of alkaline earth

*Most of the phospholipids seem to dissolve in hydrocarbons and other additives only in the crude sense, but they appear to do so only as large aggregates of molecules and give negligibly small concentration of single molecules. However, for most of other amphipaths the c.m.c. is at much higher lipid concentration; it is only above the c.m.c. that BLM is much more stable. The concentration of lipid in BLM-forming solution would vary from lipid to lipid depending upon the c.m.c. of the mixture. High lipid concentrations may however be undesirable for other reasons to be discussed later.

salts, organic ions and buffers, etc.) have been used at various concentrations (up to $4M$ in some cases) and pH (3–11). Up to 50-fold gradients for electrolytes and 100-fold gradients for nonelectrolytes can be used across BLM. However, traces of surface-active impurities have an unstabilizing effect on BLM. Also, the chaotropic ions such as perchlorate, thiocyanate, etc., have a destabilizing effect. Small amounts of antioxidant such as butylated hydroxytoluene in aqueous solutions retard lipid oxidation and increase BLM stability. It has been claimed that the addition of heparin in the aqueous solution raises the stability of BLM with time and electric fields.[115]

FORMATION OF BLM

Methods for the formation of BLM involve the application of a small amount of lipid solution to the aperture submerged under an aqueous solution, and the progress of BLM formation is followed optically. Various methods have been used for the application of lipid solution as described below:

1. The glass, polyethylene, or *Teflon* O-ring is dipped into bulk lipid solution are transferred to the aqueous medium via air. This method is well suited to simple visual studies involving the mode of BLM formation or for testing the efficiency of BLM-forming lipid solutions.

2. Application of the lipid solution with the tip of a fine brush dipped in lipid solution, as originally suggested by Mueller and co-workers,[96-99] has been successfully used with a variety of lipids. The possibility of contamination between different applications can be eliminated by using a pointed Teflon or polyethylene spatula or a hypodermic syringe for delivering the lipid solution to the rim of the hole; these devices can be easily cleaned and they last longer. A small volume of the lipid solution can also be delivered at the aperture with the help of a small-diameter tubing attached to a syringe. If the lipid solution does not contain a low-boiling solvent, blowing air bubbles with a Pasteur pipet (Dispopipet), previously coated inside with the lipid solution, in the vicinity of the hole is very convenient. With this latter technique the amount of lipid applied to the aperture is very small and the chamber usually stays clean.

3. In the marginal suction method, a modification of the method reported for the formation of soap films,[57] a small amount of lipid is delivered to fill the lumen of the ring around the aperture. The bulk lipid thus applied is then withdrawn back into the syringe causing the drop to become a biconcave disc. As the lipid solution is further withdrawn by suction on the margin of the disc, the concave surfaces collapse and form a planar membrane supported by an annulus of bulk phase lipid solution.

4. Nonsupported BLM in the form of spherical bubbles can be prepared by discharging a droplet of aqueous electrolyte solution through a fine nozzle coated (dipped) with lipid solution in a continuous or discontinuous density gradient. The detached droplets fall through the gradient until they become

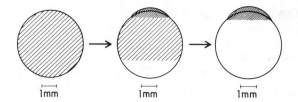

Fig. 3-3. Three successive stages during the formation of spherical bilayer. The thinned black lipid membrane is surmounted by a lens of bulk-phase lipis solution.[116]

isodense with their surrounding solution.[116] It is important that the solutions inside and out of the bubble be isoosmotic since solvent flux due to osmotic gradient would continuously change the volume of the bubble and the area of the film. These membranes show properties similar to a planar membrane, although the effect of the "lipid cap" on the membrane (Fig. 3-3) is manifested in some of its properties. The electrical properties of these membranes can be studied after impaling them with glass microelectrodes.

5. A method claimed to be specially suited to making BLM of up to 50 mm^2 in area has been reported.[117] A screen (5 mm thick) with a hole up to 9 mm in diameter is carefully machined to give sharp edges without irregularities, so as to permit diagonal fixation in a chamber of suitable size. The sides of the screen and the inner walls of the chamber are previously coated with an apolar liquid such as silicone oil; this treatment helps bring the lipid solution into contact with the wall of the screen. When the screen is dipped into the aqueous phase passing through the lipid-water interface of the contents of the chamber, a film is formed on the hole. To enhance free drainage of the film the hole is not completely lowered into water phase, but its upper rim is left in the lipid solution phase, thus allowing free drainage of the film into it. However, extreme care must be taken to shock mount the chamber to protect it from vibrations.

6. A method to "coalesce" two monolayers to form a bilayer, which permits control of the surface pressure, has been described.[106] With this technique, the bilayer is formed by the opposition of two phospholipid monolayers, each at an oil/water surface, at least one of which is curved. An adsorption film is formed on the interface between a solution of protein in water and a solution of lipid in heptane. A similar film is brought into contact with this interfacial film from the heptane side. The two films then form a thin bimolecular protein lipid membrane separating two aqueous solutions after replacement of heptane by an aqueous solution. Similarly a method has been reported which allows the formation of a BLM from two monolayers of different lipid types at constant surface pressure separated by the partition. When the partition is slowly lowered into the aqueous phase until the hole comes down under the subphase level, both the monolayers coalesce to form BLM.[118]

After application of the lipid solution to the aperture by any of the methods described above, the formation of BLM can be followed by observing the reflected light from the membrane through a low-power microscope. The formation characteristics of soap films in air were first described by Robert Hooke in 1672,[87] and the mode of formation of BLM in aqueous solutions displays similar characteristics. This similarity is not unexpected since, from the structural point of view both the soap and lipid bilayer films involve identical forces during their formation. When the lipid solution is spread on the aperture the resulting film is usually thick and colorless; its structure is pictured as being similar to that of a sandwich consisting of an organic phase between two adsorbed monolayers separated from each other by about 0.1-1 μ (Fig. 3-1). This film thins continuously under the influence of diffusion and drainage. If the thinning film is illuminated with white light, the interference colors soon appear as horizontal bands in the reflected light either at the top or the bottom of the aperture depending upon the density of the lipid solution used. Initially, only faint pinkish greenish bands appear. These bands usually increase in breadth and brightness until the purer colors of the lower orders of Newton scale appear. As the thickness approaches 1000 Å the film becomes silvery gray, also known as primary gray.[134] The attraction of the monolayers at the two interfaces due to van der Waals forces is small, and therefore the two interfaces are independent of each other. The thickness of these membranes is expected to be a unit multiple of bimolecular membranes, and their rate of thinning before appearance of black spots follows the $T \cdot t^2 \simeq$ constant relation, where T is the time and t is the thickness of the membrane.

Under appropriate conditions of diffusion, evaporation and drainage of the solvent from the film, the attractive London-van der Waals forces between the sandwiching interfaces leads to a spontaneous isothermal phase transition, presumably to the lowest stable free energy state which is the unit bilayer. At the beginning of this stage black spots are seen either singly or in clusters, all having circular boundaries. These spots coalesce or grow at a fairly rapid rate, which depends upon a variety of factors including temperature, nature of the lipid, and lipid composition. Under suitable conditions, the whole film becomes black in 10-1000 sec depending upon the experimental conditions. This film is referred to as a *secondary black film* since it has a small reflectance and therefore has only a gray sheen in reflected light. The film looks black as it is so thin that the reflections from the front and back are very nearly in counterphase. Like primary black, the secondary black appearance of films has also been noted for soap films.[57] It should be noted that a black or grayish black film may not necessarily be bimolecular in thickness. In principle, under suitable conditions transformation from primary black to bimolecular thickness should occur through several stages of thickness, which may be distinguishable under suitable illumination. Some of these intermediate stages may persist for considerable lengths of time. However the last stage in their thinning is almost always the bimolecular lamellar film before the film is disrupted.

It has been suggested that thinning and spontaneous phase transitions of primary black films occur due to Plateau-Gibbs border suction, gravitational flow, and diffusion.[56,78,135] In some cases the Plateau-Gibbs border which supports the black region can actually be seen[135] and it has been suggested that its presence is essential for the maintenance of the structural integrity and the stability of BLM. Due to curvature of the surface the Plateau-Gibbs border is always a seat of low pressure (hence suction) with respect to the adjacent flat film. This suction probably acts more strongly on a thick film than on a thin film, and consequently, there is tendency for the thicker film to be sucked in and the thinner films to be generated at these borders.[56] The border between colored and primary black films is known as the welt, and it is obvious that the foregoing remarks on Plateau-Gibbs border suction may be extended to the suction at the welt.

As the thinning of the thick lipid layer proceeds, it seems probable that contact of the monolayers at opposite interfaces in the primary black films may occur. This may be due to a variety of causes, such as thermal motion, mechanical vibration, impurities (such as dust particles), and local variation of interfacial tension. Should any one or a combination of these factors be operative, chance contact may bring van der Waals forces into play, and consequently the appearance of the black spots follows. The border suction at the welt as described above gives rise to a zipperlike mechanism leading to fast growth of this black film after the appearance of the first black spots. Thus, under a given set of conditions, the formation of black film, once initiated, apparently proceeds at a constant rate until the whole film has reached the lowest possible free energy configuration (see below). This film is most prone to rupture due to vibrations, transmembrane hydrostatic pressure changes, and effects of stirring during the transition to the black state. This may be due to the fact that the rapidly advancing boundary between the black area and the thick film (that is the welt) is a seat of high concentration stress and turbulent flow. If the viscosity of the BLM-forming lipid (in the thick membrane) solution is low, turbulence above a critical value may break the membrane, while high viscosity may slow down the process of secondary black formation. Thus during spontaneous growth van der Waals forces will favor disproportionation of the film into thin and thick parts. Although the surface tension will act against such unevenness of the interfaces, it will be more than compensated by the total decrease in the van der Waals energy. These fluctuations and corrugations at times show up in the light scattered by the thinning films. It can be shown that corrugations having wavelengths larger than a critical wavelength will grow spontaneously because of the action of van der Waals forces. These corrugations may help in the formation of BLM if sufficiently strong repulsive forces keep two interfaces of the film separated; otherwise the film may become thinner rapidly and break.[136] The critical wavelength (λ_{crit}) for such a corrugation is given by:

$$\lambda = \left(\frac{-2\pi^2\,\gamma}{d^2\,G/dh^2} \right)^{\frac{1}{2}}$$

where γ = interfacial tension; G = Gibbs energy of interaction among molecules per unit area; and h = thickness of the film. The rate at which the fluctuations above a critical wavelength grow depends on the viscosity of the liquid.

Thus the effect of additives such as n-alkanes and α-tocopherol may be important in the sense that they increase the viscosity of the lipid solution and thus confer stability to the thinning membrane in addition to acting as one of the structural components. However, the rate of membrane (BLM) formation also seems to depend upon several factors which have been noticed experimentally. Thus rate of BLM formation depends on rate of stirring, application of voltage,[137] addition of compounds such as acetyl choline to the medium, poking the thick membrane with a fine object, tapping the bench when the film is in primary gray stage of thinning, and the nature and amount of trace impurities and divalent cations in the aqueous medium. Some rationale for these observations may be seen in the semitheoretical discussion on the mode of thinning. Moreover, the rate of thinning should also depend upon the solubility of the lipid solvent in water. This is particularly important since the life expectancy of BLM should depend on the solubility of the lipid solvent in the aqueous phase, the viscosity of the lipid solution, and its density. The last variable (density) is important since it seems that besides supporting the film, the Plateau-Gibbs border also serves as a reservoir for the black film. Drainage of the border will occur more rapidly with solutions of lower viscosity and higher density. Thus a a mixture of higher hydrocarbon solvents with chloroform and methanol would give rise to lipid solutions with suitable solubility, viscosity, and density.

It is interesting and probably pertinent to recall that a variety of cells can reconstitute a new and functioning plasma membrane from the cytoplasmic lipoproteins in a few seconds following its total destruction (Chapter 7 in Ref. 2). It is particularly relevant to note that the phenomenology and kinetics of this process is similar to BLM formation as described in the following quotation (p. 9 in Ref. 2):

When the surface film is torn by rapid and repeated thrusts of the microneedle, a wave of disintegration sweeps around the protoplasm, and its complete destruction ensues. This disruption occasioned by small tears tends to be localized. A surface film quickly re-forms, closing the gap and walling off the healthy part of the protoplasm from the cytolyzed region. Repair of a localized tear of the protoplasmic surface requires the presence in the external medium of the appropriate proportional concentrations of monovalent (sodium and potassium) and divalent (calcium) cations.

On page 102 of the same book these authors further point out that conditions for the formation of new surface films are at an optimum when the electrolytes (Na, K, and Ca) are present in the external medium in the proportional concentrations characteristic of the cell's natural habitat. In the case of the marine starfish eggs, film formation readily occurs following small tears of the surface of the eggs immersed in solutions containing the same concentrations of Na, K, and Ca as seawater. If Ca is omitted from the mixture, repair from the tear is definitely impeded. For electron microscopic observations on the mode of new plasma membrane formation see Ref. 1199.

As pointed out earlier in this chapter, the kinetics of BLM formation is strongly dependent upon external conditions. BLM have been prepared at 15 to 50°C in a variety of aqueous solutions of pH 5-9. The rate of growth of black films varies with pH, the nature and concentration of ions in the aqueous phase, and the temperature. Thus, for example, the rate of formation of black films from egg lecithin at high and low pH in sodium chloride solution is found to be slower compared to the rate at pH 7. However, at higher pH (around 9) thinning could occur very slowly in the presence of calcium chloride (about 10^{-3} M). At pH 10-11 in the absence of calcium chloride complete black films could not be formed, and in the presence of calcium chloride the films tend to become rather rigid.[138]

The rate of BLM formation decreases at lower temperatures; thus BLM formation from egg lecithin in n-tetradecane takes about 30 min at 20°C, 4 min at 36°C, and only 1 min at 39°C.[139] The membranes formed at higher temperatures are generally less stable. These results are consistent with the hypothesis that phase transition (at the Kraft temperature) is involved in the formation of the black film. This is consistent with the observation that saturated lipids form BLM much more slowly than the corresponding unsaturated lipids at the same temperature. In general, stable membranes can most satisfactorily be prepared by choosing suitable transition times, i.e., long enough to give the membrane sufficient stability for the experiment but short enough so that the transition time to the secondary black stage (when it is susceptible to breakdown) is convenient. This can be achieved by a suitable choice of chain length and unsaturation of lipids, viscosity and solubility characteristics of additives and solvents, temperature, and the concentration of divalent ions. The addition of certain additives such as tetradecane, mineral oils, silicone oil, caprylic acid, α-tocopherol, squalene, and certain carotenoids increases BLM stability, prevents solidification, and protects from critical turbulence during drainage. However, these additives have considerable influence on membrane properties including permeability, electrical conductivity, dielectric breakdown, and electrokinetic and excitability characteristics (to be discussed in detail in Chapters 6 and 8). Thus for example, addition of 2% cholesterol to crude brain lipid extracts stabilizes BLM conductance to dielectric breakdown. Similarly, it has been noted that tocopherol and β-carotene make crude brain lipid films electrically excitable; however, tocopheryl acetate, cholesterol, and the products of lipid oxidation make the films less excitable and sometimes completely inexcitable. The reasons for these peculiar observations are not clear. However, the following rationalization may be considered provisionally. The concentration of solutes in the BLM is of the order of 10^{-9} to 10^{-10} mole-cm^{-2}. At these concentrations the rate of formation and the stability of BLM would be greatly influenced by the presence of trace amounts of particulate[1198] or surface-active impurities either in the aqueous phase or in lipid solution; in either case, incorporation of these impurities would lead to localized changes of interfacial tension of BLM to the extent of affecting not only some of the specific membrane properties but even their stability.[1060] An alternative way of interpreting these observations is that these "impurities" cause a change in surface potential which is likely to alter the electrical charge profile within and around the membrane and thus lead to reorientation of charged and dipolar groups.[140]

The nature of the lipid components also determines the kinetics of BLM formation. A BLM can be prepared from a rather large variety of lipids and amphipaths (see Table 3-1). The prerequisite which the lipid molecule must possess in order to form BLM are difficult to specify, but some limited generalizations can be made from the available data. The membranes have been prepared from pure phospholipids (containing single or mixed chemical species) usually dissolved in single or mixed solvents with or without any additive. During the formation of BLM at least a major portion of the solvent diffuses away. Structural requirements for the lipid molecule do not seem to be critical, as BLM have been prepared from phospholipids containing saturated, unsaturated, and branched hydrocarbon chains. Although Mueller, Rudin, and co-workers[99] prepared BLM from partially purified steer brain lipids which contain at least seven components,[141] the nature of different molecular species in the lipid mixture also seems to have little significance as far as their ability to form BLM is concerned. It appears that almost any amphipath when mixed with some suitable additive and filler can form BLM. It has already been pointed out that the major role of these additives is to modify the properties of the lipid solution so as to stabilize the film during and after thinning. One of the particularly important features of lipid composition arises from the consideration of the storage properties of the lipid solution. Thus, highly unsaturated lipids tend to be easily autooxidized and polymerized at room temperature and/or in the presence of oxygen and light.

Ability to form a stable BLM is not the sole criterion for preparation of a lipid solution. In biological membranes there appears to be some specificity associated with the degree of unsaturation, chain length, the nature of polar groups, and the relative abundance of various molecular species. Besides this, there is ample evidence suggesting that the nature and composition of lipid in cellular membranes is a function of the metabolic and functional state of the cell. Thus it becomes particularly interesting to correlate lipid composition with the properties of BLM. In fact, several modest attempts described later in this book have given significant insight into the functional significance of lipids in natural membranes. Also there seems to be considerable specificity regarding the nature of lipids present in BLM when the BLM is modified with proteins. These and various aspects are considered in greater detail in Chapter 9.

Once formed, BLM can withstand gross manipulations and vibrations, and are stable with respect to strong convection currents, are self sealing when punctured, are mobile, and are viscoelastic with low rigidity.[99,139] A small glass pipet can be pushed through and withdrawn without destroying the membrane. The BLM is liquid in the plane of the membrane and its low viscosity is indicated by the fluid motion of particulate material or of islands of thicker lipid under the influence of convection, vibration, or stirring. Depending on the lipid type the film may shatter violently or may rupture when broken. Such a behavior is indicative of differences in interfacial tension; shattering occurs in BLM of higher surface tension. However, in conclusion, under suitable conditions and when prepared from appropriate lipids, BLM have life times of more than twenty-four hours.

COMPOSITION OF BLM

The composition of BLM formed by thinning of a thick lipid film is not necessarily the composition of the bulk lipid solution used for their formation. Radiotracer analysis of fixed BLM has been reported.[126] The membranes formed from isotopically labelled components were stabilized and fixed with lanthanum nitrate and potassium permanganate.[142] A sample withdrawn by punching the stabilized membrane with a syringe needle (sample collected in the hole of a hollow needle) was analyzed for the radioactive components (Table 3-2). The thickness of these fixed membranes has been found to vary from 38 to 116 Å. It is possible that this variation is artifactual.*

The bilayer region is expected to be in continuous exchange equilibrium with the adjacent monolayers constituting the surface of the thickened lipid torus on the supporting torus. Thus a large decane concentration in the substance composing the membrane could also reflect the contribution of these areas rather than the true composition of the thin bimolecular lamellar structure.

---

*In the BLM sometimes there are highly mobile and light-reflective regions (small isolated lenses of thickened lipid or included particles) which can be seen to move continuously on the surface of the membrane. This could affect the computed composition of the BLM. If, however, the variations in the thickness of BLM are not artifacts from fixing, it has been speculated that the variations seen in the electron micrographs reflect the time-dependent fluctuations of thickness rather than a static structural characteristic.[142]

TABLE 3-2 THE COMPOSITION OF SOME BLACK LIPID FILMS

Film Components	No. of Molecules per cm^2 Film $\times 10^{14}$	Moles Decane/ moles polar lipid	Vol. Decane/ vol polar lipid chains	Method
Glycerol monooleate	6.8			Surface chemical
(> 3 mM)		0.57	0.36	method[143]
n-decane	3.9			
Egg yolk phosphatidyl	3.2			Surface chemical
choline (7 mM)		1.4	0.45	method[143]
n-decane	4.5			
Egg yolk phosphatidyl	3.0			Surface chemical
choline (7 mM)		(< 0.3)	(< 0.1)	method[143]
cholesterol (14.5 mM)	1.2			
n-decane	(0–1.3)			
Phosphatidyl ethanol-		11.4 ± 3.4		Radiotracer
amine (< 5 mM)				method[142]
n-decane				
Phosphatidyl enthanol-		10.5 ± 1.8		Radio tracer
amine (< 5 mM)				method[142]
cholesterol (< 10 mM)				
decane				
Isosorbide monobrassidate	1.7	3.3	1.7	Surface tenison measurement[144]

The radiotracer method for determining the composition of BLM is the most attractive method. It is, however, important that the radioisotopes in the film should be detected and estimated without in any way perturbing the black film. A variety of experimental difficulties must be overcome to achieve this end, and so far it has not been possible to do so. Other indirect methods based on properties of interfaces and behavior of amphipaths at these interfaces have been described.[143] One of the methods depends upon the establishment of a relationship between the composition of the interfaces of BLM and of the bulk lipid solutions with which the BLM is in equilibrium. It has been suggested that the difference in composition of these two types of interfaces is quite insignificant for films of glycerol monooleate and decane, and that this is also likely to be true for other lipid BLM and for the films of similar thickness. This is based on the experimental observation that the interfacial pressure of BLM (π^F) and bulk interface (π^S) are effectively equal and similarly dependent on the composition of their respective monolayers. Thus the amphipath content of BLM may be determined from a knowledge of either π^F or π^S and of the number of amphipathic molecules required to produce this interfacial pressure at the bulk interface. This latter information may be obtained from conventional surface balance techniques used for the study of monolayers. Yet another method for determining composition of BLM utilizes the Gibbs adsorption isotherm. The composition of BLM is calculated from measurements of the interfacial tension of the membrane and the composition and interfacial tension of phospholipid mono-layers at oil/water interfaces. Thus, the method is applicable only when the lipid is soluble in water and oil phases in the molecular sense.[144] It may be noted that both of these indirect methods for determining BLM composition are applicable to BLM prepared from one or two components only. For impure systems, or systems embodying complex mixtures, the radioisotope method seems the only one feasible. The composition of BLM prepared from various mixtures and as determined by these three methods is given in Table 3-2.

It is difficult to generalize from these results. However, the data do suggest a relationship between BLM composition and stability.[144] Furthermore, the con-centration of lipid in BLM is much higher than its concentration in the lipid mixture. Considerably lower ratios of both decane to phospholipid and choles-terol obtained by surface chemical methods are consistent with the interpreta-tion that a high decane content obtained by radiotracer method must be due to lenses of trapped hydrocarbon.

The concentration of organic solvents appearing in the aqueous medium bath-ing BLM is usually less than 0.01% as found by direct analysis. At this concen-tration the solvents generally used in lipid mixtures do not have any effect on various membrane properties and phenomena.[111]

OTHER FORMS OF BIMOLECULAR LAMELLAR LIPID MEMBRANES AND OTHER MODEL LIPID MEMBRANES

The very small areas of stable planar BLM which can be generated limit their applicability as experimental tools. For several experimental purposes, it is how-ever convenient to use a large number of small membrane-bound vesicles. These structures bounded by single or an array of lamellar layers can be prepared in a variety of ways and they offer a high surface/volume ratio. A large number of vesicles with cell-like geometry can be produced under suitable conditions by dispersing phospholipids in aqueous salt solutions above the phase-transition temperature of the lipid.[145-153,1061,1062] These vesicles are generally termed liposomes, spherules, smectic mesophase, and sometimes *Bangosomes* (after Dr. Bangham). The weight-average molecular weight of the egg lecithin aggregate sonicated for 25 min in water, is approximately 10^7 (see Ref. 146). The precise geometry of the aggregate depends upon the relative concentrations of lipid and water, the temperature, composition and nature of lipids, the salt concentration, and pH.[84,147,1061] Similarly, the size and the shape is determined by the method of dispersion:[1062] mechanical agitation promotes fragmentation and produces liposomes of 5-50 μ diameter, whereas ultrasonication induces the formation of smaller sacs averaging 250 Å across and usually bound by a single bilayer of lipid.[143,149] The physical characteristics of the smaller vesicles may be summarized as follows:[147b] sedimentation velocity, $S_{20,w} = 2.15$; diffusion coefficient $D_{20,w} = 1.87 \times 10^{-7}$ cm^2-sec^{-1}; partial specific volume, $\bar{v} = 0.9885$ ml-g^{-1}; intrinsic viscosity, $\eta = 0.041$ dl-g^{-1}; vesicle weight = 2.1×10^6 daltons.

The surface area of the vesicles can be computed by measuring the pellet volume of the centrifuged liposomes, and also by titrating the suspensions with uranyl nitrate.[147] The surface area of the dispersion as a function of the phos-

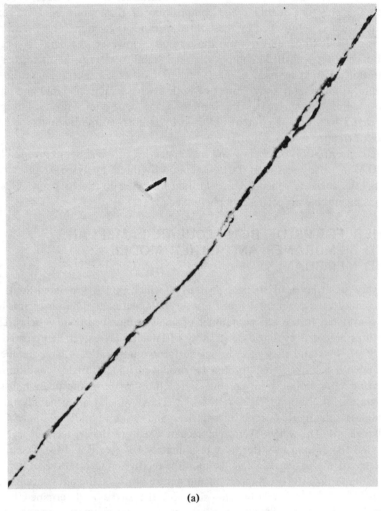

(a)

Fig. 3-4. Electron micrograph of (a) BLM prepared from phosphatidyl ethanolamine-decane fixed with La(NO$_3$)$_3$ and KMnO$_4$ ($\times$ 45,000) and, (b) liposomes prepared from lecithin/cholesterol/dicetyl phosphoric acid (240,000 $\times$).

pholipid concentration can be computed from the area of one uranyl ion, and from the amount of uranyl acetate required for the partial titration of the surface charge. The values obtained by both of these methods are in close agreement and the information can be used to determine the degree of association of liposomes in the suspension. In general, the liposomes can be handled like suspensions of cells, that is they can be centrifuged, resuspended, and counted.

Low-angle X-ray diffraction studies,[148] electron microscopy, and birefringence studies[152] suggest that the most common structural form of liposomes consist of many concentric lamellar bimolecular layers of lipid each separated by an aqueous compartment. Every bilayer presumably forms an unbroken membrane. and every aqueous compartment is discrete and isolated from its neighbors. The structures thus closely resemble myelin (compare Fig. 1-5 and Fig. 3-4b). The

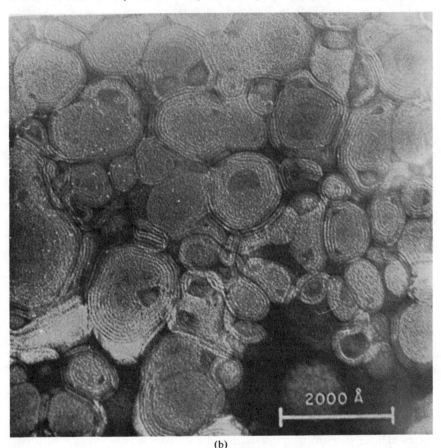

(b)

Fig. 3-4 *continued*

integrity of the multilayer membranes has been shown by permeability studies in which they act as perfect osmometers (see Chapter 5). The number of bilayers in liposomes usually varies from 8 to 30. Electron micrographs of these membranes also show differences after incorporation of various additives. Thus using egg lecithin liposomes, the mean thickness of the hydrophilic regions is increased in the presence of retinol and tocopherol. The relevance of these observations to membrane structure is not obvious.

It is possible to prepare vesicles bounded by a single bimolecular lipid layer, usually termed cellules.[111,154,1063,1200] A suspension of red cell hemolysate (or egg white or cytoplasmic constituents of a living cell) is shaken with a solution of lipids in hexane or octane. Working up the emulsion by centrifugation and recovering the aqueous phase gives cellules. Triolein or ethanol can substitute for hexane, and both purified and crude natural lipids can be used. Aggregation and suspension of the cellules can be controlled by varying the amount of divalent ions and of EDTA in the medium. Cellules can also be prepared from suitable lipid emulsions (for example those formed by using deoxycholate of dodecyl sulfate as emulsifier) by removal of emulsifier by dialysis or dilution. A large variety of single or mixed lipids in the presence of various proteins and RNA can be used for the formation of cellules. The action of proteins and RNA is nonspecific and apparently serves only to prevent the formation of the multiple concentric lipid bilayers (liposomes).

Cellules are freely permeable to water and show osmotic swelling. Electron micrographs show a vesicular structure bounded by a single bilayer of about 100 Å thickness. These vesicles can trap various electrolytes and neutral solutes in their bound aqueous phase and are thus excellent models for the living cell membrane. In fact, membrane-bounded vesicles have also been reconstituted from the dispersed subunits of natural membranes. Bacterial and mitochondrial membranes dissolved in lauryl sulfate[155] or deoxycholate[156] reform membrane fragments and membrane-bounded vesicles upon removal of the dispersing agent by dialysis, or upon dilution or addition of divalent cations.

The major disadvantage of dispersion systems, especially liposomes, is their structural complexity, particularly nonuniformity of surfaces, which makes interpretation of data difficult.

A very interesting type of bulk lipid membrane can be prepared by UV irradiation of a drop of unsaturated glycerides spread upon an oxidizing solution (one percent potassium permanganate). After a few hours a thin flexible membrane is obtained. It shows bright interference colors corresponding to a thickness of about one micron. The membranes can be prepared from both crude lipids such as linseed oil or tung oil, or from pure glycerides of unsaturated fatty acids such as glycerol monooleate mixed with oleic acid (5:1) deposited on a moist cellophane sheet in the form of a thin film. The properties of these films however do not depend upon the procedure used for their initial oxidative polymerization, nor on the interface on which they are prepared (water/air interface or cellophane film).

The oxidatively polymerized lipid films can be floated on saline solution and can support a drop of another solution, thus separating two distinct aqueous phases.[61] They show very interesting electrical properties such as cationic conductance, intercationic selectivity, rectification, and excitability. Some of their physical properties may be summarized as follows: capacitance, 4×10^{-9} farad-cm^{-2} (semi-circular Cole-Cole plot); resistance, 1.2×10^{10} ohm-cm^{-2} ($Q_{10} = 4$ at 18–28°C); thickness, 3 μ average. These membranes probably have a cross-polymerized structure and certainly do not resemble BLM in structure or form. Some of their electrokinetic characteristics are described further in Chapter 8.

Yet another type of bulk lipid membrane can be prepared on porous solid supports.[55,60] In a typical procedure, filter paper or Millipore filter discs (150-μ-thick tablets of cellulose nitrate and acetate) are dipped into a benzene solution of the lipid. After evaporation of the solvent, the coated discs can be mounted on suitable support. These lipid-coated discs show electrical resistance of the order of 10^5 ohm-cm^{-2} which depends upon the nature of the electrolytes present in the bathing solution. The curve relating film resistance to pH shows inflections and peaks at the same pH values as does the titration curve for the membrane-forming lipid. The osmotic permeability of these membranes is also significantly altered by multivalent cations present in the bathing solution. All these properties indicate that these membranes behave as cation exchangers. Their properties will not be discussed in subsequent chapters of this book.

MEMBRANE PHENOMENOLOGY

The organization of the membrane, both in its static structural and probably in its dynamic functional aspects, is governed by the general physical principle of minimum free energy. This implies that the overall process of BLM formation proceeds in the direction which leads to an overall decrease in free energy, and its rate is determined by the activation energy required for the transition. We have discussed earlier that the propensity of lipid molecules to form aggregate structures exists because these molecules can best attain their free energy minimum by orienting their polar groups towards water and bringing their hydrocarbon chains together. The bimolecular lamellar structural form is one of the forms which lipids can assume. The energy of formation of such systems for practical purposes can be defined as the isothermal work required to increase both interfaces of the membrane by one unit area. This can be measured experimentally by determining the radius of curvature of the membrane as a function of the pressure difference across it. Such pressure-induced changes in area are dependent upon the interfacial tension (surface free energy) of the membrane. Some general aspects of the energetics of BLM formation are considered in this section.

From the studies on monolayers it is obvious that there is a significant difference in the surface free energy at the interface between random and oriented hydrocarbon chains. Thus, in principle, it is possible to calculate the free energy of BLM formation in a semiempirical manner if electrostatic free energy terms are ignored. This assumption is justified later in this section since the capacity of lipid to form bilayers does not significantly depend upon charges on the polar groups. Based on these assumptions, from a theoretical standpoint alone it can be shown that the bimolecular lamellar structural form is the lowest free energy form for the organization of lipids.[157]

Consider the system water/lipid/water. The lipid molecules at the water/lipid and lipid/water interfaces will be oriented such that the polar groups are directed toward water and the hydrocarbon chains are in the lipid phase. The lipid molecules bound between these two oriented lipid layers will be more or less random. The system may be represented as follows:

Water	Oriented lipid molecules	Random lipid molecules	Oriented lipid molecules	Water
Surface energy $E_{w/p}$	$E_{o/r}$	$E_{r/o}$	$E_{p/w}$	

Such a system may be characterized by four interfaces with characteristic surface energies (given at the interfaces). Thus the total free energy of the system is

$$E = E_{w/p} + P + E_{o/r} + E_{r/o} + P + E_{p/w}$$

where P is the difference in the cohesion pressure between the oriented chains and the chains in the bulk liquid phase. Since $E_{w/p} = E_{p/w}$ and $E_{o/r} = E_{r/o}$, the total energy of the system is:

$$E = 2 \left[E_{w/p} + P + E_{o/r}\right]$$

As the thickness of the lipid layer is reduced until oriented interfacial layers come into contact, the term $E_{o/r}$ becomes zero, and when the layer is strictly bimolecular in thickness the total surface energy is:

$$E = 2 \left[E_{w/p} + P\right]$$

If the thickness of the lipid layer is decreased further a rise in surface free energy will occur since it will create a new surface consisting of a hydrocarbon/water interface favoring the transition bilayer $\longrightarrow$ micelle. If it is assumed that the new hydrocarbon/water interface thus formed has an interfacial energy (E_{CH_2/H_2O}) of $\simeq 50$ erg-cm^{-2}, the approximate free energies of formation of cylindrical and spherical micelles will be 6000 and 12000 cal/mole, respectively. Thus, in principle at least, it should be possible to calculate the proportion of material present as bilayer and as micelle (at equilibrium) under any given set of conditions.

The relationship between free energy change and the thickness of the film is shown in Fig. 3-5. The fall in free energy $(2\,E_{o/r})$ can be obtained by one of the following methods:

1. From the contact angle between a drop of randomly oriented hydrocarbon fluid resting upon a completely oriented hydrocarbon layer, such as a monolayer.

2. From the value of interfacial tension at the inverse point for the equilibrium: oil dispersed in water $\rightleftharpoons$ water dispersed in oil (where $E_{p/w} = E_{o/r}$).

3. From the difference in the polarizability of hydrocarbon chains in the direction of C–C bonds compared with the direction perpendicular to C–C bonds.[158]

4. From the critical surface free energy measurements for penetration of an oil droplet into a bilayer membrane.

The fall in free energy $(2\,E_{o/r})$ for the system under discussion has been determined by method 1, and found to be approximately 20 erg-cm^{-2}.

The conclusion from such crude thermodynamic considerations, i.e., the process of bilayer formation should proceed spontaneously, comes of course as no surprise. More sophisticated treatments also lead to similar conclusions.[159] Such analysis further suggests that practically all materials that form condensed mono-

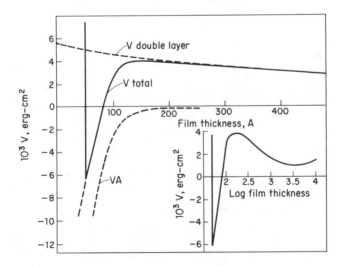

Fig. 3-5. The energetics of lipid bilayer formation. Estimates of the electric double layer, London–van der Waals, steric repulsion, and gravitational free energies have been combined to give an overall curve of free energy as a function of thickness.[164]

layers at liquid-liquid interfaces could (thermodynamically) form bilayers.* Experimentally it has not been proved so; the failure probably arises from mechanical and dynamical causes to be elaborated later in this chapter.

Spontaneous formation of BLM has several evolutionary implications for the biological membrane. In the primitive atmosphere, lipids are believed to have evolved quite late on the scene. However, basic mechanisms to fulfil membrane functions, that is compartmentalization or even concentration and accumulation, must have developed at a fairly early stage in the primordial atmosphere. Jeewanu,** proteinoid microspheres, coacervate or even air bubbles in seawater have apparent morphogenicity and the ability to locally concentrate material.[161] These and other observations, thus suggest that inherent self-organization is observed at several levels of biological order. It is the spontaneity of generation of order and consequently of organized structure that characterizes processes normally associated with *life*. Although we shall restrict our discussion to membranes formed from lipids, the underlying principles can be extrapolated to other systems.

In molecular terms the results can be interpreted as follows. Since the polar groups have a strong tendency to leave the lipid phase and the hydrocarbon phase has a strong tendency to keep water away, the hydrocarbon chains should should be effectively close-packed. Since the cross-sectional area of the polar head of a phospholipid molecule is approximately the same as that of the two hydrocarbon chains, in the absence of any other external force the continuous bimolecular lamellar leaflet is the most favored aggregate structure since the

---

*It is of course implicit in the above discussion that there are no specific forces between unlike molecules, across the interface, which do not exist between like molecules within one or both separate phases. An example of such specific forces is the hydrogen bonding between a substituted hydrocarbon chain with the species in aqueous phase. If such specific interactions exist between the molecules partitioned in two phases, $E_{o/r}$ might well be positive; and the bilayer would not form spontaneously. This may offer a rationale for the observed inability of aged and oxidized lipids to form BLM; these lipids are expected to contain epoxide, peroxide, and other oxygen functions which, besides altering the geometry of orientation of hydrocarbon chains, may also act as hydrogen acceptors in a hydrogen bond with water.

**Jeewanu (Sanskrit word for "particles of life") are formed by exposing an aqueous solution of molybdic acid, paraformaldehyde, and ferric chloride to bright sunlight for about 4–8 weeks.[160] Several other recipes are also available. Jeewanu are spheres of 0.3 to 1.5 μ diameter, they exhibit an external membranelike structure, and have a dark interior. These microspheres provide another means for establishing a delimiting boundary from the external environment since it was found that the constitution of the interior differed from that of the external medium. Jeewanu seem to undergo growth, "multiplication," and have the ability to carry out simple localized catalytic processes (they show phosphatase activity). Apparently, these processes are not due to microbial contamination since other aspects of cellular function are missing.

interchain interaction can occur most efficiently in this configuration. Complications could, however, arise due to factors such as mutual repulsion of neighboring charged polar groups. Such effects, at least in part, can be overcome by specific adsorption of cations or dilution of charge by interspacing them with neutral additives, such as cholesterol, tocopherol, decane, etc. Thus the continuous lamellar lipid bilayer structure is the lowest free energy and geometrically preferred configurational form which amphipaths such as phospholipids can assume in aqueous solutions.

In this connection it may be pertinent to recall that biological macromolecular systems exist in characteristic three-dimensional structures which are specific for each type of molecule under a given set of conditions. These structures are self-organizing since the conformation which the macromolecule assumes is thermodynamically the most stable under biological conditions of temperature, pH, and ionic strength. These aggregated structures are not dependent on the input of energy, nor are they the result of covalent crosslinking. The phenomenon of their formation and stability is simply the automatic outcome of severe limitations on the freedom of rotation around single bonds and of intermolecular interactions. Such limitations are produced by the occurrence of weak bonding, both specific and nonspecific, between different functional groups of the molecules and in part by the geometric rigidity of various bonds themselves. Almost all of these factors are encountered in the formation and stability of BLM. Thus, in common with other biological *supramacromolecular* structures, the distinctive feature of the membranes is that they have evolved (and been selected ?) for specific functions as dynamic and self-reproducing systems. The organization is a consequence of the coordinated control of the interactions between specific structures, and the control mechanisms are themselves intrinsic properties of the organized structures. Therefore, it may be concluded at this stage that the bimolecular lamellar lipid structure is a natural unit of structure, arising spontaneously just as the helixes in protein or the double helix of DNA. The nature of the interactions and the fine structure of the associated membrane-forming molecules will be dealt with in greater detail in Chapter 9. In the remainder of this chapter we shall elaborate some of the thermodynamic parameters which determine formation and stability of BLM.

THERMODYNAMIC PARAMETERS

Based on the foregoing discussion the reasons for the stability of BLM may be summarized as follows:

1. The amphipathic nature of the phospholipids which results in a large free energy term for the *removal* of the hydrocarbon chain from the aqueous solution, and the term for the *removal* of polar groups from the hydrocarbon phase.

2. The anisotropic orientation and the consequent anisotropy of polarizability

of the hydrocarbon chains of the lipids. It gives rise to a surface free energy at the interface between oriented and random hydrocarbon chains.

3. The high surface free energy at the hydrocarbon/water interface presumably due to structuring of water.

4. Electrostatic energy of interaction between charged groups in the plane of the membrane, as well as across the thickness of the membrane. Structuring of water as a consequence of electrostatic forces at the interfaces of BLM may also have a stabilizing effect on the BLM.

There are many ways in which the contribution of these and various other factors can be evaluated and the BLM stability interpreted correspondingly. Thus far, however, the absolute or relative contribution of each of these factors has not been evaluated. Some information regarding the magnitude of various effects may be obtained through the study of BLM formation under different conditions of temperature, salt concentration, pH, etc. Some of these results, their interpretation, and implications will be presented in the following sections; the nature and the contribution of factors 1–3 for membrane stability are described next and the electrostatic energy factors are treated subsequently.

The total internal surface energy, U, of a system is the sum of the changes in surface free energy (ΔF) and the entropy (S) terms:

$$U = \Delta F + TS = \gamma_{BLM} + TS$$

By differentiation

$$S = -\left(\frac{\partial \cdot \Delta F}{\partial T}\right)_{n,V} = \left(\frac{\gamma_{BLM}}{\partial T}\right)_{n,V}$$

In physical terms, the total surface free energy is the total potential energy of the molecules that form 1 cm^2 of the surface in excess of that which the same molecules would possess in the interior of the bulk liquid. The formation of a BLM biface may be imagined to take place in three experimentally distinguishable steps. These are (a) the creation of a water/oil/water biface; (b) the adsorption or migration of interfacially active molecules to the biface; and (c) the formation of the BLM from the monolayers situated at the biface. The energy changes in these individual steps are termed ΔF_1, ΔF_2, and ΔF_3. It is evident that the overall reaction of BLM formation is

$$\text{Liquid hydrocarbon + Water + Lipid} \longrightarrow \text{BLM} + \Delta F_i$$

where ΔF_i is the sum of the three free energies. Thus

$$\Delta F_i = \gamma_{BLM} - \gamma_{o/w}$$

where $\gamma_{o/w}$ is the interfacial tension of the oil/water interface, and γ_{BLM} is the interfacial tension of BLM. Thus the interfacial tension gives the surface free energy, and the temperature coefficient of interfacial tension gives the entropy term.*

As a specific example, the experimental data are available for BLM formed from the lecithin-dodecane system: $\gamma_w \simeq 72$, $\gamma_{dodecane} \simeq 22$, $\gamma_{o/w} \simeq 72$, $\gamma_{bulk\ solution\ of\ lecithin} \simeq 6.5$, and $\gamma_{BLM} \simeq 0.9$ erg-cm^{-2}. Thus, the free energy change in the formation of BLM from lecithin at a clean $W/O/W$ biface is about 50 erg-cm^{-2}. Assuming the area occupied per lecithin molecule in BLM to be 50 Å^2, the molar free energy change is calculated to be -3.6 kcal. It may be noted that the bulk solution of lecithin in n-dodecane ($\gamma \simeq 6.5$) can produce a BLM with $\gamma_{BLM} \simeq 0.9$ erg-cm^{-2}. It has been suggested that the two monolayers in a thick film of lipid solution possess a higher interfacial free energy at the $W/O/W$ biface than the BLM, i.e., the formation of BLM from lecithin monolayers involves a reduction of free energy. This is probably reflected in the kinetics of BLM formation where a thick membrane (colored) undergoes spontaneous phase transition to BLM after appearance of black spots.

Using BLM generated from dodecyl acid phosphate/cholesterol/dodecane the variation of interfacial tension with temperature has been determined. From the equations presented above the entropy, free energy, and enthalpy of BLM formation can be calculated. The results of one such study are presented in Table 3-3.

The implication of this information in relation to the molecular organization in BLM may be elaborated as follows. The molecules of a liquid in the bulk phase have other liquid molecules as their nearest neighbors. However, at an

TABLE 3-3 INTERFACIAL TENSION, FREE ENERGY, ENTROPY, AND ENTHALPY OF FORMATION FOR THE BILAYER PRODUCED FROM CHOLESTEROL—DODECYL ACID PHOSPHATE IN n-DODECANE IN 0.1 N NaCl SOLUTION[163]

Temp ($°C$)	γ_{BLM} (erg-cm^{-2})	$-\Delta F$ (erg-cm^{-2})	$-S$ (erg-cm^{-2} deg^{-1})	$-H$ (erg-cm^{-1})
25	1.1	50.6	0.008	53.0
28	1.2	50.5	0.018	55.9
32	1.3	50.4	0.036	61.4
34	1.4	50.3	0.051	65.9
36	1.5	50.2	0.077	74.0
38	1.7	50.0	0.125	88.8
40	2.0	49.7	0.190	109.2
42	2.5	49.2	0.264	132.4

*The energy of formation of BLM relates not only to the interfacial tension but also to the interaction energy between the two bulk aqueous solutions that it separates. However the latter energy appears to be of the order of 0.06 erg-cm^{-2} (Ref. 162), therefore the value of the energy of formation of BLM may be roughly considered to be equal to twice the value for interfacial tension.

interface the adsorbed molecules have a different environment on one side relative to the other side. Thus the molecules may occupy a position either in the immediately subjacent bulk phase or at the interface. In BLM such a possibility of randomness (in terms of "choice") does not exist; hence the surface entropy term on this count must be zero. However, the loss of randomness for lipid molecules in the bilayer phase (which the molecules had in bulk phase) may be partly compensated by reorientation of the water lattice at the aqueous interface. At the interface, some of the molecules of water may be oriented by the adjacent methylene groups into an icelike form. Hydrogen bonds and the tendency of water to form "icebergs" around dissolved molecules may partly be responsible for such changes. Unfortunately, little information is available on the contribution of such factors to the entropy terms in the surface energy changes during membrane formation. The net entropy changes during BLM formation may, however, be expected to be small compared to the net free energy change. Indeed, it has been found to be so; for BLM prepared from dodecyl acid phosphate/cholesterol/dodecane in 0.1 N sodium chloride solution the entropies of formation have been found to be negative and range from 0.008 at 25°C to 0.264 erg-cm^{-2}-deg^{-1} at 44°C (Table 3-3). The enthalpy of BLM formation (H_i) may be calculated from the Gibbs-Helmholtz equation:

$$d\left(\frac{\Delta F_i}{T}\right) = -\frac{H_i}{T^2}\,dT$$

The value of H_i for membrane formed from DAP/cholesterol/dodecane has been found to be 53 erg-cm^{-2} at 25°C. The values of entropy and enthalpy terms imply that the latent heat of formation of BLM is also negative, that is evolution of heat takes place during the formation of BLM.[162,163] This is specially significant at higher temperatures. One of the explanations of these values lies in the change of the orderliness of constituent molecules at the water/oil/water biface; it is also consistent with a higher degree of organization in the BLM phase.

It was pointed out earlier that the free energy change during the formation of an interface is equal to the interfacial tension. In a thinning lipid film the total interface of the film remains almost unchanged since the bulk solution withdrawn from the enclosed bulk of the film is added either to the bulk phase of a torus or some such energetically equivalent state. Therefore, the forces balancing the metastability of thin films are gravity, capillary suction from the borders, van der Waals forces, dissolution of the film, steric effects (or Born repulsion) due to crowding of molecules, charge density, order in molecular packing, and the repulsive electrostatic forces at the electrical double layer. Similarly the breaking of the film may involve conversion of surface free energy into kinetic energy of the film's free edge, dissipative energies resulting from the environmental drag and internal liquid viscosity, and the surface energy of drop formation. Thus the relative contributions of these factors should be adjudged to ensure stability of BLM; it is necessary that the net free energy changes during thinning of the thick film be positive.

The relative contribution of various factors determining BLM stability and breakdown depends to a large extent on factors such as the composition of the mixture forming BLM, the composition of the aqueous phase, and various mechanical disruptive forces. From the discussion of these various factors governing the kinetics of BLM formation it emerges that additives significantly affect the stability of BLM. It was also noted that the major contribution of additives towards BLM stability may be through the lowering of interfacial tension and by interspersing and diluting the charge at the interface of the BLM. The contribution of these factors to the total surface free energy of BLM formation is difficult to evaluate; however, circumstantial evidence is provided by several observations. A solution of pure cholesterol does not form BLM.[135] However, stable BLM can be prepared by adding small amounts of a surfactant to the aqueous medium or to the lipid mixture. It should be noted that stable BLM can only be formed when the surfactant concentration is below the critical micelle concentration; if the surfactant concentration is above c.m.c. the resultant micelles will solubilize the lipids, resulting in the destruction of the BLM. The interfacial tension of hydrocarbon/water interface is about 50 dyne/cm, whereas, introduction of cholesterol in dodecane reduces the value to 9.6 dyne/cm. Both of these systems are apparently incapable of forming stable BLM. If, however, some surfactant (below the c.m.c.) is added, stable BLM may be prepared; the interfacial tension of these membranes is less than 4 dyne/cm.[120] Thus a small but finite interfacial tension is a necessary but not a sufficient condition for the formation of stable BLM.

Less direct evidence for the role of surface energy on the stability of BLM has been obtained through their interaction with polyene antibiotics (to be discussed in detail in Chapter 6). Filipin and nystatin (at $4 \times 10^{-5} M$) have no effect on the stability of BLM containing only lecithin. At the same concentration these antibiotics disrupt films prepared from an equimolar solution of lecithin and cholesterol.[129] These results have been interpreted to suggest that the disruptive effect of the polyene antibiotics is due to an increase in the surface pressure of the membrane which could be due to specific interaction of antibiotics with cholesterol. The effect of these antibiotics may be localized such that only a part of the total membrane undergoes phase change due to increased surface pressure, thus promoting rupture. A similar explanation may be found for the destabilizing influence of various trace impurities in the medium, particularly the ones which tend to adsorb on the surfaces.

Besides the cohesive forces arising due to interaction of hydrocarbon chains, the aqueous interfaces of BLM are also the site of several electrostatic interactions. The problem may be treated simply as follows. Consider two hydrocarbon layers of semiinfinite thickness having electrolytes on both sides. The

BLM may be treated as a hydrocarbon layer sandwiched between two electrolyte solutions and considered to be formed by bringing these two semiinfinite layers toward each other. Thus the relevant electrostatic energy of the formation of BLM is the difference between the electrostatic energy of the hydrocarbon layer with electrolytes on both sides and the sum of the electrostatic energies of the semiinfinite hydrocarbon layers.

Haydon[164] has calculated the net free energy change for bimolecular leaflet formation using the theory of Verwey and Overbeek[165] for electrostatic free energy and a two-dimensional equation of state[166] for the surface interaction energy. The data on the adsorption of normal-chain lipid molecules at hydrocarbon/water interfaces suggests that in many systems the adsorbed molecules are oriented normally to the interface and the chains can be regarded statistically as rigid rods. Furthermore it is suggested that the repulsion of two monolayers is not infinite as soon as they touch each other provided the lipid molecules have more than about twelve carbon atoms in their chains and form monolayers from dilute solutions that are more than about half close-packed.[107] The potential energy curve for a hydrocarbon film as a function of its thickness has been computed[167] and is shown in Fig. 3-5. A similar curve is obtained from the considerations described earlier (p. 75). The results clearly suggest that a free energy minimum is achieved when the film thickness is about 60–70 Å, that is, when the film has a bimolecular lamellar configuration.

Parsegian developed a statistical thermodynamic theory[168] to calculate the free energy change in the phase transition between neat and middle phases of fatty acids and water. The electrostatic free energy component is derived by the Debye-type charging process as applied to cylindrical polyelectrolyte molecules by Lifson and Katchalsky.[169] To account for short-range forces between hydrophobic parts of the lipid and aqueous medium (surface interaction energy) a macroscopic thermodynamic representation is used. The resulting statistical thermodynamic theory has been applied to account for the X-ray diffraction data on lamellar liquid crystals of pure cell membrane lecithin in water.[83,170] The data has been interpreted as evidence for a long-range potential field around the electrically neutral BLM which has a field strength comparable to forces within the leaflet.[171] This field acting to repel the leaflets causes an increase in the thickness of the lipid aggregates as the water fraction decreases. The X-ray diffraction data also suggest that the chains of lecithin molecules tend to aggregate in bilayers with maximum disorder, and this disorder itself acts as a thermodynamic force when perturbed.[172] In fact, studies on the membrane interfacial properties suggest that the hydrocarbon chains are free to rotate internally.[173] Though this study is useful, it is not directly applicable to the problem of the stability of BLM since the boundary conditions on the electrostatic potential for two cases are entirely different.

Ohki and Fukuda[158] have calculated the stabilization energy of the model bilayer system by a quantum mechanical quantitative refinement. The system has been treated as composed of two oriented anisotropic molecular layers. The interlayer interaction is divided into three parts: dispersion interaction between two oriented monolayers, dipole-induced dipole interaction between one oriented monolayer, and a permanent dipole layer. From the second-order perturbation theory of intermolecular forces between asymmetric molecules, a general expression for these interactions is obtained and their numerical values computed. Interaction energies between two oriented, and an oriented and a random monolayer are 33.1 and 24.4 erg-cm^{-2}, respectively. These values indicate that the energy from dipole-dipole interactions is repulsive for thin membranes of finite size. If the dimensions of the bilayer film are small, there is a net repulsive force between the permanent dipole layers which may be greater than the sum of the attractive dipole-induced and dispersion forces. Further evaluation of the theory may throw more light on the underlying basis for the stability of BLM.

Vaidyanathan and Goel[174] have attempted a quantitative evaluation of the electrostatic energy responsible for BLM stability by approximating this term with the energy of interaction between ions on either side of the lipid film. Such an approximation involves the assumption that other electrostatic interactions remain essentially unchanged when two semiinfinite layers of electrolyte are brought together. Preliminary calculations suggest that for the BLM with the same electrolyte on both sides, there is no *extremum* in the electrostatic energy of interaction as a function of film thickness.

In the foregoing paragraphs we have seen that the dispersion forces and the electrostatic energy terms are of paramount importance in determining stability of BLM, although quantitative appraisal is not yet possible. The evaluation of the intermolecular forces between membrane components, especially in relation to their orientation and the corresponding lattice structures, and in relation to the establishment of the dynamic equilibria resulting from the balance between the cohesive and disruptive forces (e.g., thermal agitation, border suction, etc.) is probably needed.

In conclusion, based upon the above theoretical and experimental considerations, some of the requirements for the stable BLM formation may be summarized as follows:

1. The solute molecules in the hydrocarbon core must be strongly adsorbed, that is, the standard chemical potential change must be large and negative. For long-chain molecules which are adsorbed from hydrocarbon to water the value of the standard chemical potential change is determined chiefly by the size and polarity of the head group as this is the only part of the molecule which actually transfers from one phase to another.

2. The amphoteric species used to stabilize the BLM must be able to lower the interfacial tension to about 5 dyne/cm.

3. If BLM-forming substances bear net charge, neutral or co-polar molecules must be present to serve as spacers (additives or fillers) so that the repulsive forces between the charged groups can be reduced and the van der Waals interactions between hydrocarbon chains enhanced.

4. Since the surface tension of the film is finite, complete collapse will occur if a hole of greater than a critical size is formed. On the basis of a simple model, DeVries[175] has calculated that the activation energy for formation of the critical size hole is:

$$E_{act} = 11.7\, d^2 \gamma$$

For a thickness (d) of about 50 Å and tension of 1 dyne/cm, the activation energy is 7.3×10^{-13} erg or about 18 kT for a hole of 1 mm. The effect of a slight increase in surface tension of the film may thus limit the size of the hole.

5. The tendency of the lipid film to dewet and become detached from the supporting material might be a serious handicap.

6. The concentration of lipid in the BLM is determined by several factors, including c.m.c. and solubility of lipid in water and organic solvent, viscosity of the mixture, and possibility of specific intermolecular interaction between different components of lipid mixture.

7. The concentration of multivalent ions is critical in maintaining the structural integrity of the membrane; both low and very high concentrations may affect the life and formation characteristics of the BLM.

8. X-ray diffraction data points out that the interactions between chains are weak arising from van der Waals and dispersion forces, and the chains are more or less ordered. Thus any factors which disturb such an arrangement would affect membrane stability. Branching of the chains, presence of cis double bonds, and formation of epoxides and peroxides would be expected to do so.

Thus formation of stable BLM should be possible from most amphipaths which have apolar straight hydrocarbon chains. However, disruptive forces due to micelle formation, droplet formation, and very low or very high viscosity must be overcome.

4/ Intrinsic Properties of Black Lipid Membrane

That function emerges from structure is one of the basic axioms of molecular biology. The structure of the lamellar bimolecular lipid membrane and its relationship to the surrounding medium gives rise to its characteristic physicochemical properties, which in turn may be manifested in various phenomena characterizing biomembranes. However, before the properties of the protoplasmic surface film can be discussed, it is necessary to differentiate clearly between the film and the extraneous coats present on the surface of the cell. In nature, the protoplasmic surface film is seldom, if ever, exposed. Extraneous material adheres so closely to the external surface of the film that it is frequently difficult to differentiate the one from the other. Thus, as a prelude to the discussion of more complex membrane systems and associated phenomena, in this chapter we shall discuss some of the basic physicochemical or intrinsic properties of black lipid membranes. See Refs. 177, 189, 359, 1064, 1201 for reviews.

We have seen in the last chapter that the major driving force for the formation of BLM is the net free energy gain of interface formation, reflected in its low interfacial tension. The liquid crystalline character of the core of BLM is manifested in its stability and in its surface mechanical properties such as viscosity, elasticity, and mechanical strength. However, few general remarks can be made with regard to these characteristics.

Anisotropic orientation of molecules forming BLM and the consequent requirement for an overall cohesion pressure resulting from dispersion forces may imply that the physical state of the lipid molecules ranges between fairly narrow limits; therefore, even with substantial variation in lipid composition, most of the physical properties of the membrane remain essentially invariant. As the hydrocarbon core has a relatively coherent structure, it will define the basic level of dielectric and associated properties such as refractive index, electrical capacitance, and permeability characteristics. Hydrocarbons have low dielectric constants and consequently low ionic permeabilities. Thus, the lipid core provides an ideal system for the generation of an electrical potential difference, which could arise in any of the following situations:*

1. A static potential difference is set up when charged particles accumulate on charged interfaces. Since the interfaces of BLM are composed of charged groups, the counterions in the vicinity of the interface would give rise to a diffuse double layer (Fig. 2-4) and the consequent surface potentials and related electrostatic and electrokinetic characteristics.

2. A diffusion potential is set up by different mobilities of ions in the aqueous and the membrane phases. This potential can only be maintained by replenishment of ionic gradients.

3. An ohmic potential drop can be produced if current is passed from an external source between the compartments and across the BLM.

These various potentials give rise to a variety of phenomena, as we shall consider in later chapters.† In the remainder of this chapter, we shall describe some of the elementary properties of simple lipid bimolecular membranes. It may be ambiguous as to whether a given property of the BLM is due to the black portion or to the residual thick area as, for example, the torus border. Evidence that a given phenomenon occurs in the black region and is unattainable in films prior to thinning, and that the properties dependent upon the area show direct proportionability to the area of the black membrane, is generally taken as a strong implication that the given phenomenon occurs on the BLM. Since the precise membrane composition is not known, the composition referred to is always of the lipid solution employed for the formation of BLM. During the discussion of the properties of BLM, comparable data for biomembranes is also provided so as

---

*Although these various electrical characteristics of membranes and their interfaces are discussed at reasonable length in this book, an excellent discussion of basic electrochemistry involved in movement of ions and effect of the dielectrics of the medium may be found in Ref. 180.

†Before discussing specific aspects of membrane properties it might be pertinent to note contributions of yet another structural factor regulating membrane properties. Water is an important constituent of the polar region. The thickness of this region is largely determined by the electric charge density at the interface and by the ionic strength of the medium. Thus, depending upon the nature of the environment, one may envision various types of organized water including (a) water in direct contact with the ordered macromolecular membrane surface; (b) water layered between closely packed hydrocarbon chains; and (c) probably interconnected water channels within the lamellar lipid layers. These permanent and transient substructures and the cooperative ordering of water can play an important role in membrane function. Unfortunately, the precise macromolecular nature of ordered water in membranes is still a matter of speculation and is expected to depend on the nature of the protein and other constituents present therein. We do not intend to delve into such aspects of membrane structure-function correlations.

to note the similarities between the two, which may reflect the contribution of the lipid portion (presumably the lipid core) to the observed characteristics. There are significant differences in the properties of BLM and biomembranes which may be attributed to some components not present in BLM or may be due to differences in structural organization. These differences are elaborated further in Chapter 6.

SURFACE MECHANICAL PROPERTIES

The mechanical properties of BLM have not been measured, however see Ref. 1202. Qualitatively BLM appear to be resilient, liquidlike, and fragile. Detailed studies of rigidity, elasticity, extensibility, surface viscosity, etc., which may play a critical role in the function of membrane as a structural element, are certainly warranted.

Measurements of the mechanical properties of the plasma membrane have been greatly hindered by the smallness and complexity of natural vesicular structures. It is even more difficult to interpret these results due to the presence of relatively rigid structural elements—mitochondria, endoplasmic reticulum— just beneath the surface of these membranes. In spite of these difficulties attempts to measure the surface mechanical properties of biomembranes have been made from time to time. The intact erythrocyte membrane has a Young's modulus of elasticity of 10^6-10^8 dyne-cm^{-2} and a viscosity of 10^7-10^{10} poises.[1065] Comparable data for BLM are not available, but a pure lipid layer is expected to be much less viscoelastic. Studies on monolayers reveal that the surface viscosity remains approximately constant over the whole range of composition of cholesterol-lecithin monolayers at a surface pressure of 10–30 dyne-cm^{-1}, and that they show liquid character.[181] Even though the surface pressure of BLM is much higher it is expected to behave in a manner similar to monolayers. For such reasons the mechanical strength which a cell envelope may possess is unlikely to arise from the BLM structure. The reason for this may be ascribed to other components, such as proteins and polysaccharides, present in significant quantities in biological membranes, which are certainly expected to contribute to the mechanical properties. In fact, a direct role of proteins in the determination of the mechanics of the plasma membrane may be inferred from the observed effects of proteolytic enzymes on the deformability of the cell surface.* It has been found that trypsin reduces the forces required to deform

---

*When a cell changes its shape, its plasma membrane conforms through some combination of elastic and plastic deformations or through exchange of membrane components with an intracellular pool. From the uniformity in the thickness of membranes of any one type, it follows that changes in shape at constant surface area will be associated with conjugate patterns of displacement and/or two-dimensional flow. Substantial changes in total area which are not compensated by an exchange of components will result in wrinkling or rupture. An exchange of components may proceed as a passive response to local changes in tension or as a result of any one of a number of processes which are under local biological control.

the surface of various cell types, and that the cells are unable to repair tryptic lesions.[182a]

It has been noted that the mechanical properties of cell surfaces change as the forces acting on them vary; that is, they have non-Newtonian viscous properties. In fact, the surface of some cells show a remarkable fall in viscosity in the first few seconds after exposure to shear.[183] This thixotropic behaviour indicates that exposure to shearing stress can decrease or loosen the molecular packing of cell surfaces. The extent and direction of viscosity changes with shear, however, depend strongly on the shearing force and the rate of change of shear.[184] Short periods of low shear may decrease molecular packing in surfaces and lower surface viscosity whereas long periods of strong shear can produce increased molecular packing and increased surface viscosity (rheopexy). Finally, the effects of shear depend upon the molecular organization as well on the homo- and heterogeneity of the membrane.

Such observations and others to be mentioned later suggest that cell membranes are not static systems; they continually respond to the dynamics of their environment, whether it be to concentration gradients, fluid pressure gradients, temperature gradients, or mechanical forces. The analysis of viscoelastic responses to time-dependent strain may provide information of a fundamental nature.[178] The role played by lipid regions in determining viscoelastic properties may be insignificant; however, their liquidlike character and resiliency may contribute significantly toward the dynamism of their response to change in environment.

ADHESION AND COHESION

In principle, the factors maintaining two surfaces in adhesion are the forces of attraction and forces of repulsion between them, and, at equilibrium, they balance each other. Besides this the forces due to interdigitation of the hydrocarbon chains and viscosity may keep the surfaces in adhesion although these are not adhesive forces in the usual sense.[179] Because of the magnitude of the distances involved, the attractive and repulsive forces are usually due to covalent bonding, hydrogen bonding, coulombic attraction or repulsion, charge fluctuation forces, dispersion forces, and image forces due to the tendency for ions to move away from the regions of low dielectric constant, such as the BLM core. Thus the charged surface groups and ionic environment of the medium assume special significance. The driving force for adhesion of two separate bilayer interfaces may be the decrease in the free energy of mixing; however other factors may also be involved.

Adherence properties of BLM formed from brain lipids, as well as of thick lipid films (colored films), in various electrolytic solutions at different values of pH have been reported.[185,1066,1205] The results have been interpreted from the standpoint of the theory of interaction of charged surfaces. Since the fixed negative charge on the surface of the BLM is reduced by the cations in the medium, the probability of van der Waals forces assuming increased importance, and, consequently, the possibility of a decrease of adhesion and of shortening the time of adhesion is increased. The method employed is essentially measurement of the time of collapse of two membranes in surface-to-surface contact. The typical membrane adherence histogram in 100 mM potassium chloride solution showed an average arithmetical mean value of 21 sec, a dispersion of 185 sec^2, and an asymmetry coefficient of 1.4. In distilled water, membranes show no adherence for more than 30 min. In contrast, increasing cation concentration reduces markedly the adherence time. When 10^{-4} to 10^{-5} M calcium or magnesium chloride is added to the medium, the adherence time is reduced to about 10 sec. In the presence of higher concentrations of calcium and magnesium, more than about 1mM, the adherence time is increased considerably. Potassium ions antagonize the effect of calcium ions. A similar antagonistic effect of potassium ions on adsorption of calcium ions has been noted on stearic acid monolayers.[186] In these experiments, the type of anion does not seem to have any effect on the adherence characteristics of BLM. The adherence time tends to decrease with temperature or with increased contact area. Results obtained from BLM and thick colored films are very similar.

Based on these experiments, it has been suggested that the effect of physicochemical variables on adhesiveness of BLM is that expected on the basis of its bimolecular anisotropic structure, and that the role of divalent ions such as calcium in regulating aggregation of various lipidic vesicular structures, including natural membranes, may be attributed to their chelating action with the polar groups of the lipids. Various cellular properties such as vesicle formation, protoplasmic streaming, endocytosis, invagination, secretion, and adherence, among others, depend upon intercellular interactions or on the phenomenon of adhesion and coalescence between two continguous membranes.[187] The fusion processes probably are more complex and may require specific macromolecules,[1183] or appropriate physical[1203] and energy (metabolic)-dependent[1204] factors while allowing merger and reorganization of two membranes.[188]

INTERFACIAL TENSION

As discussed earlier interfacial tension is a consequence of the surface-contracting tendency of liquids and has the same significance as the free energy of the molecules at the aqueous interface of BLM. The interfacial tension of BLM is of fundamental importance since it is determined by the configuration of molecules forming the interface.

A method for measurement of interfacial tension of BLM based on the maximum-bubble-pressure principle has been devised.[163] When a bubble of one of the interface-forming liquids is blown into the other liquid (say aqueous solution) through a capillary, the shape assumed by the bubble during its growth is such that although it is always a section of a sphere, its diameter goes through a minimum when it is exactly hemispherical. Since the diameter is at a minimum, the pressure difference (P) across the membrane is maximal and the diameter of the bubble under these conditions is equal to the diameter of the capillary. The technique involves measurement of the pressure difference across the BLM, and the interfacial tension (γ_i) is calculated according to the following relationship:

$$\gamma_i = Pd/8$$

where d is the diameter of the capillary. The values of interfacial tension observed for various BLM from several lipid types are collected in Table 4-1. The values are between slightly above zero to 5 dyne-cm^{-1}. The corresponding values for biological membranes are in the range of 0.03 to 1 dyne-cm^{-1} (Table 4-1).

Certain inferences regarding the molecular packing in BLM can be drawn from the magnitude of the interfacial tension. The low values of interfacial tension of BLM suggest that their surface pressure is almost the same as the interfacial tension between pure bulk lipid and water. This implies that the molecules in the BLM are highly compressed and probably packed as densely as in crystals, and thus occupy their limiting molecular area. The surface pressure due to the kinetic energy of electrically neutral molecules adsorbed at the oil-water interface is approximately given by the relation:

$$\pi = kT/(A - A_0)$$

where k is the Boltzman constant, T is the absolute temperature, A is the limiting area of the molecules in the adsorbed layer, and A_0 is the area occupied by the molecules at the surface pressure, $\pi = \infty \cdot A_0$ determined at the air-water interface for a monolayer of phospholipids is approximately 58 Å^2 per molecules. If the value of surface pressure for BLM is taken as 48–49 dyne-cm^{-1}, the values of A are computed to be 73.7 Å^2 for the former value, and 66.1 Å^2 per molecule for the latter.[167] The molecular significance of these values is discussed in Chapter 9; it is indicated that the adhesive forces between the molecules acting laterally and the perpendicular attraction between hydrophobic portions of two monolayers forming BLM is significant.

The effect of external variables on interfacial tension of BLM has also given some information about their formation and stability. The thermodynamic parameters of BLM formation have been derived from the relationship of temperature and γ_{BLM} (Table 3-3). The results indicate that the free energy of formation of BLM decreases slightly with increasing temperature (in the range 25–44.5°C). Similarly some tentative conclusions may be drawn from the effect of pH change on interfacial tension. The interfacial tension of BLM formed from lipids having anionic groups is found to be larger at higher values of pH than the values for other ionizable lipids at higher pH.[189] For membranes prepared from

TABLE 4-1 INTRINSIC PROPERTIES OF BLM AND THE CONSTITUENTS[a]

Film	Surface Tension (dyne/cm)	Capacity (F/cm²)	Resistance (ohm-cm²)	Dielectric Str. (V/cm)	Reflec.	Water Permeability (cm/sec)	Ref.
Decane	50	2.0	5×10^{15} b	10^6		10^{-4}	
Octadec-1-ene	50	2.14	5×10^{15} b	10^6		10^{-4}	
Dodecyl acid phosphate + cholesterol	1.1	0.69	10^8	10^6–10^7	10^{-5}	10^{-3}	120
Sorbitan esters + alkanes		0.33–0.57	10^9				164
Egg lecithin + alkane	0.5	0.39	10^6–10^7	5×10^5	5.9×10^{-5}	10^{-3}	192, 210
Egg lecithin + cholesterol + alkane	1	0.39–0.7	10^7–10^8	10^6		10^{-3}	135, 194
Phosphatidyl inositol + alkane		0.29	10^7				218
Glyceryl distearate + alkane	1.5	0.39	10^8	10^6	10^{-4}		78, 120
Brain lipids + cholesterol + alkane	1	0.4–0.6	10^6–10^7	10^4–10^5		10^{-3}	96–99, 141
Red cell lipids + alkane	1–2	0.55	10^8	10^5		10^{-3}	107, 279
Chlorophyll (crude) + alkane	3.5–4.5		10^6	10^5		10^{-4}	
Tocopherol + alkane			10^{-8}	10^5–10^6			216
Biomembranes	0.1–1.5	1.0	10^2–10^6	10^4–10^6		10^{-3}–10^{-4}	

[a]The values are approximate.
[b]Given in ohm-cm^{-1} for a film of 1-cm thickness.

neutral lipids, changing pH and salt concentration have much smaller effects on interfacial tension. The effect of salt concentration on interfacial tension of BLM prepared from charged lipids is much more noticeable.* These results thus seem to suggest that the nature of polar groups and the magnitude of charge on these groups in BLM may play some yet unknown role in modulating interfacial tension.

The surface tension at the electrolyte/metal phase boundary is a function of the potential across the interface; this phenomenon is known as electrocapillarity (for a review see Ref. 191). The effect on surface tension of the applied potential is attributed to the repulsion forces of the charge of the electrical double layer which tend to counteract the forces which help to keep the surface area to a minimum. The surface-spreading effect of this potential is independent of charge sign, and the greater the surface charge, the greater is the effect. The surface tension of the charged surface is decreased by the amount of energy per cm^2 which is required to charge an interfacial double layer of capacity C_d. For a potential change dE, this energy dF is given by

$$dF = -d\gamma = EC_d dE = QdE$$

where Q is the charge density per cm^2 on the interface whose interfacial tension is γ. Thus,

$$\frac{d\gamma}{dE} = -Q$$

which on differentiation yields

$$\frac{d^2\gamma}{dE^2} = -\frac{dQ}{dE} = C_d$$

which is the double layer capacity.

The dependence of surface pressure on the surface potential of BLM is not known. However, the above relationship strongly suggests that some phenomenon analogous to electrocapillarity observed at most charged surfaces should be discernible at BLM interfaces. Studies on monolayers do, however, suggest that the relationship between surface pressure and surface potential may be used to evaluate the effects of ions, chelators, additives, and other components which might affect the surface structure by adsorption on the interface and/or in the hydrocarbon region.

---

*It may be of interest to recall that the surface tension of the aqueous solution of lecithin is reduced by about 5 dyne-cm^{-1} on addition of 1 mM acetylcholine. On the basis of this observation it has been suggested that interaction of acetyl choline with phospholipid may change the permeability properties of synaptic membranes and thus initiate the postsynaptic nerve impulse.[190] The effect of pH on this change of surface tension has been interpreted to suggest that the effect is mediated by the electrical charge on the substrate.

OPTICAL PROPERTIES

The optical properties of BLM are basically the same as those of very thin transparent solid or soap films. In principle, the BLM in aqueous solution can be regarded as a thin dielectric plate of refractive index n_i immersed in a medium of refractive index n_0. It can be shown that at a certain angle β the intensity of the component of the reflected light which has its electric vector in the plane of incidence will be zero. It should be noted that the reflected light is completely plane polarized. The angle β is known as Brewster angle and for a homogeneous isotropic film the refractive index n_i is given by the relationship:

$$n_i = n_0 \cdot \tan \beta$$

The value of refractive index for egg lecithin BLM for white light as $1.66 + 0.03$[192] or $1.60 + 0.03$.[193] This value is much higher than the expected value for egg lecithin (1.49) or of any component of the solution from which the BLM are prepared. Furthermore, these values are not consistent with the value 1.37 obtained by other coworkers.[194] The latter value can, however, be reconciled by taking into consideration the anisotropic structure of BLM.

In the reflectivity measurements, complications resulting from birefringence can be avoided by making all measurements using the light perpendicular to the plane of incidence. From the value of reflection coefficients at various interfaces the thickness of the membrane can be computed as 55–65 Å.[78,192,194–196] These calculations however require some assumptions regarding the membrane structure, for example, whether it is single layered or triple layered (Fig. 4-1). As shown in Table 4-2 reflectances computed from both of these models by assuming various values for the thickness of the layer (core) are almost identical if the polar groups of the lipid are sufficiently small.[193,194] It is however

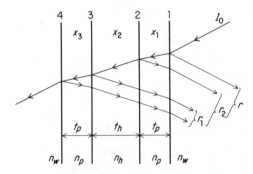

Fig. 4-1. Reflection of light from a triple-layered membrane model. $I.$ is the incident light wave; x refers to a phase difference due to a different layer; $r, r_1,$ and r_2 are amplitudes of the reflected light at various interfaces; d denotes thickness; and n is the refractive index. The subscripts w, p, and h refer to the aqueous phase, the polar layer, and the hydrocarbon portion of the membrane, respectively.

TABLE 4-2 CALCULATION OF THE REFLECTANCE $(R \times 10^4)$ FOR DIFFERENT VALUES OF n_h AS A FUNCTION OF t_p[189a]

t_p(Å) [1]	Triple-Layer Model			Single-Layer Model	t(Å) [6]
	$n_h = 1.42$ [2]	$n_h = 1.44$ [3]	$n_h = 1.46$ [4]	$n_h = 1.56$ [5]	
		$d = 48$Å			
1	5.129	7.477	10.309	32.325	50
2	6.208	8.769	11.816	34.949	54
3	7.388	10.162	13.425	37.675	56
4	8.671	11.658	15.136	40.500	58
5	10.056	12.255	16.948	43.426	60
		$d = 50$Å			
2	3.959	4.660	5.428	6.707	54
6	5.623	6.452	7.348	8.824	62
8	6.561	7.454	8.451	9.988	66
10	7.571	8.527	9.552	11.222	70
14	9.801	10.883	12.036	13.899	78

[a]See Figure 4-1 for explanation.
Other values used: $n_w = 1.33$, $n_p = 1.56$; $\lambda = 4350$; d is thickness of the hydrocarbon core.

evident that the reflectance of BLM is very sensitive to the thickness of the polar groups and the refractive index of the membrane. Although the values of membrane thickness determined by various authors are in a fairly narrow range, the values of refractive index and reflectivity differ considerably (compare Refs. 193 and 194). Theoretical analysis suggests that these measurements do not permit a conclusive determination of membrane thickness.[197,1067]

The effect of changing pH and salt concentration has not been studied extensively. It has been reported that the thickness of the BLM formed from glycerol distearate depends upon the salt concentration in the medium.[78] One of the major difficulties is to ascertain whether a certain effect is due to change in the thickness of the membrane as such or to change in reflectance and/or refractive index. In fact, experimental data strongly suggest dependence of Brewsters angle and reflectivity of BLM on the refractive index of the aqueous phase.[194]

DIELECTRIC CONSTANT AND ELECTRICAL CAPACITY

The electrical capacity of BLM in aqueous solution is due to its dielectric core between two conductors. The insulator core is essentially a mixture of n-alkane and the hydrocarbon chains of the phospholipid fatty acids as indicated by exceedingly high values for the electrical resistance and dielectric breakdown voltage. Thus, a BLM may be treated as a parallel plate condenser.

The capacity, C, of a nonconductor between two conductor plates is defined by the charge $\pm Q$ on the plates and a potential difference between them by the relationship:

$$C = \frac{Q}{V} = \frac{4\pi LQ}{\epsilon A}$$

where L is the thickness of the dielectric of dielectric constant ϵ and area A. As the potential across the insulator changes, a charging current, I, flows.

$$I = \frac{dQ}{dt} = C\frac{dV}{dt}$$

If the applied field is alternating.

$$I = \bar{I}\cos\omega t = C\omega V\cos\omega t$$

where V is the sinusoidal potential, and $\omega = 2\pi f$, f being the frequency. From these relationships can be derived the relationships of R_m (membrane resistance) and X (reactance) as a function of ω:

$$R(\omega) = \frac{R_m}{1 + \omega^2\tau^2} \quad \text{and} \quad X(\omega) = \frac{-\omega\tau R_m}{1 + \omega^2\tau^2}$$

The locus of R versus $-X$ as a function of frequency is a circle of radius $R_m/2$ and center at $(0, r_m/2)$. The angular frequency for which the real and imaginary parts of the impedance are equal corresponds to the reciprocal of the membrane constant. Plots of R vs $-X$ are known as Cole-Cole plots as shown in Fig. 4-2.[198]

Capacitance of BLM prepared from a variety of lipids, as well as that of cellular membranes, has been measured (Table 4-1). For various membranes, values range from 0.3 to 0.65 $\mu F\text{-cm}^{-2}$. Higher concentrations of cholesterol impart higher values.[204] The time constant for charging and discharging the membrane

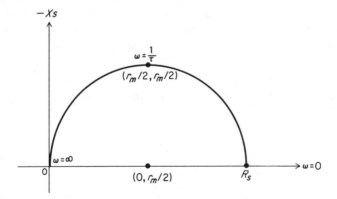

Fig. 4-2. Impedence locus for an idealized capacitor. Corresponding experimental plots for BLM are also semicircular.

(by the dc transient method) is about 10 sec; therefore, most of the determinations of capacitance have been made by the ac method. The Cole-Cole plots between 10^2-10^7 cps are semicircular, suggesting that membrane capacitance is independent of the frequency in this range.[110] Similarly, it has been shown that capacitance is independent of the nature of the lipid polar groups, electrolyte concentration (0.001 M to 0.5 M NaCl), and the applied voltages up to 50 mV.[104,105,109,110,164,199,206] Subjecting the membrane to mechanical vibrations causes a capacitance change which follows the wave form of the vibration.[178]

Treatment of BLM capacitance as that of a parallel-plate condenser gives reasonable values for the membrane thickness, or, alternately, for the dielectric constant of the hydrocarbon core. For the calculation of BLM thickness, the value of the dielectric constant is assumed to be 2.07 as the average of the values 2.0 for n-decane and 2.14 for octadec-1-ene. The thickness of the insulator core of BLM is computed to be 48 Å for a BLM prepared from lecithin-decane mixture.[110] However, this value for the dielectric constant may not be correct. By definition, the dielectric constant is a characteristic of the medium between two charges. If we consider a lipid bilayer to be composed of oriented hydrocarbon molecules, its dielectric constant must be anisotropic since the relation of the dielectric constant of a substance to the structural properties of its constituent molecules depends upon the average local field acting on the center of a molecule. Therefore, the dielectric constant of the bulk liquid hydrocarbon must be different than the value for a bilayer core. A molecular theory of a nonpolar dielectric for an oriented system in accordance with Kirkwood's theory of dielectric polarization has been developed.[200] The value for the dielectric constants derived from this theory are given in Table 4-3. The calculations suggest that dielectric constants parallel and perpendicular to the interface differ significantly. Since the hydrocarbon chains in the core are parallel to each other and perpendicular to the interface, the average dielectric constant pertaining to an electric field across a BLM is ϵ. Using these values, the thickness of a BLM core varies between 52.970 and 60.097 Å, depending upon the value chosen for the

TABLE 4-3 THEORETICALLY CALCULATED VALUES OF DIELECTRIC CONSTANT OF ANISOTROPICALLY ORIENTED HYDROCARBON LAYERS[200]

	Dielectric Constant[a]		
	Bulk	Monolayer	Interchain Distance
$\epsilon_{\parallel}$	2.4897	2.5910	4.472
$\epsilon_{\parallel}$	2.2706	2.2706	5.000
$\epsilon_{\perp}$	1.7435	4.0277	4.472
$\epsilon_{\perp}$	1.5143	2.2762	5.000

[a]$\epsilon_{\parallel}$, Dielectric constant parallel to hydrocarbon chains.
$\epsilon_{\perp}$, Dielectric constant perpendicular to hydrocarbon chains.

interchain distance. The limiting values are 4.472 and 5.00 Å for close-packed and liquid structures, respectively. Therefore, the thickness of the BLM core might change depending upon the degree of compression of the molecules forming BLM. The higher value for the dielectric constant of hydrocarbon layers oriented anisotropically as compared with the value for bulk liquid hydrocarbon, implies that strong intermolecular forces operate in the oriented hydrocarbon layers.

Based on the model just described treatment of membrane dielectric as a layer of hydrocarbon sandwiched between regions of a mixture of polar groups and aqueous solution is rather an oversimplification, not only in terms of anisotropy of orientation of the hydrocarbon layer but also due to the effect of the electrical double layer. If the polar groups are ionized, an electrical double layer will be present at the film interfaces. The resulting model of the membrane may be presented as shown in Fig. 4-3. The behavior of this model has been analyzed in terms of the equations for a three-layer dielectric consisting of hydrocarbon, polar, and aqueous phases.[167,201,202] The results are simplified under the conditions of low potentials or when the applied potential is much smaller than the electrical double-layer potentials. For zero net charge the results are the same as those obtained by neglecting the effect of the electrical double layer. The membrane capacity shows a dependence on surface charge density and consequently on electrolyte concentration.[110] Thus at low electrolyte concentrations, the electrical double layer should cause a decrease in the membrane capacitance. However, at more than $10^{-3} M$ electrolyte, the charge density becomes sufficiently high to nullify the effect of the electrical double layer. The effect of the double layer is expected to be of significance at higher electrolyte concentrations for membranes formed from strongly polar lipids, as, indeed, is observed experimentally. Under an applied ac potential of about 40 mV, the capacitance per unit area of lecithin + n-decane BLM in 4.18 M sodium chloride is approximately 1000 times higher at 5×10^5 cps than in 0.1 or 0.001 M sodium chloride solution.[110,137,201,203]

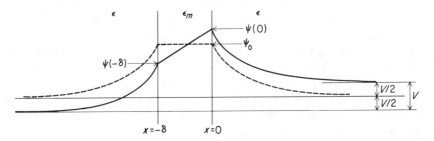

Fig. 4-3. The triple-layer model for BLM dielectric. The dashed line represents the potential before, and the full line after, a potential V is applied across the membrane.[167]

On the application of a dc potential the capacitance of BLM changes.[137,199,204] The initial change following application or withdrawal of the current appears to follow an exponential time course which attains a constant value in 3-4 min; the time constant is approximately 30 sec. Maximum changes of up to 15% of initial values have been recorded and these appear to be proportional to the square of the applied voltage. The time constant has a large temperature coefficient, $Q_{10} \sim 3.5$, and is strongly dependent on the ionic mobility. However, the capacitance changes in the media of low ionic strength and mobility are rather small. Similar changes in membrane capacity are elicited by altering temperature.[205] These effects of electric field or temperature on membrane capacity may be associated either with a decrease in the thickness of BLM and an altered dielectric coefficient, or with the extension of its surface,[137] or are a consequence of changes in composition and geometry of the diffuse layer. It has been demonstrated that there is a correlation between the values of surface tension and the change in capacity under the influence of the applied voltage. Such changes are ascribed to some special effect of the electrolytes on the BLM interface.[199] The electrolyte concentration versus capacitive change plots show a pronounced dip at the salt concentration for which the Debye length* is just equal to the dipole length of the polar group in the lecithin molecule. The occurrence of a minimum is found to be strongly dependent upon the composition of the membrane; in the presence of cholesterol or in membranes prepared from aged lipids, the curves tend to be flattened out. These results suggest that the applied potential alters the ionic distribution near the interface in an energetically reversible manner. It has, however, been emphasized that these effects are secondary and the primary variable is change in membrane thickness.[205,1068] This conclusion is supported by square dependence of C_m on applied potential, since an electric field exerts a compressive force on a parallel-plate capacitor which is proportional to the square of the applied potential.

The effects of other factors on BLM capacitance may be summarized as follows (see Ref. 204, 205 and 1068):

1. Capacitance as a function of pH shows a minimum around pH 4 (for lecithin BLM), and increases symmetrically toward low pH or high pH. The pH minima are different for different lipids, and the shape of the curves changes with Ca ion concentration.

2. The breakdown potential is function of pH; e.g., it is 200 mV in solutions in the pH range 4-8. However in solutions of pH 3 or 9 the breakdown potential is less than 140 mV, and is even less than 60 mV in the solutions at pH 2.

3. BLM modified by addition of various substances does not affect the capacitance even though its conductance increases by several orders of magnitude (see Chapter 6 for further discussion).

---

*The Debye length refers to the effective thickness of ionic cloud.[180]

Most of the measurements of the cell membrane capacity yield values some-what higher,[207] in the neighbourhood of one μF-cm^{-2}, than those found for BLM. The capacity of biomembranes is found to change with the frequency; in contrast, BLM capacitance is independent of the frequency of ac current. The low and high frequency capacitance of a squid giant axon membrane has been estimated to be 0.8 and 0.6 μF-cm^{-2}, respectively.[208] This means that if the greater part of the membrane structure is based on a lipid layer, a certain percentage of the area must consist of material with a higher frequency-dependent dielectric constant. At the moment, not much is known about the origin of such differences although some speculations have been made in this regard. In the case of striated muscle (frog), values of capacitance of 5 to 8 μF-cm^{-2} are observed. A careful study by Falk and Fatt[209] has resulted in a modified membrane model involving two time constants. An electrical repre-sentation of the shunt admittance per unit length (reciprocal impedance) of muscle fiber is shown in Fig. 4-4a. In addition to the parallel capacitance and re-sistance, C and R_m, respectively, there are introduced the series elements C_s and R_s. For an ideal two-time-constant network, such as illustrated in Fig. 4-4b, where the capacitances are widely separated in value, two nonoverlapping circu-lar loci result. Thus, in the low-frequency range one branch remains virtually open-circuit so that a typical single-time-constant behavior due to the second branch remains. As the frequency is further increased, the latter branch behaves like a pure resistance with negligible capacitative reactance, whereas in the other branch the admittance is no longer zero. Again the behavior is characterized by that of a single-time-constant circuit.

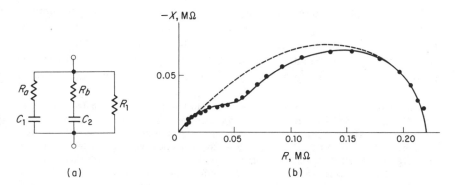

(a) (b)

Fig. 4-4. (a) Electrical circuit representation of a two-time-constant network as a model to explain the capacitance of muscle fibers as a function of frequency. When the capacitances are widely separated in value, two nonoverlapping circular loci result. Otherwise varying de-grees of overlap between two loci can give rise to dispersion effects. (b) Impedence locus for frog sartorius muscle fiber over the frequency range 1 to 10,000 cps. Dots are corrected ex-perimental points. The theoretical curve for the two-constant model is solid, whereas the single-constant curve is shown slashed. Both figures are after Ref. 209.

When the capacitances are not greatly different, overlap occurs and a locus such as the dotted curve in Fig. 4-4b results. The latter is typical of results obtained from measurements on muscle fibers which yield two dispersions. The ratio of transmembrane voltage to current, the so-called characteristic impedance, is the quantity normally measured. The impedence plot of Z corresponds to the dotted curve in the figure. The results still require a satisfactory explanation in physical terms.

The results of studies concerning capacitance measurements on biological membranes and those on BLM would imply that the surface charges of the BLM are rather localized,[1206] and that the stereochemistry of the interface must allow considerable ionic movements, more than would appear to be possible if the lecithin dipoles were firmly or permanently bound in the plane of the membrane. As the capacitance can be regarded as an index of molecular packing and polarizability,[1207] it seems reasonable to postulate that many natural and artificial lipid-based membranes are similar in this respect and the relationships deduced for one system should have relevance to others.

ELECTRICAL CONDUCTIVITY*

The hydrocarbon core of BLM separates the two media of high dielectric constant. In order to transfer a charge from a medium of high dielectric to a medium of low dielectric constant, a certain amount of electrostatic energy is needed. It may be calculated from the following relationship:

$$U = e^2/2r \, (1/\epsilon_1 - 1/\epsilon_2)$$

where U is the energy needed to transfer a charge e of radius r from a medium of dielectric constant ϵ_1 to a medium of dielectric constant ϵ_2. If we consider the transfer of a potassium ion (radius approximately 2 Å) from water (dielectric constant 80) to the membrane phase (dielectric constant 2), the electrostatic energy needed will be approximately 1.5 ev. The concentration of potassium

---

*A conductor is defined as a material in which a substantial number of charges are free to move. An electric field exerts a force on every charge and those free to move are accelerated simultaneously. However, each charge travels only a short distance before colliding with some molecule of the conductor and giving up some kinetic energy to it. Thus the random kinetic energy of all the molecules (temperature) is increased and the potential energy of each free charge is decreased. The acceleration-collision process goes on continuously for all free charges and the net result is that the charges move through the medium at an unvarying, rather slow, average speed. Thus current flow involves the circulation of charges in a closed path, and Ohm's law describes current flow in a conductor if the average speed of charge movement is directly proportional to the electric field and hence to the applied voltage. That is, friction is proportional to speed. Thus Ohm's law follows from the reasoning that the steady-state flux of ions is proportional to the driving force of an electric field (cf. Chapter V).

ions in the membrane phase $[C_m]$ can be correlated with its concentration in the aqueous phase $[C_w]$ by the Boltzman distribution:

$$[C_m]/[C_w] = \exp(-U/kT) = \exp(-60) = 10^{-25} \text{ approx. (at } T = 300°K)$$

Thus at room temperature ($T = 300°K$) an extremely small fraction of the total charges should be present in the membrane phase. It is also evident from this approximation that the membrane conductance should be extremely small.

The electrical conductance of BLM is determined by applying a current and dividing the measured change in membrane potential by the intensity of applied current (Ohm's law). It is difficult to find a truely representative value of membrane conductance. For BLM prepared from most lipid types, the resistance is not constant for the life of the membrane[108,110,117,124,210] but characteristically passes through several phases of resistance drop. The stability and the absolute value of the membrane resistance seem to depend upon the composition of the aqueous media as well as on the nature of the lipids. Membranes prepared from hydrocarbon solutions almost invariably seem to have their resistance higher (about one or two orders of magnitude) than those formed from a chloroform-methanol solution of amphipaths. The values for membrane resistance range from 10^5 ohm-cm^2 to about 10^9 ohm-cm^2 (Table 4-1).

The rise time of the membrane current on the application of an applied dc pulse is usually 0.1 to 10 sec. The fact that the voltage signal has a slower rise and fall than the applied current pulse suggests that the BLM has a characteristic charging-up time. Thus it has the property of a leaky condenser whose voltage lags behind the current passing through it with a time constant depending on the capacity as well as the resistance of the BLM.

The current-voltage relationship of a BLM is shown in Fig. 4-5. Typically, the membrane resistance up to 50 mV of applied potential is ohmic and for potentials greater than 60 mV the BLM resistance increases less rapidly.[101,138,141,211,1069] This decrease indicates that some type of reversible resistance change is taking place in the BLM. This deviation from ohmic behavior is strongly dependent upon the BLM composition and, in general, is relatively more significant for membranes having low resistance.

Temperature change has anamolous effects on BLM resistance. For BLM prepared from lecithin-decane mixture, the resistance shows a maximum near 23° and a marked transition at 29-30°C;[212] however, sphingomyelin + tocopherol membranes show transition at 48°C.[1069] Such anamolous temperature effects have been observed in a variety of natural and artificial membranes including the n-butanol liquid phase membrane, cellulosic type membranes, plasma membranes of the green alga *Valonia*[213] and BLM modified with a proteinaceous material EIM[214] (see also Chapter 8). It has been proposed that these transitions are possibly associated with the onset of increased motion in the hydrocarbon chains[1069] (see also Chapter 9), or to a change in water structure in or

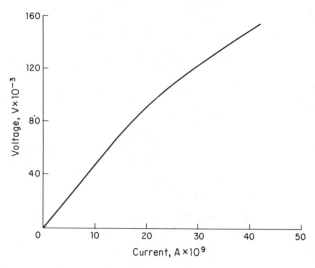

Fig. 4-5. The current-voltage relationship for a BLM (lecithin–cholesterol–dodecane). The current–voltage relationship is linear to about 60 mV, then exhibits a gradual decrease in the voltage developed slope with further increase in current until dielectric breakdown.

adjacent to the BLM. The Arrhenius plot (log conductance vs reciprocal of absolute temperature) is linear between 10–55°C for BLM showing no temperature anomalies. The activation energy for conductive processes is found to be 14–18 kcal-mole^{-1}. However, BLM showing large temperature-dependent transitions in conductance have activation energies in the range 20 kcal-mole^{-1} at higher temperature, but the value at lower temperatures is usually less than 15 kcal-mole^{-1}.

The change in membrane conductance with the area of BLM is as expected from considerations of the membrane as an insulator.[104,105] During the formation of BLM, spot checks for conductivity show that the conductance increase is proportional to the growth of the black film. The conductance through the bulk lipid torus or through the septum has been found to be negligible.[108] Constant activation energies over a large range of conductances also suggests that the ion penetration mechanism is similar in all cases, supporting the thesis that the variation in conductance of different membranes is dependent upon the structure of BLM rather than the result of membrane faults or edge leakage.

The precise value of BLM conductance is dependent on the pressure[1070] and concentration of electrolytes in the aqueous medium, and probably on the nature of the polar groups of the lipid. The relationship between electrolyte concentration and conductance is shown in Fig. 4-6. For various electrolytes, up to a certain concentration, the conductivity is constant and usually independent of the nature of the anion or the cation. Similarly, above 0.1 M concentration the conductivity of lecithin BLM increases linearly with the salt concentration[130]

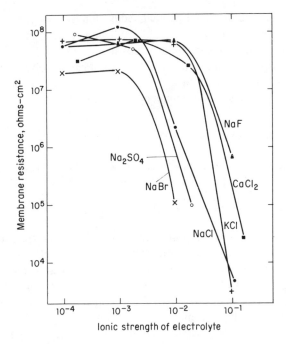

Fig. 4-6. Resistance of BLM (cholesterol–decane in 0.008% hexadecyl trimethylammonium bromide) in various salt solutions (pH 6.85–6.95).[124]

but is independent of the nature of the univalent cation species. When chlorides of Cd, Mn, or Cu were added in low concentrations the BLM conductance decreases by a factor of two to nine. Ferrous and ferric ions markedly increase the conductance, whereas magnesium and calcium ions have little effect.[108]

The decrease in BLM resistance at relatively high salt concentrations has been attributed to the equilibrium reaction:

$$R^+ Br^- + X^- \rightleftharpoons R^+ X^- + Br^-$$

occurring at the surface of the membrane.[124] This exchange of anions could affect the BLM conductivity in more than one way. It is possible that R^+X^- may salt out of the membrane or form an ion pair. Since R^+Br^- is a membrane component, in either case the conduction would be affected. It is also possible that, at high salt concentrations, BLM may undergo phase transitions which may be reflected in alteration of the interfacial tension. For the electrolytes studied with hexadecyl trimethylammonium bromide cholesterol BLM the order of increasing effectiveness for reducing BLM resistance is the same as the Hofmeister series:

$$I^- > Br^- > SO_4^{2-} > Cl^- > F^-$$

This implies that the smaller hydrated radius and greater polarizability of the anions run parallel; both factors should result in a stronger interaction with the cationic head of the lipid membrane. Consistent with this hypothesis is the sequence of transference numbers for alkali metal cations observed for development of diffusion potentials (a measure of relative ionic permeability) across lecithin BLM.[108] Transference numbers for a series of cations were found to be in the order:

$$H^+ < Li^+ < Rb^+ < Na^+ < Cs^+ < K^+$$

The bulk aqueous solution values listed below show that the transference numbers within the membrane are quite different.

$$Li^+ < Na^+ < K^+ < Rb^+ < Cs^+ < H^+$$

Clearly, the ion discrimination process in the membrane is different from that in bulk water. It appears that ionic selectivity for these membranes is determined by the polar group of the lipid: Cationic head groups permit passage of anions, and vice versa. The selectivity sequence observed for various membranes are not necessarily those predicted theoretically (compare selectivity isotherms predicted on p. 166). In the anionic selectivity sequence observed for lecithin BLM the permeability of protons is small compared with their permeability in bulk water. This may be attributed to the fact that large values of the proton transference number in aqueous solution depend upon the proton jump mechanism, and it is likely that this mechanism may not operate for proton transfer across BLM. Thus it is probable that the membrane phase through which conduction occurs has a lower dielectric constant than bulk water. Consistent with this hypothesis is the observation that cationic transference numbers for bilayers formed from a variety of phospholipids do not depend on the type of phospholipid, thus suggesting that the polar group on the surface of the bilayer may not in itself be the major factor governing cation permeability.[215] It may also be noted that BLM prepared from a variety of apolar compounds, such as tocopherol and cholesterol without any polar lipids, have conductances similar to the BLM prepared from polar lipids.[216]

The effect of pH on the conductivity of egg lecithin BLM in the presence of Na and Ca ions has been studied.[138,217] The resistance in NaCl medium shows a maximum at pH 4, whereas the resistance in calcium chloride solution increases continuously with increasing pH (Fig. 4-7). This difference is probably due to dissociation of surface polar groups at extreme pH values which will tend to open the lattice and consequently its permeability to ions. At higher pH and in the presence of calcium ions, this expansion is prevented by chelation. A similar explanation has been forwarded to account for the effect of calcium on the surface pressure and surface potential curves of egg lecithin monolayers as a function of pH.[51]

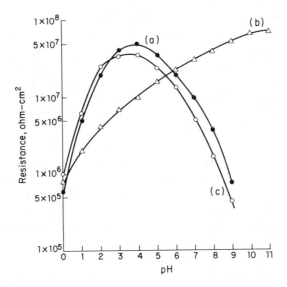

Fig. 4-7. Resistance of phosphatidyl choline BLM as a function of pH (*a*) in 0.1 *N* sodium chloride; (*b*) in 0.1 *N* calcium chloride, and (*c*) in 0.1 *N* magnesium chloride.[217]

Abnormal changes in BLM conductance have been reported by the presence of iodide ions[218] in aqueous solution, or by irradiation with X rays. In both of these cases, the results have been shown to be artifactual. Conductance due to iodide is attributed to the presence of traces of iodine; and this phenomenon is discussed further in Chapter 6. Increased conductance on exposure to X rays has been attributed to improper controls.[219]

Of the two possible mechanisms of conductance—ionic and electronic—across BLM, ionic conductance seems to account for the observed data. None of the membrane components could be expected to participate in electronic conduction, although extremely small amounts of impurities capable of supporting electronic conduction may bring such a mechanism into play.* However, the ob-

---

*The possibility of electron tunneling in BLM can at least partly be resolved by calculating Christov characteristic temperatures (T_{ch}) for an idealized barrier.[220] This is the temperature at which the probabilities of tunneling and hopping are equal. Below T_{ch} electron transport is by tunneling and above T_{ch} hopping is the prevalent mode. T_{ch} depends critically on the rate of curvature at the top of the barrier (potential), its height, and width. The dimensions of BLM suggest that the possibility of electron transfer occurring by tunneling is justified. However, the magnitude of the resistance (10^7–10^8 ohm-cm^{-2}) suggests that the passage of current through BLM must occur predominantly either by charge transfer complex formation or by an ionic mechanism. A BLM resistance of the order of 10^9 ohm-cm^{-2} for cholesterol membranes might be explained by an electron tunneling effect. However, for the lack of any direct evidence in favor of such a mode of charge transfer in these systems we shall restrict our discussion to the ionic charge-transfer mechanism across BLM.

served diffusion potential, individuality of transference numbers and dependence of BLM conductance on salt concentration (usually above 0.25 M) all point to electrolytic conduction as the sole mode of charge transfer across BLM. The nature of the charge carriers is not very well established but the dependence of conductance on salt concentration and water permeability characteristics point to the participation of electrolytes as the principal charge carriers. Participation of H^+ and OH^- ions in conductance may not, however, be ruled out, especially at extreme values of pH.

As pointed out earlier, the observed value for the resistance of BLM is consistent with the ionic mechanism of charge transfer. If only the hydrocarbon core is assumed to contribute to the resistance of the BLM, the resistivity of the membrane would be 10^{15} ohm-cm^{-1}, as calculated from the observed conductance of bulk liquid hydrocarbon, taking 50 Å as the thickness of the hydrocarbon core. However, studies on thick films containing phospholipids also show a limiting resistance of 10^8 ohm-cm^{-2} when expressed as surface resistivity.[221] This is consistent with the view that the electrical resistance of BLM is due to two lipid monolayer resistances in series. Thus from the data available thus far it is difficult to discern whether the hydrocarbon core is the permeability barrier for charge transfer or whether the surface layer forming the electrical double layer is the primary barrier for charge transfer. Considerable inferential support may, however, be gathered for the hypothesis that ion transfer occurs by a solubility-diffusion mechanism; that is, partition of ions between the water and membrane phases gives rise to current carriers.[130]

The non-ohmic conductance of BLM is not very well understood. It has been suggested as being due to the existence of ionic association within the conductance phase of the membrane.[108] It could be due to a field modified barrier conduction mechanism[222,1069] or simply due to the dissociation field effect operating within the membrane.[130] Quantitative analysis shows that the influence of the Field dissociation effect (Wien effect) on the membrane conductivity is appreciable only if the energy barrier at the interface is sufficiently high, that is, if the rate-limiting step for the ion transport is the passage of the ion across the interface.[223] Similarly, a relatively sophisticated theory has been developed from the assumption that the rate of flow of charge-carrying species is proportional to the gradient of its electrochemical potential.[224,225] The theory describes conductance under conditions of stationary flow and predicts that under conditions of high level bipolar flow the current varies directly with the third power of the applied transmembrane potential difference.[226] This asymptotic cube law has been applied and is supposedly obeyed by the non-ohmic part of the current voltage curve of lecithin BLM in sodium iodide and thus modified with iodine.[218] The treatment does not throw any light on the mechanism involved in ionic conduction across the BLM. Conductance properties of modified BLM and biomembranes are described in Chapters 6 and 8 in relation to ion-transport mechanisms involved in these systems, which differ significantly when compared to that of BLM.

DIELECTRIC STRENGTH

BLM are stable to large current flux, as great as 2×10^{-6} amp-cm^{-2}, and usually rupture at applied potentials ranging between 80 and 800 mV, depending upon the composition and nature of lipids constituting the membrane. Dielectric strength can be calculated from the value of the membrane breakdown voltage. Assuming a thickness of the insulator hydrocarbon core of 50 Å, a value of 50 mV for the BLM breakdown voltage would give a dielectric breakdown voltage of 10^5 V-cm^{-1}. BLM containing cholesterol have relatively high dielectric breakdown voltages. The breakdown voltages are usually independent of the nature of the electrolytes in the medium although very low and very high values of pH make BLM rather unstable at higher voltages.

The breakdown mechanism of the liquid dielectrics is not known, although for pure liquids it could be due to massive ionization of the dielectric caused by emission of electrons from the surface of the metallic electrodes.[227] This mechanism cannot be expected to operate in the BLM. It is also possible that both dielectric breakdown and nonlinearity of current-voltage curves are due to a structural reorganization in the membrane induced by high fields. It has, in fact, been suggested that field-dependent polarization distortion may well be a triggering process leading to several structural changes in membranes.[228] Such a change is expected to have a time constant which might be large enough to give rise to a "dispersion" region in the breakdown voltage-frequency curve. No such region has been found in the region from 50 to 5,000,000 cps.[110]* It is possible, however that a critical concentration of ions in the core of the membrane under the influence of high field strength causes the breakdown of the membrane. Acceleration of ions in the membrane could also result in thermal stimulation.[229] If only a small number of sites carry ions across the membrane, one can visualize a local increase in temperature caused by the dissipation of about 10^{-5} cal-min^{-1}-cm^{-2} for an applied potential of about 100 mV. Since the BLM is a poor conductor of heat, the energy thus released may raise the temperature locally and induce a convective turbulence in the membrane. Such fluctuations could also arise due to movement of the lenses of the bulk lipid on BLM.

INTERFACIAL CHARGE PROFILE AND ELECTRICAL PROPERTIES OF BLM AND BIOMEMBRANES

By way of recapitulation of the observed phenomenology of ion transport and charge profile across BLM, some general remarks and a comparison with biologi-

---

*Both conductance and capacitance show anomalous dielectric dispersion due to the presence of a thin (about 40,000 Å thick) layer adjacent to the membranes where the electrical properties differ from those of bulk water.[215]

cal membranes may be pertinent. We have noted that conductance of BLM is significantly lower than the conductance of plasma membranes in general even though the electrical capacity of both of these systems is almost the same. Thus it can be concluded that differences in conductance cannot be ascribed to the dielectric of the membrane phase.

As we suggested earlier in this chapter, separation of bulk aqueous phases by BLM gives rise to a variety of potentials, both static, due to fixed charges at the interface, and diffusion potentials, due to difference in the rate of movement of various ions in the two phases. The presence of an electrostatic field at the interface is expected to have serious effects on the transfer of matter, particularly ions, across such interfaces, and on the molecules which are in this field. This situation may give rise to a variety of phenomena observed in biological systems. Not much is known concerning the actual contribution of such field effects to interfacial properties of membranes, but both theoretical and experimental studies do suggest that their contribution to the overall process may be significant.[65,140]

The electrostatic field at the interface can influence adsorption onto the surface, which is pertinent to adhesion, coalescence, permeability characteristics, ionic selectivity, and mobility among other effects. Such effects are of considerable importance since they can affect a variety of cellular functions, the most notable being ionic selectivity and contractile processes. The permeability changes are manifested in volume and size regulation, and such changes are induced not only by the osmolarity of the suspending medium but also by specific stimuli such as ionic strength and the presence or absence of certain macromolecules in solution as demonstrated for nuclei.[230] Also experiments with intact *Amoeba proteus* indicate that contraction can be brought about at the periphery of the cell by local application of a polyelectrolyte to an advancing pseudopodium[231] (see also Chapter 7), suggesting that changes in surface charge can give rise to large and specific changes in the permeability of the membrane. Thus, in principle, it can be seen *how* a cell could be "aware" of surface potential changes. In this respect, the influence of surface potential on surface pressure, surface tension, the kinetics of surface reactions such as adsorption and adhesion, permeability, interfacial pH, frictional coefficient, or ion binding to the interface assume added importance. On the basis of a theoretical model it has been suggested that an appreciable rise in electrostatic potential can occur when charged interfaces approach to within 20 Å or less of each other.[65] It has been suggested that such electrostatic changes might be responsible for initiating changes in membrane permeability such as those occurring at the electrotonic synapse,[232] in changes leading to ionic intercellular communication,[233] and for the increased permeability of the membrane of the amoeba *Chaos chaos* during pinocytosis.[234] These observations may thus imply that the surface potential may act as mediator between the action of environmental effectors and cellular responses via changes in membrane properties. Moreover electroosmosis may result due to movement of the medium past a charged surface in an electric field. Such a process underlies various hypotheses forwarded to explain the mechanism of nerve impulse propagation.* Similarly, fluids may induce streaming potentials on vesicular structures floating in an electrolytic solution.

Yet another interesting aspect of membrane structure and consequent properties may lie in the relationship of surface potential and transmembrane potential. If the cell surface is a completely reversible (nonpolarizable) electrode, then surface and transmembrane potentials should be completely independent of each other. However, if the surface acts as a polarizable electrode, the change in transmembrane potential due to metabolism of the cell would alter the surface charge.[236,237] The influence and implications of such changes may be far reaching. Unfortunately, there are few studies concerning such aspects on either living or artificial systems.

MEMBRANE THICKNESS

One of the most interesting and biologically most relevant structural feature of BLM is their extreme thinness. A good estimate of the thickness of the BLM is important because it provides dimensional limitations on the ways in which lipid (and proteins, if any) may be packed in the membrane phase. The thickness of BLM of roughly known composition has been determined by several methods given in Table 4-4. The value of 70 ± 3 Å obtained by adding the length (eighteen carbon atoms) of hydrocarbon chains and the polar heads of two lipid molecules arranged tail-to-tail is in excellent agreement with the values obtained from these various methods. This implies that the thickness of BLM is constituted by two lipid molecules placed lengthwise at full length; however, the polar head groups do not appear to be fully extended or stretched into the aqueous phase. Under certain conditions the polar groups may be completely extended, and, thus, the thickness of BLM determined by optical and X-ray diffraction methods may vary. In fact, the thickness of BLM prepared from glycerol distearate at 32°C has been found to be dependent upon the salt concentration in the medium.[78] The results also suggest that the molecules are laterally compressed, giving an area per molecule (hydrocarbon chain) near 22–25 $Å^2$, which is in between the values for crystalline and liquid phases (see Chapter 9 for the significance of these results).

The thickness of plasma membranes is very much dependent on the type of surface layer present. The cellular membrane in general has a layer that appears to be 80–90 Å thick as seen in electron micrographs whereas its surface layer of mucopolysaccharides and proteins has thicknesses up to 5000 Å. Such surface

---

*In fact, a theory of memory has been forwarded on the basis of such a phenomenon.[235]

TABLE 4-4   THICKNESS OF BLM AND PLASMA MEMBRANES

Membrane Type	Thickness (Å)	Remark
BLM	65–85	Electron microscopy [99,126,1071]
	65–73	X-ray diffraction[84,238]
	67–70	Optical reflectance[78,137,192-196,239]
	48–56	Capacitance[99,104,110,204,239,1206]
	70 ± 5	Theoretical—obtained by adding the lengths of two phospholipid molecules (chain length C_{18}).
Myelin	150–180 (periodicity)	Electron micrographs shows multilayered structure (Fig. 1-5)
Plasma membrane		
Intestinal epithelium	95–110	EM, thickness of luminal side
	80–90	EM, thickness of lateral side
Golgi apparatus	70–80	EM of proximal convoluted tubules
Mitochondrial	50–60	EM of outer membrane
	50–60	EM of inner membrane
Retinal rods and cones	50–53	EM and X-ray diffraction

coats have been shown to be present on a large variety of cells, including red blood cells, ascite tumor cells, bacterial, and plant cells. Irregular projections seen by negative staining of the endoplasmic reticulum and red blood cells is also associated with similar structures located on plasma membrane interfaces. Myelin appears to have no surface coat. In spite of these significant differences in the thickness of surface coats, the plasma membranes seem to differ very little in their thickness. In general the lipoprotein layer constituting the plasma membrane appears to occur in two thickness ranges: myelin and related plasma membranes have a repeat thickness of about 90–150 Å, and the mitochondrial and endoplasmic reticulum membranes appear to be 50–80 Å thick (Table 4-4).

Although the values for thickness of biological membranes quoted in Table 4-4 are those obtained by electron microscopy, various other methods for determination of plasma-membrane thickness have been used. Low-angle X-ray diffraction and scattering measurements give approximately the same values for the membrane thickness. However, low-angle reflections show appreciable differences in average chain lengths of fatty acid residues in the lipid components and due to incorporation of proteins into the lamellar structure. The external variables such as tonicity, pH, etc., seem to modify the structural parameters and membrane thickness, which might give further information regarding the physicochemical characteristics of the membrane components.[1206]

By way of recapitulation, it is obvious from the foregoing discussion that the intrinsic properties of BLM are not only consistent with their bimolecular lamellar structure, but also there is significant correspondence in the properties of BLM and biomembranes. There are significant differences in these systems; their elaboration may however be postponed till Chapter 6.

5/ The Permeability Characteristics of BLM

The concept of the membrane as a barrier implies that at least some substances can pass across it unimpeded, and that the *flow* of these species is causally related to forces. That such characteristics are implied in the concept of the membrane as a barrier is extremely important; they seem to be the major features of membrane phenomenology. In fact, it would not be an exaggeration to say that "every cell of every organism, at every moment of its existence, is dependent upon diffusion processes for supplying it with necessary materials from its surroundings, for distributing these and other materials within its boundaries, and for removing to a safe distance metabolic products which, if allowed to accumulate, would be injurious."[240] Needless to say protein synthesis, transduction and translation of genetic information, morphogenesis, volume regulation, growth, nervous conduction, and numerous reactions of intermediary metabolism are among those manifestations and consequences of the controlled movement of metabolites, ions, water, and other species across the boundary of the living cell. This is due to the fact that the organization of the reactions within the cell depends not only on the reactivities of the substances involved, but also on their access to one another. It is evident that, without a system of barriers, without segregating and desegregating processes, and without compensation (replenishment) of loss of metabolites from the vesiculated structures, the ordered flow and metabolism would be impossible. In fact, diffusion is a prerequisite to steady-state control in an open system. Indeed, the simplicity of control of metabolic events through modifying access between reactants is observed at several levels of cellular organization, as well as for the whole organism. Moreover, the control of specificity and the modification of permeability-dependent

111

processes are achieved at several different levels, including genetic regulation and metabolic feedback. In this chapter and the subsequent chapters of this book we shall consider the properties and mode of action of the many specialized systems that move metabolites across cell membranes. As we shall see in the following chapters, a living cell employs many mechanisms to translocate and transport numerous substances that it requires. These various modes of solute transport are: (1) simple diffusion across the lipid layer; (2) solvent drag through aqueous pores; (3) pinocytosis and phagocytosis; (4) facilitated transport—by channel or carrier mechanisms; (5) exchange diffusion and co-transport; and (6) active transport and related processes.

It is implicit in the definition of transport as mediated and controlled by membranes, and as observed experimentally, that there are three major conditions which affect the movement of solutes:

1. The characteristics of the membrane—its chemical nature and the presence of catalytic proteins in the matrix.

2. The characteristics of the permeant species being transported—size, shape, solubility, charge polarity, hydrophobicity, and degree of ionization which may reflect the characteristics of the membrane in a complementary fashion.

3. The driving force on the permeant molecules which may or may not be independent of the properties of the membrane and the solute. For the convenience of argument the characteristics of diffusion processes shall be discussed first.

DIFFUSION

It is taken to be axiomatic that the prime mover of chemical transport is diffusion. All such processes have in common the flow of a substance due to thermal energy when its density is nonuniform. For example, a molecule that reacts to form two or more fragments by a unimolecular process is ordinarily excited to an activated state which makes decomposition to products possible. This excitation is a result of a succession of random activating and deactivating collisions with neighboring molecules. Similarly, diffusion occurs because of random forward and backward displacements of the diffusing molecules. Examples of diffusion include the diffusion of gases in regions where unequal concentrations exist, the flow of heat in a region of nonuniform temperature distribution, the flow of ions in aqueous solutions, and the flow of current in a semiconductor where an unequal distribution of charge carriers exist. Diffusion as commonly observed is therefore a one-sided spontaneous, irreversible rate process which, in a homogeneous system, must, at equilibrium, bring about uniformity of concentration everywhere within that system. For an excellent discussion of diffusion processes see Ref. 240.

Permeation is a consequence of diffusion and may be viewed as a *reaction* that equalizes the activity of the solute (and solvent), which is free to move in the solution and across the barrier. The term permeability as used here has the clear connotation of a rate and refers to the permeant and the membrane besides the variable parameters such as time, direction of flow, and the concentration difference or gradient. The phenomenological or descriptive basis of diffusion theory may be traced back to the theoretical work of Adolf Fick published in 1855.[241] For an isothermal, isobaric binary system in which diffusion occurs in only one direction, Fick's first law states that the quantity of diffusing substance which passes through unit area of a plane at right angles to the direction of diffusion per unit time is linearly proportional to the concentration gradient of the diffusing substance. If dc is the concentration difference over a distance dx, the number of moles dn crossing area A in the interface in time dt is given by:

$$dn/dt = -D \, A \, dc/dx \tag{5-1}$$

The negative sign on the left-hand side of the equation indicates that diffusion occurs in the direction of decreasing concentration. Now if one considers diffusion over a period of time, it is evident that the initial concentration, c_0, and hence dc_0/dx (or the corresponding complete differential) must vary, and in fact, decrease with time. Thus it becomes necessary to know c as a function of distance from the diffusion barrier and of time. For diffusion in one dimension only it can be described as:

$$\frac{\partial^2 c}{\partial t^2} = D \frac{\partial^2 c}{\partial x^2} \tag{5-2}$$

This equation, known as Fick's second law, is a concise mathematical statement of what all diffusion processes in one dimension possess in common. The above equation states that the rate of change of concentration at any position is proportional to the rate of change of the concentration gradient at that position. Thus, for calculation of the concentration-distance profile, suitable boundary conditions characterizing the system must be used.

In equation (5-1) the term $(1/A) \, dn/dt$, the total number of permeant molecules crossing a unit area of the membrane, is defined as the flux J:

$$J = -D dc/dx \tag{5-3}$$

The proportionality factor D is known as the diffusivity or *coefficient of diffusion*. It has the dimensions of $L^2 T^{-1}$ and is usually given in $cm^2 sec^{-1}$. The flux J has the dimensions of $ML^{-2} T^{-1}$ (flow of mass per unit time per unit area).

At first sight, Fick's postulation of a direct linear relation between the rate of diffusion and the concentration gradient appears reasonable, but is, in fact, only the form assumed for *infinitely* dilute solutions by a much more complicated law or series of laws. In fact, it bears the same resemblance to some more general laws that Boyle's gas law does to those governing actual gases, or that van't Hoff's law of osmotic pressure does to those actually obeyed by concentrated solutions. This similarity between the laws of Fick, Boyle, and van't Hoff is more than a superficial one. In reality these three laws are related to one another, and similar phenomena may be responsible for deviations from all of them.[240]

The phenomenological basis of Fick's law is shown elsewhere,[240] that

$$D = u'kT = \frac{RT}{N} u' = \frac{RT}{N} \frac{1}{6\pi\eta r}$$ (5-4)

applies for any case to which Stokes law* is applicable. This equation is commonly known as the Stokes-Einstein or, more appropriately, the Sutherland-Einstein-Stokes equation. It permits experiments on diffusion to be linked up with other phenomena involving the mobility of solutes; i.e., phenomena in which there are forces that produce drift velocities. Two such forces are the force experienced by a solute when it overcomes the viscous drag of a solution and the force on an ion arising from an applied electric field. A significant extrapolation of the Stokes-Einstein relationship lies in the approximation generally used in structure-permeability correlations across membranes. It is assumed that the radius of a molecule is related to the square root of the molecular weight for molecules of low molecular weight, and to the cube root of the molecular weight for molecules of high molecular weight. It follows that[242]

$$D = \text{constant} \times 1/r$$

or $$DM^{1/2} = \text{constant}$$

or $$DM^{1/3} = \text{constant}$$

Viscosity is assumed to be constant for any given type of membrane. This relationship, as we shall see in subsequent sections, has been used extensively to predict the behavior of a variety of solutes.

In actual practice the measurement of steady-state diffusion typically has a membrane separating the permeant at two different concentrations. After a lapse of time, a steady-state is achieved; that is, the rate at which the permeant enters on one side of the membrane is equal to the rate at which it leaves the membrane on the other side. Thus the flux J becomes independent of time and of position within the membrane. Under the assumption that D is also constant

---

*According to Stokes law, for a sphere of radius r (which must not be small in comparison with that of the surrounding molecules) and a viscosity η of the incompressible fluid, the frictional resistance to motion is $6\pi\eta r$ (=1/mobility). The approximate nature of Stokes law as used in diffusion studies must be emphasized. Apart from the validity of extrapolating from the macroscopic sphere-continuum-fluid model of Stokes to an atomic nonsphere in a molecular liquid, other reasons for the limited validity of Stokes law arises from equations concerning the radii which should be substituted in any application of the law. Further, the experimental value of viscosity is the bulk average viscosity of the whole solution. whereas it is the local viscosity in the neighborhood of the permeant which should be considered for the calculation of mobility. Finally, some modification may be required when considering the mobilities of charged particles which tend to orient solvent dipoles and may thus affect the viscous drag.

it follows that the concentration gradient is constant and hence may be written as $\Delta c/\Delta x$ or $(c_1 - c_2)/\Delta x$, where c_1 and c_2 are the concentrations of the permeant on each side of the membrane and Δx is the thickness of the membrane. Thus equation (5-3) reduces to:

$$J = - D\frac{\Delta c}{\Delta x} = - P\Delta c \qquad (5\text{-}5)$$

The new term P, generally known as the *permeability coefficient*, is introduced to avoid complications arising in the measruement of the thickness of the permeability barrier. It may be noted that P has the dimensions LT^{-1} (length per unit time); that is, velocity. To obtain the full significance of the term P, consider the case in which the initial concentration of the permeant is equal to the steady-state concentration; that is, $c_1 = C$ and $c_2 = 0$ at $t = 0$. Then the equation (5-5) is:

$$J = - PC = - d\bar{c}/dt$$

By integration:

$$\bar{c} = C \exp (-Pt) = C \exp (-t/\tau)$$

in which P is the reciprocal time-constant (τ) of the diffusion process by which the uniformity of concentration is attained.

The situation of equation (5-5) has been oversimplified but may be corrected by the following considerations. In order for a permeant to traverse a membrane from one side to other, it must enter the membrane at one interface, diffuse within the membrane to the other interface, and leave the membrane at this second interface. Thus, the relationship $P = D/\Delta x$ should also contain an equilibrium distribution term K (partition coefficient), the equilibrium ratio of permeant concentration in the membrane phase to that in the external phase. Thus the permeability coefficient relating flux to concentration gradient in the aqueous phase is best expressed as:

$$J = - P\Delta c$$

and

$$P = KD/\Delta x$$

This relationship is developed further later in this chapter.

EXPERIMENTAL METHODS

The value of the permeability coefficient, P, can be determined by several direct methods, such as isotope exchange, osmotic flux measurements, electrical conductance, or fluorescence or absorbance measurements. Indirect methods are usually based on induction of a streaming potential or osmotic swelling by solutes which have permeability less than or equal to that of the solvent (to be discussed later in this chapter). The choice of the methods and the precise experimental conditions rests heavily on the nature of the permeant and the magnitude of the permeability coefficient. Measurement of isotope fluxes is perhaps the ultimate simplification in diffusion studies. In such a case, the chemical potential gradient is zero and the driving force is the isotopic gradient, the activity coefficients remaining constant throughout the system. Consequently, the free energy change in such a process is zero, except of course for a very small term reflecting the entropy of mixing.

Values of permeability coefficients determined by various methods range over several orders of magnitude. Thus before we discuss the molecular basis of diffusion across a biomembrane, it would be pertinent to discuss the permeability characteristics of BLM and biomembranes for water and nonelectrolyte.

WATER PERMEABILITY

In BLM, the permeability coefficient of water is several orders of magnitude larger than the permeability coefficient of most solutes. This value has been determined by isotope exchange using tritiated water and also by measurement of osmotic flux. In contrast to their extremely low electrical conductance, the BLM show significant water permeability, in the range of 0.5×10^{-3} to 1×10^{-2} cm-sec^{-1}. Earlier reports showed considerable differences in the values obtained by osmotic and tracer flux methods. Under the conditions of zero net volume flux values in the neighborhood of 4.4×10^{-4} cm-sec^{-1} were usually reported.[103,104] The presence of calcium chloride, urea, sucrose, and sodium chloride on each side of the membrane has no effect on the permeability coefficient. The values obtained by osmotic measurements were considerably larger,[103] 17.3 to 104×10^{-4} cm-sec^{-1}. From the estimate of the outer surface area of liposomes, their osmotic permeability coefficient[147a] has been calculated as 0.8×10^{-4} cm-sec^{-1}, however the osmotic permeability coefficient for single walled vesicles[1072] is 44 and 70×10^{-4} cm-sec^{-1} at 25° and 37°C.[1072]

A difference in osmotic and tracer permeability coefficients for water has also been observed in measurements on several biomembranes (Table 5-1). This discrepancy has given rise to considerable controversy over the mechanism involved in these processes. The inequality of the two types of permeability coefficients may be interpreted in several ways. The phenomenon may be due to some basic difference in the mechanism of water flux in the membrane under different conditions of diffusion exchange or volume flux, or due to salt effects, or to lack of proper mixing of the permeating species, such as tritiated water, in the vicinity of the membrane interface. Although the controversy is more or less resolved, some of the earlier theories and experiments will be discussed here at some length since it has brought out a number of interesting arguments which have paved the way for the explanation now generally accepted.

TABLE 5-1 THE PERMEABILITY COEFFICIENT FOR DIFFUSION OF WATER ACROSS VARIOUS MEMBRANES

Membrane	Permeability		"Effective" Pore radius (Å, approx)
	Osmotic	Diffusional	
	$(cm\text{-}sec^{-1} \times 10^4)$		
Self diffusion	$(= D^a)$	2.41×10^{-5}	
Through hexadecane[255]	$(= D^a)$	4.16×10^{-5}	
BLM[b]	10–12 at 25°	9–11	
Liposomes	44 at 25°		
Valonia	2.4	2.4	
Amoeba	0.37	0.23	2.1
Frog skin	3.3	1.1	4.3
Frog ovarian egg	89.1	1.28	30
Frog body cavity egg	1.30	0.75	2.8
Dog erythrocyte	400	63	7.5
Ox erythrocyte	152	51	4.3
Human erythrocyte[c]			
adult	115–130	45–50	3.5–4.0
fetal	61	23	3.9
Squid axon	250	30–35	8.5
Toad bladder	4.1	0.95	8.5
Toad bladder + vesopressin	230	10.8	18
Goat ventricular walls	270	2.8	36
Dialysis tubing	380	10.9	23

[a]This value is in $cm^2\text{-}sec^{-1}$. Considering a thickness of the water slab of 100 Å, the permeability coefficient can be obtained as 24.1 $cm\text{-}sec^{-1}$.
[b]The permeability of BLM is dependent upon its composition; see text for further discussion.
[c]It may be noted that thiol reagents cause a substantial decrease in osmotic permeability and the ratio of osmotic to diffusional permeabilities.[1083] This may imply that the pores are modified in the treated cells.

It is implied in Fick's relationship that, under conditions of permeability measurements, there is only a membrane-solute interaction, and no consideration is given to solvent-membrane and solute-solvent interactions. Thus, if two systems are separated by a membrane which is permeable to solutes as well as to water, then the chemical potential of water is not the sole driving force on water transport; the moving solute can exert a frictional drag on the water and these drags constitute extra forces. Such effects can, however, be quantitatively accounted for within the theoretical framework of the thermodynamics of irreversible processes as shown elsewhere.[243,244] Without going into discussion of these accounts, we shall present some of the salient conclusions. From these theoretical considerations it emerges that a solute concentration gradient will cause osmotic flow at a rate proportional to $\omega RT \Delta C_s$ in the particular case of zero volume flow. This relationship refers to the simultaneous passage of a single solute and of solvent across the membrane. The solute diffusion coefficient ω, which is of the same form as a conventional permeability constant, denoted by the term P, is equal to $(L_D - \sigma^2 L_P)C_s$, where σ is the reflection coefficient, L_p is the pressure filtration coefficient, L_D is the solute permeability coefficient, and C_s is the solute concentration. The reflection coefficient (σ) is defined as the ratio of the osmotic coefficient (or the ultrafiltration coefficient) to the pressure filtration coefficient for a leaky membrane. It accounts for various drag factors and is amenable to experimental measurement. For an ideal impermeable solute σ is equal to unity, and it is zero for a solute as permeable as the solvent itself. Negative values of reflection coefficients indicate anomalous osmosis.[244] The reflection coefficient is not an arbitrary correlation factor but can be related by thermodynamic equations to other rate constants, such as sieving coefficients, filtration coefficients, and permeability coefficients. A clear-cut physical explanation or description of any of these coefficients is difficult to elaborate; however, the meaning of σ is best grasped by the following relationship:

$$\sigma = -J_D/J_v$$

σ is the ratio of the flow of water relative to solute compared with the total volume flow under the influence of the pressure gradient alone.* If the membrane does not distinguish between solute and solvent there is no relative flow and σ is zero. If the membrane selects absolutely, $\sigma = 1$. Thus σ is a measure of semipermeability of the membrane.

It must be emphasised that ω, σ, and L_p are three independent parameters, all of which are obtainable experimentally, which suffice to characterize the membrane.

Values of the reflection coefficient less than one for permeant molecules arise from two different factors: the rate of osmotic flow from the dilute to the concentrated solution is reduced by the volume of solute diffusing from the concentrated to the dilute solution, and the diffusing solute will drag along some water from the concentrated to the dilute solution, if and only if the solute and water interact while crossing the membrane. Based on these considerations the following expression has been derived:[245]

$$\sigma = 1 - \frac{\omega - V_s}{P'} - \frac{K f_{sw}}{f_{sw} + f_{sm}} \tag{5-6}$$

where P' is the osmotic permeability coefficient, ω is the solute permeability coefficient, and V_s its partial molar volume, K is the partition coefficient, and

---

*The physical significance of the reflection coefficient is best understood in terms of osmotic flow across a leaky membrane. Staverman pointed out that in such membranes the osmotic pressure never reaches the value it would if the membrane were ideally semipermeable. This is because the tendency of an *impermeable* solute to diffuse through the pore will partially counterbalance the force causing water to enter from the other side. Thus the ratio of the initial osmotic pressure of a leaky membrane to the classical pressure developed by an ideal membrane would be a measure of the ability of the pores to discriminate between the *permeant* and solvent (water).

f_s are frictional coefficients between solute and water (f_{sw}) or solute and membrane (f_{sm}). The term ($\omega - V_s$)/P' represents the reduction in σ due to the volume of diffusing solute, and the term $Kf_{sw}/(f_{sw} + f_{sm})$ arises due to dragged water. These thermodynamic factors thus result in lower rates of osmosis for permeant molecules, although these may not be the only factors.

The experimental values of σ have also given rise to the term "equivalent pore radius,"[246] which is the pore size required to account for the σ values of the probing molecules.* Thus a ratio of the osmotic and tracer permeability coefficients greater than unity is interpreted to suggest that the biological membrane is characterized by water-filled pores through which an osmotic pressure gradient permits bulk flow of water. Thus the net flux of water under an osmotic gradient may consist of two components: one due to diffusion through pores in the membrane and the other due to filtration. A quantitative development of this thesis leads to an analysis of the isotopic and osmotic permeability data for a given membrane in terms of an equivalent pore radius and the fractional pore area.[247-250] The value of these constants involved in the permeation of water across the membrane have been calculated (Table 5-1). The pores appear to be 5-16 Å in diameter and appear to involve a very small fraction (usually less than 0.1%) of the total membrane area.[251-254].

It is interesting to note that the ratio of isotope and osmotic flux for water across BLM and quite a few biomembranes is almost the same[103,104] (however, see below) and the absolute magnitudes are comparable, as well. However, some of the experimental observations can be interpreted to argue against the presence of water-filled pores in BLM. The value of electrical conductance, for example, calculated from the computed values of pore radii and total membrane area involved in the diffusion of water, is at least one order of magnitude larger than observed values.[244] Moreover, BLM have small but finite interfacial tensions; the stability of any permeant pore, it may be argued, would be difficult to rationalize, although it can be attributed to the presence of some external component, such as protein in the biological membranes. At best it seems that if pores or their equivalent do exist in the BLM they must be transient, a result of local fluctuations in molecular order within the membrane, and the product of the frequency of occurrence of such pores and their life time would be expected to be small. A simple alternative might be statistical fault formation in the bilayer. This

---

*In order to ascertain the existence of pores in the membranes, the following reasoning is usually employed. The friction per molecule of water during water flow across a membrane is greater if each single molecule has to travel separately, that is to diffuse, than when it travels as bulk through the pores in the membrane. Similarly a substance which is not soluble in the lipid part of the membrane passes the membrane by filtration with a rate proportional to the total area of the pores, but when the substance is pressed through the pores by a hydrostatic pressure or by an osmotic pressure difference the rate depends not only on the total area, but also on the average radius of the pores. From a difference in the diffusion permeability coefficient the total area and the average radius can be calculated.

mechanism has been proposed to account for water and gas permeation in mono-layers at water-air interfaces.[256]

It has been suggested that a difference in isotopic and osmotic permeability co-efficients might be the consequence of isotope effects as observed in silica mem-branes.[1208] A model has been suggested to account for the isotope interaction in phenomenological terms.[257] The osmotic flux of D_2O across BLM measured in sodium chloride, gives results which dismiss this hypothesis completely. The pos-sibility of isotopic effects is also ruled out by the close agreement among the values of self-diffusion coefficients of water, DHO, D_2O, and THO.[258-260]

The "long pore effect"[261] indicating membrane-water interaction offers yet another possible explanation for the differences in permeability. Unfortunately, it is difficult to test (p.40 in Ref. 262). The possibility of a salt effect on osmotic fluxes can also be eliminated since the nature of the salt has no effect on the value of the permeability coefficient.[104]

Finally, the possibility of stagnant layers of water adjacent to the membrane interface have been given serious consideration. The usual assumption for mea-surement of water permeability is that, at each *moment*, the permeant is dis-tributed uniformly in both compartments. However, in order to have a flux across the membrane, the solute must diffuse to it. This implies that the concen-trations of the solute at the membrane interfaces are always different from the bulk concentrations and that the rate of permeation is lower than it would be without these concentration gradients near the interfaces. The slowing down of permeation will increase with time and will be more pronounced for rapidly per-meating substances. The quantitative importance of such unstirred layers is diffi-cult to evaluate. According to the usual treatment, a stationary thin layer of solution in contact with the interface is assumed. Although a gross oversimplifi-cation, within this layer diffusion alone controls the transfer of a substance to the interface. The diffusion boundary layer has no exactly defined thickness; it is simply a depth which it is convenient to define as the region within which the maximum change in concentration occurs.* It can be shown that this thickness is proportional to the physical properties of the solution (rate of stirring, vis-cosity, etc.) and geometrical constraints,[1210] as well as to the value of the dif-fusion coefficient. Thus, if the moving fluid contains a solute, and if conditions are such that the concentration of this solute varies with time, the effect of mo-lecular or ionic diffusion is added to the situation. The largest change of concen-tration occurs in the thin unstirred boundary layers.

Based on these arguments it might be expected that convective stirring close to

---

*The fundamental problem in stirred solutions is that one is dealing with a fluid moving past the stationary membrane. Exactly at the surface of the membrane the fluid flow is zero. Some distance away, the flow velocity has a value characteristic of the bulk of the solution as determined by rate of stirring. Such boundary layers always exist with all real liquids at ordinary flow velocities.

the membrane occurs during the osmotic flux of water. No such convective stirring could result from the minute amount of THO influx in the isotope-exchange studies. This would imply that the lower values of isotope-exchange coefficients reflect the presence of unstirred layers in the aqueous solution adjacent to the membrane.[263-265] The apparent permeability (P) is related to true permeability (P_t) for the membrane system with stagnant layers by the following relationship.[266]

$$\frac{1}{P} = \frac{1}{P_t} + \frac{\bar{\ell}_1}{D_1} + \frac{\bar{\ell}_2}{D_2} \tag{5-7}$$

Where $\bar{\ell}_1$ and $\bar{\ell}_2$ are the thickness of these stagnant layers and D_1 and D_2 are diffusion coefficients of the species in flux in two aqueous phases bounded by the stagnant layers. Thus in human erythrocytes for example, the thickness of the unstirred layer has been calculated as 5.5μ since the permeability coefficients differ by a factor of 2.4.[267]

In the BLM, if the largest value obtained in the osmotic experiment gives the best estimate of the intrinsic membrane resistance to water flow, the thickness of the stagnant layer can be calculated. The earlier estimate of 0.03 cm[104] for the thickness of unstirred layer was somewhat higher than the values reported recently: 0.017 cm[282] or 0.011 cm.[1076] It may however be pointed out that the thickness of the unstirred layer is a function of the area of the membrane, the thickness of the support, and the viscosity of the medium.[268-272,1076] Furthermore the thickness of the unstirred layer was found to be the same in the absence and in the presence of amphotericin[1076] which induces hole of 5-6 Å in the BLM (see pp.125 and 199). Thus by taking suitable precautions to eliminate the effect of unstirred layers the value of the permeability coefficient is found to be about 1×10^{-3} cm-sec^{-1} for water flow across BLM. The actual value for P has been claimed to be within 20% of this observed value.*[268-272] Furthermore, when the unstirred layer is perturbed as for example for tracer diffusion in the presence of osmotic flow, higher tracer permeability coefficient is observed.[78,104,268,1209] This is apparently due to higher membrane permeability under these conditions, which arises due to additional stirring of the layer of water adjacent to the membrane during osmotic flow.

The permeability coefficient has also been found to depend upon the lipid composition of BLM.[271] This is probably best illustrated in Fig. 5-1, where the permeability coefficient of water for phospholipid BLM is plotted as a function of cholesterol content. The results clearly indicate that at higher concentrations of cholesterol the permeability of water decreases considerably. This has been

---

*In a theoretical analysis of osmotic and tracer permeability in a homogeneous liquid membrane the ratio is found not to reduce to unity, but be given by a more complicated expression. The effect of unstirred layer is only a part of total deviation, other factors need be considered (1073).

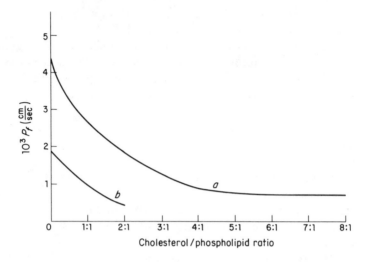

Fig. 5-1. Water permeability coefficient (P_f) as a function of cholesterol/phospholipid molar ratio at 36° in solutions either of unbuffered 100 mM NaCl or of 100 mM Nacl + 0.2 mM $MgSO_4$ + 5 mM histidine (pH 7). After the membrane had become completely black, the concentration of NaCl was increased on one side, and the rate of water movement determined. Curve a is for lecithin membranes, and curve b for "mixed lipids" from brain (ox.)[271]

attributed to an increase in the viscosity of the hydrophobic core as a function of cholesterol content.[271,1211]

The equality of the two permeability coefficients is consistent with the concept of a continuous lipid phase without channels or pores. It is also interesting to note that the barrier (resistance) offered by the bilayer to water transport is of the same order of magnitude by which monolayers reduce the rate of water evaporation.[273] This is consistent with the diffusion theory subject to partitioning of water in the hydrophobic phase (see below). Finally, similarity in the permeability characteristics of water in BLM and biomembranes suggests that in biological membranes the protein component has at best only a minor role in water permeation. The question of difference in osmotic and isotope fluxes of water in biological membranes is still unresolved. However, as suggested by several workers the contribution of unstirred layers may be predominant.[274] At least in some cases, such as *Valonia*, the two permeability coefficients have been found to be the same.[275] The results obtained from other biological systems do however indicate the difference in osmotic and tracer flux; their significance is discussed later in this chapter.

PERMEABILITY OF NONELECTROLYTES ACROSS BLM

The permeability coefficients of various solutes across BLM have been determined by a variety of methods (Table 5-2). A theoretical analysis of the factors involved in such measurements may be found elsewhere.[268,269] It may, however,

TABLE 5-2 PERMEABILITY COEFFICIENTS FOR VARIOUS SOLUTES ACROSS BLM

Permeant[a]	Permeability, coefficient $(cm\text{-}sec^{-1})$	Lipid Composition[a]	Method[a]	Ref.
Water	4.2×10^{-3}	Egg lecithin— tetradecane	T,O	271
Urea	4.2×10^{-6}	Egg lecithin— tetradecane	T	269
	5.9×10^{-7}	RBC lipids	T	279
Thiourea	4.5×10^{-6}	RBC lipids		281
Acetylocholine	6×10^{-6}	Egg lecithin		101
Salicylamide	29×10^{-6}	Egg lecithin		101
Procaine	16×10^{-6}	Brain lipid	F	277
PABA	11.6×10^{-6}	Brain lipids	F	277
Salicylic acid	10.8×10^{-6}	Brain lipids	F	277
Pyridoxamine	6.1×10^{-6}	Brain lipids	F	277
Pyridoxine	3.6×10^{-6}	Brain lipids	F	277
Quinidine	8.2×10^{-6}	Brain lipids	F	277
Quinine	9.5×10^{-6}	Brain lipids	F	277
Indole	$150\text{-}250 \times 10^{-6}$	Brain lipids	F	277
5-OH-Indole	200×10^{-6}	Brain lipids	F	277
Indole-3-ethanol	230×10^{-6}	Brain lipids	F	277
Indole-3-acetic acid	0.59×10^{-6}	Brain lipids	F	277
Tryptamine	4.0×10^{-6}	Brain lipids	F	277
Serotonine	0.66×10^{-6}	Brain lipids	F	277
Glycerol	4.6×10^{-6}	RBC lipids	T	269
Erythretol	0.75×10^{-6}	RBC lipids	T	269
Mannitol	3.3×10^{-8}	RBC lipids	T	269
D-Glucose	$2\text{-}6 \times 10^{-8}$	RBC lipids	T	101,279
Sorbitol	$< 1.7 \times 10^{-7}$	RBC lipids	T	279
D-Fructose	$0.4\text{-}0.6 \times 10^{-8}$	RBC lipids	T	276

[a]RBC, red blood cell; PABA, p-aminobenzoic acid; T, tracer diffusion, O, osmotic, F, fluorometric methods.

be noted that water flow diminishes the permeant flow; in fact, the drag influence of water flow can be demonstrated and determined by observing the isotope exchange against and in the direction of osmotic flux. The effect of such factors is usually small. For example, urea is entrained to about 2% of the maximal possible value.[270] It has also been noted that the values for the permeability of various solutes such as glucose, fructose, and mannitol are initially high and gradually decrease for the first 30–40 min, reaching a steady-state flux which lasts for more than 2 hr.[276] In contrast, permeability of p-aminobenzoic acid increases more than fivefold during a 3-hr experimental period[277] accompanied by a simultaneous and proportional increase in the membrane conductance. Similar conductance increases have been observed with sphingomyelin + tocopherol + cholesterol BLM using indole-3-acetic acid as the permeant.[277]

Several factors seem to govern the permeability characteristics of BLM as adduced from the permeability studies of several chemically related organic solutes, for example the indole derivatives. The permeability coefficient for these solutes decreases with increasing polarity: indole > 5-OH indole > indole-3-acetic acid > serotonin > tryptophan, all at pH 7.0.[277] Indole, the substance most nearly approximating a simple aromatic hydrocarbon nucleus shows the highest permeability. The neutral substituted indoles permeate at slightly smaller rates. The hydrophilic groups of these compounds do not appear to decrease permeability substantially. An ionizing group however introduces a change of several orders of magnitude in permeability coefficient, although the ionizing groups do not entirely control the permeability. The effect of lipid composition and pH of the medium has also been studied for several membranes.[271,277,284]

A series of benzene derivatives show similar behavior. The effect of changing pH has been interpreted to suggest that the carboxylic acids are permeable only in the un-ionized form. The value of the permeability coefficient calculated for the free acid was constant with varying pH. An interesting observation was that the permeability of most amines appears to be independent of their degree of ionization, and the concentration-dependent membrane diffusion potentials of the amine anaesthetics procaine, and related substances, are proportional to the logarithm of the amine concentration. Although diffusion potentials are normally used to indicate specific permeability, they do not necessarily reflect the magnitude of permeability, since a potential may arise even from the transport of immeasurably small amounts of charged species. Thus the development of a diffusion potential simply implies that the membrane may be permeable to the cationic form of the drug even when it is completely impermeable to inorganic cations. Furthermore, for the development of a diffusion potential the substrate need not actually diffuse through the membrane in the charged form. Ionization may be suppressed by entry into a region of low dielectric constant, leading to the formation of the neutral, diffusible species by exchange reactions at the interface.[277]

Permeability studies across liposomes show graded permeabilities for various permeants. Solutes such as ethylene glycol and methyl urea diffuse more readily than glycerol, urea, malonamide, and erythritol. Sodium acetate and sucrose were found to be poor permeants.[147] The permeabilities of glycerol and erythritol show strong temperature dependence[278] but activation energies show little sensitivity to the degree of unsaturation and presence of cholesterol in the BLM.[1074] Thus for a rise in temperature of about 30° P increases ten times. Similarly introduction of olefinic lipids increases the permeability.[1075,1080]

The permeability coefficients for various solutes given in the Table 5-2 are found to lie in the same range as the data reported for living cells; however, some significant differences may always be noted from cell to cell (see below). Thus, for example, permeability of glucose across red cell membrane varies over four orders of magnitude in various organisms. In general the experimental data on permeability coefficients and the effect of temperature on permeability may be used to divide these species into two groups.[280] Erythrocytes from rabbit and related species show relatively low permeability for glucose ($P = 5 \times 10^{-8}$ cm-sec^{-1}), the temperature coefficient for permeation is small, and the effect of temperature change is independent of pH. In contrast, erythrocyte from human, beef, and hog show considerably larger permeability;[240] they show a large temperature coefficient, which is strongly dependent upon the pH of the medium. In the first group the behavior is comparable to the one observed in BLM, even those prepared from lipids of human red cell ghosts.[1084] Thus the mechanism of glucose transfer may be the same in group I and BLM. However, large permeability of polar solutes like glucose in the latter group has been attributed to the presence of polar pathways in the biomembranes. This is further substantiated by the observation that incorporation of several agents such as enniatin,[281] nigericin,[276] nystatin, and amphotericin B[282,283] makes BLM significantly more permeable to polar solutes, both ions and nonelectrolytes. For a detailed discussion of the effect of these agents see Chapter 6. It may however be noted that in the BLM modified with nystatin or amphotericin *B* (*polyenes*) the osmotic permeability coefficient for water is about three times greater than the value obtained by tracer flux. Similarly, these modified membranes are much more permeable to small molecules whose Stokes-Einstein radii are less than 4 Å (Table 5-3) as manifested both in tracer permeability and in reflection coefficients for various solutes. All these results are probably best rationalized by the assumption that these agents induce pores of about 4 Å diameter across the BLM.* Thus the presence of pores offers a rationale not only for the observed difference in the osmotic and tracer permeabilities of

TABLE 5-3 PERMEABILITY OF LOW-MOLECULAR-WEIGHT SOLUTES ACROSS MEMBRANES CONTAINING PORES[282]

| | BLM modified with | | | | Red blood cell |
| | Nystatin | | Amphotericin B | | |
Permeant (mol-radius, Å)	$P \times 10^{-4}$	σ	$P \times 10^{-4}$	σ	σ
Water (3)	12.0		6.0		
Urea (2.03)	0.95	0.55	0.68	0.57	0.62
Thiourea (2.4)	0.95	–	–	–	–
Ethylene glycol (2.24)	0.45	0.67	–	–	0.63
Glycerol (2.74)	0.115	0.78	0.075	–	0.88
Glucose, sucrose, NaCl (> 4)		1.0		1.0	0.99
Water (osmotic/tracer flux)	3.3		3.0		2.5

*Quantitative differences in the results reported by these various authors have been attributed to unstirred layers.[282,1076]

water but also for higher permeability of polar molecules of low molecular weight across biomembranes. Furthermore, as expected the *porous* membranes (modified BLM) exhibit quasilaminar water flow during osmosis[1076] and coupling of solute and solvent flows within the membrane phase.[1077] Similar features of biomembranes are further elaborated in later sections.

PERMEABILITY OF NONELECTROLYTES ACROSS BIOMEMBRANES

Quite a few studies of nonelectrolyte permeability and selectivity have been reported (Table 5-4). In spite of the diversity of the organisms and tissues studied, the qualitative patterns that emerge regarding permeant specificity are rather general. Undoubtedly there are some significant deviations if one makes a comparison of a pair of nonelectrolytes with approximately the same size, molecular weight, free solution diffusion coefficients, and empirical formulas. Such aspects of nonelectrolyte permeability are summarized in this section. For excellent reviews of this topic see Refs. 243, 290 and 300.

The classical studies of Overton[11] suggest that the rate of permeation of most nonelectrolytes across almost all cell membranes is governed by their lipid solubility relative to their water solubility. This concept has been extensively

TABLE 5-4 LANDMARKS IN PERMEABILITY STUDIES OF NONELECTROLYTES ACROSS BIOMEMBRANES

Tissue/Organism (Ref.)	Number of Substances Examined	Major Conclusions	Authors
Nerve, muscle, etc. (11, 25, 235, 286)		Permeability across cell is governed by lipid solubility	Overton (1895-1900)
Chara ceratophylla (24)	45	Permeability is proportional to partition coefficient and to square or cube root of molecular weight	Collander and Barlund (1933)
Nitella mucronata (289, 290)	70	Exception to general lipid solubility pattern noted	Collander (1954)
Chara spp. (243)	49 (data from Ref. 289)	Effect of hydrogen bonds in solute membrane interaction emphasized to postulate membrane interface as the rate-limiting barrier in permeability	Stein (1967)
Rabbit gall-bladder (291-293, 300, 1213)	206	Deviations from the general pattern of lipid solubility are explained in terms of molecular structure of the membrane as a continuous anisotropic lipid layer. Results from plant and animal tissues appear to be in general agrement.	Wright and Diamond (1969)
Red blood cell, plant cells (247-254, 294-299, 308)	more than 100 (total)	Presence of pores in various biomembranes	Solomon *et. al.* (1956-58)

confirmed and modified since. The concept of lipid solubility implies further that if permeability were governed solely by lipid solubility, a regular increase in permeability coefficients and decrease in reflection coefficients must accompany an increase in the lipid water partition coefficient (K).* It further implies that the rate-limiting step in the permeation of a solute is the diffusion in the membrane phase and that the mechanism of nonelectrolyte diffusion is the same through cell membranes as through a bulk lipid phase. Thus the following expressions should hold:

$$P = KD/d$$

and

$$K = \exp(-\Delta F/RT) \qquad (5-8)$$

where d is the membrane thickness, $P(= \omega RT)$ is the permeability coefficient, which may be interpreted as the velocity the particles assume under the influence of unit driving force; D is the diffusion coefficient for the solute in the membrane interior; and K is its lipid water or membrane water partition coefficient. The second expression relates the partition coefficient to thermodynamic parameters. As pointed out earlier, for diffusion in bulk solvents the relation $DM^{1/2} =$ con-

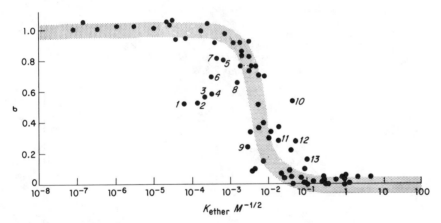

Fig. 5-2. Reflection coefficients of non-electrolytes as a function of the ether/water partition coefficient (K) times the reciprocal square root of molecular weight (M) for each non-electrolyte across rabbit gall bladder. Points referring to small solutes and branched solutes are numbered: small solutes, 1 = urea, 2 = methyl urea, 3 = formamide, 4 = acetamide, 5 = ethylene glycol, 6 = dimethyl urea, 7 = ethyl urea, 8 = propionamide, 9 = dimethyl formamide; branched solutes, 10 = pinacol, 11 = isovaleramide, 12 = 2-methyl-2, 4-pentanediol, 13 = triacetin. The shaded band is drawn to indicate the general pattern of the other points and has no theoretical significance.[291]

---

*The partition coefficient is the equilibrium ratio of the test solute's concentration in a lipid solvent to its concentration in water. The two phases are in contact but mutually immiscible. See Ref. 1212 for an excellent review.

stant, where M is the molecular weight, holds approximately for small molecules, and $DM^{1/3}$ = constant holds for very large molecules. Thus the permeability coefficient (P) and the reflection coefficient (σ) should correlate closely with $KM^{-1/2}$ in a homogeneous, bulk lipid membrane. In fact this relationship has been the guideline for most of the studies on nonelectrolyte permeability. The main characteristics of such permeability processes may be summarized as follows.

(1) As shown in Figs. 5-2 and 5-3, plots of σ vs K, and plots of $PM^{1/2}$ vs K suggest a systematic correlation: P increases conspicuously with increasing K, but also increases somewhat with decreasing M at constant K. Thus orders and ratios of permeability coefficients in biological membranes are similar to those for permeation through a bulk lipid phase.

2. Measured reflection coefficients do not become negative even when the partition coefficient is increased 10^3 times over the values at which σ is already

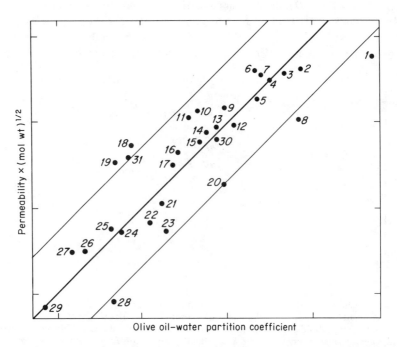

Olive oil–water partition coefficient

Fig. 5-3. The permeability of cells of *Chara ceratophylla* to nonelectrolytes of different oil solubility and different molecular weight. Permeants have been assigned the following numbers: *1*, Triethyl citrate; *2*, urethan; *3*, trimethyl citrate; *4*, antipyrin; *5*, valeramide; *6*, ethanol; *7*, urethylan; *8*, diacetin; *9*, butyramide; *10*, cyanamide; *11*, propionamide; *12*, chlorohydrin; *13*, glycerol ethyl ether; *14*, propylene glycol; *15*, succinamide; *16*, glycerol methyl ether; *17*, dimethyl urea; *18*, formamide; *19*, ethylene glycol; *20*, monoacetin; *21*, ethyl urea; *22*, thiourea; *23*, diethyl malonamide; *24*, lactamide; *25*, methyl urea; *26*, hexamethylenetetramine; *27*, urea; *28*, dicyandiamide; *29*, glycerol; *30*, diethyl urea; *31*, acetamide.[289]

virtually zero. It may be recalled that, under a given set of conditions, σ is 1 for an impermeable solute and 0 for a solute as permeable as the solvent itself. Thus for solutes with very high lipid solubilities, diffusion through the unstirred layers probably becomes rate-limiting (Table 5-5).

3. There are significant exceptions to the main pattern of nonelectrolyte permeability. Small molecules permeate relatively more rapidly (points to the left of the shaded curve in Fig. 5-2), and branched molecules permeate relatively more slowly (points to the right of the shaded curve). It has been found that small polar molecules of low lipid solubility permeate more rapidly than one would expect on the basis of their molecular weight. They also show other abnormalities. The inverse relation between σ and temperature is less steep for them than for other solutes, and is little affected by a decrease in pH, which increases the value of σ for other solutes.[292] These anomalies, however, do not show up in any of the bulk properties such as partition coefficient. Two possible explanations have been considered: (a) these solutes may follow the same route through

TABLE 5-5 REFLECTION COEFFICIENT OF SOLUTES ACROSS RABBIT GALL BLADDER

Substance	Mol wt	K_{ether}	K_{oil}	σ^a
Methanol	32	0.14	0.0078	0.04
Formamide	45	0.0014	0.00076	0.57
Acetamide	59	0.0025	0.00083	0.59
Urea	60	0.00047	0.00015	0.53
Methylurea	74	0.0012	0.00044	0.53
Thiourea	76	0.0063	0.0012	0.98
Diethylurea (asym)	116	0.019	0.0076	0.77
Hydroxyurea	76	—	—	1.06
Diethylthiourea (sym)	132			0.16
Ethanol	46	0.26	0.032	0.05
Ethyleneglycol	62	0.0053	0.00049	0.81
Propanol	60	1.9	0.13	0.02
1, 3-Dihydroxy propane	76	0.012		0.92
Glycerol	92	0.00066	0.00007	0.95
Butanol	74.1	7.7	0.53	0.01
isobutanol	74.1	4.5	0.25	0.03
tert-butanol	74.1	2.2	0.23	0.03
Diethylether	74.1	10	3.8	0.01
sec-butanol	74.1	6.9	0.44	0.05
1, 4-Dihydroxy butane	90	0.019	0.0021	0.86
1, 5-Dihydroxy pentane	104	0.055	0.0061	0.71
1, 6-Dihydroxy hexane	118	0.12	0.0068	0.34
1, 7-Dihydroxy heptane	132			0.17
1, 8-Dihydroxy octane	146			0.06

[a]Values taken from Ref. 291.

membrane lipids as do other solutes, and their enhanced permeation may be an expression of the sieving properties of oriented membrane lipids;[290] or (b) these solutes may follow a separate route through the cell membrane, interacting with membrane polar groups rather than with the hydrophobic chains, for example through pores.[246] Diamond and Wright[300] have reviewed the evidence in favor of the latter mechanism.

The low permeability of branched solutes is attributed to their partition co-efficients, which are lower than those of isologous straight chain molecules in the bulk phase. The fact that cell membranes discriminate more sharply against branched solutes than do bulk lipid solvents is probably attributable to the anisotropy of orientation of membrane lipid chains. Lipid-lipid forces are largely very short range van der Waals forces, which are effectively proportional to surface contact area. A branched molecule has less area of close contact with surrounding molecules than does a nonbranched molecule. This difference is amplified in an environment where "solvent" molecules are rather rigidly oriented. Also a branched-chain molecule passing through the membrane will, on the average, have a larger disruptive effect on the local structure, and encounter greater steric hindrance than a straight-chain molecule. The latter can present a smaller cross section by assuming an end-on configuration parallel to the hydro-carbon tails. Permeability to a branched molecule will therefore be reduced below the value expected from the bulk partition coefficient in an isotropic solvent.

4. Yet another point of departure from the main selectivity sequence may be a consequence of the rate-limiting step in the overall process of permeation. The main selectivity sequence is consistent with either of two different mechanisms: the rate-limiting step may be diffusion through the membrane interior, or it may be the passage through the water membrane interface. This dilemma is difficult to resolve since the intermolecular forces governing expected selectivity patterns are the same in either case (see Ref. 300 for discussion). As an argument in support of interfacial transfer as the rate-limiting step, it has been pointed out that interfacial resistance will depend largely upon the number of solute-to-water hydrogen bonds (Fig. 5-4). The limiting maximal $PM^{1/2}$ for a non-hydrogen bonding solute, as obtained by extrapolation, has been suggested to be the diffusion coefficient in the membrane interior. However, this has no general validity since the same arguments which predict low interfacial resistance to a hydrocarbon explain its high partition coefficient and hence low resistance for passage through the membrane interior.

5. The transmembrane flux of molecular species that can undergo association-dissociation reactions, such as acids and bases, is generally proportional to the concentration of uncharged species, which may be a function of pH. For mole-cules having large lipid moieties, such a correlation is not observed. In fact,

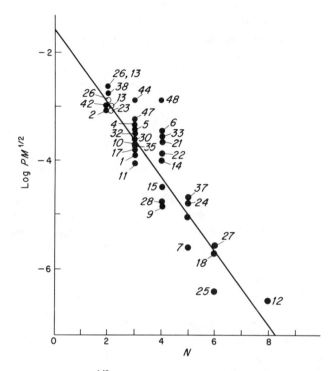

Fig. 5-4. Variation of log $PM^{1/2}$ (P in cm-sec^{-1}) with the number (N) of hydrogen bonding groups in the permeant in *Chara* spp. *1*, acetamide; *2*, antipyrin; *3*, butanol, *4*, butyramide; *5*, cyanamide; *6*, diacetin; *7*, dicyandiamide; *8*, diethylene glycol; *9*, diethyl malonamide; *10*, diethyl urea; *11*, dimethyl urea; *12*, erythritol; *13*, ethanol; *14*, ethylene glycol; *15*, ethyl urea; *16*, ethyl urethane; *17*, formamide; *18*, glycerol; *19*, glycerol + CO_2; *20*, glycerol + CO^{2+}; *21*, glycerol ethyl ether; *22*, glycerol methyl ether; *23*, isopropanol; *24*, lactamide; *25*, malonamide; *26*, methanol; *27*, methylol urea; *28*, methyl urea; *29*, monoacetin; *30*, monochlorohydrin; *31*, propanol; *32*, propionamide; *33*, (α,β)-propylene glycol; *34*, (α,γ)-propylene glycol; *35*, succinimide; *36*, tetraethylene glycol; *37*, thiourea; *38*, triethyl citrate; *39*, triethylene glycol; *40*, trihydroxybutane; *41*, trihydroxybutane + CO^{2+}; *42*, trimethyl citrate; *43*, urea; *44*, urethane; *45*, urethylan; *46*, urotropin; *47*, valeramide; *48*, water, *49*, 2,3-butylene glycol.[243]

changing pH has little effect on the permeability of higher fatty acids. At low pH there is significant deviation in the permeability of even neutral molecules, possibly because of altered charge and packing of membrane lipids.

6. Neurohypophyseal hormone extracts increase the osmotic flow of water and other small molecules through the outer membrane of the epithelial cell layer of several secretory tissues, such as frog skin, by severalfold. The diffusion rate as determined by isotope exchange is also changed.[301] Discrepancies reported in the earlier literature appear to be due to unstirred layers. These observations are

consistent with the assumption that neurohypophyseal hormones increase the pore size in some layer of the skin without increasing the total area available to diffusion. It is quite possible that these effects of hormones are secondary; that is, that the primary site of their action is not the membrane (also see p. 228). However, vasopressin (10^{-9} M) has been shown to increase water permeability across BLM by a factor of two.[1214]

MECHANISM OF PERMEATION*

What emerges from the discussion in the foregoing pages is the view that both BLM and biological membranes in their closed configuration act as highly efficient molecular assemblages for partitioning the aqueous phase. Since kinetic theory invokes diffusion as a natural consequence of atomic or molecular movement, it should in principle be possible to predict the behavior of molecules in this system. One would expect the diffusivity in the membrane *microphase* to depend on a variety of factors which would normally operate in any bulk phase. Thus, intermolecular forces are expected to be prominent. Thus a rational explanation of molecular events, even in phenomenological terms, would be expected to have its origin in classical kinetic theory of gases and solutions, according to which diffusion is considered to be a *random-walk* process. The diffusion coefficient D has been thus far treated as a phenomenological coefficient. But how does this coefficient depend on the structure of the medium and the interatomic forces which operate? To answer this question one should have a deeper understanding of this coefficient than that provided by the empirical first law of Fick (equation 5-1) in which D appeared simply as the proportionality constant relating flux and concentration gradient. To search for a more realistic basis of the diffusion coefficient one has to begin with classical kinetic theory.

In relating the macroscopic aspects of diffusion to the mobility of individual species, attention is given to the statistical features of the large number of possible paths that can be traced out by the molecules in a given time. It may be recalled that a net diffusive transport of a molecular species occurs, in spite of the completely random zig-zag motion of individual species, because of unequal numbers of molecules in different regions. It can be proved that for such a random-walk model the diffusion coefficient is related to the mean jump distance, that is, the mean free square path (ℓ) (in fact, to its component along the

---

*At the outset, it is of value to clarify the sense in which we employ the term mechanism. The ultimate aim of the study of membrane phenomena is to define completely the interaction between the membrane and the transported species at the molecular level. However for the reasons developed below the approach in this section is phenomenological. This approach, as we shall see, permits a convenient classification of biological transport processes and provides considerable insight into the nature of several of these phenomena in spite of our ignorance of the underlying molecular and biochemical mechanisms.

diffusion direction) and mean jump time (τ) by the relationship:

$$D = \tfrac{1}{2}\, \ell^2 / \tau$$

or

$$D = \tfrac{1}{2}\, \ell^2 \, \vec{k} \tag{5-9}$$

where $\vec{k}$ is the jump frequency, i.e., the number of jumps per second. This equation shows that the rate of diffusion depends on how far, on the average, a molecule jumps and how frequently these jumps occur.

In elementary treatments like the present one, the assumption is frequently made that each individual jump takes place in a direction independent of that of the preceding ones. This assumption of random displacements is an oversimplification. The mechanics of the process of collision between molecules is such that after a collision a molecule is more likely to have a component of velocity along the direction it was moving before the collision than in the opposite direction. Consecutive flights in the *same* direction occur with a larger than random probability, so that the diffusion coefficient (even in an isotropic medium) is greater than given by equation 5-9. This may be formally taken care of by rewriting equation 5-9 as:

$$D = 1/2 \cdot \ell^2 \vec{k} f \tag{5-9a}$$

where f is called the "correlation factor and it will have a value greater than one.

To go further one has to examine the factors which govern the mean jump distance ℓ and the jump frequency $\vec{k}$. For this, the picture of the medium in which diffusion is occuring as a structureless continuum is inadequate. For this purpose liquids may be treated as having structures which may be local in extent, transitory in time, and mobile in space. The details of such structures are not necessary for the present discussion. What is important is that molecules zig-zag in a random walk and, for any particular jump, the molecule has to jump *out of* one site in the medium *into* another site.* This jumping process may be represented as:

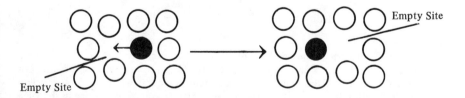

Empty Site

Empty Site

*These vacant sites may be associated with transient pockets of free volume or mobile structure defects (kinks) arising from thermal fluctuations of hydrocarbon chains of membrane-forming lipids. For models based on such assumptions see Refs. 1078 and 1079.

where a molecule goes through one step of the random walk by jumping to the empty site.* The mean jump distance ℓ is seen to be the mean distance between sites, and its numerical value depends upon the details of the structure of the medium.

The jump of a permeant molecule from site to site always occurs at a finite rate. Thus, diffusion is a rate process like a chemical reaction. In fact a cursory observation of the site-jump model just described suggests a strong analogy between these two rate processes, i.e., diffusion and chemical reaction. Thus the basic theory of rate processes should be applicable to both the processes.

We shall not go into a detailed discussion of the basis of this theory. However, an oversimplified discussion may facilitate further discussion. The basis of the theory of rate processes is that the potential energy (and the standard free energy) of the system of particles involved in the rate process varies as the particle moves to accomplish the process. Very often, the movements crucial to the process are those of a single particle, as is the case with the diffusive jump of a molecule from site to site. If the free energy of the system is plotted as a function of the position of the crucial particle, e.g., the jumping molecule, then the standard free energy of the system has to attain a critical value—the activation free energy, ΔG^0—for the process to be accomplished (Fig. 5-5). This is usually identified as an *energy barrier* for the rate process. The number of times per second the barrier is crossed, k, i.e., the jump frequency in the case of diffusion, is given by:

$$\vec{k} = \frac{kT}{h} \exp\left(-\frac{\Delta G}{RT}\right) \tag{5-10}$$

The $\vec{k}$ on the left-hand side is the jump frequency; the k in the kT/h term is the Boltzmann constant.

To obtain the diffusion coefficient in terms of atomic or molecular quantities, one has to insert the expression for the jump frequency into that for the diffusion coefficient. Thus from equations (5-9a) and (5-10) one obtains:

$$D = \frac{f}{2} \, \ell^2 \, \frac{kT}{h} \, \exp\left(\frac{-\Delta G^0}{RT}\right) \tag{5-11}$$

---

*The possible directions for the next jump of the molecule are clearly not of equal or random probability. The vacancy is immediately available to effect a second jump of the atom in a direction the reverse of the first, and the probability that this will take place is clearly greater than the probability that the vacancy will migrate through the lattice in the same direction as the first. Consecutive pairs of jumps in the same direction therefore occur with less than random probability, and pairs in opposite directions, leading to no net displacement, with greater than random probability. The net effect must be to reduce D below the value calculated assuming the jump directions are uncorrelated. The correlation factors will therefore be less than one in this case.

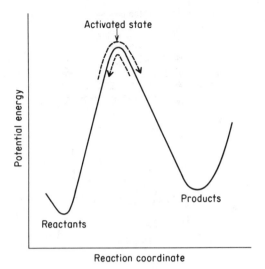

Fig. 5-5. A schematic potential energy profile for the system

The numerical coefficient $f/2$ has entered here only because the relationship (5-9) was derived for a one-dimensional random walk. In general, it is related to the probability of the permeant's jumping in various directions, not just forward and backward. For convenience the coefficient will be taken to be unity,* in which case:

$$D = \ell^2 \frac{kT}{h} \exp\left(\frac{-\Delta G^0}{RT}\right) \tag{5-12}$$

This relationship has fundamental theoretical importance. The term ℓ^2 can be correlated to the viscosity of the medium and the radius or mass of the molecule undergoing the random walk. Thus a relationship closely related to the Stokes-Einstein relationship may be generated.[302] Also, by evaluating D as a function of temperature, an estimate of thermodynamic parameters such as entropy, enthalpy, and activation energy can be obtained, and correlated to molecular or atomic parameters such as frequency factor, mean square free distance, etc. Although this treatment provides a molecular basis for the permeability phenomenon discussed earlier in this chapter, this approach has not been exploited experimentally, and alternative approaches have been devised.

---

*Obviously the value of f depends on the relative jump rates of the diffusion and host species. When these are equal, as in self-diffusion, f becomes a geometrical constant dependent only on the structure.

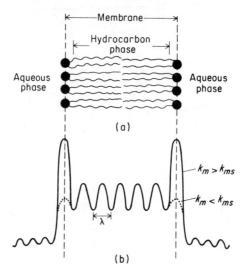

Fig. 5-6. (a) A schematic representation of membrane interfaces and the phases involved in solute transfer. (b) Possible potential energy profile for a solute molecule diffusing through a membrane.[1079]

For the experimental evaluation of the thermodynamic parameters characterizing diffusion across a membrane, the permeability coefficient (P) can be correlated with the diffusion coefficient (D) by the following considerations. The total resistance encountered by a permeant molecule during diffusion may be viewed as due to five resistances in series, as shown in Fig. 5-6: two resistances at the two interfaces of the membrane, one due to the membrane phase, and two due to the bulk solvent phases. It is generally assumed that diffusion in the bulk solvent phase is much faster than in the membrane phase or at the membrane/bulk solution interfaces. Thus, if D_{sm} and D_m are the diffusion coefficients of the permeant at the membrane/bulk solution interface and in the membrane phase, respectively, they can be related to P by the following relationship:[303]

$$\frac{1}{P} = \frac{2\ell}{D_{sm}} + \frac{t}{D_m K} \tag{5-13}$$

where t is the thickness of the membrane, K being the partition coefficient or the equilibrium constant of the permeant, K is also defined as k_{sm}/k_{ms}, where k_{ms} is the rate constant for diffusion from membrane to solution, and k_{sm} for solution to membrane. Thus equation (5-13) can be solved for the three possible cases: (a) membrane limiting, $k_{ms} \gg k_m$; (b) interface limiting, $k_m \gg k_{ms}$, and (c) the membrane-interface limiting.

An idea of the mechanism of permeation of nonelectrolytes can be arrived at

TABLE 5-6 EXPERIMENTAL ACTIVATION ENERGIES AND DERIVED
QUANTITIES FOR MOVEMENT ON WATER IN VARIOUS SYSTEMS

System	Activation (Energy (E_a) (kcal/mole)	Entropy (cal/mole/degree)	Enthalpy (kcal/mole)	Free Energy (kcal/mole)	Ref.
Bulk water (self-diffusion of HDO)	4.6	2.5	4.0	3.9	
n-Hexadecane	3.4	−0.2	2.8	3.4	
Liposomes (lecithin)	8.25				1072
BLM/lecithin/cholesterol	14.6				121
Oxidized cholesterol*	6.8	−13 to 15	6.2	10 to 11	
Nitella translucens	8.5		7.9		
Turtle bladder					
No vasopressin	9.8		9.2		
With vasopressin	4.1		3.5		
Ascites tumor cell	9.6		9.0		
Bovine erythrocyte	5.4		4.8		
Sea urchin egg	14.7 to 16.6		14.1 to 16.0		

*Calculated values corresponding to ℓ = 1-3 Å (mean free path).

from thermodynamic data. For water permeation such data is collected in Table 5-6. Very little can, in fact be said about the mechanism involved in the transfer of water molecules across the membrane. However, some meaningful generalizations have been arrived at from the study of thermodynamic parameters characterizing nonelectrolyte permeation. Before discussing specific examples it may be instructive to relate thermodynamic parameters to various experimentally observed quantities. The following expression relates the lipid/water partition coefficient (K) to the free energy change (ΔF_t) occuring while transferring one mole of solute from water to a lipid solvent:

$$K = \exp\left(- \Delta F_t/RT\right) = Pd/D \qquad (5\text{-}14)$$

If the nonelectrolyte permeation through the interior of the membrane is considered so that the equation (5-14) holds, the physical chemistry of the aqueous and organic phases should determine intermolecular forces* and be reflected in the free energy changes. Thus the free energy change resulting from a change in structure of the solute can be related to the incremental thermodynamic quantity by which the group or substituent alters the partition equilibrium between membrane lipids and water. Thus the "increments in free energy" ($\delta \Delta F_t$), enthalpy ($\delta \Delta H_t$) and entropy ($\delta \Delta S_t$) of transport can be obtained. If nonelectrolyte per-

---

*It is implied in this assumption that forces governing intermolecular interactions are essentially the same in bulk (isotropic) and membrane (anisotropic) phases.[1081,1082] Similar assumptions have been made for structure-activity correlation of pharmacological agents where contributions of hydrophobicity and stereoelectronic factors in a given set of molecules may be developed.

meation is limited by the membrane/water interface the incremental quantities have the significance of activation parameters. Obviously, it is not possible on a purely thermodynamic basis to decide whether the rate-limiting step in transfer of the solute is interfacial transfer or diffusion across the lipid layer.

Incremental free energies of transfer (or solution) associated with various organic substituents are summarized in Table 5-7. The significance of these values is best understood in terms of the following considerations. Consider the transfer of solute from water to the membrane to occur in two steps: Vaporization of solute from an aqueous phase at infinite dilution into a vacuum, followed by condensation and solution of solute from the vacuum into the lipid phase. Thus, $\Delta F_t = - \Delta F_w + \Delta F_\ell$, where ΔF_w and ΔF_ℓ are the partial molar free

TABLE 5-7 INCREMENTAL FREE ENERGIES OF SOLUTION FOR VARIOUS ORGANIC SUBSTITUENT GROUPS[a] (CAL/MOLE)[300]

	$\delta\Delta F_t$ $(= -\delta\Delta F_w + \delta\Delta F_t)$				$\delta\Delta F_w$	$\delta\Delta F_\ell$			
	Iso-butanol	Ether	Olive Oil	Nitella		Iso-butanol	Ether	Olive Oil	Nitella
—OH	1000	2100	2800	3600	−7000	−6000	−4900	−4200	−3400
—O—	600	1400	1400	800	−4000	−3400	−2600	−2600	−3200
O ‖ —C—	—	2100	2200	2500	−6100	—	−4000	−3900	−3600
O ‖ —C—OH	1100	1700	2800	—	−8600	−7500	−6900	−5800	—
O ‖ —C—O—R	1200	1200	1400	1400	−5400	−4200	−4200	−4000	−4000
O ‖ —C—NH$_2$	1700	4900	4800	6200	—	—	—	—	—
O ‖ —NH—C—NH$_2$	1900	5500	5300	7300	—	—	—	—	—
C≡N	—	1100	—	—	−5800	—	−4700	—	—
—NH$_2$	1100	3500	—	—	−6600	−5500	−3100	—	—
—CH$_2$	−530	−670	−660	−610	160	−370	−510	−500	−450

[a]The second through fifth columns give for each group listed in the first column the average values of $\delta\Delta F_t$, the amount by which the group changes the difference between the partial molar free energy of solution for a nonelectrolyte in water and its partial molar free energy of solution in isobutanol, ether, olive oil, or *Nitella* membrane. Each 1000 cal/mole corresponds to a 5.4-fold reduction in partition or permeability coefficients. The sixth column is $\delta\Delta F_w$, the amount by which the group changes the partial molar free energy of solution for a nonelectrolyte (out of vacuum) in water. The seventh column contains $\delta\Delta F_\ell$, the amount by which the group changes the partial molar free energy of solution for a nonelectrolyte (out of vacuum) in isobutanol, ether, olive oil, or *Nitella* membrane. The more positive the value of $\delta\Delta F_t$, the more effective is the group at reducing partition coefficients or permeability coefficients. The more negative values of $\delta\Delta F_w$ and $\delta\Delta F_\ell$ imply greater solubility in the given solvent.

energies of solution at infinite dilution in water and in lipid solvent respectively. Similarly, $\delta\Delta F_t = -\delta\Delta F_w + \delta\Delta F_\varrho$ for any specific substituent group by which the parent molecules differ. The values are collected in Table 5-7, from which the following conclusions may be drawn:[300]

1. Values of $\delta\Delta F_t$ associated with various substituent groups are positive for all substituents in all four phases (isobutanol/water, ether/water, oil/water, Nitella/water) except for the methylene ($-CH_2-$) group. That is, every substituent reduces the permeability and partition coefficients, except $-CH_2-$ which increases them.

2. Virtually every substituent has a negative value of incremental change in partial molar free energy of solution $\delta\Delta F_\varrho$ or $\delta\Delta F_w$ at infinite dilution in all the five solvents, that is, the change in net attractive intermolecular forces does not favor transfer from water to the organic phase. The sole exception is $-CH_2-$ in water, which has a positive $\delta\Delta F_w$, and thus decreases water solubility.

3. For every substituent (except $-CH_2-$) in all four systems, the value of $\delta\Delta F_w$ is considerably greater than $\delta\Delta F_\varrho$. This implies that solute/water forces are much stronger than solute/organic solvent forces for nonelectrolyte substituent groups. Thus, most substituents decrease permeating power because they increase the energy required to break the solute loose from water.

4. Resolution of these free energy terms into enthalpy and entropy shows that the effects of OH, NH_2, CO, COOR, and $-O-$ groups (and probably CN and COOH too) in increasing water solubility are enthalpy controlled, that is, these substituents strengthen solute/water intermolecular forces, even though this reduces the system's degrees of freedom and causes a small loss of entropy. On the other hand the effect of the $-CH_2-$ group proves to be entropy controlled.

Qualitatively different selectivity patterns are thus expected to be determined largely by differences in ΔF_w between different substituent groups. The quantitative differences between different systems must be due to differences in $\delta\Delta F_\varrho$ for a given substituent. Of the four lipid solvents compared, *Nitella* provides the greatest selectivity, in the sense that it has the largest range (6000 cal/mole) of $\delta\Delta F_t$ among substituents other than methylene. Hence it has the largest spread of permeability coefficients and partition coefficients. For instance, the ratio of P for methyl acetate to P for glycerol is 780,000 in *Nitella*, whereas the K ratio in olive oil is only 6100, in ether only 4100, and in isobutanol only 26. This is attributable to both the smallest range (800 cal/mole) and the lowest absolute values of ΔF_ϱ's for *Nitella* which makes it the "weakest solvent."

Thus an interpretation of thermodynamic data in molecular terms can be extended to include intermolecular interactions. The principal forces between nonelectrolytes and water depend upon hydrogen bonds, whereas the principal forces between nonelectrolytes and lipids are short-range van der Waals forces.

As noted earlier in this chapter (p.130), selectivity in partition and permeation is dominated by interactions in the aqueous phase, and these aqueous phase interactions are enthalpy, rather than entropy controlled (p.131). Thus non-electrolyte permeability should be explicable in terms of number and strength of hydrogen bonds.*[243] Specific effects of hydrogen bonds in a closely related pair of solutes having similar molecular environment are predictable from qualitative considerations of the strength of hydrogen bonds, as discussed elsewhere.[243] Attempts to predict permeability ratios for different and unrelated nonelectrolytes by attributing a certain number of hydrogen bonds to each group regardless of its molecular environment have had a limited success, as shown in Fig. 5-4. The scatter is attributable to the molecular environment which may affect intermolecular hydrogen bonding by intramolecular hydrogen bonding and by inductive effects. The former affects the effective number of hydrogen bonds and the latter alters their strength by delocalizing electrons (hence proton donating and accepting ability). Similar effects have been observed in the permeability of polyhydric alcohols across liposomes.[1074] In these cases the activation energy is found to be independent of the degree of unsaturation or the presence of cholesterol in the lipid barrier, but seem to depend only on the number of hydroxyl groups in the solute. It therefore appears that the activation energy for dehydration may be the rate-determining factor in the transfer of hydrated solutes across lipidic barrier.

ION TRANSPORT ACROSS BLM

As discussed in the previous chapter, the most important principle governing the distribution of charged particles across the BLM is that these particles must respond not only to their chemical concentration gradients but also to electrical gradients in the system. Thus the permeability and equilibrium distribution of a particular ion is determined by the combined electrochemical potential of counterions in the medium. The electrical gradient itself may be partly or entirely the result of asymmetries in the overall ion distribution. Analysis of such intricate systems in terms of strictly physical and chemical forces, sometimes operating together and sometimes in opposition, has generated an exten-

---

*It may be recalled that a hydrogen bond is a bridge formed by hydrogen between two electronegative atoms acting as proton acceptor and proton donor respectively. The strongest bridges involve oxygen atoms as the donors and oxygen or nitrogen as acceptors; dipole forces resonance stabilization, and dispersion forces also contribute to the bridge strength. The approximate donor potency sequence for biological systems is:

$$COOH > CONH_2 > OH >> NH_2$$

and the potency sequence for acceptors is:

$$NH_2 > COOH > CONH_2 > OH > -O- > CHO > CO, COOR > CN > NO_2$$

sive and highly complex literature; however, new insights have come very slowly. The classical view is based on the assumption that transfer of charges across a phase boundary leads to the development of an electrochemical potential. When the equilibrium potential is reached, the apparent charge transfer ceases as in the case of a chemical reaction at equilibrium. However, on a molecular scale, a constant exchange of charge occurs continuously. Thus in the absence of an electrochemical process at a phase boundary, the equilibrium potential must be attainable, which is a measure of ionic permeability and selectivity (see Chapter 6).

Movement of ions across the membrane should be governed by the same rules which govern movement of other solutes; however one has to take into consideration the properties of both cation and anion, and their charged character. From the considerations of the electrical conductivity of BLM (Chapter 4) we might expect that the membrane model we have been considering for transport of nonelectrolytes would have to be modified to account for the properties of biological membranes. The basic predictions of the lattice or continuous lipid layer model are rather satisfactorily confirmed by the electrical properties of BLM and by the data on hyperfilteration through uncharged membranes.[1216] Also the observed rates of ionic diffusion through BLM, under steady-state conditions are immeasurably slow. Thus, for example, the flux of sodium or chloride ions could not be detected in isotope tracer flux experiments.[108,130,269] The lower limit for ionic permeability set by these experiments is of the order of 5.6×10^{-9} cm-sec^{-1} for sodium and 8.3×10^{-8} cm-sec^{-1} for chloride ions. The electrical conductance of BLM, although small but finite, could arise due to "solubility diffusion" of ions. Diffusion potentials developed across BLM in salt gradients suggest that the membrane is generally more permeable to one ion than to its counterion. Membrane potentials and consequent transference (transport) numbers determined from such experiments are given in Table 5-8. Permeability coefficients calculated from such data are 4.8×10^{-9} and 2.1×10^{-9} cm-sec^{-1} for sodium and chloride ions respectively.[108,227] The mode of transfer of these ions is unknown. Such movement of ions would, however, involve work against the electrostatic forces, the surface tension, the electrical potential, and the cohesive forces in the layer. Thus, the activation energy of ion movement across the continuous lipid layer of BLM would be large. It is for such reasons that permeability across BLM has been studied by indirect methods, including the measurement of equilibrium diffusion potentials.

Thus on introducing a salt gradient, BLM show potential changes which depend upon both the nature of the lipid forming BLM, as well as the nature of the ions. Such potential changes occur in two steps: a rapid transient one caused by a change in the boundary potential due to the altered ionic strength, and a slow one resulting from the establishment of a new equilibrium potential.[304] The results obtained from such studies (Table 5-8) show that there is a correlation between the charge on the polar head group and the relative ion permeability.

TABLE 5-8 EQUILIBRIUM DIFFUSION POTENTIALS AND CATIONIC TRANSFERENCE NUMBERS IN BLM[a]

Lipid	Electrolyte	Change in Potential per Tenfold Increase in Salt Activity (mV)	Net Charge on Lipid	Cationic Transference Number (t^+)
Lysyl phosphatidyl	NaCl	+52 to +54	+	0.05
glycerol	KCl	+47 to +55		0.04
	HCl	+48 to 49		0.09
Phosphatidyl glycerol	NaCl	−27 to −54	−	0.95
	KCl	−39 to −54		0.95
	HCl	−47 to −54		0.95
Diphosphatidyl glycerol	NaCl	−27 to −43	−2	0.86
	KCl	−19 to −44		0.87
Diglucosyldiglyceride	NaCl	−3 to −13	0	0.61
	KCl	−6		0.55
	HCl	−38 to −39		0.82
Phosphatidyl ethanol-	NaCl	−6	0	0.55
amine (E. coli)	KCl	−10 to −14		0.62
	HCl	−25 to −35		0.79
Phosphatidyl choline	NaCl	−5.6	0	0.55
(egg)	KCl	−10.1		0.52
Erythrocyte lipids	NaCl	−36		0.82
(high-K sheep)	KCl	−36.6		0.84
Erythrocyte lipids	NaCl	−31.5		0.78
(low-K sheep)	KCl	−33.9		0.80

[a]The anionic transference number is $1 - t^+$. The sign of potential corresponds to the side containing higher concentration of the salt. Membrane potentials are usually determined for 0.1/0.01M ratio of the salt or one pH unit difference for HCl in the neighborhood of 5-7.[304]

Positively charged membranes are highly anion selective ($t^+ =$ less than 0.1) whereas negatively charged membranes are highly cation selective ($t^+ =$ more than 0.85). BLM prepared from uncharged lipids show only slight cation selectivity ($t^+ = 0.55$-0.62). Similar results have been obtained with BLM prepared from other fixed charge membranes[305] and liposomes.[148,149,151,306] However, liposomes prepared from phosphatidylserine and phosphatidylglycerol show a K^+/Na^+ diffusion rate ratio of approximately ten.[1217] Other phospholipids do not show such intercationic selectivity.

As indicated earlier conductance of biomembranes is several orders of magnitude larger than the conductance of BLM; thus implying a higher rate of ionic movement across biomembranes, which also show selectivity for transport of closely related ions. In multicellular membranes such as epithelia the "tight junctions" between cells may provide the main route for passive ion permeability.[1292] The specificity and selectivity is generally characteristic of the tissue and the metabolic conditions of the cell. Some of these features become apparent from the data collected in Table 5-9. Such observations suggest that the ionic permeability characteristics of biomembranes are due to physiological specializa-

TABLE 5-9 PERMEABILITY[a] OF VARIOUS MEMBRANES TO WATER AND INORGANIC IONS

Membrane	Water	K^+	Na^+	Cl^-
BLM (phosphatidylcholine)	1.1×10^{-3}	3.4×10^{-12}	10^{-12}	10^{-9}-10^{-12}
Human erythrocyte	1.1×10^{-2}	2.4×10^{-10}	10^{-10}	2×10^{-4}
HK Sheep erythrocyte	—	0.84×10^{-10}	4.4×10^{-10}	—
LK Sheep erythrocyte	—	5.7×10^{-10}	3.7×10^{-10}	—
Squid axon (resting)[b]		5.6×10^{-7}	1.5×10^{-8}	1.0×10^{-8}
Squid axon (excited)		17×10^{-5c}	5×10^{-6}	10^{-8}
Frog sartorius muscle (resting)[b]	1.3×10^{-2}	1.6×10^{-7}	1.4×10^{-7}	9.5×10^{-8}

[a]All values are given in cm-sec^{-1}.
[b]Values correspond to passive flux only.
[c]Conductance corresponds to completely depolarized state under voltage-clamp conditions.

tion, and, therefore, that such permeability characteristics may have some functional significance and may be under metabolic control. It is interesting to note in this connection that disturbance of intermediary metabolism is frequently associated with changes of the electrolyte distribution between the cells and the medium. In fact, as we shall discuss in Chapter 7, significant parts of ionic fluxes are associated with energy-dependent transport across the membrane (*active transport*). To complicate matters further, there is an increasing body of experimental evidence suggesting that alteration of metabolic conditions also affects passive ionic movement. For example, human erythrocytes are permeable to potassium ions, whereas the permeability of other ions is relatively small. In the presence of metabolic inhibitors, such as fluoride ions or iodoacetate, the permeability of potassium ions across the erythrocyte membrane decreases some twentyfold. The sodium permeability remains largely unaffected.[308] Such observations therefore, are generally interpreted to suggest the presence of a specialized permeability mechanism since simple solubility-diffusion mechanisms cannot account for such large discrepancies between the observed permeabilities of BLM and biomembranes.

Summarizing the permeability characteristics of a continuous lipid bilayer and that of a biomembrane (generally speaking of course), it is fair to say that permeability characteristics of nonelectrolytes suggest a general solubility-diffusion-mechanism operating through a lipid layer. Such a mode of mass transfer can be formalized within the framework of the theory of rate processes, and the thermodynamic parameters further substantiate a mechanism in which a continuous lipid core is implicated. The solute permeability coefficient depends upon size, steric and electronic configuration, and the partition coefficient between the membrane phase and the bathing solution. This somewhat idealized picture has some serious limitations when one considers permeability of small molecules and ions. These differences may be summarized as follows:

1. Extra acceleration of permeability of small molecules, such as water, urea, etc., across biological membranes seems to suggest a sieving effect.

2. Differences in osmotic and diffusion permeability of water (if it is proved to be real) are generally rationalized by postulating the presence of pores (or regions of discontinuity) in membranes.

3. The effect of hormones further substantiates the presence of pores.

4. The excessive permeability of certain ions and metabolites cannot be rationalized by a lipid solubility-diffusion mechanism.

5. The temperature coefficient for various diffusion processes as collected below further attest to difference in their mechanism.

Process	Q_{10}	Activation energy (kcal/mole)
Ion transport: (for K^+		
free diffusion in water	1.26	4.2
in ion-exchange resins	1.4	6.5
in ion-exchange resins (highly crosslinked)	1.6	8.62
in passive diffusion across most biological membranes	2 to 4	11 to 25
Water transport (BLM)	1.4	6.5
Water transport (biological membranes)	1.5-2.5	7 to 20
Nonelectrolyte (BLM and biological membranes)	1.35-1.5	6 to 10

All these exceptions taken together seem to suggest the existence of alternate pathways in biomembranes when compared with those present in BLM. Indeed, as we shall see in the next chapter, modification of BLM by suitable "agents" leads to the induction of permeability characteristics similar in several important respects to those shown by biological systems.

6/ Extrinsic Properties of Bimolecular Lipid Membranes

Thus far we have considered the properties of BLM prepared from a variety of lipids. In spite of a wide variation in lipid composition, the characteristics (*intrinsic*) of BLM fall within a fairly narrow range. However, as noted in the last chapter and summarized in Table 6-1, there is a multiplicity of functions associated with biomembranes which mitigate against the acceptance of a simple continuous lamellar lipid layer structure for biological membranes. As was pointed out in the first chapter, it was realized quite early in the study of biomembranes that the protein components of the cellular membranes play functional roles besides having structural importance. From a functional point of view, the cell surface seems to consist of several specific components such as blood group substances, transport proteins, catalytic proteins, and so forth. Such membrane-bound molecules seem to be responsible for surface charge characteristics, immunological characteristics, and transport properties besides having their role in DNA replication, protein synthesis, etc. Some of these complexes have been shown to consist of proteins, mucopolysaccharides, complexes of sialic acid, aminosugars, and amino acids. The question of their functional identity and structural significance is far from solved. To complicate matters further, a considerable amount of experimental evidence has accumulated suggesting a high degree of specificity of interaction of pharmacological agents and other biologically active compounds with biomembranes; these include toxins[310,311] and pheromones.

These and various other observations thus seem to imply that the lipid bilayer in biomembranes, in contrast to BLM, contains regions of discontinuity. Such a discontinuity could arise due to lipid-bound proteins which modify the prop-

145

TABLE 6-1. A COMPARISON OF THE PROPERTIES AND PHENOMENA ASSOCIATED WITH BLM AND BIOMEMBRANES

Property or Phenomenon	BLM	Biomembrane
Thickness (Å)	50–70	50–150
Mechanical properties	Relatively unstable	Stable
Surface tension (dyne/cm)	2–0.1	0.5 to almost 0
Capacitance ($\mu F/cm^2$)	0.33–1	0.5–2
Dielectric breakdown voltage (mV)	80–500	80–500
Electrical resistance ($\Omega \cdot cm^2$)	about 10^8	less than 10^5
Permeability (cm/sec)		
Water (tracer)	10×10^{-4}	4–40×10^{-4}
Water (osmotic)	10×10^{-4}	Usually more than tracer flux
Nonelectrolytes	As predicted from lipid solubility	As predicted from lipid solubility except for small molecules which show much higher permeability
Electrolytes	As predicted from lipid solubility	Show much higher permeability than that predicted from lipid solubility
Ionic selectivity	Little: as determined by polar head groups	Show considerable ionic selectivity both for ions of different charge and for related ions
Specialized functions Effect of hormones and external stimulus, photosynthesis, oxidative phophorylation, immunological characteristics, etc.	Do not show any	All these functions can be observed in membranes of various specialized cells

erties of the hydrophobic barrier inherent in the continuous lipid layer. Such modified structures would thus impart the dynamic properties of the catalytic proteins to the hydrophobic barrier. The behavior of the modified membrane would thus be reflected in the modulation of the levels of the metabolites, altered electrokinetic properties, morphological changes, and in altered antigenicity. Also there may be more distal effects arising from altered metabolic relations of the system as a whole.

Specific aspects of lipid-protein ineraction are discussed in Chapter 9; until then, however, we shall correlate the properties of *modified* BLM with the dynamic and functional characteristics of biomembranes. As it turns out, experimental alteration of the permeability characteristics of a continuous bilayer structure can account for some of the dynamic aspects of biomembranes. Thus the purpose of this chapter is twofold. In the first part, experimental evidence concerning the distinctive permeability characteristics, selectivity patterns, and the underlying molecular basis of these processes is developed with particular reference to studies on modified BLM. The second part is concerned with extrapolation of this information to characterize the permeability characteristics of biomembranes with special reference to facilitated transport. The extrinsic properties of (modified) BLM, as expected, can be characterized in terms of their thermodynamic parameters, stoichiometry of constituents, and spatial density of the modifying agent. The overall picture is far from complete, although considerable insight may be drawn from the work already done in this field.

Several modes of mass transfer across biological membranes have been considered from time to time. As we noted earlier, certain ions and molecules penetrate living cells at a rate much faster than would be predicted from the consideration of simple solubility-diffusion mechanism. In contrast, such membranes show much smaller permeabilities to most of other solutes. An apparently abnormal permeability for specific solutes has been attributed to the presence of specific sites on the membrane which allow the passage of those permeants across the membrane. Some of these sites are, however, characterized as mobile within the membrane phase whereas the characteristics of other systems are taken to implicate the presence of pores or channels across the membrane, that is, the sites are fixed in the membrane matrix and may or may not be accessible from both interfaces. It must be realized that the existence of channels or carriers cannot be proved directly and their existence may appear hypothetical; however, the evidence for their presence in membrane systems may be inferred from experimental data such as kinetics of transport processes, influence of external factors, substrate specificity, and related phenomenon. In fact, considerable experimental evidence has accumulated over last fifty years suggesting the presence of facilitated transport mechanisms in biomembranes.

In relation to the above hypothesis, it may be mentioned that the incorporation of various substances* into BLM gives rise to considerable differences in their permeability characteristics and such systems are excellent models for detailed study of facilitated transport phenomenon. In the following sections we shall mainly restrict ourself to a phenomenological description of transport through channel or carrier-mediated transport systems, mainly as characterized from the study of modified BLM. No attempt has been made to provide a detailed mathematical treatment, although some generalizations regarding quantitative relationships are mentioned.

There is considerable confusion in the.literature regarding the terminology as

---

*A large number of molecules (anaesthetics, steroids, and drugs) seem to interact with the membrane components by virtue of their molecular geometry or charge. Their incorporation into BLM may have little effect on the packing order of lipids and may thus have little effect on many of the membrane properties, but the effect on viscosity drag may become significant and thus change the solubility diffusion characteristics. Very little can be said about such interactions and their effects at a macromolecular level. Therefore, we shall not consider such cases in any detail, except for some brief remarks in the text at relevant places.

definition of "mobile carrier," and to some extent of the "channel." For the present purpose these two systems are characterized as follows: The term *mobile carrier* implies the existence of a *particle* (a single molecule or an aggregate) which forms a stoichiometric complex with the permeants either on the membrane interface or in solution. In all the models for carrier transport, complex formation is followed by translocation of the substrate-carrier complex across the membrane, where the substrate is released, and finally the carrier system returns to its original state. Such translocation may be achieved by free diffusion, by rotating molecules, or by a propelled shuttle mechanism where energy is expended. However, there is a tendency to describe carriers on the basis of the relatively restricted conformational changes that may be permitted in the membrane system. This may imply a change of accessibility of the translocation site from one side of the osmotic barrier to the other. Therefore, little or no movement of the translocator or the carrier may occur. The translocation step may be achieved simply by changing the bonding relationship in the carrier site of the translocator. Such a situation becomes particularly important due to the fact that the dimensions of the membrane are small and comparable to those of large molecules which may be associated with them.

The terms *"channel"* and *"pore"* imply fixed sites on the membrane or in the membrane phase through which a permeant can pass. In pore formation the active molecules affix to the surface of a membrane, providing an area which offers a lowered energy barrier for the penetration of the permeant. In channel formation several active molecules stack together forming a channel which completely breaches the membrane. The permeant species pass through the channel, forming intermediary complexes progressively with successive molecules of the channel. In principle, channels or pores may be treated as expanded lattices in the membrane phase. In narrow pores or channels, the diameter of which is comparable to the molecular dimensions of the permeant, the permeant molecules will line up in a column. It follows that all those molecules which enter the pore on the left will always be to the left of the molecules which enter from the right of the solution. In such a system the encountering molecules cannot separate, and the displacement of the set of molecules in the pore is due to the effective hits of the molecules in solution against the edge of the set. It may, however, be noted that the driving force for a given permeant passing through a pore or channel will depend not only on the concentration gradient but also on the permeabilities of all substances passing through the pore. Thus the efflux of chloride from muscle cells slows down in the presence of internal nitrate ions. The movement of permeant may not necessarily involve any change in the binding relationships—chemical or conformational—at the active site. The permeant may, however, be recognized on the basis of physicochemical properties such as charge, size, degree of hydration, shape, nature of the atoms or groups of the permeating molecule which determine weak intermolecular interactions. In the following discussion no attempt has been made to distinguish pores from channels, and these terms have been used almost interchangeably.

The two modes of facilitated transport, as mediated by mobile carriers or by channels, are probably extreme cases of what is generally encountered in various membrane systems—both biological and artificial—and the distinguishing features of these two modes are not easy to establish in some cases. For example, it is conceivable that surface-bound macromolecules or macromolecular aggregates may form part of the membrane structure, and movement of only a part (say a side chain) may close the access of the binding site to one side and open the other side. Such systems would be expected to share the salient features of both of these models. Some of the common as well as the distinguishing features of mobile carrier and channel models are collected in Table 6-2.

TABLE 6-2 CRITERIA FOR IDENTIFICATION OF FACILITATED TRANSPORT SYSTEMS

Mobile Carrier	Pore-Channel Flow
1. Operates on the electrochemical gradient of the permeant; the rate of movement of permeant is faster than would be predicted from the oil/water partition coefficient; the temperature coefficient is lower than would be predicted if the rate-limiting step for the movement across the membrane were breaking the hydrogen bonds with water.	Same as for mobile carrier
2. The initial rate of entry of permeant depends upon its concentration but reaches a maximal value; also, the rate of penetration of permeant shows saturation kinetics with increasing concentration of permeant; thus the rate of diffusion is not given by Fick's law	Same as for mobile carrier, although some differences may be noted
3. The permeant must show a very high degree of specificity with regard to molecular parameters such as shape, charge, size, sterochemistry, or any other criterion employed for the recognition of the permeant by the carrier; thus, like enzyme substrate binding, this system allows for stereoelectronic specificity, high turnover rates, and sensitivity toward inhibitors and protein reagents	Relatively poor discrimination on the basis of various physicochemical factors is expected; the most general criterion for selectivity is size
4. The rate of penetration may be markedly reduced by the presence of molecules structurally analogous to the permeant; the blocking agent would have a tendency to compete for the same carrier sites; however, certain irreversible blocking agents can be conceived	The channel may be blocked completely by certain molecular species, and the effect is of the *all-or-none* type

Table 6-2 (cont.)

Mobile Carrier	Pore-Channel Flow
5. In certain cases it is possible to link the facilitated movement of species down its electrochemical gradient with the movement in the opposite direction of structurally analogous molecules; thus, this second species may be driven up its gradient	Counter-transport is not possible
6. In biological transport systems proteins appear to be involved invariably since they are the only molecules which possess the required degree of specificity to discriminate between possible substrates; thus, transport may be inhibited by reagents which interact with proteins or may be neutralized immunologically, or may be suppressed genetically	All these possibilities exist in this case too; however, the effect of protein reagents may not be as dramatic
7. As the carrier molecules move across the membrane plane, there would be considerably lesser degree of specificity expected in lipid-carrier interaction; at times this may be reflected in the lipid composition of the biomembranes	Since the orientation of the molecules forming the channels requires a considerable degree of specificity, it may be reflected in the composition of the membrane and certain types of lipids may always be needed for specific transport activity
8. The concentration of carrier vs. permeability plots are almost always exponential	The concentration of channel forming agents vs permeability plots show sigmoidal characteristics
9. The relationship between log membrane resistance and log concentration of charge-carrying ion in the medium is sigmoidal	The log membrane resistance vs log concentration of charge-carrying ion is linear
10. The current-voltage curves are characterized by limiting current	The current-voltage curves are characterized by finite limiting conductances approached in the limits of high positive and negative fields
11. Inhibition is not anisotropic	Partial inhibition of entry or exit is possible if the entry or exit sites are physically different (interface)
12. Fixed charges on the mobile carrier would not give rise to electrokinetic phenomena	Fixed charges on or in the channel would give rise to streaming potential and other electrokinetic phenomena

The most outstanding feature of the two models of facilitated transfer is the energy of transfer. This property is a consequence of the lipid phase being effectively discontinuous, and the transfer from the aqueous phase to the *modified* membrane may no longer be the rate-limiting step. Thus the process of facilitated transfer, even though occurring under the driving force of thermal agitation leads to equilibrium far more rapidly. This feature is particularly obvious when one compares transport of ions in modified and unmodified bimolecular lamellar lipid membranes as discussed in the part I of this chapter.

Theoretical Expectations

ION TRANSPORT ACROSS MODIFIED BLM

The low permeability of ions across BLM is thought to be the consequence of the extremely low solubility of ions in the hydrocarbon phase of the membrane core (see p. 101 and Ref. 312). Alternative explanations have been sought from time to time to account for the high permeability characteristics of biomembranes for ions. In principle, the passage of charged polar species through BLM is possible only if the charge is neutralized through interaction with a counterion, or if the charge is buried inside a nonpolar hydrophobic macromolecule. Thus the permeant species are usually counterions which neutralize the charge of a fixed or bound site on the membrane. Alternatively, the permeant ion may form a suitable complex with an uncharged carrier species such that the surface of the complex is nonpolar and hydrophobic. Thus, whatever may be the mechanism of transport of these complexes across BLM, the nature of the ion-binding or -complexing sites on the membrane, and the concentration of counterions and coins on or in the vicinity of the site becomes critical in determining the selectivity pattern and mobility of ions across the membrane. The model based on this general principle in its simplest form postulates formation of ion-carrier complexes (refer to Fig. 6-1 for details):

$$XI \rightleftharpoons X^- + I^+$$

$$I^+ + S \rightleftharpoons [IS]^+$$

$$X^- + [IS]^+ \rightleftharpoons [IS]^+ X^-$$

These reactions are assumed to occur in the aqueous as well as in the membrane phase. However, the complex formed by combination of carrier with one of the ions (in this case the cation I^+) has considerably different solubility properties than those of the ion itself, and thus under suitable conditions the complex may pass through the hydrocarbon core of a lipid membrane. In fact, the concept of the ion pair was introduced by Bjerrum in 1926 to account for the behavior of ionophores in solvents of low dielectric constant. Such a pairing of ions* often leads to the formation of new thermodynamically distinct species possessing properties strikingly different from those of free ions. As would be expected, the dielectric constant (D) of the medium will have a strong effect on the dissociation constant (K_{dis}). For an ideal dielectric medium** (hypothetical structureless, continuous but polarizable):

$$-RT \ln K_{dis} = \Delta F_d - T\Delta S_d$$

$$= \frac{e^2}{D(r_1 - r_2)} + BD - T\Delta S_d \qquad (6\text{-}2)$$

electrostatic)
binding)

where $\Delta F_d \simeq \Delta H_{dis}$, the energy of dissociation; ΔS_d is the standard entropy of dissociation per ion pair; r_1 and r_2 are the radii of ions treated as charged spheres (for the sake of simplicity); and BD is the binding energy of the pair due to forces other than coulombic ones.

Thus an increased effective radius or formation of pores or channels lined with polar groups lowers the energy barrier significantly. The passage of an ion accompanied by passage of the carrier results in a lowering of the conductance which may give rise to a variety of phenomena. Consequently, before discussing more specific features of ion transport across BLM, a discussion of some of the more general aspects of ion translocation phenomenology is relevant.

For theoretical treatment of membranes containing mobile carriers† one of

---

*It has been assumed that ions can only bind electrostatically and that polarization of charges should be complete to accomplish this. This is not a realistic picture; however, induced polarizability of the carrier may give rise to a significant stabilization of the complex formed between an ion and a neutral carrier. These cases are discussed in detail later in this chapter.

**The situation in real solvents differs from that for an idealized medium. They have discrete molecular structures and their molecules undergo random motion. Ions and ion pairs imbedded in such a medium orient solvent molecules. The degree of orientation is greater around free ions than in the vicinity of ion pairs. Hence the dissociation of ion pairs into free ions decreases the entropy of the system. Also the energy of two ions imbedded in a solvent and separated by a distance r depends on the average configuration of the surrounding molecules. The average configuration varies with temperature, and thus the potential energy curves become a function of temperature.

†As discussed earlier a mobile site may be regarded as one which differs from a fixed site only through the relative ease of rearrangement of its sites. In a typical case it may be noted that although these sites are confined to the membrane phase, they are free to move within that phase; however, two new properties are encountered in such a system. First, the sites can cross the membrane-solution interface and participate in boundary reactions; the sites thus are redistributed between the aqueous solution and the membrane phase. Second, in response to external forces, the concentration of sites will rearrange within the boundaries of the membrane. Such rearrangement could lead to changes in permeation characteristics. Further differences would arise from the possible formation of associated sites, and interactions with counterions giving diffusible molecular species.

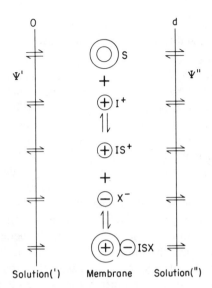

Fig. 6-1. A schematic representation of a membrane interposed between two aqueous solutions whose electric potentials are designated by ψ' and ψ''. The species I^+, S, IS^+, X^-, ISX refer to the free ion, neutral molecular carrier, the complexed cation, the free anion, and the neutralized complex, respectively. Although these species are illustrated within the membrane phase, the arrows at the membrane solution interfaces indicate that equilibria exist between these species and their counterparts in the aqueous solutions.[314]

the simplest models is that described in Fig. 6-1. Although various molecular species are illustrated within the membrane phase, equilibria exist between these species and their counterparts in the aqueous solutions. Several simplifying assumptions can be made to develop the system. Thus, it is generally assumed that: (*1*) The effect of polar head groups on transport is negligible. (*2*) The low conductance of BLM is due to a small number of ions expected to be present in the dissociated form in the core of membrane. (*3*) The charge carriers are mobile. (*4*) The membrane is permeable to co-ions. (*5*) No flow of bulk solution occurs. (*6*) All the ions are completely dissociated, behave ideally, and have constant mobilities throughout the membrane.

From the model illustrated in Fig. 6-1, an explicit expression can be derived for the behavior of the system with regard to three separately measurable properties: the potential difference across the membrane at zero current, the limiting value of the resistance of the membrane at low currents when interposed between identical solutions, and the solvent extraction of salt into a bulk organic phase such as hexane if such a phase is considered to be similar to the hydrocarbon core of BLM.

In general three fundamentally distinct approaches have been used in characterizing phenomena associated with ion permeation across membranes, which show differential permeability towards different ions:

1. Nernst first showed the nature of the equilibria of ions in electrolytes and described the motions of the ions under electrical and diffusion forces. This theory was soon applied by Planck to a constrained junction between electrolytes. The Planck formulation has since been frequently extended. As it has grown, this theory has become widely applied to the interpretation of electrochemical data on liquid junctions, and on artificial and natural membranes. The Nernst-Planck flux equations, the relationship between the membrane potential and the concentration asymmetry for any given ion, are derived in the form of a set of differential equations relating concentration to membrane potential.[313] The thermodynamic properties of the membrane are assumed to remain constant. These equations (in principle at least) can be solved for any given set of boundary conditions. However, the derivation itself requires detailed knowledge about the structure and thermodynamic properties of the membrane.

2. Application of the theory of irreversible thermodynamics permits a phenomenological description of the membrane by regarding it either as a continuous phase or as an interface of a discontinuity. With this type of approach one can proceed to interrelate ionic fluxes, water transport, and electrical potentials without specifying the ion permeation mechanism involved. This approach is limited in its ability to describe only specific steady-state properties (for example current-voltage relationships) and kinetic phenomenon (time-dependent relations).

3. The transition-state theory for transport based on the statistical mechanics of rate processes was described in the last chapter. According to this treatment the transport of an ion or molecule occurs by a series of discrete jumps, where each ion or molecule has to cross an energy barrier between adjoining equilibrium positions. Although this approach has been very successful in giving a quantitative account of transport properties of nonelectrolytes, it has not been developed sufficiently to explain ion permeation.

It is obvious that all the three approaches assume a macroscopic set of variables concerned with the membrane. This may range from concentration differences between ions in the internal and external milieu to macroscopic potential measurements across the membrane. However, none of these theories can claim complete success in explaining ion transport across biological membranes.[1218] Recently a relatively complete description of the behavior of BLM modified with macrocyclic antibiotics has been possible within the framework of classical electrochemistry.[314-316] Similar, although not as explicit, expressions can be arrived at from approximate solutions of the Nernst-Planck flux equations or from postulates of irreversible thermodynamics. For an extensive and terse theoretical analysis the reader is referred to the trilogy of papers just mentioned.

For the present discussion, we shall develop some of the major conclusions arrived at by these authors.

Now, referring to the model of a carrier-mediated transport system described in Fig. 6-1, the membrane potential (V_0) at zero current is given by:

$$V_0 = \frac{RT}{F} \ln \frac{\sum\limits_{i=1}^{n} \left[\frac{u_{is}^* k_{is} K_{is}^+}{u_{js}^* k_{js} K_{js}^+} \right] a_i'}{\sum\limits_{i=1}^{n} \left[\frac{u_{is}^* k_{is} K_{is}^+}{u_{js}^* k_{js} K_{js}^+} \right] a_i''} + \frac{RT}{F} \ln \frac{C_s^{\text{Tot}'}}{C_s^{\text{Tot}''}}$$

$$+ \frac{RT}{F} \ln \frac{1 + \sum\limits_{i=1}^{n} K_{is}^+ a_i'' + \sum\limits_{i=1}^{n} \sum\limits_{x=1}^{m} K_{is}^+ K_{isx} a_i'' a_x''}{1 + \sum\limits_{i=1}^{n} K_{is}^+ a_i' + \sum\limits_{i=1}^{n} \sum\limits_{\lambda=1}^{m} K_{is}^+ K_{isx} a_i' a_x'}$$

(6-3)

where a_i and a_j are the activities of two monovalent cations, and a_x that of an anion in solution (either $(')$ or $('')$). Note that V_0 depends on such parameters as the mobility (u_{is}^*, u_{js}^*) in the membrane of charged complexes, their partition coefficients (k_{js}, k_{is}), and their association constants in the aqueous solution (K terms with appropriate subscripts). C_s^{Tot} refers to the total concentration of carrier in the appropriate solution. The number of species of cations and anions is given by n and m, respectively.

In the derivation of this relationship no assumption as to the constancy of the individual fluxes, or of electroneutrality, has been made. The first term of equation (6-3) is recognized as the equivalent of the classical Goldman-Hodgkin-Katz equation.[313]

It is convenient to define the bracketed term (a product of mobility and equilibrium parameters, ukK) as the permeability ratio:

$$\frac{P_i}{P_j} = \left[\frac{u_{is}^* k_{is} K_{is}^+}{u_{js}^* k_{js} K_{js}^+} \right]$$

This term determines the relative effects of the ionic species on the membrane potential and is dependent upon the nature of membrane and carrier.

The second term indicates that the membrane potential depends linearly on the logarithm of the ratio of the total carrier concentrations in the two aqueous compartments. This term becomes zero when the carrier concentration is kept the same on both sides of the membrane.

The third term in equation (6-3) results from the possibility of significant formation of IS^+ and ISX complexes in the aqueous solution. For sufficiently dilute solutions this term reduces to zero. It is also zero at high salt concentrations provided the constants K_{is}^+ and K_{isx} are sufficiently small in the aqueous solution. This condition seems to be fulfilled by macrocyclic antibiotics which do not form stable complexes in aqueous solutions. When this term is not negligible its effect is to flatten, and even to reverse the slope expected on the first term. Thus, for the same concentration of carrier on both sides of the membrane and over a wide range of salt concentrations, the equation (6-3) reduces to the simple form:

$$V_0 = \frac{RT}{F} \ln \frac{\sum_{i=1}^{n} \frac{P_i}{P_j} a_i'}{\sum_{i=1}^{n} \frac{P_i}{P_j} a_i''} \tag{6-4}$$

which for the usual experimental situation of a mixture of only two species I^+ and J^+, can be written:

$$V_0 = \frac{RT}{F} \ln \frac{a_j' + \frac{P_i}{P_j} a_i'}{a_j'' + \frac{P_i}{P_j} a_i''} \tag{6-5}$$

Membrane conductance under the limiting conditions of zero current across a membrane interposed between identical solutions is given by:

$$G_0 = \frac{F^2}{d} \cdot \frac{C_s^{\text{Tot}} \cdot \sum_{i=1}^{n} u_{is}^* k_{is} K_{is}^+ a_i}{1 + \sum_{i=1}^{n} K_{is}^+ a_i + \sum_{i=1}^{n} \sum_{x=1}^{m} K_{isx} a_i a_x} \tag{6-6}$$

From this expression it is apparent that the membrane conductance (G_0) is proportional to the total concentration of carrier in the aqueous solution regardless of the ionic concentrations. Moreover when association in aqueous solution is negligible, the denominator of equation (6-6) can be approximated by unity and the expression for conductance reduces to:

$$G_0 \simeq \frac{F^2}{d} C_s^{\text{Tot}} \cdot \sum_{i=1}^{n} u_{is}^* k_{is} K_{is}^+ a_i \tag{6-7}$$

which becomes

$$G_0(J) \simeq \frac{F^2}{d} C_s^{\text{Tot}} u_{js}^* k_{js} K_{js}^+ a_j \tag{6-8}$$

when only one cationic species J^+ is present in the solution. Equation (6-8) predicts a simple linear dependence of $G_0(J)$ on ionic activity.

If successive measurements are made of G_0' and G_0'' in solutions of differing composition designated by (') and ('') and the ratio of conductance expected from equation (6-8) is compared with equation (6-5) for the electric potential, the following relationship results:

$$V_0 = \Psi'' - \Psi' = \frac{RT}{F} \ln (G_0'/G_0'')$$

This relationship predicts an identity between the ratio of the membrane conductances for two different solutions and the membrane potential measured between these solutions. In particular, when (') and ('') refer to single cation solutions of I^+ and J^+ at the same ionic activity and in the presence of the same total concentration of the carrier, equation (6-8) reduces to

$$\frac{G_0(J)}{G_0(I)} = \left[\frac{u_{js}^* k_{js} K_{js}^+}{u_{is}^* k_{is} K_{is}^+} \right] = \frac{P_j}{P_i}$$

when association is negligible in the aqueous solution. $G_0(I)$ and $G_0(J)$ denote the zero current value of the conductance under such experimental conditions.

At this stage the theoretical expectations of the model described in Fig. 6-1 may be summarized as follows:

1. An expression for the membrane potential in ionic mixtures in terms of the aqueous concentrations of ions is identical in form to the Goldman-Hodgkin-Katz approximation of the Nernst-Planck equation.

2. Conductance at the limit of zero current depends on the same parameters as does membrane potential.

3. The ratio of conductance measured in single salt solutions for two different ions should be identical to their permeability ratio.

4. The membrane conductance is proportional to the concentration of salt for dilute solutions but may become independent of concentration at high concentrations.

5. The membrane conductance is directly proportional to the total concentration of the modifying agent.

It is not inconceivable that similar expectations follow from a channel or pore model. Thus, potential and conductance measurements cannot, by themselves, provide a unique test for the carrier hypothesis. In fact, both fixed and mobile site ion-exchange membranes give the same potentials when they separate aqueous solutions of the same concentration, each containing one univalent cation whose counterion does not go through the membrane.[317-319] The current-voltage relationships, on the other hand, were found to be different and were governed by the condition in which the sites existed in the membrane, mobile or fixed, and the extent to which the sites are dissociated. Such aspects are elaborated below.

Theoretically it is possible to distinguish between fixed-site and mobile-site transport systems. Perhaps the clearest distinction between these two types of transport mechanisms lies in the marked discrepancy between the diffusion constant for the membrane as measured electrically (D_e), and the diffusion constant (D_d) expected from diffusion (flux) studies in the absence of electric current. Theoretically the ratio of two diffusion coefficients is given by:

$$\frac{D_d}{D_e} = \frac{u_i}{(u_i + u_s)} + \frac{u_{is}}{(u_i + u_s)} \cdot \frac{(1 - \alpha)}{\alpha}$$

where α is the degree of dissociation of the ion-carrier complex, which is also related to K_{is}, the dissociation constant and the total concentration of the carrier (C_{is}^{Tot})

$$K = \frac{\alpha^2 \, C^{Tot}}{(1 - \alpha)}$$

where u_i and u_s are the mobilities of the ion and the mobile carrier in the uncomplexed form and u_{is} is the mobility of the carrier in the complexed form. When $\alpha \longrightarrow 0$, the ratio between the two diffusion constants becomes inversely proportional to α. This is the classical example of exchange diffusion, which may occur as a consequence of the fact that association leads to a transport of counterion in an electrically neutral form due to combination with the oppositely charged mobile carrier.[1085] This is only a limited example of exchange diffusion, the phenomenon is developed further later in this chapter (see p. 223).

Yet another aspect of ion transport is the current-voltage (I-V) relationship. The charge-carrying species in the BLM determine the net charge transferred across the membrane under the influence of an electrochemical gradient, and, consequently, the electrical conductance characteristics. It is therefore, useful to determine the membrane potential during the course of a sustained charge transfer process at the membrane interface; that is, while a net current flows through the electrochemical cell. The steady-state I-V relationships of membrane with mobile carriers are characterized by finite limiting currents.[317] This is because the existence of a saturating current depends on the reduction of site and counterion concentrations at one end of the membrane to zero, and dissociation becomes essentially complete regardless of the value of the dissociation constants. The situation is analogous to that of polarizable electrodes used in voltammetry where development of a high degree of concentration over potential leads to limiting currents. More generally, the steady-state current voltage characteristics depend on the mobility of the undissociated species. Thus, the magnitude of the limiting current is a function of the degree of dissociation, the mobility of the undissociated carrier-ion complex, and the mean mobility of the dissociated species.[317,318] The expectations from the theoretical analysis of these factors are in good accord with the experimental data obtained from mobile-site ion-exchange membranes.

The properties of membranes having fixed sites is also influenced by a variety of factors as are the properties of membranes having mobile sites. For the sake of comparison some of these factors may be summarized as follows:

1. Available pore area or the concentration of the carrier.

2. Geometrical factors such as interaction with pore walls, ionic affinity of the carrier, interaction energies of the ions, co-ions, and counterions.

3. Nature of the ion-binding site.

4. Ion-matrix interactions which imply that chemical properties depend on the counterion mole fractions in these systems in which mobilities and standard chemical potentials are functions of hydration or unit cell size.

5. Buffer capacity and pH of the medium.

6. Solvent drag.

The relative contributions of these factors to ionic transport depends on the diameter of the channel, ionic strength, and polarizability of the groups forming the channel, which may be considered as membrane factors. Solvent factors such as the nature of the binding ions, buffer capacity, and the pH of the aqueous phase are also expected to influence ion transport. Thus, for example, in membranes having pore diameters sufficiently large, the majority of the permeants can move across the membranes as through free solution except for the constraints on the concentration of counterions and co-ions implied by the charged sites on the walls of the pore. Thus, the mobilities and the activation energies for diffusion in such systems would have values characteristic of diffusion in aqueous solution, implying that factors (1), (2) and (5) are relatively more important in this situation.

Perhaps the most distinctive feature of transport across membranes with large diameter pores is the possibility of solvent drag and phenomena associated with it, such as streaming potential. The fixed charges on the pore walls create electric double layers giving rise, in the presence of an applied field, to electro-osmotic water flow and to such related phenomena as anomalous osmosis and ion sorting.[320] Yet another interesting effect exhibited by such membranes is the existence of a region of differential negative resistance in the current-voltage curve.

Most of these characteristic phenomena of large pore membranes are exhibited by silica or porous glass membranes. In these membranes the current-voltage characteristic is a function of fixed charge density in the membrane which imperfectly excludes co-ions in contrast to membranes with small pore diameters, which exclude co-ions efficiently and in which the current-voltage relationship is independent of the fixed charge density. The differential negative characteristics exhibited by a variety of artificial and biological membranes are discussed in Chapter 8. However, it may be noted that this feature of modified BLM and probably of biological membranes is certainly not due to the presence of channels of large diameter.

In small channels, less than 10 Å in diameter, there is no longer any liquid water, but only bound water. In such cases specific interaction between the mobile species and the sites and matrix of the channel occurs in such a way as to alter the mobilities and standard potentials from the values characteristic of aqueous solutions. Thus, for example, the presence of narrow pores lined with ionogenic groups of high charge density gives higher values for transport numbers and reflection coefficients.

When the pore diameter is reduced sufficiently, a limit is reached at which there is no free solution in the channel—all the water is bound. In such cases only site-to-site migration occurs, essentially as would occur in an anhydrous matrix. The co-ions are also completely excluded. In such a situation Donnan boundary conditions must be replaced by boundary conditions characteristic of an ion-exchange equilibrium in which differences among counterions of the same charge, for example sodium and potassium, become important. The mobilities in such systems are much smaller than those in free solution, whereas the activation energies are much larger. Moreover, whether mobilities and activation energies are constant or are dependent on the mole fraction is a function of the mechanism of ion migration. If a defect mechanism, for example, a simple vacancy or hole migration, is involved, the mobilities will be dependent on mole fraction since the concentration of vacancies is a function of the degree of exchange. If a defect mechanism is not involved, mobilities should be constants independent of the degree of exchange. However various complications could arise[318,319] due to effects of varying partial molal volumes, the metastability of the membrane itself due to changes in applied electric currents and through alteration of the composition of the external solutions.

This has been an oversimplified representation of the phenomenology associated with ion transfer across channels or pores lined with polar or polarizable groups. Little can be said by way of comparison with biological membranes since these conclusions have been reached from studies done on inorganic ion-exchange membranes. However, the physical mechanisms are expected to be the same.

In the phenomenology of ion transfer across membranes as described thus far, selectivity of ion binding is the factor governing selectivity of ion transport process. This assumes an added significance if one considers the role of specific ion-binding processes in enzymatic activation, relationship of selectivity of ion transport in cotransport, active transport and related phenomena (see Chapter 7), and changes in cationic permselectivity in transport mechanisms giving rise to bioelectric phenomena. The physicochemical factors governing interionic permselectivity are described in the following section.

Interionic Selectivity

Even in the simplest cases our understanding of the factors and mechanisms governing ionic selectivity in exchange and transport processes is imperfect. The exact physical basis of ionic permselectivity may not be known until macromolecules involved in biological translocation mechanisms are characterized. However, one would expect the physical contact of the ion with the carrier to be of critical importance. The selectivity patterns observed in biological and nonbiological systems show striking and detailed similarities so that it is likely that the underlying physical mechanism of interionic discrimination is the same in both cases. Ionic permselectivity appears to be a function of interaction between a fixed charge or polarizable group with the ionic species. This would imply that the relative selectivity of two ions is nothing but the ratio of the activity coefficient of the ions in the ion exchanger. (The term ion exchange is used here in its widest sense so as to include ion-ion, ion-dipole, dipole-dipole, and ion-induced dipole interactions between a polarized and a polarizable species.) Thus, intuitively, one would expect to find the physical basis of ionic permselectivity in Coulombic forces, ionic radii, coordination number, and valence of ions, and in those factors, both thermodynamic and kinetic, governing the behavior of electrolytes in solution.

The simplest case of solution formation is that of the so called intrinsic salts, that is, those salts which are already completely ionized in the crystal lattice. On dissolution of the crystal lattice each ion simply gains an increase in freedom of translational motion. No chemical processes are normally required, except solvation for ions present in the salt lattice to become free in solution upon physical dissolution of the lattice structure. Thus solvation has been recognized as of key importance in determining solution behavior of an ion.

Solvation is a very general term which has been used with many liquid and gaseous systems wherever deviations from ideality are observed. However, this term is usually limited to the short-range interactions between a solute and the surrounding solvent molecules. Solvation will occur when the solute-solvent interactions are different from the normal solvent-solvent interactions. When the solute and solvent molecules differ only in their respective size, dipole moment, or polarizability, the difference in interactions are usually small; but when molecules capable of forming hydrogen bonds are involved, solvation may become quite significant, as with sugar in aqueous solution. With ionic solutes in polar solvents solvation effects are always important and in many cases are strong enough to immobilize completely a certain number of solvent molecules, which then behave as one entity with the ion or molecule. Except for a few water molecules adjacent to small multivalent ions, the water

TABLE 6-3 PHYSICAL CONSTANTS AND STANDARD THERMODYNAMIC FUNCTIONS FOR HYDRATION OF ALKALI METAL IONS[a] AT 25°

Function	H^+	Li^+	Na^+	K^+	Rb^+	Cs^+
Atomic radius (Å)	0.375	1.52	1.86	2.25	2.43	2.62
Ionization energy (kcal-mole^{-1})	313	124	118	100	96	90
Ionic radius, unhydrated (Å)	—	0.6–0.75	0.95–1.0	1.34	1.48	1.68
Ionic radius, hydrated (Å)						
Stokes	0.253	2.36	1.80	1.21	1.16	1.15
Effective	>0.5	5.2	3.8	2.5	2.2	2.2
Hydration number	1 ± 15	6 ± 1	4 ± 1	2 ± 1	2 ± 1	1
a_0 (Å)[b]	1.545	2.09	2.29	2.97	3.12	3.28
Coordination number	4 ± 1	4	4, 6	6	6	6
Hydration heat (kcal-M^{-1})	−258	−118	−94.5	−75	−69	−62
Molar entropies (cal/ g-ion-degree)	0.0	4.7	15.0	24.2	28.7	31.8
Partial molar volume (ml/g-ion)	+0.2	−1.0	−1.6	+8.45	+13.55	+20.95
Partial molar heat capacity (cal/g-ion-degree)	−15.9	−2.4	−9.1	−14.3	—	—
Ionic conductance at infinite dilution (ohm^{-1}-cm^2/g-eq)	316.6	33.3	43.4	64.4	67.5	67.7
Ionic mobility at infinite dilution (cm-sec^{-1}/esu of potential)	9.61	1.03	1.35	2.0	2.095	2.1
Transport number (as chloride)	0.825	0.3289	0.3918	0.4902	—	—
Ionic mobility in ice (as chloride)	≈10^4	<10^{-6}	<10^{-6}	<10^{-6}	<10^{-6}	<10^{-6}
Reorientation time (10^{-11} sec)	—	2.3	1.4	0.7	0.5	—

[a] Values reported in the literature vary over a wide range (usually ±10%). The values reported here have been collected from various sources.[6,321,322]
[b] a_0 is separation of solvent molecules from the center of the ion, corresponding to the state of minimum free energy.

molecules in the hydration shell are in continuous exchange with the water in the bulk of the solution, and it is only on a time-average basis that a hydration shell may be said to exist around most ions. It follows from this discussion that the free energy of ion-solvent interaction can be defined as the work done when one transfers an ion from vacuum into a position deep inside the solvent.

The process of solution may be considered in terms of hydrated ionic radii, or volumes of the solvated ions, or energetics of hydration, or variations of the effective field strengths of the fixed groupings as a consequence of hydration. The thermodynamic properties such as heats of solution and partial molal quantities of electrolytes are related to ion-solvent interactions. A compilation of these quantities for alkali metal cations is given in Table 6-3. Similar values for some monovalent anions are collected in Table 6-4. The most characteristic feature of the values of various thermodynamic parameters presented in these tables is their monotonic variation as a function of ionic radii.

The interpretation of these functions in terms of ion-solvent interactions is not simple, since it is not easy to devise experiments capable of measuring the thermodynamic properties of individual ions. However, it may be generally stated that since the solubilities of many salts are appreciably large in water, it is clear that the standard free energy of these dissolution processes must be

TABLE 6-4 PHYSICAL CONSTANTS AND STANDARD THERMODYNAMIC FUNCTIONS FOR HYDRATION OF HALIDE ANIONS[a] AT 25°C

Function	F^-	Cl^-	Br^-	I^-	OH^-
Ionic radius, unhydrated (Å)	1.36	1.81	1.95	2.16	—
Ionic radius, (Å)					
Effective	3.4	2.4	2.3	2.2	—
Stokes	1.67	1.2	1.16	1.16	0.443
Hydration heat (kcal-M^{-1})	−88	−59	−54	−33	−80
Hydration number	4 ± 1	2 ± 1	2 ± 1	1 ± 1	—
Molar entropies (cal/g-ion-degree)	−2.3	13.3	19.7	25.3	−2.5
Partial molar volume (ml/g-ion)	−1.9	+18.0	+25.04	+36.6	−5.23
Partial molar heat capacity (cal/g-ion-degree)	—	−14.3	−15.4	−15.7	−16.1
Ionic conductance at infinite dilution (ohm^{-1}-cm^2/g-eq)	46.7	65.4	67.4	67.4	176.6
Ionic mobility (cm-sec^{-1}/esu of potential)	1.45	2.03	2.09	2.09	5.49
Reorientation time[b] (10^{-11} sec)	1.8	0.7	0.6	0.4	—

[a]See Table 6-3 for explanation.
[b]Reorientation time is the time in which the coordinating interaction of water to the ion relative to a given initial direction is lost. These values are measure of rigidity of complex; smaller values imply that complex is not rigid.

small or negative. There is a wide range in the corresponding equilibrium constants but the standard free energy values are numerically small compared with the magnitudes of lattice energies or ionization energies of metal ions. Thus the standard free energy of solvation must be a large negative quantity for ions.

At this stage we may ask for possible explanations for the specificity of ion-site interactions. Specificity in ion binding is certainly not a consequence of a peculiar electron configuration as that which distinguishes transition metal ions from main group metals, for example Ni^{2+} from Ca^{2+}. For ions of noble gas-like electron configuration the models are therefore based on factors such as solvated and unsolvated ionic radii, and peculiarities of the structure of the complexing agent. Complex formation involves the replacement of one or several solvent molecules from the inner hydration shell of the ion. The energetics of solvation and complexation of the ion and site are obviously to be considered as factors determining ion selectivity.

The simplest, although not necessarily the most profound, picture of ionic hydration arises from considerations of ionic mobilities in dilute aqueous solution. It is well known that the mobilities of alkali metal cations in aqueous solution increases in the order (Table 6-3):

$$Li < Na < K < Rb < Cs$$

The usual interpretation of this phenomenon is in the form of Stokes law which suggests that, qualitatively, these ions exist in aqueous solution in a form such that Cs behaves as if it were the smallest and the Li were the largest, which is opposite to their ionic radii in crystals. The most natural way of accounting for this anomaly is to suppose that lithium has many more water molecules stuck to it than to cesium, and thus unhydrated lithium which is smallest becomes the largest ion when it is hydrated, and thereby has the smallest mobility.

Now referring back to the mode of ligand substitution for solvent in the coordination shell of ions, we should expect two monotonic size dependencies for ion binding.

1. If the incoming ligand is more tightly bound than the solvent molecule to be substituted, the stability of the complex will decrease with increasing radius of the metal ion, the order of nonhydrated radii. The smaller the metal ion the larger will be the gain in binding energy for the ligand.

2. If the incoming ligand is less tightly bound than the solvent molecule the size dependencies of complex stability will be reverse of the order (increasing) of apparent hydrated size, since now the smallest metal ion will prefer the solvent molecules.

Thus two monotonically varying but completely opposite selectivity sequences can be generated depending upon whether the ligand is more or less tightly bound to the ion. These arguments are of general nature and can be applied to any set of ionic species since their physical basis arises from relative polarizabilities of the fixed sites. More information regarding the stability constants of ion-site complexes and the selectivity sequence arising therefrom can be obtained from intermediate polarizabilities, and additional assumptions regarding the mode of interaction and origin of required interaction energy can be made. Eisenman and co-workers have considered differences in electrostatic energy of interaction of ions and sites to be of primary importance in determining selectivity isotherms and have come up with reasonably general predictions.[317,319,323]

In Eisenman's theory, the hydration of ions is recognized as of key importance. However, it is considered primarily in terms of its energetics rather than in terms of ionic radii or volume. The quantitative treatment of the theory is based on two specific assumptions. (a) Equilibrium cation specificity depends, by definition, upon the free energy difference between ion–site and ion–water interaction. (b) Free energies of interactions depend largely upon electrostatic forces. Binding of cations may be considered as a system in which the fixed binding groups are anionic with cations as bound counterions. Both cations and anions may, for the sake of simplicity be considered as nonpolarizable point charges each at the center of an incompressible sphere. If the fixed grouping has radius r_A and the cation has radius r, the electrostatic energy* of interaction is given by $e^2/(r_A + r)$. where e is the electronic charge.

Since free energy is required to remove from the fixed site and from the counterion as many water molecules as are necessary to permit the contact or close approach of the fixed groupings and counterion, such free energy would be closely related to the standard free energy of hydration of the fixed ion-binding site and the counterion. Therefore during the exchange of cations, the overall gain in free energy of the system would determine the typical activity

---

*The simplest basis for an electrostatic treatment of the free energy of solvation is that of Born.[324] Considering the difference of energy required to charge ions of radius r and charge ze in vacuum and in a solvent of dielectric constant ϵ, it is easy to show that, per ion:

$$\Delta F = -\frac{(ze)^2}{2r}\,[1 - 1/\epsilon]$$

where ΔF is the electrostatic part of the total free energy change associated with the solvation of ions from the gas phase. Contribution from the configurational free energy term (free energy of mixing), will depend on the concentration in solution, and is usually small compared with ΔF. Also, for water and most polar solvents that dissolve ions appreciably, $\epsilon \gg 1$, ΔF corresponds to a large extent to the loss of energy required for charging in a vacuum. Physically, this loss corresponds to energy stored in the dielectric as a result of the creation of a charge with its corresponding distance-dependent field. The charging energy is a free-energy quantity and its heat and entropy components readily follow if ϵ is a function of temperature. This relationship is based, of course, on the electrostatic energy of charging a sphere of radius r. Although the Born equation has been used as a basis for considering solvation energy, there are a number of limitations for its applicability.[325]

sequence for cationic selectivity. Thus, if r_1 and r_2 are radii of two cations competing for the anionic fixed site of radius r_A, the gain of free energy, ΔF, by the system during the exchange is given by:

$$\Delta F = \frac{e^2}{(r_A + r_2)} - \frac{e^2}{(r_A + r_1)} - (\Delta F_2 - \Delta F_1)$$

where ΔF_1 and ΔF_2 are the free energies associated with the hydration of ions 1 and 2, and of course

$$\Delta F = - RT \ln K$$

where K is the equilibrium constant for ion binding.

Both of the above equations generate predictions. Consider the situation where r_A is large; that is, the fixed grouping has low field strength. In this case the electrostatic energy term would be small and consequently the ΔF term would be determined by the difference of the standard free energies of hydration of the counterions. Thus the selectivity isotherm would be:

$$Li > Na > K > Rb > Cs$$

In contrast, consider the case where r_A is fairly small; that is, the case of a fixed grouping of high field strength. In this case the first electrostatic term will be more important. Consequently, the exchanger will give preference to the ions of smaller radius. Thus small r_A gives rise to a completely reversed selectivity isotherm:

$$Cs > Rb > K > Na > Li$$

In ion exchange resins, the sulfonate group having low field strength because of its large size, exhibits the first selectivity sequence; the carboxylate group, having high field strength owing to its small size, shows a completely reversed preference. That is, the second selectivity sequence. In fact, transition from one sequence to another can be brought about by varying the radius of the fixed site. The net free energy change in the exchange process, as is apparent from the foregoing discussion, is either equal to or proportional to the difference in the standard energies of hydration of two counterions. From these values and with assigned values for hydrated cationic radii, one can calculate the theoretical consequence of varying r_A. In this way one can predict eleven sequences:

I	Cs > Rb > K	> Na > Li
II	Rb > Cs > K	> Na > Li
III	Rb > K > Cs	> Na > Li
IV	K > Rb > Cs	> Na > Li

V	K > Rb > Na > Cs > Li
VI	K > Na > Rb > Cs > Li
VII	Na > K > Rb > Cs > Li
VIII	Na > K > Rb > Li > Cs
IX	Na > K > Li > Rb > Cs
X	Na > Li > K > Rb > Cs
XI	Li > Na > K > Rb > Cs

These reversals in cationic selectivity have actually been observed with liposomes prepared from phosphatidylserine,[1217] lipoprotein complexes,[1220] natural and synthetic ion exchangers including alumino silicates, polymeric ionic resins, colloidian membranes, modified BLM and the biomembranes.[326] Each sequence differs from the preceding and following sequence only in inversion of the position of one pair of cations. Sequences II through X constitute transition sequences linking I, the lyotrophic series defined by increasing apparent hydrated size, to sequence XI, defined by increasing nonhydrated size.

The underlying reason for these selectivity patterns is that cation-site interaction forces decrease more slowly with increasing distance than do cation-water interaction forces. The forces between ions in Coulombic or monopole-monopole interaction decrease with the inverse square of the distance. The forces between hydrated cations are largely dipolar and quadripolar forces which vary with the third and fourth power of the distance, respectively. Thus, the difference between the attraction of a small ion, Li for example, minus the attraction for a large ion, for example, Cs, is less marked for a monopolar site than for water molecules. As site strength is increased, the ion-site attraction gradually begins to increase in strength relative to ion-water attraction. Thus, the free energy term becomes negative most rapidly for Cs and most slowly for Li. Hence inversions in order affecting the largest ions occur first.

It may be noted that these sequences are those predicted for considerations of Coulombic forces alone and that the occasional deviations from these patterns are those expected from considerations of the most important non-Coulombic forces. It would be a serious misunderstanding, however, to assume that there must be some totally different explanation for specificity in systems lacking fixed ionic charges; specificity will still be controlled by these factors if proper electrostatic or induced dipole or multipole interactions are involved. For example, in a molecule with carbonyl groups or ether linkages, the carbon-oxygen bonds will have permanent dipole moments. Thus interaction of cations will still depend upon the magnitude of the formal negative charge on oxygen and the distance between the cation and the oxygen atoms in the complexing molecule. If the field strength falls slowly with respect to the field of water molecule the same selectivity sequence as predicted above may be observed.

It would be pertinent to consider various factors implicit in the elementary atomic origin of equilibrium ionic specificity as elaborated above. Some of these factors may be summarized as follows:

1. An increase in the amount of water in the vicinity of the mobile ion and a site of opposite sign reduces the magnitude of selectivity ratios without altering the selectivity sequence. This is because only those ions in immediate contact with fixed charges without intervening water molecules contribute significantly to selectivity patterns differing from those of free solution.

2. On a site in which the replacement of one ion by another leads to no change in the structure or hydration state of the fixed site, the entropy change should be given by the difference in entropies of hydration.

3. For ions of the same charge, variation in site spacing or in ion coordination number of sites which the ion nearly touches does not affect the pattern of the transition sequence; closer site spacing or higher coordination number merely yields a sequence corresponding to higher field strength.

4. The steric effects (Born repulsion forces) may assume additional importance under particular circumstances; for example, in rigid and close frameworks, steric effects may reduce the mobility of larger ions giving rise to ion sieving. Thus, if an ion has ionic radius less than the channel radius, but a hydrated radius greater than the channel radius, the cation may be unable to enter the channel, especially if the free energy of hydration is large.

5. The selectivity of H^+ relative to group I_A cations is also expected to depend upon anionic field strength, H^+ being markedly preferred at the high field strengths while alkali cations are preferred at lower field strengths.

6. Various ionic species will be partitioned in a site-free media in accordance with their solvation energies in water and the non-aqueous phases. Thus, when electrostatic interactions dominate the solvation energy those species having the larger charge and smaller size (and hence higher hydration energies) tend to be excluded from solvents having dielectric constants lower than that of water.

Considerations identical to those outlined above for discrimination among five alkali metal cations can be extended to the four halide anions (F^-, Cl^-, Br^- and I^-), except that one must consider interactions of negative mobile ions with positive sites instead of positive mobile ions with negative sites. Thus from experimental free energies of formation of alkali halides as crystals and in aqueous solution, using different alkali cations as models of positive sites with different field strengths, the following seven selectivity sequences are predicted and observed experimentally in biological systems:[326]

I	$I > Br > Cl > F$
II	$Br > I > Cl > F$
III	$Br > Cl > I > F$
IV	$Cl > Br > I > F$
V	$Cl > Br > F > I$

$$\text{VI} \quad Cl > F > Br > I$$
$$\text{VII} \quad F > Cl > Br > I$$

Sequence I, the order of the free solution mobilities, is associated with weak binding sites, while sequence VII in which the ion with the smallest nonhydrated radius is preferred, is associated with strong binding sites. Five transition sequences link I to VII and the first inversion involves the weakly hydrated I^- and Br^- which have the largest ionic size. The inversions II → III and III → IV involve the more strongly hydrated Cl^- and the remaining inversions involve F, the smallest and most strongly hydrated anion.

Although most of the experimentally observed intrinsic equilibrium selectivity sequences for cations and anions, whether in biological or artificial systems, correspond to theoretically predicted selectivity isotherms, it may be emphasized that in certain ways Eisenman's treatment is rather crude. Although these calculations are based on variations of the radius of the fixed site, producing different field strengths and hence different selectivity sequences, the cation-binding sites in most biological systems will be either carboxyl or phosphoric acid groups. At first glance this would imply that only two different values of site radius should occur in biological systems despite the wide range of the observed selectivity sequences. However, even without considering non-Coulombic modes of ion-site interaction, the field strength (hence selectivity pattern) of carboxyl and phosphate groups will vary greatly, depending upon whether inductive effects exerted by the site's immediate molecular environment increases or decreases electron density at the site.

This influence of the environment upon the field strength of a chemical grouping is reflected not only in counterion selectivity patterns but also in the much more familiar variation in the acid dissociation constant or pK_a of the binding site. The pK_a depends upon field strength because an ionic site with high field strength binds protons more firmly than one with low field strength; hence the protons dissociate less readily and the stronger site has a lower dissociation constant and higher pK_a (a weaker acid). Examples of dependence of pK_a on environment are provided by the large differences between the pK_a's of free amino acids and their values in different proteins.

These considerations require greater emphasis on the positions, numbers, orientations, and freedom of motion of water molecules, counterions and sites, and other structural elements of the system. Also contribution to ion selectivity from non-Coulombic forces, the geometry of the hydrated ionic species, site spacing, entropy effects and nonequilibrium effects such as ion-sieving must be taken into account while predicting ionic selectivity isotherms. Theoretically, it is not easy to accommodate all these factors in the calculations. However, a more sophisticated and more general approach to evaluate factors governing interionic selectivity isotherms has been devised by Ling,[6] who calculated site-mobile ion and water–mobile ion interactions for a linear model, taking into account all nonexchange forces including such non-Coulombic effects as polarization of the ion by the site, polarization of the site by the ion, and Born repulsion forces. Over a range of values for the site polarizability (up to 2 Å) these detailed calculations yield the same sequences as those predicted from Coulombic forces alone, although high polarizabilities yield different ones. For an alternative treatment see Ref. 326a.

Thus far we have developed some physicochemical principles to explain the problem of equilibrium ionic selectivity. However, movement of ions across a a membrane under an electrochemical gradient contains a nonequilibrium component (as a mobility or an interfacial activation energy) in addition to the equilibrium affinity. To a large extent the forces governing nonequilibrium selectivity are the same as those governing equilibrium selectivity, acting in the opposite direction. Note that while considering various selectivity isotherms it is very important that in each series of decreasing ion size, the cavity of the ligand coordination shell can just as freely contract to the ion size as can the solvent coordination shell. However, this is often not possible; steric hindrances as well as ligand-ligand repulsion may prevent contraction below a certain size. Thus, in a rigid and close framework, steric effects may reduce the mobility of larger ions, giving rise to ion sieving. For all ions larger than this critical size the complex stability will increase with decreasing metal ion radius, assuming the ligands are more tightly bound than the solvent molecules. Below the critical size, however, this behavior will be reversed, since now there will be little or no gain in ligand binding energy with decreasing metal ion radius (the size of the cavity is fixed), and there will be an increased requirement of energy for desolvation. Note that this optimal fit will produce a maximum in the thermodynamic stability constant, but that this does not mean the optimally fitted complex is absolutely the most stable complex as demonstrated in Fig. 6-2. The two upper curves (b and c) represent the free energy of binding of two different chelating agents. They consist of multidentate ligands, which for complexation have to substitute for the entire solvation sphere of the metal ion in order to enclose that metal ion in a cavity. Due to the fact that the free energy of ligand binding (curves b and c) approaches saturation more quickly, the difference of the ligand binding and solvation energies (curves a–c and a–b) may well show a maximum at a given size of the metal ion and it will correspond to the optimal fit of the metal ion to the ligand. It is just this difference which determines the stability constant of the complex. Two outstanding features of this set of curves may be noted. First, as long as the increase in free energy of solution exceeds the change in $RT \ln K$, the free energy of ligand binding will change monotonically regardless of whether stability constants increase or decrease as expected from any simple electrostatic picture. Second, the maximum of the stability constant (K) can occur at any position depending upon how soon

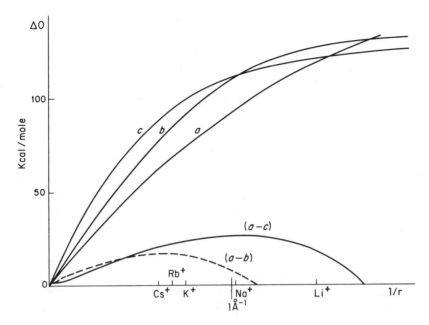

Fig. 6-2. Dependence of free energies for ligand binding (curves *b* and *c* for two hypothetical cases) and solvation (curve *a*) on the reciprocal radius of alkali metal cation. Curves *a–c* and *a–b* show the difference in binding for these two ligands. See text for details.[327]

the ligand binding reaches saturation, and that this position does not necessarily coincide with the $1/r$ value at which saturation is reached.

By way of recapitulation, the factors governing ionic selectivity in ion exchangers are fundamentally the free energies of interaction of the mobile ion with water on one hand and with fixed sites on the other. Although the energy necessary for desolvation is a major contributory factor, desolvation need not be complete. The selectivity sequences are for equilibrium affinity. However, one would intuitively expect that selectivity involving nonequilibrium parameters should not be different. This will be so because the competing interactions of mobile ions with water molecules versus ion interactions with fixed sites will control not only the equilibrium ion-exchange properties (and consequently the phase boundary potential) but these interactions will also be importantly involved in the activation energies governing ion migration from solution into and through the phase containing the fixed site. Finally one may summarize the thermodynamic, kinetic, and geometric constraints of a size-specific ion carrier as follows:[314,319,327]

1. As many solvent molecules of the inner hydration shell as possible should be replaced by the coordinating sites of the carrier molecule. The difference in

free energy thus gained will contribute towards stability (at room temperature 1.4 kcal/mole corresponds to an order of magnitude in stability constant).

2. The optimal fit of ion to cavity is related to the specific architecture of the carrier in which the difference in the free energies of ion binding, of ligand-ion conformation, and of solvation of ions and the site is important. This difference involves appreciable entropy increments in favor of the chelating ligand. Thus, optimal ion size for a cavity does not necessarily occur at the minimum size of the cavity.

3. The carrier molecule should possess sufficient flexibility in order to allow stepwise substitution of the solvent molecules. If complete or substantial desolvation is required for the ion to slip into the cavity, the activation barrier would be high for single step desolvation.

4. The electrophilic groups should point toward the ion binding site, i.e., cavity, whereas the lipophilic groups should be located at the surface in contact with the core of the membrane.

5. If the overall size of the complex is approximately the same regardless of the particular ion of the sequence bound, as is likely for the situations we shall be concerned with, the mobilities of the complexes will be the same for all ions. In this event, the permeability and conductance ratios are expected to depend only on equilibrium selectivity parameters.

6. In general, the rate of substitution of a given solvent molecule in the inner coordination sphere depends mainly on the nature of the metal ion rather than that of the substituting ligand. Thus the main energy barrier results from loosening of an ion-water bond rather than from formation of a new bond between ion and incoming ligand. In fact, despite the high solvation energy values, substitution is an extremely rapid process (rate constant = 10^9 $\sec^{-1}$) involving only a few kcal/mole of free energy of activation. Also the rate constants for formation and dissociation of complexes in aqueous solutions for most alkali metal cations are expected to approach the diffusion-controlled limit and so these steps are unlikely to be rate limiting in the aqueous phase.

7. Constraints imposed by the membrane, particularly those based on interfacial charge profile and viscosity and mobility of the chains in the lipid core would be similar to those expected for passive diffusion of nonelectrolytes (see p. 122 and Ref. 316a).

Carrier-Mediated Transport Across BLM

Transport of Protons Across BLM. A large number of compounds which can undergo association-dissociation equilibria of the type

$$R-H \rightleftharpoons R^- + H^+$$

are capable of transporting protons across BLM by a simple carrier mechanism (Table 6-5). In its simplest form this is reflected in increased specific conductance of the membrane as a function of the pH of the aqueous solution, and the

TABLE 6-5 EFFECT OF PROTON CARRIERS ON BLM

Agent[a]	Concn. for Half Maximal Effect (conductance)	Optimum pH	Ion Transported	Ref.
m-Nitrophenol	1.5 mM	8.2	H$^+$	329
2,4-Dinitrophenol	0.34 mM	4.3	H$^+$	329-333
Picric acid	70 μM	1.0	H$^+$	329
2,4-Dichlorophenol	0.3 mM		H$^+$	329
2,4,5-Trichlorophenol	30 μM		H$^+$	329
Pentachlorophenol	30 μM		H$^+$	329
FCCP	0.9 μM	6.2	H$^+$	329, 330
TTFB	3 μM	6.0	H$^+$	329
CCPH	2.6 μM		H$^+$	329, 330
S-13	1 μM		H$^+$	331
Dicoumarol	1 mM	5.5	H$^+$	333
Salicylic acid	1 mM	4	H$^+$	333
Acetoacetic ester	1 mM		H$^+$	333
Decachloroborene	0.1 μM	2-12	H$^+$	338, 339
Dimethyldibenzyl ammonium chloride	1 mM		Anion	329
Tetraphenylboron	1 μM		Anion	329, 335
Methyl borenyl mercury	1 mM		Anion	339

[a]TTFB = Tetrachlorotrifluoromethylbenzimidazole; FCCP = carbonyl cyanide-p-trifluoromethoxy phenylhydrazone; CCPH = carbonyl cyanide-m-chlorophenylhydrazone; S-13 = 3-t-butyl-5-chloro-2'-chloro-4'-nitrosalicylanilide.

development of a transmembrane potential in the presence of a proton gradient (that is pH gradient) across the membrane. As expected, pH vs conductance curves show a maximum at a pH characteristic for each agent—its acid dissociation constant, pK_a.[328-337]

The current passing through the BLM under a fixed constant potential difference decreases with time. This is connected with a rise in the proton gradient inside the membrane when the concentration of the charge-carrying species is the factor that limits the rate.[1221] In solutions of high buffer capacity, this phenomenon, as expected from association-dissociation considerations, is not observed. The bell-shaped pH vs conductance curve for BLM is also rationalized in terms of dissociation of the charge-carrying species. On the alkaline side of the pK_a, where these agents exist in anionic form, there are few H$^+$ ions in the membrane phase (bound to the carrier) and the conductance is very low. On the acidic side of the pK_a, the dissociation of the carrier is repressed, and hence it loses its ability as a mobile proton carrier even though its concentration in the membrane phase is relatively high. At any given pH, the effect of concentration of the nonionic form of these agents on the conductivity of BLM is linear, thus implying once again that the uncharged agent is the proton carrying species. In fact, it has been suggested that the primary charge carrier is a dimer formed between the undissociated and dissociated form of the weak acid.[328,1086,1087]

The transport (transference) number for transport of protons across modified

BLM is very close to unity. The slope potential of the pH gradient-transmembrane potential curve is 56–59 mV. This value is little affected by the presence of other ions, thus suggesting that these membranes function as perfect hydrogen electrodes.

The effect of the agents which increase proton conductance (Table 6-5) across BLM is, in general, only slightly dependent on the nature of the phospholipids, especially the hydrocarbon chains. However the effect of polar groups is significant. For example, the pH of maximum conductance for each proton-carrying agent is dependent on both the agent and the lipid polar group. It is lower for each agent in a positively charged lipid such as lysyl phosphatidyl glycerol and higher in negatively charged lipid such as phosphatidyl glycerol.[331] The difference in conductance of differently charged lipid bilayers at equal concentrations of the agent, as well as the change of pH optimum of conductance with lipid charge, can be explained in terms of an electrostatic energy contribution of the fixed lipid charges to the distribution of the agent-anion (the carrier) between the aqueous and the membrane phases.

In comparison to the results just described it must be mentioned that, mechanistically, the results are consistent with an electronic charge-transfer mechanism. Rosenberg and co-workers have relegated a considerable body of experimental evidence to this effect.[332] However, most of this evidence refers to charge-transfer complex formation of these various agents with lecithin and oxidized cholesterol. The evidence for the interaction of DNP has been obtained from the UV spectrum of the lipid-DNP ($1:1$) complex in carbon tetrachloride solution. For egg lecithin the UV maximum is at 310 nm (ϵ, 12650) and $K = 720$ l/mole, and for synthetic lecithin the absorption maximum is at 300 nm (ϵ,11375) and $K = 49.3$ l/mole. With increasing temperature, the conductivity in bulk and BLM phase has been shown to increase exponentially, yielding a value for the activation energy for the charge transfer across BLM of 0.2 eV, and across BLM modified with DNP of 1.2 eV. These authors prefer to interpret these and various other observations to suggest that the mechanism of charge transfer across these modified membranes may be electronic rather than protonic. The evidence, however, is not conclusive in these cases although BLM modified with iodine does show electronic charge transfer mechanisms as described in the next section.

Before concluding this section it may be noted that a number of other charge carriers which can be incorporated into BLM are also included in the Table 6-5. We shall not discuss them in the text; however, their mode of action is consistent with the carrier mechanism.

Charge Transfer Across BLM in Iodine-Iodide Solution. In some early experiments on the determination of conductance of BLM it was noted that the iodide ion has a considerably higher mobility across BLM than other halide anions and alkali metal cations.[340] This abnormality was traced to the presence of small amounts of iodine present in the iodide solutions. Once iodine was reduced with sodium thiosulfate, the mobility of the iodide ion was found to be in reasonable accord with that for other halides.[340] Iodine when added to the medium bathing BLM does not significantly change its conductance unless iodide ions are present in the medium. The log-log plot of conductance as a function of iodine concentration is linear. If other ions are present in sufficient concentration, the presence of iodide (above 10 mM) has little effect on conductance. However, when an iodide gradient is present across the BLM, the presence of even small amounts of iodine in one or both compartments gives rise to a diffusion potential. This varies linearly with the logarithm of the ratio of the iodide ion concentration with a slope of 60 mV, positive on the side containing the higher concentration of iodide.[340-343] These results suggest that the charge transfer is mediated by an ionic species resulting from iodide-iodine.

The high permeability of iodide in the presence of iodine across BLM has been generally attributed to the tendency of iodine and iodide to form polyiodides. The following reactions are known to occur in aqueous solutions:

$$I^- + I_2 \rightleftharpoons I_3^-; \quad I_3^- + I_2 \rightleftharpoons I_5^-; \quad I_5^- + 2I_2 \rightleftharpoons I_9^- \quad \text{etc.}$$

and

$$2I^- + 6I_2 \rightleftharpoons I_{14}^{2-}$$

The equilibrium constant for the formation of triiodide is 6×10^2 at $40°C$.

The polyiodide ions thus formed have large radii and their charge is delocalized; consequently, the electrostatic field at the surface of polyiodide species should be relatively small. Thus polyiodide species may have a considerable solubility in a medium of low dielectric constant. For example, if the radius of the iodide ion is increased from 2 Å to 6 Å after formation of a polyiodide ion, the solubility of the iodide as polyiodide in the membrane phase will be increased by a factor near 10^{17} [343] (also see p. 101). Such theoretical considerations combined with the observed second-order dependence of conductance on iodine concentration and the first-order dependence on iodide concentration has led to the postulation that the species mediating charge transfer is I_5^-. However, it remains undecided whether I_5^- itself shuttles across the membrane interface or simply acts as the mediator in charge transfer by some other mechanism.

An electronic charge-transfer mechanism may operate through a charge-transfer complex formation of the type lecithin. I_2, and the BLM may act as an electronic conductor:

$$I^- \cdot \text{Lipid} \rightleftharpoons \text{Lipid} + 1/2\ I_2 + e$$

Such a mechanism is known to operate in crystalline complexes of organic compounds with iodine.[344-346] The transport of charge in such organic *semicon-*

ductors (charge-transfer complexes) may result from the migration of either electrons or holes (see Ref. 347 for an excellent discussion of semiconduction processes in organic compounds).

Kallman and Pope[348] studied the dc electrical conductivity of thin (10–20 μ) crystals of anthracene, utilizing electrolyte solutions (usually $0.02M$ NaCl) as electrodes. They obtained specific resistivities of the order of 10^{15} ohm-cm^{-1}, the current being ohmic up to 10^6 V/cm. When one of these electrodes was replaced with a solution of I_2 in $1.0M$ NaI, the current increased about fourfold when the polarity of the adjacent crystal interface was positive but approximately eighty-fold when it was negative. Upon removal of iodine, the behavior of the crystal returned to what it was prior to iodine addition, showing that no permanent chemical change had occurred. Similar current increases were observed when the solutions contained another strong oxidant, ceric ion. An extrinsic mechanism of carrier generation, i.e., hole injection, was proposed to account for these results. Similar results have been obtained for solid complexes with naphthalene, tetracene, pyrene, chrysene, phenanthrene, diphenyl, p-terphenyl, and p-quaterphenyl, when iodine is replaced by N-benzylnicotinamide, riboflavin, and other pyridinium compounds in the presence of light.[349]

The criteria for semiconductivity may however be summarized as follows: (a) the semiconductor should have ordered crystalline structure; (b) conductivity must be electronic rather than ionic; (c) pure conductor should insulate at low temperature and the temperature coefficient is nonlinear; (d) electrical conductivity can be induced by external agents because of high sensitivity of the electrical properties to lattice imperfections and the possibility of freeing the charge carriers in the lattice.

It follows that the rate of such electronic charge transfer and of migration of electrons and holes should be considerably affected by perturbation with external energy sources such as irradiation with light of a suitable wavelength.*

Such photoconductivity studies are not applicable to strongly interacting charge-transfer complexes, since photoconduction cannot be distinguished from dark conductivity. In fact, light-induced changes in the conductivity of BLM in the presence of I_2–I^- have been observed at low concentrations of iodine.[350,351] See also Ref. 1222 for a review. When BLM prepared from ox-brain lipids is irradiated with pulses of light of wavelength 3371 Å, the membrane resistance falls reversibly by less than 2%; the magnitude of this response varies linearly with light intensity. The effectiveness of the nitrogen wavelength (3371 Å) and a xenon

---

*When a molecule in its ground singlet state absorbs a photon of light one of its electrons in the highest filled π molecular orbital is raised to one of the lowest unfilled π molecular orbital. The normal photophysical processes by which the molecule gets rid of this electronic energy are competitive with the process of charge carrier generation. Thus external agents, including light, that deliver energy to a close packed lattice may excite valence electrons of molecules to create an electrical conductivity.

flash lamp (pulses of 10 μ-sec) correlates with the strong absorption of poly-iodides near 3600 Å. Similarly, the lesser effectiveness of an argon laser (wave-length 4880 and 5150 Å) accords with approximately ten-fold decrease in absorp-tion of polyiodides in this region. A similar mechanism may underlie the photoelectric responses of BLM in iodide solution (on one side only), develop a photovoltage of up to 160 mV, and currents up to 1600 pA. The action spectra for the photoelectric effect compares well with the absorption spectra of the membrane or their constituents.[350]

These experiments and others[1089,1090,1206] suggest that at least a fraction of the conductance induced by iodine–iodide is not due to a carrier-mediated mechanism in which the polyiodide species shuttle across the BLM. The flux measurement of iodine–iodide is complicated because of high diffusion rates of iodine alone even without the passage of current.[352] However as shown in Fig. 6-3, little or no flux of I^{131} could be detected across BLM whose resistance was only three to five times more than that of the hole on which it was formed. Furthermore, application of a dc current in the direction which would have aided the flux of anionic iodide species had no effect on the flux of $I.^{131}$ Thus, the pos-sibility of a simple carrier mechanism may be ruled out. Such an assertion is further supported by the observation that, in a gradient of iodine with equal concentrations of iodide on both sides, a flow of electrons toward the compart-ment containing the higher concentration of iodine is observed. The potential across the membrane developed under these conditions is of the order of 3–4 mV negative on the side containing the higher concentration of iodine. This can only be explained in terms of a redox reaction occurring at the interface:

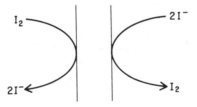

The reason why only a small potential develops in an iodine gradient is not clear. This could however be due to almost a complete depletion of iodine gradi-ent in the unstirred layers. These observations coupled with the photoelectric re-sponses shown by BLM in the presence of iodine and iodide strongly suggests that charge transfer is mediated by polyiodide species. At the same time, the evidence does tend to rule out the possibility that polyiodide itself carries the charge by a shuttle mechanism. The electrolyte contact experiments on anthracene and other semiconducting materials do, however, provide an important insight into the mechanism of charge transport. Thus, if an anthracene ion were placed into an otherwise neutral lattice, electron or hole transport would not be expected to occur unless there was sufficient intermolecular overlap be-tween the wave functions of the ion and the neutral molecule. In the gas

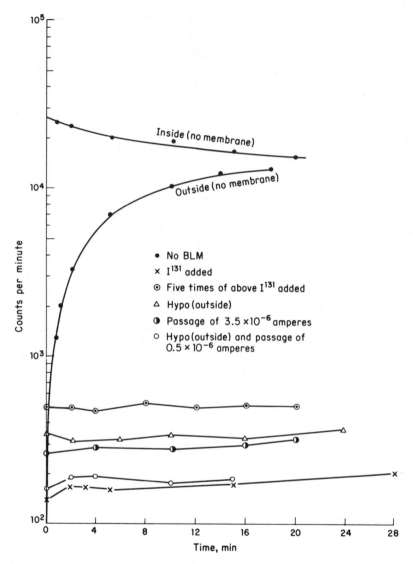

Fig. 6-3. Flux of I^{131} across BLM under various conditions. Concentration of KI in both compartments in 5 mM in all experiments.[352]

phase or in solution where the molecules are still well separated the inter-molecular interactions are at a minimum and the resulting absorption spectra and photophysical processes are similar to those of free molecules. In the solid state, however, where the molecules are separated by only a few Å, intermolecular interactions become significant. Electrons can thus jump from one molecule to next.

Polyiodides have unfilled d orbitals at low energy which make them suitable for an electronic charge-transfer process provided a close approach between donor and acceptor molecules is envisioned. One of the more attractive mechanistic possibilities is formation of a charge-transfer complex between a membrane component and iodine. Formation of stable charge-transfer complexes of iodine with a variety of molecules is well known; these complexes are frequently nonstoichiometric and their electrical resistivity may show a strong dependence on the halogen content.[345] Because of the strong charge-transfer interaction in the ground state of these complexes they show strong ESR signals. In liposomes, I_3^- and high concentrations of I_2 were found to displace a hydrophobic probe.[353] From this observation, it appears that iodine-iodide species are not simply incorporated into the interior of liposomal membranes, but they interact with the membranes in a definite way. ESR signals resulting from direct interaction of iodine with phospholipids have not been reported.

It is also possible that iodine-iodide species do not interact with any of the membrane components, but that they aggregate to form *excitons*. Since electrons may be considered as an array of interacting molecules in an excited state, it is useful to imagine the electron transfer as hopping from one molecule to the next. The diffusion length of this exciton energy transfer will depend strongly on the lifetime of the excited molecular states involved.

Both of these models for charge transfer across BLM containing iodine–iodide complex are capable of explaining the observed data, although some questions remain to be answered. Charge transfer by electronic mechanism has been proposed from time to time in biological processes,[354,355,1088] especially in photosynthesis, receptor response, and other membrane-related processes to be described in Chapter 7. It is pertinent to note here that conductance change induced by iodine has been brought out in a mechanism advanced to account for the hormonal action of thyroxin.[356] Evidence in support of the proposed mechanism includes consideration of the structure and degradation of thyroxin, the ability of iodine to exhibit thyromimetic actions, and the interaction of iodine with model systems.

It is pertinent to note here that BLM and liposomes modified with ferrocene and its homologs show electron transfer across the lipid bilayer.[1191] Furthermore the uncoupling agents of oxidative phospholrylation (proton conductors) can increase the rate of an oxidation reaction mediated by transmembrane electron transfer. These results have strong implications on theories of mitochondrial function.

Transport of Alkali Metal Cations Across Modified BLM. The electrical resistance of BLM bathed in alkali metal salt solutions is considerably decreased in

TABLE 6-6 SOME CHARACTERISTICS OF ION-TRANSPORTING MACROCYCLIC COMPOUNDS BOUND TO MEMBRANES

Macrocyclic Compound (Ref.)	Ion Specificity	R_m drop at 1 M in KCl soln. (0.1 M)	Antibacterial Activity[a] (γ/ml)	No. of Ring Atoms	Charge at pH 7
Enniatin A (374)	Rb > K > Cs > Na > Li	10^2	1.5 (366)	18	0
Enniatin B (374, 375)	Rb > K > Cs > Na > Li	10^4	9 (366)	18	0
Enniatin C (374)	Slightly K selective	10^2	50 (366)	18	0
Beauvericin (1091)	Rb $\gtrsim$ Cs > K $\gg$ Na > Li		2–4	18	0
Valinomycin (211, 358, 376)	H > Rb > K > Cs $\gg$ Na > Li	10^6	0.8 (366)	36	0
Nonactin (314–316, 375)	NH_4 > K > Rb > Cs > Na $\gg$ Li	10^5		32	0
Monactin (358, 377)	NH_4 > K > Rb > Cs > Na $\gg$ Li	10^6		32	0
Trinactin (1224)	NH_4 > K > Rb > Cs > Na	10^6		32	0
Nigericin (1092)	K > Rb > Na > Cs	10^3		"18"	−1
Monensin (1092)	Na > K > Cs	10^4		"18–19"	−1
Dianemycin (1125)	Na > K > Cs > Rb > Li	10^4		"18"	−1
Dicyclohexyl-18-crown-6 (378)	K > Rb > Cs > Na > Li	10^3		18	0
"Transport peptide" (381)	K > Rb > Cs > Na > Li		Yes	?	0
Gramicidin A (380)	Rb > Cs > K > Na > Li		Yes		0
Gramicidin B (379)	K $\gtrsim$ Na	10^2	Yes	"18–19"	0
Gramicidin S	K $\gtrsim$ Na	10^2	Yes	30	0
X-537A	K > Rb > Na > Cs > Li	10^2	Yes	"18"	−1
Antamanide (373a)	Na > K	10	Yes	30	0

[a]Minimal growth inhibiting concentration for *Staphylococcus aureus*.

the presence of certain macrocyclic compounds such as valinomycin, actins, enniatins, etc. (Table 6-6). The group includes macrocycles of peptide, depsipeptide, depside, and ether residues interspersed with suitable methylene and methine residues (Fig. 6-4). In addition to these, the group includes compounds which are linear (Fig. 6-5) but behave similarly; that is, they can also act as ion carriers in the hydrophobic phase of the lipid bilayer. This is because, as we shall see later, these molecules "wrap around the cation like a collar, buttoning shut with a pair of hydrogen bonds, and presenting a highly hydrophobic exterior."

The efficacy of both these groups of compounds to induce ion transport in model membranes can be correlated with their ability to induce cationic permeability in a variety of plasma membranes,[357–367] which gives rise to an uncoupling action on oxidative phosphorylation, respiratory stimulation, mitochondrial ATPase activation,[357–359,1093] and antibiotic activity.[363–365,1094] However, all these various effects are due to a selective increase in the permeability of these macrocycle-treated membranes to monovalent alkali metal cations, especially potassium ion. Divalent cations and most of the anions show little change in their permeability.

The effect of macrocyclic compounds is best studied by their addition to the aqueous medium after the formation of the black film is complete. Since these

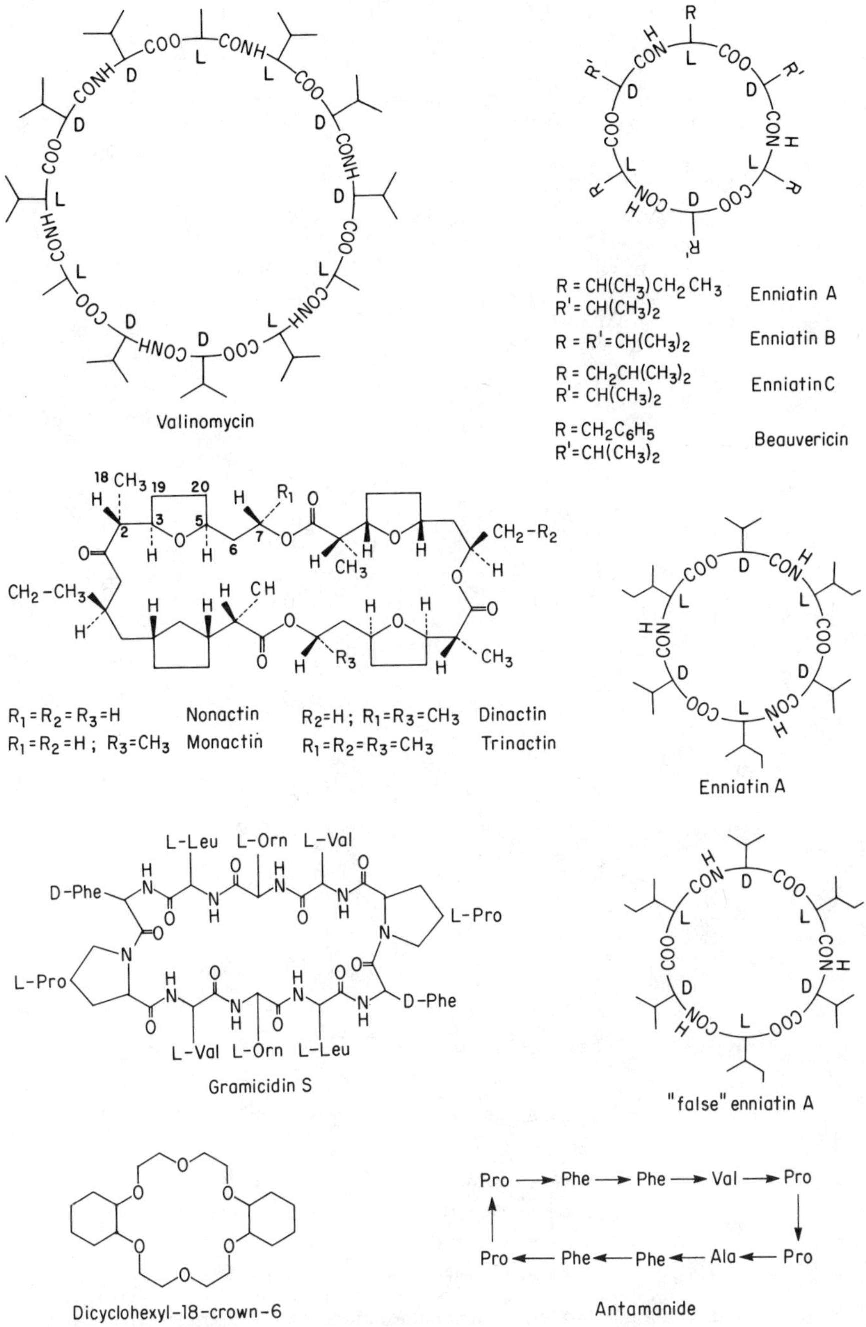

Valinomycin

R = CH(CH₃)CH₂CH₃
R' = CH(CH₃)₂   Enniatin A

R = R' = CH(CH₃)₂   Enniatin B

R = CH₂CH(CH₃)₂
R' = CH(CH₃)₂   Enniatin C

R = CH₂C₆H₅
R' = CH(CH₃)₂   Beauvericin

R₁ = R₂ = R₃ = H   Nonactin
R₁ = R₂ = H ; R₃ = CH₃ Monactin

R₂ = H ; R₁ = R₃ = CH₃ Dinactin
R₁ = R₂ = R₃ = CH₃   Trinactin

Enniatin A

Gramicidin S

"false" enniatin A

Dicyclohexyl-18-crown-6

Pro → Phe → Phe → Val → Pro

Pro ← Phe ← Phe ← Ala ← Pro

Antamanide

Fig. 6-4. Structure of some macrocyclic antibiotics.

I 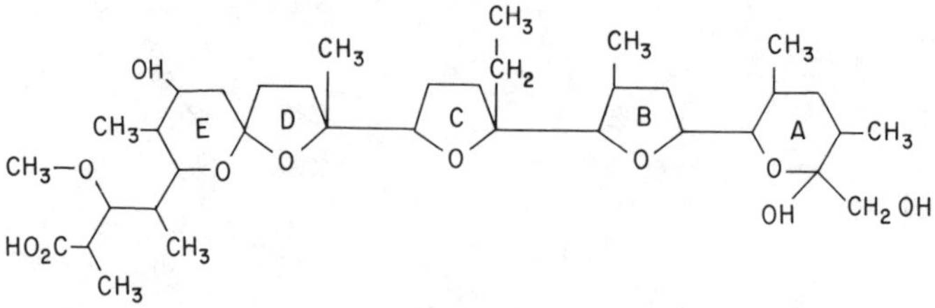

II 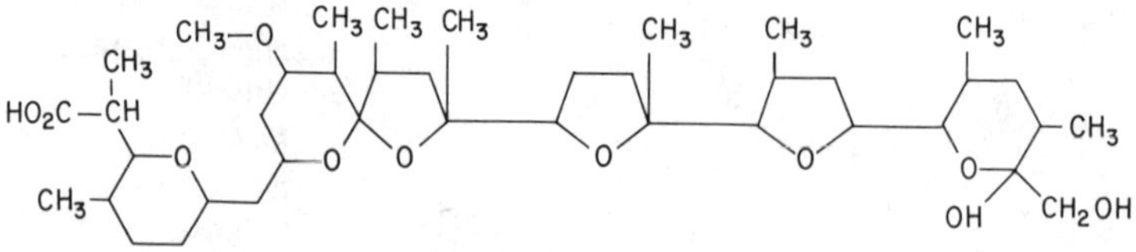

III 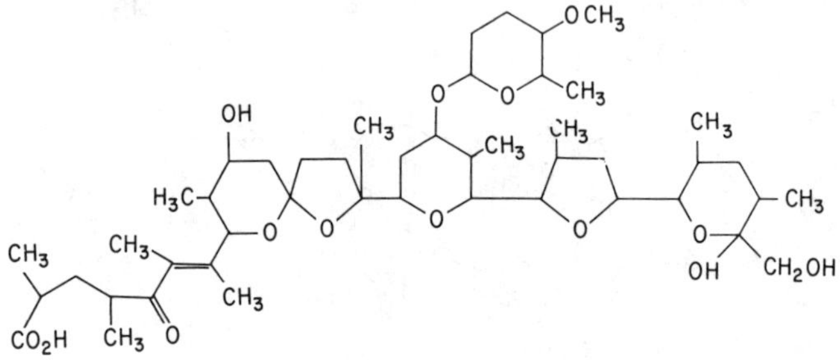

IV 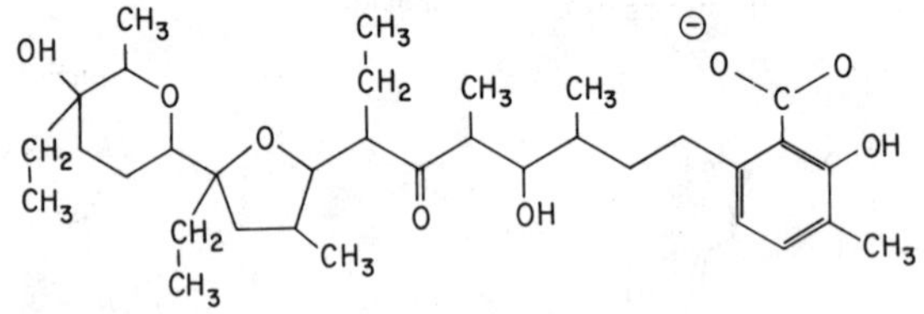

V 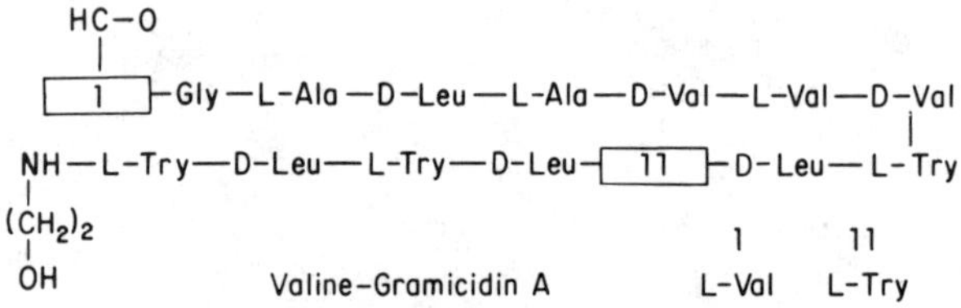

	1	11
Valine-Gramicidin A	L-Val	L-Try
Isoleucine-Gramicidin A	L-Ileu	L-Try
Valine-Gramicidin B	L-Val	L-Phe
Isoleucine-Gramicidin B	L-Ileu	L-Phe
Valine-Gramicidin C	L-Val	L-Tyr
Isoleucine-Gramidicin C	L-Ileu	L-Tyr

Fig. 6.5. Molecular formulas of "open" ionophores/monensin (I), nigericin (II), dianemycin (III), the antibiotic X-537A (IV), and gramicidins (V).

agents act as ion carriers, their incorporation into BLM can be conveniently followed by monitoring the resistance drop of BLM. The rate of black film formation is significantly affected by the presence of some macrocycles; for example, formation of BLM is almost inhibited in the presence of valinomycin in the aqueous phase. The experimental results obtained from the study of the effect of various macrocyclic compounds on BLM suggest a definite parallelism between the role of these compounds on model and biological membranes as cation carriers. These observations may be summarized as follows:[357-370,1223]

1. Macrocycle-treated BLM show a significantly lower resistance than untreated BLM. Generally speaking, the effect of these agents is instantaneous and the membrane resistance becomes steady a few minutes after the addition of these agents to the medium. Most of these agents cause an increase in the conductance of BLM, thick lipid films, and an increase in the ionic permeability of liposomes,[368] bulk organic solvents,[369] and organic polymers.[370]

2. The effect of macrocyclic agents on ion transfer across BLM formed from various lipids and additives is as would be expected to result from effect of changing viscosity of the medium on the mobility of the carrier. Thus, for example, higher temperature[1294] or addition of tocopherol to the membrane-forming lipid solution[316] increases the membrane responsiveness, whereas lower temperature or higher cholesterol depresses it. Although there are some quantitative changes on conductance of BLM which can be correlated with the nature of lipid, the effect of these agents is otherwise quite independent of the nature of the lipid. The effect of polar groups is as would be expected from the electrostatic consideration of the surface charge profile, which would affect the concentration of cations on the membrane interface.

3. Generally speaking, the current-voltage curves of these modified membranes are linear; however, in some cases slight rectification at higher voltages has been noted. As expected, the electrical capacity and dielectric breakdown voltage of these modified BLM remain essentially unchanged. This is presumably because only a small fraction (of the order of one in 10^5) of the membrane area is occupied by these agents.

4. If salt solutions of the same alkali metal at different concentrations are placed on two sides of BLM containing valinomycin or any of the macrocyclic compounds listed in Table 6-6, an open-circuit potential develops. It is negative on the side containing the higher salt concentration and has a magnitude of 54–58 mV per decade of salt concentration gradient at room temperature. This implies that valinomycin *confers* cation selectivity, and that the selectivity with respect to anions may approach infinity at relatively low salt concentrations and for small gradients. Similarly if two compartments separated by the membrane contain the same concentration of sodium chloride and potassium chloride, respectively, biionic potentials up to 150 mV negative, usually, on the side containing potassium ions are observed. These observations strongly imply that the

macrocycle-trated BLM are permeable to alkali metal cations, and intercationic permeabilities may differ by several orders of magnitude. These intercation selectivity sequences are also collected in Table 6-6. The values of potentials developed in single-salt gradient, as well as the biionic potentials, are predicted by equation (6-3).[316]

5. The intercationic selectivity sequences deduced from potential measurements at zero current flow (open-circuit potential) are manifested in the conductance of modified BLM in various salt solutions as predicted by equation (6-3). Thus, for example, the cation transference numbers for both sodium and potassium ions (t_{Na} and t_K, respectively) are approximately 0.85 for BLM prepared from sheep red cell lipids or from lecithin, and these membranes do not distinguish between sodium and potassium ions. However, in the presence of macrolides, t_K for these membranes is unity, and the transport number of sodium is virtually unaffected at low salt concentrations. Addition of the aromatic anion trinitrocresolate to these modified membranes markedly reduces their K^+/Na^+ selectivity.[1295]

6. The conductance of BLM is a function of macrocycle concentration and nature, and of the concentration of salts. Temperature has little effect on these changes in membrane permeability (Q_{10} = 1.2–1.4) however see Ref. 1294. For any given salt, the membrane conductance is a linear function of macrocycle concentration. The effect of increasing concentrations of valinomycin on BLM conductance in 0.1 M sodium chloride and 0.1 M potassium chloride solutions is shown in Fig. 6-6. With over a hundredfold change in the valinomycin concentration only a small decrease in membrane resistance is produced in sodium chloride solution, whereas in potassium chloride the membrane resistance drops by several orders of magnitude. Increase in membrane conductance is the same when the macrocycle is present on one or both sides of the BLM; it is the concentration in the vicinity of BLM that determines the amount of macrocycle adsorbed by the BLM. For reasonably permeable ionic species, usually the first two or three members of the observed selectivity sequence, an increase in salt concentration beyond 0.1 M has a negligible effect on the BLM conductance.

7. The complexed cation is solubilized in the bulk organic phase, but only in the presence of suitable anions which accompany the charged complex. Measurements of equilibrium partitioning of various cation salts show partition coefficients which can be correlated with equilibrium selectivity isotherms and thus found to be dependent upon the nature of macrocyclic agents.[314,315,1095] These results further indicate that the nature of the complex formed is independent of the nature of the bulk solvent, and that the size and shape of the complex formed between macrocycle and a given cation is the same regardless of the species of cation bound. Isotope exchange and flux studies have also given information regarding rate of movement of the complex and its stability constant in the bulk organic phase which can be correlated with the corresponding values in thin lipid membranes.

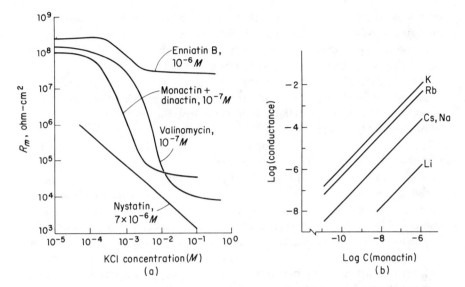

Fig. 6-6. (a) The proportionality between the electrolyte concentration and conductance of modified BLM.[361] (b) The relationship between membrane conductance and the concentration of monactin. The BLM was prepared from purified soya bean lecithin; and the concentration of monovalent cation chloride in the aqueous phase was 0.1 M. Points indicate the experimentally observed values. The solid lines have slope of one.[316] It may be noted here that the conductance of modified BLM is function of the nature of lipid chains, additive, nature of the polar groups, temperature, the nature of the modifying agent (ionophore), and of course the ions in the aqueous phase.

8. When current is driven across a bilayer containing monactin-dinactin or valinomycin with equal concentrations of KCl in the two bathing solutions, more than half of the current is carried by some component other than K^+. This unknown charge carrier is not the anion present with K^+ and it requires presence of O_2. Addition of superoxide dismutase and catalase eliminates this unknown charge carrier—presumably a superoxide anion.[1295]

These experimental observations are in complete accord with the model suggested for carrier transport (Fig. 6-1), and confirm the theoretical expectations of this model. With such a phenomenological correspondence between the model and experimental data, it is no surprise that a considerable amount of work has been done to elaborate the precise molecular changes undergone by a carrier during the course of ion complexation, its migration in free and complexed form, and to elaborate various partial reactions and processes implied in the model. These studies may be discussed in terms of the following methods:

1. Correlation of structure and conformational parameters with their physicochemical ion-complexing properties in different media with their behavior towards the most varied type of membranes and with their antibiotic activity. Such studies are particularly warranted since, thus far there is no available

macrolide which shows higher affinity for sodium than for potassium. The biological significance of this pursuit is obvious.

2. Use of physicochemical techniques to characterize the mode of ion binding and conformational states of carrier in free and complexed forms.

3. Characterization of various thermodynamic and kinetic parameters involved in ion binding and complex translocation steps. These are particularly important by way of characterization of carrier-transport phenomenology as a whole, and interfacial processes and rate-determining steps in particular.

The macrocyclic structure of ion-carrying molecules has engendered the idea that the ions are sequestered in the central cavity of the molecule; the relative sizes of the cavity and of the naked or hydrated cations serve to explain the observed ionic specificity. From the behavior of various compounds listed in Table 6-6 as ion carriers, it is evident that the cause of their effect on membranes lies in the unique molecular structure of these compounds which confers on them the ability to complex with alkali metal cations in lipophilic media. All of these compounds are characterized by large rings (Fig. 6-4) or by the capacity to form ring structures by noncovalent ring closure, as evidenced in the case of gramicidin, nigericin, monensin, etc. (Fig. 6-5).

The effectiveness of both synthetic and natural macrocyclic compounds of different ring sizes (namely 36 membered for valinomycin, 32 for actins, 18 for enniatins), of different patterns of amino and hydroxy acid residues, their nature and configuration on ion-binding specificity, and interaction with membranes correlates with their biological activity.[360] In a given series either increasing or decreasing the ring diameter from the optimum size seem to affect their ionic selectivity and membrane binding capacity adversely. Increasing the number of hydrocarbon residues in the lipophilic part increases their ability for incorporation into hydrophobic phases. Thus, not all the cyclic compounds of a particular ring size show their ionic selectivity or other effects to the same extent, which bears evidence of the importance of conformational factors. Corey-Pauling molecular models of macrocyclic compounds and considerations based on ion-dipole interactions involved in complex formation suggest that the complexing capacity is highly conformation specific, since binding of a cation of a given radius, charge, and coordination number by polar groups far removed along the chain forming the ring will depend on both the number of such groups and their location and orientation in the macrocyclic compound. Studies with synthetic analogs have brought out several structural features of interest involved in complex formation. It has been found that changing the configuration of a single α-hydroxyl group in valinomycin destroys 98% or more of its activity; in contrast, analogs of valinomycin[371] and enniatin B[372] in which the amide group has been replaced by an ester function show activity comparable to the parent compound. Similarly, adding or removing two monomers from the hexameric enniatin structure abolishes the antibiotic activity, but the inverted configuration of each monomer component, so as to form the enantiomorph of enniatin, results in a compound with the full biological activity of naturally occurring enniatin. In contrast, "false" enniatin A, differing from the parent macrocycle by a *shift* in the position of the side chains (Fig. 6-4), does not bind ions detectably and its conformation differs from that of "normal" enniatins, and is retained in solvents of various degrees of polarity.[366] These observations in general suggest that the absolute configuration of components in macrocyclic compounds is as critical in determining their activity as is relative orientation and chemical relationship of groups in cation binding and hydrophobic interactions in conferring lipid solubility to the complex after formation. It may be noted that both of these features are completely preserved in the enantiomorphs. It follows that the geometry of complex and ligand gives rise to yet another possible means of intercationic discrimination due to differences in the coordination number of various cations. Sodium only forms structures with coordination number six and eight with respect to oxygen, whereas coordination numbers of potassium-oxygen compounds have been found to be 6, 7, 8, 9, 10, and 12.[373] Thus, there is a relatively wide array of geometries which can accommodate potassium, and far fewer which can accommodate sodium. In fact, most of the experimentally observed selectivity sequences for macrocyclic compounds are those which can prefer K to Na (Table 6-6).

Physicochemical studies on complexed and uncomplexed macrocyclic molecules have been extensively persued by conductometric, potentiometric, and spectrophotometric methods. The results obtained from these studies have been confirmed by direct single crystal analysis of various complexed and free compounds. The results may be summarized as follows.

The physicochemical studies on uncomplexed macrocyclic compounds, in general, suggest that in solution the molecules exist as an equilibrium mixture of different forms, and that the equilibrium point is strongly dependent upon the polarity of the solvent. In nonpolar media (CCl_4, hexane, etc.) molecules like valinomycin prefer a conformation in which all NH groups form intramolecular hydrogen bonds; in contrast, in more polar solvents, a portion of the NH protons are involved in hydrogen bonding with solvent molecules. However, none of these conformations correspond to fully open ring structures. Manipulation of molecular models suggests that, in nonpolar solvents, irrespective of the ring size, an active macrocyclic compound forms a rigid system of condensed rings resembling a bracelet or donut about 6 to 10 Å in diameter, and 4 Å in height. It may be particularly emphasized that the highly stable compact conformation of valinomycin (36-membered ring) and actins (32-membered ring) makes the effective size of the inner cavity the same as that of the 18-membered enniatin. The conformation of oxygen atoms bound to the metal-cation is similar in these cases; however, small-ring molecules form less compact structures than their large-ring analogs. This implies that binding of ions is strongly dependent on the structure of the ligand and the polarity of the medium. Moreover, even though molecules with various ring sizes form almost identical cavities, their ion-binding capacity would be expected to differ considerably. In fact, these conclusions have been fully confirmed by various experimental studies on valinomycin,[376,383-386] actins,[387-388] enniatins,[366,389] gramicidins,[390-392] cyclic polyethers,[378] and various others.[357-366] These studies are consistent with the conclusions that the conformation of macrocyclic compounds in nonpolar solvents is suitable for ion binding by interaction with centrally directed polar groups and that the exterior surface of the complex is hydrophobic, being formed by various nonpolar groups. The physicochemical studies further suggest that the ion is bound through ion–dipole interactions with oxygen functions. Although an ion–dipole interaction of this type is not a novelty (see Refs. 393 and 394 for other examples of ion pairs and related interactions), some of the macrocyclic compounds are unique in the high efficiency and ionic selectivity in the formation of the complex, and its transport. It may therefore be instructive to examine the mode of ion binding and the structure of the complexes in detail.

It has been suggested that a complexed ion is bound to the macrocyclic compounds by means of induced ion–dipole interactions with a systematized configuration of amide carbonyl, ether, and/or hydroxyl oxygen atoms. It is well known that compounds with such groupings, for example, dioxane, dimethyl formamide, and N-methyl pyrrolidone, are capable of effectively solvating polar molecules, ion pairs, and ions. One could therefore expect that fixation of the solvating elements of the solvent by incorporating suitable polar groups in cyclic molecules of appropriate size and conformation should strengthen the solvates. Evidence for such intramolecular ion solvation by macrolides has been derived from several sources. The most characteristic feature of such compounds is that several structurally remote, strictly oriented polar groups take part in cation binding, so the complexing capacity of the macrocycles is determined by the conformational parameters of the molecule as a whole to a greater extent than by the properties of the participating functional groups taken individually. Through use of molecular models, it can be shown that most of the K^+-discriminating macrocyclic compounds assume a unique configuration which forms an eightfold oxygen-bonded complex with K and Rb ions. The smallest space that can be enclosed by a cubic array of eight oxygens in either valinomycin or the actins is too large to permit a close fit by an unhydrated sodium ion. However, it is possible to adjust six reactive oxygens of antibiotics, such as monensin, to accommodate an ion as small as sodium.

Values of stability constants obtained by various methods for various macrolide-cation complexes are given in Table 6-7. It may be noted that complexation does not occur at all in water, and that the complexes display their highest stability in nonpolar media. The complex formation constants in methanol are smaller than those in ethanol by a factor of 1 to 10. Divalent ions such as

TABLE 6-7 STABILITY CONSTANTS OF COMPLEXES OF ALKALI METAL CATIONS WITH MACROCYCLIC COMPOUNDS

Compounds	Li	Na	K	Rb	Cs	Anion	Solvent	Method	Ref.
Enniatin B		1.3×10^3	3.7×10^3	4.0×10^3	2.2×10^3	CNS	Ethanol	Conductometric	946
		2.4×10^2	8.4×10^2			Iodide	Methanol	Potentiometric	180
Beauvericin	10^2	3×10^2	3.2×10^3	3.5×10^3	3.5×10^3		Ethanol	Conductometric	1091
Enniatin C		2.5×10^3	5.5×10^3	7.5×10^3	4.1×10^3	CNS	Ethanol	Conductometric	946
(Tri-N-desmethyl)-enniatin B		2.5×10^3	2.6×10^3			CNS	Ethanol	Conductometric	946
Valinomycin		0	2.0×10^6	2.6×10^6	6.5×10^5	CNS	Ethanol	Conductometric	946
Nonactin		1.2×10^1	8×10^3			Iodide	Methanol	Potentiometric	180
Nonactin		1.3×10^2	5×10^3			CNS	Methanol	Potentiometric	179
Monactin		1.1×10^3	2.5×10^5			CNS	Methanol	Potentiometric	179
1,2,10,11-dibenz-3, 6,9,12,15,18-hexa-oxacyclooctadecane			100	47	12	Iodide		Potentiometric	368b
Antamanide		2800	270						1096
EDTA	6.3×10^2	5.3×10^2							368a
$P_2O_7^{4-}$	1.3×10^3	2×10^2	2×10^2	2×10^2	2×10^2				368a
Dibenzoyl methane	8×10^5	3×10^4	5.5×10^3	3.2×10^3	2.5×10^3				368a
Sulfate	40	51	80						368a
Substituted picrylamine anion	10	63	5×10^3	1.6×10^4	1.6×10^5			Partition	368c

Ba with an ionic radius comparable to that of K also seem to show a tendency to complex with the macrolides, but no such binding is found with Sr, Ca, and Mg. It would therefore appear that macrocyclic agents form complexes with alkali metal cations only after binding to the membrane.

Use of optical rotatory dispersion (ORD) has been successful in ion-binding studies even though the amide and hydroxyl residues of these macrocyclic molecules do not change the shape of the dispersion curves. It is generally assumed that complexation-induced changes in ORD curves are not so much the result of changes in electronic characteristics of the chromophore as of conformational changes in the molecule as a whole. The ORD curves of enniatin A and B are virtually identical in polar and nonpolar solvents, indicating that these compounds possess very similar, stable conformations.[383] However, the ORD curves for their complexes are found to differ sharply from the curves of the uncomplexed molecules in nonpolar media (compare various curves shown in Fig. 6-7). In the same figure are shown dispersion curves obtained from enniatin C, which differ very little in the complexed and uncomplexed forms. Similar changes in dispersion curves of valinomycin solutions have been noticed in various solvents, of which only that in the least polar solvent (a 10:1 mixture of heptane and dioxane) resembles the curve obtained from the valinomycin-KBr complex.[383] It is noteworthy that the conformation of the K^+-complex of enniatin B is the same as that of the noncomplexed compound in a polar solvent (trifluoroethanol), as can be seen from the similarity of curves 3 and 4 in Fig. 6-7. These results have been confirmed by fluorescence polarization[1224] and Fourier transform ^{13}C-NMR spectroscopy.[1293]

The studies just described suggest that in suitable solvents a complexed and uncomplexed macrocyclic compound retains the same configuration and conformation. Thus, in a comparison of the complexes with cations of various sizes

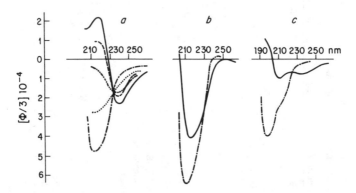

Fig. 6-7. ORD curves of enniatin B (*a*), enniatin C (*b*), (tri-*N*-desmethyl)-enniatin B (*c*), and their K^+ complexes in ethanol ([CDP] = $1.5 \times 10^{-4} M$). ---[KCl] = $0.75 \times 10^{-4} M$; ········· [KCl] = $2.25 \times 10^{-4} M$; ···[KCl] = $4.5 \times 10^{-4} M$; –··–[KCl] = $75 \times 10^{-4} M$.

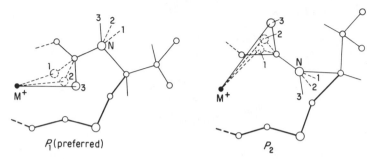

Fig. 6-8. Effect of the size of the complexed cation of orientation of amide group (1 Li⁺; 2 K⁺; 3 Cs⁺) in P_1 and P_2 conformations of (tri-N-desmethyl)-enniatin B.[360]

J_{NH-CH} Values of (tri-N-desmethyl)-enniatin B complexes
with monovalent cations

Cation	M⁺... O Distance $(r_{M^+} + r_0)$	J_{NH-CH^2}, Hz Calculated P_1 form	P_2 form	Observed
Li⁺	1.2	3.2	7.9	5.1
K⁺	1.7	5.8	7.0	7.9
Cs⁺	2.1	7.6	6.1	8.4

one would expect such complexes to differ in the size of the internal cavities formed by the oxygen atoms of the group participating in the ion–dipole interaction. Obviously, the effective size of the cavity for a given ring is determined by the orientation of the oxygen functions. The closest distance between cation and coordinating oxygen atom would be determined by factors other than the radius of the cation and the steric factors governing flexibility of the ring. Inspection of molecular models of enniatin for example, suggests that such changes in the orientation of carbonyl groups invariably lead to a simultaneous rotation of the CH-NH plane (Fig. 6-8), the ultimate result of which is to increase the dihedral angle between the H–C–N and H–N–C planes from approximately 130 to 160°, which should be reflected in the proton magnetic resonance spectra (pmr) by an increase if J_{HC-NH} from approximately 4 to 9 cps. Indeed, the pmr spectra of (tri-N-desmethyl)-enniatin B complexed with Li, K, and Cs display a regular increase in the $^3J_{CH-NH}$ constant from 5.1 to 6.5, 7.9, and 8.4 cps, respectively. These changes are shown in Fig. 6-8 by the dotted lines in the inset.

Similar evidence has been derived from pmr studies on the valinomycin-KCNS complex in deuteriochloroform solution.[360,386,1097] The spectra of free and complexed valinomycin show considerable differences (Fig. 6-9). Apart from the shift in the peaks due to amide protons (at 7.7 ppm) other peaks are sharpened and shifted. Sharpening of most of the peaks is attributed to the fact

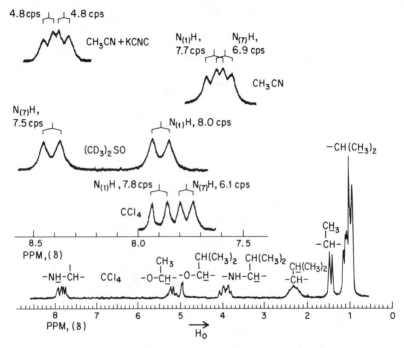

Fig. 6-9. NMR spectra of valinomycin and its K^+ complex. The $^3J_{NH-CH}$ constants should be corrected for the electronegativity of the substituents (+ 0.6 Hz). The assignment of the NH proton signals was confirmed by the NMR-^{1}H spectra of valinomycin one of whose L-valine residue was enriched with ^{15}N.[360]

that in solution these molecules exist as an equilibrium mixture of various conformers and on complexation one of these is *frozen*. The shift in the position of the peaks is interpreted to suggest a change in the environment of the protons in the complex relative to the free molecule, thus implying a conformational change resulting from complexation; such changes are, however, small. The spectra of K and Cs complexes are different, the significance of which is not clear.

Salt-induced chemical shifts have also been noted in 220 MHz pmr spectra of nonactin.[387,388] It has been noted that the protons whose resonances are shifted to the greatest extent upon complexation, namely H_7 and H_3, are those which are geometrically in close proximity to one of the four centrally directed carbonyl groups, each located at the corner of an irregular tetrahedron enclosing the central cavity of the nonactin ring. The H_5 proton also experiences a fairly large downfield shift (7 cps) upon complex formation. Since this proton is not in close proximity to a carbonyl group but is adjacent to an ether oxygen of its tetrahydrofuran ring, one might infer from the sizable H_5 shifts some participation of the ether oxygens in the coordination of the K ion. Thus approximate eight-fold cubic coordination as suggested by X-ray crystallographic studies (discussion to follow) may arise due to ether oxygens occupying the remaining corners of the cube which encompasses the tetrahedron of the carbonyl oxygens, at very similar K–O distances.

PMR spectroscopy also yields information about the rate of exchange of ions between macrocyclic ligands. A rapid exchange equilibrium would be indicated by a broadening or fusion of the peaks upon complexation. If the intensity of a resonance peak decreases from its uncomplexed position and reappears in its complexed position without detectable alteration of the peak positions of band widths, an upper limit may be set on the possible rates of cation turnover in the exchange process, such as:

$$\text{Macrolide* + KCNS} \cdot \text{Macrolide} \rightleftharpoons \text{Macrolide*} \cdot \text{KCNS + Macrolide}$$

The upper limit for cation turnover for complex formation of valinomycin in $CDCl_3$ solution, calculated from the lack of observed band broadening, is about one per second. This compares with estimates two orders of magnitude higher for the optimal turnover numbers of valinomycin observed in mitochondrial transport.[358]

The interaction of the ionophore-cation complexes with the fluorescent probes produces enhanced fluorescence emission, increased life-time and polarization, and a significant blue shift of the emission maxima. These changes have been observed even in aqueous solutions.[1224] At constant antibiotic and probe concentration in water, the intensity of the fluorescence emission was found to be function of the cation concentration. Although exact location of probes on the complex is not known, however the spectral shifts may result from binding to sites of very low polarity or from restricted solvent molecule relaxation in the region of the binding sites.

The conclusions arrived at through physical and model building studies have been confirmed by single crystal X-ray studies of various complexes (Table 6-8). The most important feature of the structures thus elucidated is that the exterior of the complex is overwhelmingly hydrophobic. This fact in itself is adequate to explain the solubility of these complexes in organic solvents and their instability in water. The cation is always located in the center of the cavity (Figs. 6-10 and 6-11). The coordination about the cation is to oxygen functions, such as carboxylate, carbonyl, ether, or alcohol. A single ligand molecule, in all cases, completely encircles the central cation, thus forming a one to one complex; but the number of the coordinating oxygen atoms and the geometry around the cation is different for different compounds. The coordination about the central monovalent cation (K in most cases) varies from fivefold in nigericin, to sixfold in monensin, enniatin, and valinomycin, to eightfold in nonactin and X-537A.

TABLE 6-8 SINGLE-CRYSTAL X-RAY STRUCTURE DATA FOR SOME ION-MACROCYCLE COMPLEXES

Complex (Ref.)	Backbone	Orientation of Polar Groups	Shape	No. and Arrangement of Oxygens	Range of K-O Distance (Å)
Enniatin-KI[a] (395)	Ring has 18 atoms made up of both peptide and ester links	Both peptide and ester groups are in trans form and planar	Disc shaped; anions do not occupy definite position	Octahedral coordination by 6 oxygens arranged 1.5 Å above and below the mean molecular plane	2.6–2.8
Nonactin-KCNS (396)	Ring has 32 atoms made up of ester and ether oxygen	All the oxygen functions are centrally directed	Approximately spherical formed by foldings like seam of tennis ball; no fixed position for anion	Eight oxygens—four from carbonyl and four from THF—in approximately cubic coordination	2.73–2.88
Valinomycin-KAuCl₄ (397)	Ring has 36 atoms made up of peptide and ester linkages	Ring is internally hydrogen bonded; the nitrogen of each of the six valine residues linked to the ester carbonyl oxygen three residues removed (O . . H 2.8–3.0 Å)	Approximately spherical; anion occupies definite position in the complex	Six carbonyl of valine residues in octahedral coordination	2.7–2.8
Nigericin-Ag[a] (398)	A linear chain which can cyclize by a pair of H-bonds to form 17-membered ring	Three THF[b] rings are approximately planar and their oxygens are centrally directed	Disc shaped (irregular)	Five oxygens coordinate; carboxyl group participates	
Monensin-Ag (399)	A linear chain which can cyclize by a pair of H-bonds to form 17-membered ring; H-bond distances 2.5–2.9 Å	Three THF rings are approximately planar and their oxygens are centrally directed	Disc shaped (irregular)	Six oxygens coordinate; carbonyl group does not participate	2.4–2.6[c]
[X-537A]₂-Ba·H₂O (400)	A linear chain	Centrally directed	Dimer (disc)	Eight oxygens	2.8–3.0

[a] Isomorphous with corresponding Na complex.
[b] THF—Tetrahydrofuran.
[c] Refers to Ag–O distance.

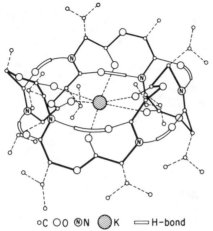

○C ○○ Ⓝ N ⬤ K ⊂⊃ H–bond

Conformation of the K⁺ complex of valinomycin

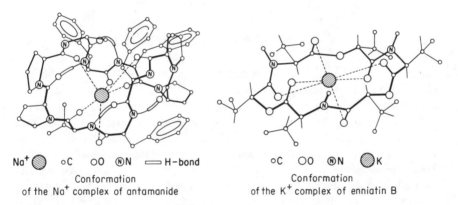

Na⁺ ⬤   ○C   ○○ ⒪   Ⓝ N   ⊂⊃ H–bond

Conformation
of the Na⁺ complex of antamanide

○C   ○○   Ⓝ N   ⬤ K

Conformation
of the K⁺ complex of enniatin B

Fig. 6-10. Three-dimensional structures of cation complexes of three macrocyclic iono-phores.

Similarly, nonactin and valinomycin show cubic and octahedral geometries, re-spectively. This three-dimensional arrangement is contrasted with the disc-shaped arrangement observed in the enniatin complex. Monensin and nigericin show more or less planer but irregular coordination patterns. The complexes formed by nigericin, monensin, and valinomycin are stabilized by hydrogen bonding (intramolecular), but the others are not. The monensin complex is quite rigid and inflexible; the nigericin complex is also rigid but can adapt to larger cations by a rotation of the carboxylate group. The dianemycin complex is quite flexible and adaptable.[1225] Finally, even though the complexed monactin, ennia-tin, and valinomycin molecules are electrically charged due to the sequestered

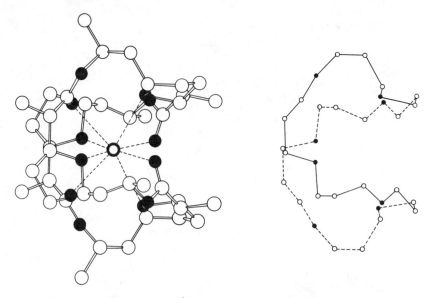

Fig. 6-11. Structure of the K^+-nonactin complex according to X-ray data.

cation, only valinomycin complex has its anion in a well-defined cavity between two cation-valinomycin molecules. In other cases, the anion does not occupy any definite position. All the complexes contain only unsolvated cations.

The structures of these ligands and their complexes as developed above lends substantial support to the hypothesis advanced earlier in this chapter that ions complexed with macrocyclic compounds can traverse nonpolar media, whether bulk organic phase or thin BLM. It must be emphasized that the principles evolved from the studies of macrocyclic compounds are rather general in nature. For example, the cavity for cation binding need not be intramolecular; it can be formed from two molecules, as is the case with the "*transport peptide*" (Fig. 6-12).

However, little can be said about the exact mode of the ion transfer across hydrophobic phases. The macrocyclic compounds which we have discussed thus far can act on membranes as channels by stacking together in columns and breaching the membrane, or as carriers by moving freely through the lipid interior of the membrane and collecting and discharging their ion at the interfaces of the membrane. Isotope diffusion and spectroscopic studies have shown that the carrier complexed with ions can move through the apolar phase. Such movement by a channel-type mechanism would imply formation of enormously large channels, a rather improbable event. Not only do X-ray crystallographic data not suggest stacking of complexed macrocyclic compounds in crystals. Furthermore, a considerable amount of circumstantial evidence favors a carrier mechanism:

1. There is no indication in the pmr and ord spectra that the macrocyclic

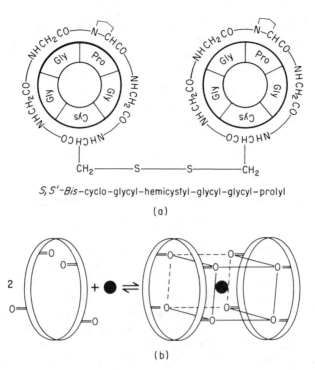

S, S'-Bis-cyclo-glycyl-hemicystyl-glycyl-glycyl-prolyl

(a)

(b)

Fig. 6-12. Structure of the "transport Peptide" (a), and mode of its complexation with a cation (b).[381]

compounds (which we have described thus far) have any tendency to associate in apolar media.

2. Exchange reactions involving cations between macrocyclic molecules (single molecular species) as studied by line broadening in pmr spectra are indetectable in apolar solvents. Thus under these conditions the complexed cation does not seem to leave the pocket or cavity.[386]

3. If the ions were moving in single file through intramolecular channels or pores formed by macrolides oriented in the membrane plane, it would be expected that the flux ratio would exceed the electrochemical activity. However it has been found that over an approximately 1000-fold range of concentration, the flux ratio is approximately equal to the electrochemical activity ratio. This suggests that cations cross the membrane singly and independently, and that most of the cations (mobile) in the membrane are present as complexes.

4. Quantitative comparison of measurements of equilibrium salt extraction by bulk organic phases with electrical studies on BLM suggests that BLM may be considered as equivalent to a bulk phase of comparable thickness[315,316] which shows phase transition characteristics.[1294]

5. If a channel mechanism were to operate, aggregation of a large number of molecular species would give rise to an exceedingly high temperature coefficient; Q_{10} of the order of at least 10^6 for membrane thicknesses near 2000 Å. However, colored membranes do not show such abnormal behavior.

6. Valinomycin-induced conductance in BLM varies linearly with the concentration of valinomycin in the medium. Higher-order relations would be expected for an aggregation reaction.

A carrier mechanism, strongly suggested by these experimental observations, predicts various phenomena, such as competitive inhibition, saturation kinetics, trans inhibition, and an inverse relationship between rate of transport and affinity at saturation concentration. The high rate of complex formation as observed in bulk solutions[327] is relevant in the present context. The carrier can be selective only if the overall transport rate is not limited by the rate of metal ion release.[1226] If the latter process were rate limiting, the high selectivity reflected by a high binding constant would be compensated by the slower rate of release. Thus, for example for Na-monactin complex the time constant for release is in the microsecond range; for the more selectively bound K^+ ion it would reach almost the msec range. The release times would be appreciably longer if the recombination process did not have a rate constant as high at 10^8 (M^{-1}-sec^{-1}). This implies that the carrier can play its selective role in alkali ion transport across BLM only due to its high–almost diffusion controlled–recombination rate. These conclusions are in accordance with the conclusions arrived at from the theoretical analysis of steady-state current-voltage characteristics of these modified membranes.[403,404] The analysis suggests that the rate-determining step in the ion transport of the ion is the translocation of the complex across the interior of the membrane. Similarly the turnover number for valinomycin mediated transfer of ions has been computed to be of the order of several hundred.[401,402]

The foregoing discussion may be concluded by a few remarks on the effect of macrolides on biomembranes and its consequences. Much evidence has now accumulated suggesting that various effects of macrolides on biological systems, such as antibiotic activity, uncoupling of oxidative phosphorylation, increased efflux of cations, inhibition of catechol amine-stimulated lipolysis in vitro,[405] modulation of metabolic reactions,[1227] inhibition of counterion fluxes,[1241] and other consequences have a common basis; that is, intensification of potassium (or cationic) ionic fluxes through the diffusion barrier. In all these cases the induced transport is passive, and ions flow down their electrochemical gradient. It may also be noted that not always an increased permeability results after treatment of a biomembrane. For example, the permeability of squid axonal membrane remains unchanged after treatment with monactin.[406] Also in mitochondria, valinomycin-induced K transport appears to be energy dependent and is accompanied by stimulation of respiration or by consumption of ATP.[357,358]

Such induced uphill (apparently) movement can be the result either of lowering the intramembrane diffusion barrier for K transport by *active* process (energized) or of direct interaction of the enzymatic system with the macrocyclic agent, or with its cation complex. These matters are still undecided, and are discussed further in Chapter 7.

At this stage one may ask for the relevance of all these studies on the functioning of biomembranes. Probably one may refer to the striking observations made by M. J. V. Osterhout in 1934:[1102] ".... the ability to distinguish electrically between sodium and potassium ... appears to exist in *Nitella* only as long as it contains a group of organic substances." This effect can be removed from *Nitella* by leaching in distilled water and thus removing a substance. This substance can be recovered from the water extract and applied to the cell and thus restore its ability to distinguish electrically between potassium and sodium ... "the *Nitella's* cell contains a substance which is soluble in petroleum ether and which is responsible for K$^+$ effect."

Although none of these ionophores have been completely characterized, however the presence of ionophores has been shown in mitochondria,[1098] brain,[1099] and electroplax.[1100] Furthermore, the physiological role of the ionophores described in this chapter remains obscure; however, some of these may be implicated in ion-transport processes. In fact, anion binding characteristics of cytochrome has led to the postulation of cytochrome c in anion translocation across mitochonrial membrane.[1101]

Pore- or Channel-Mediated Transport Across BLM

We have already described earlier in this chapter some of the general characteristics of transport through pore or channels. The assumption here is that a hydrophilic region extends through the membrane from interface to interface; permeants move or diffuse through this region rather than through the hydrophobic region. The transport characteristics may arise from the specific binding properties of the molecules which line the pore or form the channel; these systems show saturation kinetics, may allow specific translocation of a substrate, may allow a unidirectional flow of solute, and may show streaming potentials under proper conditions. The specificity of transport through pores would depend both upon the specific constitution of the molecules lining the pore and upon the specific stereochemical properties of the aggregate forming the channel. The type of terminal polar residues present at the rims and the length and bulkiness of their lipophilic segments would have direct and indirect effects upon the size, shape, structure, degree and type of hydration, zeta potential, etc., of the channel. For example, if crosslinking of many head groups could occur readily, the zeta potential of the pores would tend to be relatively low and negative, and the passage of cations would be favored. If steric restrictions were to prevent bridging the polar groups, the divalent cations might tend to have one valence free, in which case the sign of the zeta potential might be reversed, favoring the passage of anions. Several variations of this type may be conceived depending upon the nature of the ions passing through and the nature of the molecules lining the pore. (For other specialized features see Ref. 407.) More speculative systems such as oscillating pores, reorienting pores, and transient pores have also been considered from time to time.[42,256,408] However, such systems put a severe demand on the membrane structure. Since very little is known about such systems, we shall restrict our discussion to some relatively simple ones: channels across BLM formed by certain antibiotics, and pores formed in BLM and biomembranes in early stages of *lysis*.

Channel-Formation Across BLM. A number of compounds appear to form channels across BLM (Table 6-9). These compounds appear to have a wider structural variation than those acting as mobile carriers across BLM (compare Figs. 6-4 and 6-5 with Fig. 6-13). Although most of the compounds in both of these classes are macrocyclic, they are not structurally similar; the following two differences are significant: Compounds acting as mobile carriers have centrally directed oxygen atoms whereas channel forming compounds do not. Also, channel-forming compounds have a much larger range of molecular weights, ring sizes, and distribution of polar groups in the molecule.

Generally speaking, channel-forming agents do not have as dramatic an effect on BLM permeability as do the molecules which act as mobile carriers. This may be ascribed to a number of factors: channel-forming compounds show specific

TABLE 6-9 A LIST OF COMPOUNDS WHICH FORM CHANNELS ACROSS BLM[a]

Substance	Mol. Wt.	No. of Ring Atoms	No. of Molecules per Channel	Specific Lipid Required	Ion Selectivity	Ref.
Alamethicin	1,111	53	6	Cholesterol	Cations	123
EIM	~10,000	?	?	Sphingolipids	Cation	111,410,411
Gramicidin A	900		2	−	Cation	412
ATPase	250,000	?	?	?	Cation	413
Monazomycin	~1,000	?	?	−	Cation	414
Monamycin	1,000	18	6	−	Cation	415
Tyrocidin B	900	30	2–3	−	Cation	416,417
Cytochrome c	13,000	?	2–3	Oxidized cholesterol	Anion	418
Filipin	571	28	6	Cholesterol	Anion (?)	129,419–422
Nystatin	932	35–40	10	Cholesterol	Anion	129,419–421, 423,424
Amphotericin B	960	38	10	Cholesterol	Anion	129,419–421, 425
Pimaricin	681	26	?	Cholesterol	Anion	419–421
Etruscomycin (=Lucensomycin)	700	26	?	Cholesterol	?	419–421
Tetrin A	695	26	?	Cholesterol	?	119,426,427

[a]The assignments are mostly based on circumstantial evidence, however, unless some direct evidence is obtained these assignments may be considered provisional.

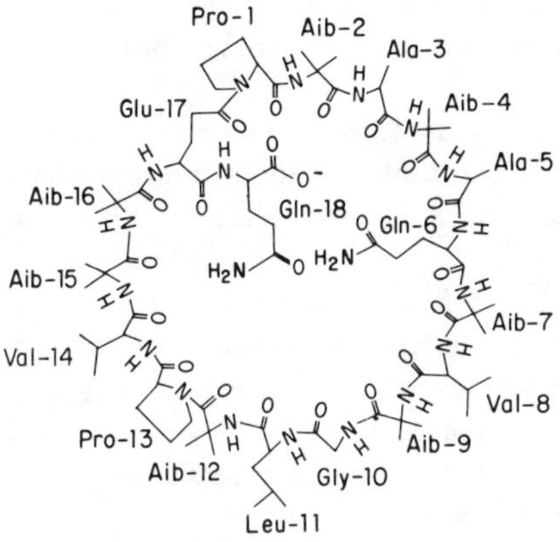

Alamethicin

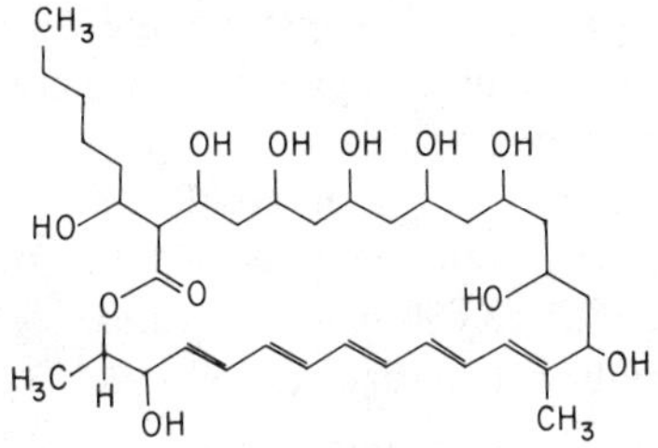

Filipin
$C_{35}H_{58}O_{11}$

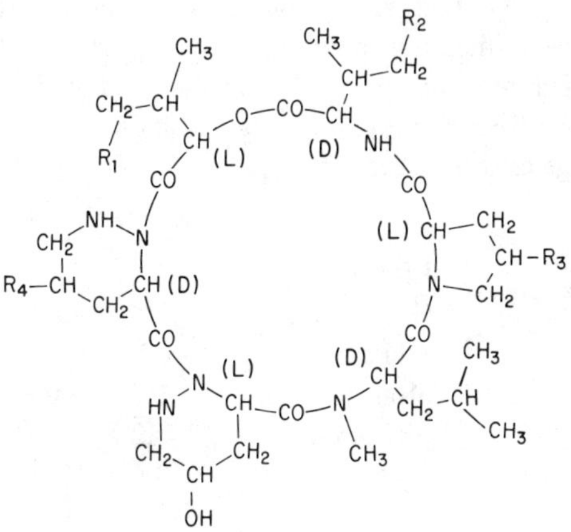

Monamycin D_1
$R_1 = CH_3$, $R_2 = H$, $R_3 = CH_3$, $R_4 = H$

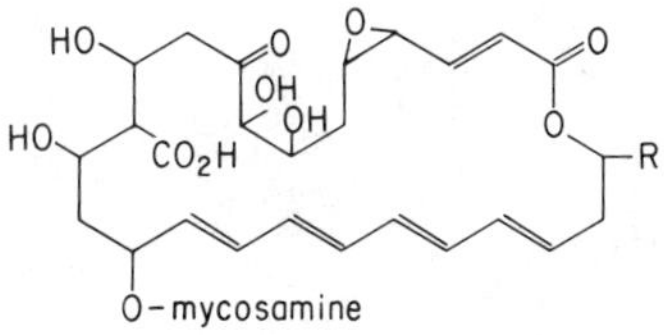

Pimaricin: $R = CH_3$
Etruscomycin: $R = -(CH_2)_3-CH_3$

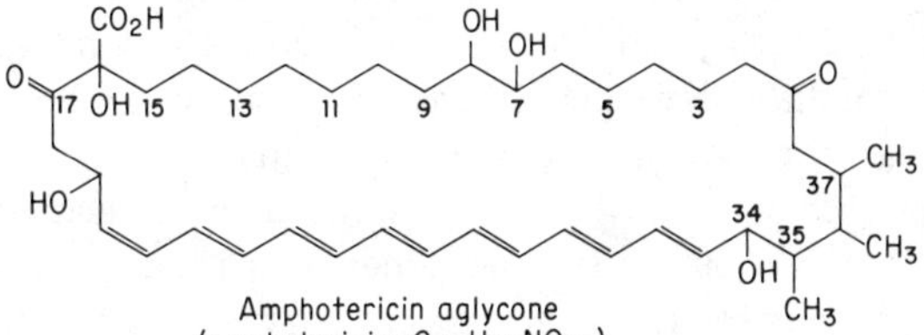

Amphotericin aglycone
(amphotericin = $C_{46}H_{73}NO_{20}$)

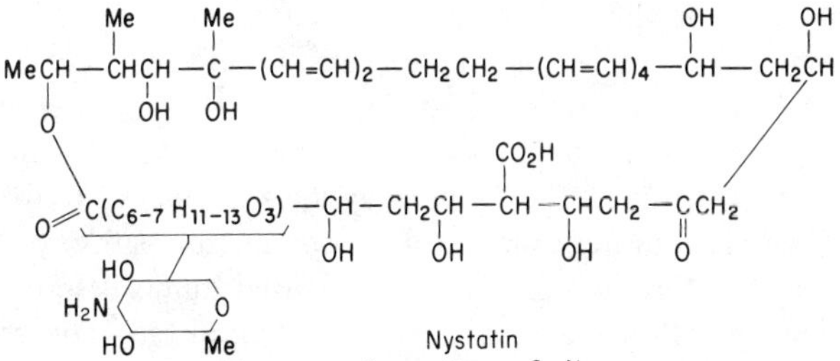

Nystatin
$C_{46-47}H_{73-75}O_{18}N$

Fig. 6-13. Structures of some macrocyclic compounds which form channels across BLM.

requirements for lipids in the membrane; they show a strong aggregation tendency in lipid and aqueous phases; and the kinetics of channel formation is multimolecular.

Characteristics of BLM and biomembranes modified with various channel forming agents and the kinetics of their adsorption into the membrane phase show some distinctive features,[359,409] (see also Ref. given in Table 6-13) which may be summarized as follows:

1. The conductance of a carefully prepared BLM is so low that, under suitable conditions, it is possible to detect and characterize the conductance corresponding to a single channel formed by adsorption and association of a modifying agent. In some cases at low concentration of the modifying agent in the aqueous medium, conductance increases of BLM may be seen as discrete jumps which are multiples of a constant value. It may be noted that such discrete, quantified, conductance changes induced by various agents fall in fairly narrow range as shown below:

		Ref.
EIM	3×10^{-10} ohm^{-1}	410, 411
Gramicidin A	2.4×10^{-11} ohm^{-1}	412
ATPase	10^{-10} ohm^{-1}	413
Monazomycin	10^{-10} ohm^{-1}	414
Rat brain	3.3×10^{-10} and 3.0×10^{-11} ohm^{-1}	1297
Electroplax	2.2×10^{-10} and 2.4×10^{-11} ohm^{-1}	1297

The observed conductance value for a single channel roughly corresponds, in saturated NaCl and for 100-mV applied potential, to transfer of about 10 to 100 million ions per second. This suggests that the effective ionic mobility across the membrane is a little less than that for inorganic ions in water. It may however be noted that for nystatin quantal fluctuations in the conductance are absent, although there are other indications that nystatin molecules aggregate in the membrane phase to form a channel (see below).

2. The plot of log conductance as a function of the log of concentration of the modifying agent in the aqueous medium is linear. However, the slope of this line varies from compound to compound. Thus, for example, conductance varies as the tenth power of nystatin concentration, as the sixth power of alamethicin concentration (Fig. 6-14), but only as the second power of gramicidin A and tyrocidin B concentration. The slope may be equated with the number of molecules of the agent aggregating to form a conduction pathway. Furthermore, as suggested by the flip-flop nature of the quantal jumps described above, the action of these agents on membranes is reversible; in fact, reversal of conductance occurs when the modifying agent is removed from the aqueous phase.

3. Some of these modified membranes show ionic selectivity as observed by the development of a trans-membrane potential as a function of salt gradients. However, selectivity, even for ions of different charges, is rather poor. Thus, for a BLM modified with nystatin, the biionic potentials obtained between several pairs of sodium salts are as follows:[361]

Side 1	Side 2	Potential difference* (mV)
0.01 M NaCl	0.01 M NaSO$_3$CH$_2$CH$_2$OH	55–60
0.1 M NaCl	0.1 M NaSO$_3$CH$_2$CH$_2$OH	55–60
0.1 M NaCl	0.1 M NaSO$_3$CH$_3$	40
0.03 M NaCl	0.03 M NaF	22
0.1 M NaCl	0.1 M NaI	3–8
0.1 M NaSO$_3$CH$_3$	0.1 M NaSO$_3$CH$_2$CH$_2$OH	20–25
0.1 M NaCl	0.1 M KCl	1–2

*In all cases, side 1 is positive with respect to side 2.

The data suggest that BLM modified with nystatin are more permeable to anions than to cations. As expected, such ionic selectivity arises from the presence of a cationic group (ammonium in this case) in the center of the ring (Fig. 6-13).

4. The temperature dependence for conductance increase resulting from the addition of these agents is usually large; for example, the temperature coefficient (Q_{10}) for the conductance increase by nystatin in cholesterol-containing BLM is approximately 10^4 (between 31.5 and 38°). Such a high value is consistent with the tenth-power dependence of conduction on nystatin concentration which implies that ten molecules of nystatin must be adsorbed to create a simple conductance channel. With this assumption, the Q_{10} for the adsorption of a single molecule becomes $\sqrt[10]{10^4} \simeq 2.5$, a reasonable value.

5. Aggregation of the molecules of the modifying agent in the membrane phase puts severe restrictions on their relative orientation and mobility. This implies that their orientation in the membrane phase may be fixed or frozen by specific interaction with some membrane component. Such an interaction may be specific and thus modification of these membranes would show an absolute requirement for the presence of the pertinent membrane components. The polyene antibiotics (the last six entries in Table 6-9), for example, show an absolute requirement for cholesterol not only for their interaction with BLM but also for their interaction with monolayers, liposomes, and a variety of biological membranes. In the absence of cholesterol, thus for example in mitochondrial membranes,[428] these agents are not incorporated into the membranes. The specificity of lipids may also be manifested in the mobility of the monomeric species which aggregate in the lipid phase, or in adsorption as influenced by the interfacial charge profile. In fact, these various factors can be discerned

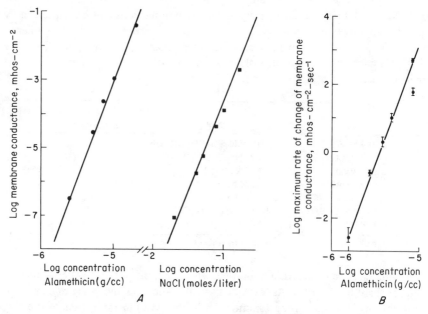

Fig. 6-14. *A*, (Left) Relation between membrane conductance (I/V) and concentration of alamethicin in the aqueous phase at constant ion concentration (0–1 M sodium chloride) and constant voltage (60 mM). (Right) Relation between membrane conductance and concentration of sodium chloride in the aqueous phase, at constant alamethicin concentration (10^{-5} g/ml), and constant voltage (60 mV). All concentrations are identical in both aqueous compartments. The straight lines have slopes to the sixth power which are Independent of membrane lipid composition and of the voltage as long as the conductances induced by alamethicin are much larger than the intrinsic membrane conductance. These curves were each obtained from a single membrane. *B*, Maximum rate of change of membrane conductance after a sudden displacement of the membrane potential by 100 mV (voltage clamp), as a function of the concentration of alamethicin. Membrane in 0–1 *M* sodium chloride, both sides, membrane solution as in Fig. 1*B*, temperature, 35° C. The slope of the straight line again indicates a sixth power function. Each point is an average of ten values obtained from ten different membranes. The bars give the total scatter.[123]

by study of the kinetics of adsorption of the modifying agent on BLM as a function of lipid composition and salt concentration in the aqueous phase.[417]

6. When a carrier-inducing agent, such as valinomycin, is added to a solution bathing BLM modified with a channel-forming agent, separate conductive pathways are formed. This can be easily demonstrated by varying the temperature. Thus, for example, when nystatin and valinomycin are present simultaneously, by varying temperature only anionic conductance is influenced significantly.

7. As expected, the channels in these modified membranes have a sieving effect; thus, for hydrophilic nonelectrolytes the selectivity sequence is based on the size of the solute. In fact, as shown by the data presented below, in BLM modified with amphotericin B, the permeability coefficients for nonelectrolytes increase in inverse relationship to solute size, and the rate of water flow during

osmosis increases about thirtyfold. The values of permeability coefficients (P) and reflection coefficients (σ) for solutes of effective hydrodynamic radius (r) as given below were determined for BLM modified with 1 μM amphotericin B.[429]

Substance	$r(\text{Å})$	$P(cm\text{-}sec^{-1}) \times 10^{-4}$	σ
Glucose, sucrose, raffinose	4.0	0.09 to 0.14	1
Arabinose	3.8	0.53 ± 0.15	0.7
Ribose	3.6	0.61 ± 0.16	0.62
Glycerol	3.1	3.28 ± 1.12	0.43
Acetamide	2.5	5.48 ± 1.42	0.15
Urea	1.8	10.4 ± 0.9	0.08
Water		300 ± 10	0.00

The effective pore radius calculated from this data is in the range of 7 to 10.5 Å; a corresponding value for membranes modified with nystatin has been calculated to a 8 Å.[430] If such pores are considered to be cylinders filled with electrolyte solution, it is very unlikely that they will have a conductance less than the lower limit for detection of single channel conductance (2×10^{13} ohm^{-1}). The fact that conductance changes greater than this are not detected indicates that if pores exist in these polyene-modified membranes they remain open for less than 100 msec.

8. In monolayers containing cholesterol, the polyene antibiotics increase the collapse pressure by about 15 dyne-cm^{-1}; also the potency of these antibiotics is correlated with their capability to reduce the surface pressure of monolayers.[409]

9. Filipin and estruscomycin cause the release of sequestered solutes from liposomes.[420] The effect is the same whether liposomes are preapred from lipid solutions containing cholesterol or not. Nystatin and amphotericin are far more effective when cholesterol is preincorporated; the cholesterol-lecithin ratio of 1:15 seems to be optimum for the release of various markers. However, the difference in these results could be due to the difference in the concentration of the antibiotics used in these experiments.[419]

10. The effect of polyene antibiotics on cell membranes can be explained in terms of their altered permeability properties, which leads to the loss of cytoplasmic constituents and ultimately culminates in the death of the organism.[421-427] However, not all the organisms and organelles show sensitivity to these agents; their efficacy can be correlated to their binding capacity to membranes which, in turn, can be attributed to their interaction with specific membrane components, like cholesterol, and the physical state of the component. As this component has been identified as cholesterol or cholesterol-lecithin complex,[431] it implies that systems containing no cholesterol will not show any affinity for these antibiotics; indeed, it has been found to be so. Insensitive bacteria or organelles do not contain cholesterol;[428] the lytic effects of these antibiotics can be reduced by addition of cholesterol to the culture medium, although, in some cases, inhibition of growth and lysis by the anti-

biotic is observed only with the cells cultured under conditions such that external cholesterol is incorporated into the membrane.

11. Further indications of interaction of cholesterol with polyene antibiotics come from the following observations. Digitonin, a sterol-specific complexing agent, has an action similar to that of polyene antibiotics on protozoa and fungi. In liposomes prepared from lipids containing 50% cholesterol and a small amount of filipin, electron microscopy has revealed interesting substructures in the bimolecular leaflets; electron micrographs show the presence of pits and rings, which resemble in form, but not in dimensions, the circular structures observed in the electron micrographs of lipid structures modified by the addition of a saponin like digitonin.[409,432] In contrast, inactive polyenes do not produce pits. Moreover, when digitonin is added along with polyenes, it markedly inhibits the polyene uptake by intact cells or isolated membrane fragments.

12. It is particularly instructive to note that the behavior of the channels formed by incorporation of various agents into BLM can be altered by external agents or stimulii and cofactors as summarized below:

Agent	Function altered	Stimulus or Cofactor
Dipicrylamine (State I) and several other ions which are soluble in organic solvents[359]	Nondielectric capacitance arises which is strongly voltage dependent; electrokinetic effects which may be due to reciprocal enrichment and depletion at interfaces	Electric field
Dipicrylamine (State II) (forms channels or bridge to translocate K and/or H ions) and TFB in the presence of Cu^{2+} [359]	Conductance which shows symmetrical differential negative characteristics	Electric field
EIM[111,410,411]	Shows various electrokinetic phenomena	Electric field
Alamethicin and monazomycin[123,414]	Same effects as for EIM	Electric field
EIM or alamethicin with protamine[111]	Shows various complex electrokinetic phenomena	Electric field and chemical agents (drugs and pheromones)
Streptolysin O[359]	Oxidation reduction reactions may be mediated across the membrane	Redox gradients
Tyrocidin B[416,417]	Selectivity ratio for Na/K changes	Amines in aqueous solution
AChase[433,1100]	Permeability change (transient)	Acetylcholine
Antigen[434,435,1113]	Permeability change (transient)	Antibody
Bovine spleen proteolipid[1299]	Permeability change	Noradrenaline and related drugs

Such changes induced by various agents give rise to more complex phenomena at the macroscopic level; these include non-Ohmic $I-V$ curves, rectification of applied pulse, generation of action potential and surface-catalyzed reactions regulated by asymmetric environments.

As diverse it may be, the experimental evidence as summarized in the preceding paragraphs suggests that the BLM modified with various agents (Table 6-9) have discrete channels. The nature of the environment in these channels, their exact shape and size, and the nature of interactions stabilizing them is a matter of speculation. Some of the experimental data discussed below may lead to development of such information but generalizations are not warranted at this stage.

The experimental data described above is compatible with the hypothesis that, in the interaction of membrane-bound sterols with polyene antibiotics, there is stereospecific hydrogen bonding between the hydrogens of 3-OH groups of appropriate sterols and indeterminate sites on the polyene antibiotics. Thus, either removal of cholesterol or its substitution by either cholesterol palmitate, dihydrotachysterol, *epi*-cholesterol, or cholest-5-en-3-one inhibits incorporation of polyene antibiotics into BLM.[436] At relatively higher concentrations, the antibiotics may, however, interact with other lipids; the properties of membranes modified in this manner are significantly different than those of membranes containing cholesterol–polyene complexes. Similarly, modification of the functional groups of amphotericin or nystatin by esterification (methylation, acetylation, etc.) significantly changes the permeability characteristics for both electrolytes and nonelectrolytes.[431] Similarly, esterified, *per*hydrogenated, irradiated, or saponifiied filipin looses all its efficacy to be associated with cholesterol and consequently with membranes.[419]

The interaction of polyene antibiotics with membrane-bound lipids such as cholesterol is also consistent with the fact that cholesterol molecules have a considerable degree of freedom of rotation and reorientation,[436] a condition which would be necessary if the polyene antibiotics were to form channels by aggregation and by assuming a relatively rigid secondary and tertiary structure. According to this view, the polyene-sterol *unit* involved in channel (or pore) formation could be multimolecular.

The tendency of various lipids, lipoproteins, and other amphipathic molecules to aggregate is of relevance when considering the formation of channels across lipid membranes. Thus, studies of the solution properties of the tyrocidines,[437,438,1103] gramicidins,[1104-1106] and other cyclic polypeptides forming channels across BLM indicate that extensive aggragation occurs in aqueous solution. The CD spectra of tyrocidine A, B, and C are almost identical. Since the tyrocidines differ in several amino acids, yet readily form homo- and hetero-aggregates (polymers), it is reasonable to assume that the aggregation process is largely a function of conformation and of the resulting orientation of the side chains.[437,1103] The nature of the side chains does not seem to prevent aggregation. This may imply that some general factor, such as hydrophobicity of the side chains, is involved.[438] If it is so, then interaction of lipid with a modifying agent assumes an added significance—stabilization of a cylindrical conformation of the channel. The aggregation tendency of alamethicin is of particular interest in this respect.[439] The average molecular diameter of alamethicin (Fig. 6-13) in the plane of the ring is 25 Å, whereas the thickness is about 4.5 Å. Alamethicin forms a lipid-soluble complex with alkali metal ions over a wide range of pH, suggesting that, as in such complexes with other macrocyclic compounds (see p. 195), the ion is held inside the ring by ion-dipole interactions. However, such interactions are not accompanied by gross conformational changes as indicated by proton magnetic resonance studies in D_2O. The ionizable carboxyl group confers a negative electrophoretic mobility, which increases as the pH is raised from 5 to 10. The properties and the amphipathic nature of the molecule make alamethicin extremely surface active and therefore confers a tendency to aggregate in solution. In fact, under proper conditions, alamethicin can form an octamer. The c.m.c. at pH 8 in NaOH solution is 2.4 μM. The average area per molecule is about 200 Å^2 in monolayers at a surface pressure of 30 dyne-cm^{-1}, which is about one-third of the fully extended alamethicin molecule (about 600 Å^2). This would suggest that the molecules must be tilted or take up some folded conformation at the interface. Physical studies in general may therefore be interpreted to suggest that localized structural changes may be involved in the interaction of alamethicin with lipid bilayers.

It may also be noted that pore and carrier mechanisms are but limiting cases of a general mobile structure model. Thus, one may expect that certain agents may occupy a position intermediate between simple carriers and highly associative channel-forming macrocyclic compounds. Tyrocidin B[417] and gramacidin A[1229] appears to be one such case. Studies on BLM doped in tyrocidine B suggest that two molecules of it interact with the BLM. The results are markedly dependent on the polarity of the lipid used, indicating that both the charge profile of the bilayer and hydrophobicity of the core influence the rate of adsorption of the antibiotic. The charge profile can be changed by adsorption of Ca, on the one hand, or acidic glycolipids on the other. The results have been interpreted to suggest that the rate of the conductance increase is governed by two kinds of dimerizations associated with the polar surface and the hydrophobic phase. However, the species in the lipid phase is not freely dispersed as a solute as it is in the bulk aqueous phase. The rate-limiting step, which is field dependent, is *autocatalysis* of lipid phase dimers to form ion-conducting channels. Since this step involves mobilization and orientation of aggregated species, one would expect that nature of the hydrophobic chains which determines mobility would also affect the rate of aggregation; this is found to be true.

Thus far we have been concerned with the incorporation of macrocyclic compounds into a continuous lipid phase to form regions of discontinuity. These agents also introduce similar *channels* into biomembranes[440] as implied in their antibiotic action and also in an increase in permeability to certain ions and non-electrolytes. Qualitatively, similar results may be achieved by a variety of agents which induce the phenomenon commonly known as lysis.

Lysis. A variety of compounds interact with cellular membranes to form large holes or pores across them. Such a process in biological or plasma membranes results in indiscriminate leakage of the cytoplasmic constituents and, at times, in complete disruption of the membrane; the term lysis is used to designate the phenomenon. In fact, cell lysis as determined by the loss of turbidity or optical density of a cell suspension is the terminal stage of a series of events leading to membrane disorganization, disruption, and autolysis. By the time visible lysis is detectable, the alteration of cell permeability is quite nonselective and irreversible damage has occurred to the permeability barrier.

Lysis may have far reaching consequences. Thus, for example, lysins are responsible for limited life of cells and for the anaemia of cancer. Some of the relatively simple naturally occurring lysins are C_{18} fatty acids, oleci acid, ricinoleic acid, 10-hydroxydecanoic acid (the *cytolysin* from royal jelly of bees), vitamin A, medium-length polypeptides, etc. They are characterized by their similarity of action rather than similarity of structure, thus implying that no specific molecules act as receptors.

Permeability changes following exposure to lytic systems may range from selective effects on ion fluxes to gross damage to membrane structure. These effects may be due to any of the following factors.

1. Dissolution of lipids from the membrane by lipid solvents, such as high concentration of detergents, by the enzymatic hydrolysis of the lipids, or by some other similar specific process.

2. Induction of phase transitions in bilayer structures as exemplified by the action of a low concentration of surfactants or photodynamic dyes.

3. Removal of a specific membrane component by the action of antibodies or complements, bacterial lysins, or metal chelating agents.

4. Partial disruption of membrane structure by high-energy radiation such as ultraviolet and X-ray photons.[469]

Initiated by any of the steps described above the overall process of lysis takes place in several steps: adsorption of lysin $\longrightarrow$ ion permeation through the complex $\longrightarrow$ breakdown of the complex $\longrightarrow$ swelling of the organelle by the development of colloid osmotic pressure $\longrightarrow$ leakage of the macromolecules accompanied by extensive damage to the membrane. The development of the colloid osmotic pressure can also occur due to metabolic maltreatment, for example, by disturbing the ionic environment causing the cell to swell and become leaky and finally leak out the intracellular components.

The initial steps in the above scheme are more or less characteristic of each type of the agent. However, the basic physical process involved in the initial changes is probably adsorption of lysins into the membrane phase and attainment of surface pressures exceeding a critical collapse level. This may occur by any of the following four mechanisms. The lytic agent could (a) overcome polar adhesive forces between phospholipids; (b) disrupt apolar cohesive forces between hydrocarbon chains; (c) break cohesive forces imposed by water in the region of charged head groups; or (d) form a complex with one of the membrane components such that the complex can no longer be a part of membrane structure. Thus the mechanism by which various lysins interact with membranes depends upon the nature of both lysin and membrane itself.

Mechanisms by which various lysins act on biomembranes remain largely unclear because the identity of target molecules is unclear. Various physicochemical methods such as X-ray diffraction and electron microscopy, however do suggest rather conclusively that lytic agents act on the membrane.[432,441] Ultrastructural evidence indicates that these agents induce not only functional but morphological changes in model and biomembranes: beading, fraying, coalescence, and the appearance of net-like areas may be observed as lysis progresses. This evidence does not permit conclusions as to whether the new structures are due to composite structures of lipoprotein with lysins or due to lipid or to protein or to lipoprotein alone. A rather convincing argument may, however, be made from the study of the effect of certain pharmacologically active drugs on membranes. Several narcotics, tranquilizers, antihistamines, antiinflammatory nonsteroidal drugs and anaesthetics at low concentrations (10^{-8} to 10^{-3} M) protect cells against hypotonic or mechanical hemolysis,[1107-1109,1230] whereas at higher concentrations they lyse the cell.[442,443] Although the precise mechanism of the drug-induced membrane stabilization at low concentrations of these drugs is not known, it is possible that it is due to incorporation of these molecules into membrane structure which may result in membrane expansion[444,1298] and stabilization.[1230] The membrane expansion that occurs increases the surface area/volume ratio of the cell, and thus the cells with higher area/volume ratios should be associated with a lower osmotic fragility. In fact, a close correlation between membrane stabilization and narcotic action has been observed.[445-447] To measure quantitatively the membrane concentration of a local anaesthetic, Seeman et al. allowed chlorpromazine to interact with hemoglobin-free ghost membranes in concentrations known to be membrane stabilizing for both nerve fibers and erythrocytes. They found that the uptake of anaesthetics, under conditions of 50% stabilization, was 0.033 mole/l, in agreement with the value (0.03 mole/l) which is predicted by the Overton-Meyer theory. The Overton-Meyer lipid solubility hypothesis is formulated as:[448] "narcosis commenses when any chemically indifferent substance has attained a certain molar concentration in the lipids of the cell ... this concentration depends on the nature of the animal or cell but is independent of the narcotic."

Such a correlation of narcotic potency of a compound with its lipid solubility has been reasonably successful in explaining the behavior of compounds which lack specific chemical groupings. The corollary of this theory is that anaesthetics act by virtue of an ability to dissolve and cause disorder in nerve membranes. Thus a variety of unrelated substances such as alcohols, steroids, secondary and tertiary amines, and even inert gases are effective, and it is thought that their action is basically to impede the rearrangement of groups in the excitable membranes and thereby block the action potential (see Chapter 8). It is true that Pauling[449] and Miller[450] suggest that the aqueous sites present in the vicinity of pores across membranes are occluded by the formation of relatively stable hydrates of anaesthetics; thus a reduced permeability leads to blockage of the action potential (Chapter 8). It is very difficult to decide unambiguously between these alternatives. However, on the simplest model the Overton-Meyer theory would predict that general anaesthetics would increase cation permeability through the membrane, whereas the Pauling-Miller theory would predict decreased cation mobility. To confuse the situation, both of these effects have been observed in model systems,[150] which could be correlated with their activities in causing nerve fiber narcosis.[451] Organic solvents such as ether, chloroform, and n-alkanols (up to C_8) all increase potassium diffusion across liposomes. In contrast, local anaesthetics such as nupercaine, tetracaine, cocaine, and procaine at concentrations known to produce narcosis reduce potassium release, perhaps by reducing the zeta potential; that is, the surface charge density of the negatively charged membrane by specific adsorption of the ionic form of the narcotic.[452] However, both procaine and cocaine at higher concentrations show reversal of this trend even though the zeta potential continues to decrease;[451] this may be attributed to penetration by the un-ionized form of the local anaesthetic into BLM. In fact, these agents are known to be lytic at these concentrations. Even the argument that expansion of the membrane indicates adsorption of lysin into the membrane phase is not without its weakness. There are several possible alternative mechanisms which can explain drug-induced expansion: drugs may replace some membrane-associated component which keeps the membrane in the condensed state, for example Ca^{2+}; or the drugs may induce conformational changes in protein and/or alterations in water structure. Similarly, even though some anaesthetics can induce increased permeability for cations across liposomes and across BLM, it is yet to be established that the mechanism of permeability change induced by lysins across model and biomembranes is the same.

Liposomes have proved to be useful in exploring the mechanisms of interaction of various types of compounds leading to changes in labilization, stabilization, and the permeability properties.[145,151,451,452] The effect of various drugs on these systems can be closely correlated with their action on biomembranes, even to the extent where a very high degree of specificity of interaction is involved. Ethiocholanone and other 5-β-H steroids but not their 5-α-H isomers increase the overall permeability of liposomes. Progesterone isomers, deoxy-corticosterone, and diethylstibesterol also have similar actions but to a lesser extent. Etiocholanone-type steroids are known to induce fever in human males; females are less susceptible. A possible rationale for this effect may be found in the following observation. As little as 0.1% 17-β-estradiol (preincorporated) in liposomes render them refractory to release induced by other steroids.[453] The androgens do not possess this property. The stabilizing effect of 17-β-estradiol is also influenced by (in the order) 17-α-estradiol, estriol, and estrone. Anti-inflammatory steroids such as cortisone have been shown to protect isolated sub-cellular particles against injury by a variety of agents such as progesterone or etiocholanone.

As a corollary to the preceding argument it follows that polyvalent cationic species should be lytic since they can interact with negatively charged membrane interfaces and induce removal of one of the membrane components. Indeed, basic proteins such as poly-L-lysine, etc., are known to lyse red cells[454] and liposomes.[1110,1111] Similarly the neurotoxic effect of snake venom,[455] bee venom[119] and bacterial toxins such as streptolysin S or streptococcal α-toxin[151] or surfactin[1114] may involve similar actions. Most of these venom toxins, as well as the physiologically active kinins, kallidins, and bradykinins are basic proteins and their attachment to appropriate sites may thus involve charge-charge inter-actions as an initial step to alter the permeability characteristics of the mem-brane. However, in all probability, the action of these peptides at physiological concentrations occurs at specific target sites which may or may not change the permeability of the membrane. It is also possible that these toxins may have synergistic lytic effects similar to the accelerative effect of halogenated aromatic compounds on the lysis induced by saponins;[456] halogenated aromatics do not have any effect of their own. Since lysis does not seem to depend on ionic inter-actions between the cationic peptide and the membrane interface, and most of the lysins have an extraordinary affinity for a hydrophobic environment, it seems probable that the surface affinity of lysins or their associations with the core of lipid bilayers accounts for their activity.

The lytic effect of snake venom is due to phospholipase activity facilitated by the basic proteins present in the venom. Cleavage of phospholipids gives lysoleci-thins, which cannot be incorporated into a bimolecular lamellar structure. These results are substantiated by studies on the effect of lysolecithin on the electrical resistance of lecithin bilayers.[457] When increasing quantities of lysolecithin are added to the BLM-forming solution containing a fixed concentration of lipid, an appreciable decrease in the membrane resistance is observed. A further increase in the proportion of lysolecithin causes a more gradual decrease in resistance un-til it prevents the formation of BLM. Perhaps the resistance drop induced by lysolecithin is related to the prelytic stage, that is leakage of cations. For details see Ref. 1232. Similar leakage can be induced by n-alkanols,[458] various surfactants,[459,1112] and anaesthetics.[460] It may be noted that the nature of the hydrocarbon chains (length, unsaturation, branching, etc.) of lytic agents, as well as of membrane lipids, determines the ratio at which the alteration in permeability occurs. Interference with membrane-ribosome association by carcinogens[1182] appears to involve interactions which are somewhat more specific than those discussed so far.

The effect of amphipathic lytic agents on biomembranes has been extensively studied, is well characterized, and has been exploited commercially. The gram-negative bacteria are highly susceptible to the action of cationic surfactants, but are usually unaffected by anionic and zwitterionic surfactants, which, however, are active against gram-positive organisms. It has been shown that organic cations alone remove somatic antigen from the cell wall of gram-negative bacteria; thus cations together with EDTA remove other cell wall components and render the cell sensitive to the lytic action of zwitterionic surfactants.[461] The order of effectiveness of various surfactants in lysing protoplasts of gram-negative bacteria is:[462]

$$RNH_3^+ > R\overset{+}{N}Me_3 > RSO_4^- > RSO_3^-$$

This is consistent with an action depending on the type and strength of hydrogen bonding; hydrogen donors are more lytic than acceptors.[463] In contrast, the order of effectiveness of different surfactants for hemolysis is the same as for penetration of a cholesterol monolayer:[464]

$$RNH_3^+ > RSO_4^- > RSO_3^- > RCOO^- > RNMe_3^+$$

These studies also suggest lower hemolytic concentrations for cationic detergents relative to anionic detergents of the same chain length.[465] The surface tension of aqueous solutions of surfactants at lytic concentrations is 40–50 dyne/cm, characteristic of concentrations much below the c.m.c. Thus, the release of phospholipid is suggested to be caused by the electrostatic attraction rather than co-micellization. However, the interaction of polar groups with specific membrane components may also be important.

The mechanism of lysis of biomembranes has attracted considerable attention. As already indicated lysolecithin does not form stable membranes and has a lytic effect above a certain concentration. It is pertinent to note that in the series of monoacylphospholipids, members with short hydrocarbon chains are practically devoid of hemolytic activity, but with increasing chain length their potency increases so as to attain a maximum hemolytic activity for the compounds having a saturated C_{18} acyl chain. Further increase in chain length is accompanied by a decrease in lytic activity, and a compound with a C_{26} acyl chain is inactive against erythrocytes. Of the diacyl phosphoglycerides studied for their hemolytic activity, compounds containing either two C_9, C_{10}, or C_{11} chains are very potent in lysing erythrocytes. The hemolytic activity disappears when the chain length of the two fatty acid constituents is increased.[466] The lytic activity of the lecithins with medium fatty acid chain lengths are of a much higher magnitude than those of their corresponding monoacyl derivatives. Similar structure-activity correlations have been observed with other amphipaths acting as lysins. These observations imply that the lytic effect of these various amphipaths is an interfacial phenomenon involving the interaction of surfactants with the lipids of the surface membrane. On these grounds, attempts have been made to correlate the hemolytic activity of synthetic surfactants with their interfacial tension characteristics, with their ability to penetrate lipid monolayers, and with their critical micellar concentration, which in turn, depend on the nature of both hydrocarbon chain and polar group (Chapter 2).

The ionic surfactants show lytic effect in two stages (Fig. 6-15): an initial exponential rise followed by a sigmoidal increase in the lytic activity. The accelerators like halogenated aromatics do not affect the initial rise but do affect the subsequent progress of lysis.[467] With anionic detergents the rapid second phase can be eliminated by repeated washing of the cells in physiological saline. Lecithin is lost under these conditions, and the cells so treated are rapidly lysed when lecithin is added. The two phases in the progress of lysis probably reflect complexing with free and bound phospholipid, respectively. Saponins do not show these two stages of activity, perhaps reflecting the state of cholesterol in biomembranes with which saponins appear to interact specifically.

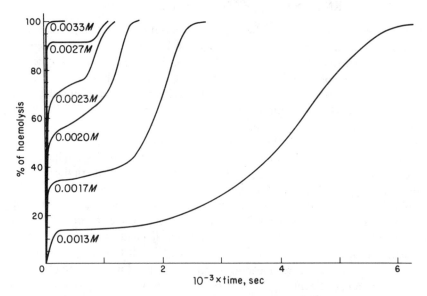

Fig. 6-15. Percentage haemolysis-time curves for lysis by various concentrations of sodium decyl sulphate at 23° C and pH 7.2 for a system containing $1 \cdot 6 \times 10^8$ cells in 3 ml.[467]

The lysins which we have described thus far do not show a high degree of target specificity; exhibition of high target specificity is, however, commonplace. The direct action of antibody and complement on a membrane produces holes,[465,468] which may be a first step in lysis. In fact, such an 'immune response' has been induced in BLM[299,435] and liposomes.[470,1113,1231] Probably, the most direct evidence for the interaction of lytic agents with specific membrane components is obtained by the treatment of red blood cells with 1-fluoro-2,4-dinirobenzene, which makes them permeable to sodium and potassium ions but not to small water-soluble molecules or to hemoglobin. Such treated cells eventually lyse in isotonic buffer. In contrast, cells similarly treated with 1,5-difluoro-2,4-dinitrobenzene do not lyse even though they do become permeable to alkali cations.[471] These reagents are known to react readily by displacement of fluorine with free amino, sulfhydryl, tyrosyl, and histidyl groups to form stable dinitrophenol derivatives. The monofluoro compound reacts with one such group, whereas the difluoro derivative reacts with two groups, provided they are about 5 Å apart, to form a dinitrophenylene crosslink. Thus the modification of erythrocyte membranes with a bifunctional crosslinking reagent makes the membrane strong enough to withstand a colloid osmotic pressure of about 0.6 atm due to hemoglobin and its counterions; the second disruptive phase is thus eliminated.

By way of recapitulation, interaction of a lysin with the membranes is the crucial step involved in the overall process; what follows after induction of permeability to small solutes across these membranes is generally a function of altered osmotic balance. The evidence at hand suggests that the initial site of interaction of most of the lysins with the membrane may be the lipid core; although polar groups and other specific membrane components can not be ruled out. However, when the amount of lysin exceeds a critical concentration, disorganization of the continuous lipid layer structure takes place, presumably by a bilayer ⇌ micelle type of phase transition. Such a rearrangement of membrane structure with or without the lysin participating as a structural unit, opens a few polar pathways across the barrier. Information about the state of lysins in the membrane phase may be obtained by physicochemical methods, especially PMR and ESR spectroscopy. Thus the nuclear magnetic relaxation rate of benzyl alcohol, anaesthetics, etc., is increased in the presence of erythrocyte membranes, myelin, synaptosomes, unmyelinated nerve, and rabbit *vagus*.[472] The results suggest that the structural changes in membranes are caused by these molecules. Similarly ESR studies have given quite detailed information about the number of binding sites, their location in the membrane, and the tightness of the binding.[473] Such studies reveal an irreversible structural breakdown of the membrane above a certain concentration of anaesthetic, even though no change is observed by electron microscopy. These observations are quite general and are common to all the membranes investigated, although their structures and functions are quite different.

FACILITATED TRANSPORT ACROSS BIOMEMBRANES

Catabolism of exogenous substrates requires transport of the metabolized molecule through the plasma membrane. Moreover, biochemical investigations of cellular processes suggest transport to be prerequisite of metabolism, both in lower and higher organisms. For example, some mutants of bacteria lose their ability to grow on certain substrates (Fig. 6-16) although the enzymes of the catabolic sequences are active inside the cell or in cellular homogenates. Further study reveals that this crypticity often results from a transport defect, which may be under genetic control. In addition, it has been recognized that transport systems are subject to induction comparable to the induction of enzyme proteins. These alleged transport proteins (rather transport systems) have varyingly been termed "permease"* or referred to by the non-committal term "membrane protein" (M-protein) or transport protein; the phenomenon of transport as catalyzed by these systems is best described as *facilitated transport*.[474-485,1115,1116] In its simplest form, a carrier protein allows the substrate to overcome the high energy of activation needed for the penetration of the lipid membrane, and the carrier may be regarded as a substance with a binding site that presents itself alternately as the two interfaces.

Facilitated transport occurs in a wide range of cell types, although only a relatively smaller range of metabolites are transported by such a mechanism (Table 6-10). Sugars, amino acids, purines, cations, and anions of a wide variety, vitamins, and hormones are some of the common type of permeants which cross the cellular membrane aided by some transport system. It may also be noted that such a transport mechanism seems to be involved in bacteria, yeast, red cells, brain, etc.—a whole range of organisms and tissues. The methods used for such studies vary considerably from one organism or tissue to another; however, tracer and osmotic flux experiments and volume distribution techniques aided by suitable analysis of the kinetic data have given considerable evidence supporting involvement of a facilitated transport system. Experimental studies suggest that a facilitated transport system may operate under the electrochemical

---

*The suffix *-ase* is usually reserved for catalytic proteins—the enzymes. Even though these transport proteins (or systems) do not catalyze a chemical reaction, nevertheless they have many characteristics in common with those that do. They facilitate transfer of polar substrates through a hydrophobic barrier. The lowering of the activation energy is quite apparent when the carrier-catalyzed process is compared with the energy that would be required to dehydrate an ion per se and bring it into the nonaqueous phase. Furthermore, the *carrier* molecules need be present in only very small numbers and are used over and over again without being consumed. In fact, the turnover number of quite a few carriers are within the range of those for enzymic reactions, and it is expected to depend on the same physicochemical principles as those governing dissociation equilibria and rates in enzymic processes. Furthermore, these carriers have a very definitive active site, the conformation of which is governed by a variety of factors similar to those governing the behavior of catalytic site on enzyme.

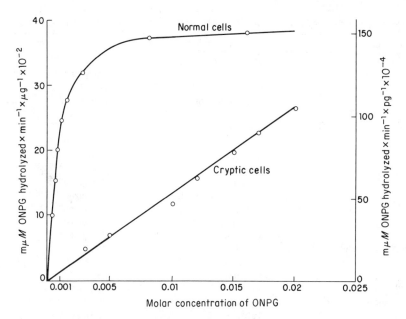

Fig. 6-16. The influx of o-phenyl-β-D-galactoside (ONPG) in two types of *E. coli* as measured by the rate of ONPG hydrolysis. Upper curve: normal type (left ordinate). Lower curve: cryptic (without permease) mutant (right ordinate).[474]

gradient of the substrate itself (*passive* facilitated transport), or it may be energized (the so called *active* transport). A discussion of both of these transport systems follows.

Passive Facilitated Transport across Biomembranes

The characteristics of these transport systems are summarized below[243,475,477,484,1115,1116] (see also Table 6-10). Although these characteristics are ascribed to passive facilitated transport, the behavior of all facilitated-transport systems is expected to be similar. The distinguishing features rest on the energy-coupling mechanism.

1. The transport systems show high substrate specificity, both with respect to the organism and with respect to the molecular structure of the permeant.

2. The transport is driven only by the electrochemical gradient of the substrate although Fick's law is not obeyed (Fig. 6-16). A nonlinear relation is observed between the rate and concentration of the permeant, and saturation kinetics is observed at high concentrations of the substrate.* Also, the rate of

---

*In the long run the maintenance of the structure of the membrane and the structure of the transport system itself may require the expenditure of metabolic energy, but this cannot be linked in time or in stoichiometry to the kinetics of any particular permeation process.

TABLE 6-10 THE FACILITATED TRANSPORT SYSTEMS IN BIOMEMBRANES

Organism	Substrate	Mode
Escherichia coli	Magnesium	Energized[486]
	Sulfate	Energized[487]
	Galactoside	Energized[488]
	Glucose-6-phosphate	Energized[489,490]
	L-arabinose	Energized[489,491]
	Maltose	Energized[492]
	Lactose	Energized[493]
	Glucuronide	Energized[494]
	Galactose	Energized[495]
	Leucine	Energized[495]
	Neutral amino acids	Energized[496-498]
	Arginine	Energized[499,500]
	Cystine	Energized[500]
	Glutamine	Energized[500]
Red blood cell[478,480]	L-leucine	Passive[501]
	Glucose	Passive[502,503]
	Purines	Passive[504]
	Choline	Passive[505]
Hymenolepsis diminuta	Acetate	Passive[506]
	Palmitate	Passive[507]
	L-methionine	Passive[508]
	D-methionine and other amino acids	Passive[508]
Kidney	Amino acids, sugars, ions	Passive and active[509-513]
Brain	Amino acids, sugars, transmitters, vitamins	Passive and active[513-516]
Muscle	Amino acids, purines, sugars, etc.	Passive and active[513,517]
Uterus	Estrogen	? (probably active)[518]

exchange of substrate in carrier-mediated transport system is usually higher (by several orders of magnitude) than the rate of net flux in either direction.

3. The temperature coefficient for most of the facilitated transport system is about ($Q_{10} \simeq 2$) for the temperature range 27-37°. This value is well above that for simple aqueous diffusion and suggests, but by no means proves, that transport involves the formation and breakage of chemical bonds.

4. It is implicit in the process of specific substrate–carrier interaction that certain chemically related species compete with the permeant species. In fact, competitive inhibition by compounds related to the permeant has been observed. Also a variety of *protein reagents,* such as thiol blocking agents, may cause irreversible blockage of transport processes.

5. A mediated transport process can, to a first approximation, be described quantitatively by Michaelis-Menten kinetics; that is, its behavior is predicted by the law of mass action. The symbols characterizing the two constants of the

Michaelis-Menten relationship for enzymic processes, V_{max} and K_m, have also been usefully adopted to characterize transport process (Fig. 6-17). The term V_{max} expresses the maximum flux that the cell studied demonstrate towards a particular permeant. Thus, V_{max} depends, under standard physical conditions, only on the number of cells present, the cell type, the number of transport sites on each cell, and the class of permeant used. K_m is the second constant which may in certain circumstances express a type of "affinity" of permeant for the transport system, but formally expresses merely the substrate concentration at which the flux is exactly one-half the limiting flux, V_{max}. Thus K_m is independent of the number of cells present but characterizes each cell type/permeant pair. Further discussion of this point is given later.

6. The transport systems show a very high degree of functional specialization; that is, they are distinct from metabolic enzymes involved in the utilization of the permeant.

7. The transport systems are located in membranes and their biosynthesis is genetically controlled.[1233] Thus a single genetic mutation leads to the loss of the capacity to transport all substrates of a given transport system without affecting any other transport system or in some cases the metabolism of the cell. Furthermore, a transport system can only be induced by one of its substrates and not by a substrate of a different transport system. In fact, transport deficient cells can be prepared by growing cells in a medium lacking the substrate and/or the one containing a suitable analog.

8. Since the transport of a substrate across the biomembrane may be considered the first step in its metabolism, it would seem reasonable to expect that transport processes are subject to regulatory control mechanisms. Reports have appeared demonstrating that various permeation systems may be regulated by induction, feed-back inhibition, catabolite repression, and genetic regulation of both protein and lipid components involved in transport system. Other mechanisms of control, such as rapid turnover of a transport component, are also observed.[519] Genetic control and regulation of transport function is not re-

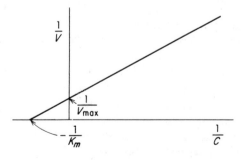

Fig. 6-17. A typical plot used for determining the transport constant V_{max} and K_t.

stricted to lower organisms. Quite a few transport defects have been implicated in inherited catabolic disorders, such as, hyperprolinemia, iminoglycineuria, and a host of others, including those listed in Table 1-2.[556] The fact that most of these defects are inherited as autosomal recessive traits suggests that clinically affected patients have a double dose of the mutant gene or are homozygous for the mutant gene.

Thus, there is a large body of experimental evidence suggesting that facilitated transport systems in biomembranes show substrate specificity, genetic specificity, as well as specificity of induction. One of the most commonly employed models for the interpretation of the data summarized in the preceding paragraphs is based on a mobile carrier. It is assumed that the permeant (S) combines with a specific carrier (C) located in the membrane in limited amounts. Thus, the overall process of-transport may be formulated as follows:

$$\text{Outside} \quad | \quad \text{Membrane} \quad | \quad \text{Inside}$$

$$S_o \; + \; C \; \underset{k_2}{\overset{k_1}{\rightleftharpoons}} \; [SC] \; \underset{k_4}{\overset{k_3}{\rightleftharpoons}} \; C \; + \; S_i$$

In most of the quantitative treatments of this model the following assumptions are made:

1. The carrier and substrate bind reversibly.

2. The free and complexed carrier are confined to the membrane phase, but are accessible at both interfaces.

3. The rate of transfer of the free and complexed carrier between the two interfaces of the membrane is independent of direction.

4. The rate of transfer of carrier across the membrane is very much greater than the rate of diffusion of substrate through the membrane phase.

5. The rate of transfer of the carrier-substrate complex is much smaller than the rate of formation and dissociation of the complex. In some cases the rate of dissociation of the complex at the inner interface is considered to be rate limiting.

6. The affinity of the substrate for the carrier is the same at both interfaces.

7. The amount of carrier in the membrane remains constant; that is, it is neither being synthesized nor destroyed.

8. The rate at which the carrier and complexes cross the membrane is proportional to the difference of their concentrations at the two interfaces.

Even within the framework of these assumptions and the model to which they apply, several variants of specific steps are possible and some of them may have physiological significance. A variety of detailed mechanisms have in fact been suggested for the specific steps involved in the *translocation* of complex and carrier from one interface to the other. Basically, most of the mechanisms require the carrier to be available at both the interfaces; this may be achieved either by rotational (diffusion) or conformation changes. Thus if we further assume that the rate-limiting step in the transport process, as visualized above, is the rate of carrier migration (a reversible first-order process with respect to the concentration of the complex), and that the rate of back and forth movement of the carrier and the complex is the same, then the rates of movement of the carrier and/or the complex will depend upon thermal agitation and simple diffusion, and therefore upon their respective concentration gradients. Thus, the net rate of transport, V, of a substrate from inside (subscript i) to outside (subscript o) may be written:

$$V = D[(CS_i) - (CS_o)]$$

where D is the diffusion coefficient of the complex within the membrane. In practice, (CS_i) and (CS_o) cannot usually be determined directly and it is necessary to determine their concentration in terms of experimentally measurable quantities. In fact, the concentration of the complex can be related to the concentration of substrate and of complex by the law of mass action:

$$K = \frac{(C)\,(S)}{(CS)}$$

Then the proportion of total carrier present as CS at each interface is given by:

$$\frac{(S)_o}{(S_o) + K_o} \quad \text{and} \quad \frac{(S)_i}{(S)_i + K_i}$$

If $K_i = K_o = K_m$ the net rate of transfer of the substrate is:

$$V = D[(C) + (CS)] \left[\frac{S_o}{S_o + K_m} - \frac{S_i}{S_i + K_m} \right] \tag{6-9}$$

The product $D[(C) + (CS)]$ can be replaced by a constant V_{max}, that is, the maximum rate of transport when $S_i = 0$. Thus the rate of transport under the conditions of unidirectional flux is

$$V = V_{max} S_o/(S_o + K_m) \tag{6-10}$$

The term V_{max} expresses the maximum rate at which permeant can be transported when the concentration of a given substrate in the medium is saturating, that is to say, beyond a certain substrate concentration the rate cannot be increased by further increasing the amount of substrate. K_m is numerically equal to the substrate concentration when $V = V_{max}/2$. Furthermore, the assumption that the transmembrane movement of the complex is rate limiting provides the basis for equating the Michaelis-Menten constant to the mass law dissociation constant. Hence, the quantity $1/K_m$ may be taken as an approximate measure of the affinity of the substrate for the carrier. Similarly, the relative affinities of a series of substrates which combine with the same carrier can be expressed in terms of their K_m values.

TABLE 6-11 CHARACTERISTICS OF SUGAR TRANSPORT SYSTEM IN RED BLOOD CELLS

Characteristic	Experimental observation
(a) Substrate specificity	Aldoses: pyranosides in chair (CI) conformation
(b) Rate of uptake (at 0.03 mM)	Glucose > mannose > galactose > arabinose
Rate of uptake (at 1.5 mM)	Sequence is exactly reversed
(c) Competitive inhibition sequence	
Monosaccharides	Glucose, mannose > galactose, xylose, arabinose > sorbose > fructose
Disaccharides	Maltose > cellebiose > isomaltose; trehlose, lactose, sucrose have no effect up to 83 mM
Based on mol. wt.	Glucose > maltose > maltotriose dextran
(d) Counterflow: exchangeable sugars	D-galactose, D-mannose, L-arabinose, D-xylose, D-lyxose, 2-deoxy-D-glucose.
Nonexchangeable sugars	D-fructose, L-sorbose, D-arabinose, D-ribose, L-fucose, L-rhamnose

(e) Kinetic parameters
Transport substrates

Sugar	K_m (mM)	V_{max} (mM/min/ cell unit)
Glucose	4–10	100–500
Galactose	20–100	600–700
Mannose	13–14	300–600
Xylose	50–70	600
Ribose	2000	500–700
L-arabinose	220–250	700
Arabinose	5500	620
L-sorbose	3100	124
Fructose	9300	124

Inhibitors (K_i), mM

Phloretin polyphosphate	1.1
Maltose	44000
Phloretin	5
Phlorizin	100

(f) Rate of Glucose metabolism	About 1/250th of the transport rate (V_{max})
Rate of net uptake	About 1/50 to 1/200th of exchange rate
(g) Energy of dissociation of carrier-substrate complex	10 kcal/mole
Activation energy	20 kcal/mole at $20°$; 7–10 kcal/mole at $37°$
Effect of pH on transport	Bell-shaped pH = rate profile
(h) Inhibitory protein reagents	Most of the thiol blocking agents, iodine, tannic acid, dinitrofluorobenzene (effect accelerated by glucose; half-maximal concentration for this activity corresponds to its K_t)
(i) Temperature dependence	K_m decreases with temperature; at $45°$, H, 6 kcal/mole and E, 22 kcal/mole

(a) From Refs. 521–523; (b) and (c) from Ref. 524; (d) from Ref. 525; (e) trom Rets. 243, 503; (f) from Ref. 503; (g) from Ref. 526; (h) from Refs. 523, 527, 1117; (i) from Ref. 541.

Equation (6-10) is comparable to the Langmuir adsorption isotherm or to the Michaelis-Menten equation for enzyme kinetics. Thus, it is obvious that the kinetic methods used for characterization of enzyme reactions must be useful to elaborate the mode of operation of simple carrier-mediated transport. It should, however, be noted that a kinetic analysis simply provides a mathematical description of the process under consideration, and while we may discard a mechanism which fails to correspond to this description, it does not follow that one which does is necessarily correct since others of the same kinetic form might be found. Nevertheless, the more varied the kinetic evidence the more stringent the mathematical requirements and the fewer the mechanistic possibilities. With this note of caution we now examine some of the experimental results.

As collected in Table 6-10 (the list is by no means complete) many organisms and tissues show carrier-mediated transport of specific solutes. However, only a few of these systems have been examined in detail. Probably the best characterized among these are the glucose transport system of erythrocytes (Table 6-11), sugar transport system of yeast,[520] and various transport systems inducible in *E. coli* (Table 6-10). Even in these diverse organisms (for others see Refs. 528–530) the facilitated transport systems show more or less similar characteristics, as shown in Table 6-11 for the sugar transport system in human erythrocyte. The most important feature of the kinetic analysis arises from the fact that the value of V_{max} is unchanged at 600 μM/min/cell unit for most of the sugars which are transported. This is just what the model of carrier-mediated transport predicts, since although the sugars may have different affinities for the carriers, their maximum rate of transport is limited in all cases by the velocity of movement of carriers themselves. This view is substantiated by exchange diffusion (or counter transport) studies as discussed below.

In the following subsections we shall discuss some of the general features of the facilitated transport systems in biomembranes.

Countertransport (Exchange Diffusion, Counterflow, Transflow, Substrate-Facilitated Transport)

All these terms designate a special case of interaction between simultaneous movements of two different substrates or two different forms of the same substance which have a comparable affinity for the carrier. Thus, in a carrier-mediated transport system, counterflow is a consequence of asymmetric competition for carrier sites. If one substrate is present at equal concentrations at both sides of the membrane, the carriers are equally distributed throughout the membrane. No concentration gradients exist and hence no net flow occurs. The addition of a second substrate reduces the number of sites available for the first one. If the second substrate is present at different concentrations on the two sides of the membrane, concentration gradients for the loaded carriers are established which lead to uphill movements of the previously equally distributed

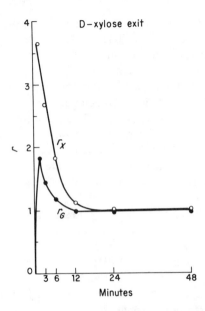

Fig. 6-18. Countertransport of D-glucose (originally at equilibrium) induced by exit of D-xylose from human red cells at 24°. The movement of glucose out of equilibrium against the gradient accompanies the decreasing gradient of xylose.[534]

substrate. However, the interaction is not limited to this special case but is also observed when the first substrate moves with a finite rate in the same direction as the second substrate; in this case the interaction leads to slowing of the movement of the second substrate.

The countertransport mechanism was suggested as a corollary of the mobile carrier model.[531] Experimentally, it has been demonstrated for a number of carrier-mediated transport systems: 3-O-methyl D-glucose and L-arabinose by D-glucose in perfused rat heart;[532] sorbose or galactose by glucose in yeast;[520] labelled β-thio-O-methylgalactoside by unlabelled β-TMG in E. coli;[533] galactose by glucose[526] and xylose by glucose[534] in human red blood cell (Fig. 6-18); xylose by galactose in rabbit r.b.c;[535] amino acids by sugars in rabbit ileum.[536] The countertransport mechanism also operates for Ca extrusion against a sodium gradient in mammalian cardiac muscle[537] and the squid axon.[538,1118]

The quantitative data from countertransport experiments is important since it can give information regarding relative mobilities of loaded and unloaded carriers.* Recent studies on facilitated transport of sugar across the red cell mem-

_____

*It may be noted that the foregoing arguments for characterizing countertransport or "exchange diffusion" are valid only when the carrier mediates one-for-one exchange, and the movement of individual solute molecules occurs independently of one another. This follows from the assumption which we made earlier, that the ratio of the fluxes of solute in opposite directions must be equal to the ratio of the electrochemical activities of the solute on each side.[539]

brane suggest that the true unidirectional flux is so enhanced by the trans concentration that the mobility of the unloaded carrier must be at least three times slower than that for the loaded carrier.[540] This implies a primary stimulation of the unidirectional forward flux by the trans solute. This would also mean that the rate constants based on the assumption of equal mobility of both loaded and unloaded carrier have to be reevaluated.

Biochemical Characterization of the Transport System

Early work on the pleiotropic effect of certain mutants indicated that facilitated transport is mediated by the cell membrane. Also certain mutants have been shown to possess the metabolic apparatus necessary for substrate utilization once it entered the cells yet the substrate does not permit appreciable growth in these mutants. Moreover, preparations of cell ghosts obtained by removal of all (at least most) cytoplasm under suitable conditions show transport activity. Spheroplasts prepared by treatment of cells with lysozyme and EDTA are capable of sustaining the transport system originally present in intact cells. For example, spheroplasts prepared from *E. coli* having a proline transport system are capable of taking up proline. In contrast, spheroplasts prepared from bacteria lacking in the proline transport system are incapable of concentrating proline.[496] Such preparations appear in electron micrographs as sacs of diameter varying from 0.1 to 1.5 μ. These sacs are usually surrounded by one to four membrane layers, each 65–70 Å thick. Chemical analysis of these preparations shows that they contain no cytoplasmic structures. Similarly, by using immunological[498] and autoradiographic methods[542] the substrate-binding components have been shown to be localized only in the cell envelope and not in the cytoplasm.

Significant progress in biochemical characterization of facilitated transport systems has been made through the observation that a protein component having a significant affinity for a specific permeant which is transported by the membrane system can be isolated (for brief reviews see Refs. 543–545). These proteins can be released without loss of cell's viability and no loss of osmotic stability by subjecting bacteria to a moderately severe osmotic shock. The method is the following:[545] Cells are harvested in the midexponential phase of growth and washed several times with 0.03 M *Tris* buffer (pH 7.0). The pellet of cells is suspended in 80 parts of 0.5 M sucrose containing 0.033 M Tris-HCl, pH 7.2, and 1×10^{-4} M EDTA. The mixture is gently agitated for 10 min and centrifuged, after which the supernatant solution is removed. The well-drained pellet of cells is now rapidly dispersed, by vigorous shaking, in 80 parts of cold 5×10^{-4} M MgCl$_2$ solution. Once more the mixture is gently stirred and centrifuged and the supernatant solution, called the shock fluid, is removed. This fluid contains, among other proteins, the *binding proteins*. The binding proteins from the cells of higher organisms are generally prepared by cell homogenization and other conventional techniques.[546]

TABLE 6-12 SOME PROPERTIES OF PURIFIED TRANSPORT PROTEINS

Substrate	Source	No. of Sites	Mol. Wt.	$K_{dis}(\mu M)$	$K_{trans}(\mu M)$	Ref.
Sulfate	S. typhimurium	1	32000	0.1	4	487
	E. coli	1		30		
β-Galactoside	E. coli	1	31000		60	547
D-galactose	E. coli	1	35000	1	4	495
	S. cerevisiae			600	3000	548
D-glucose	S. cerevisiae			17000	6000	549
	Human r.b.c.	1		18000	8–12000	550
	Human r.b.c.		45000	1000		502
L-Arabinose	E. coli	1	35000	2		491
Estrogen	Rat and rabbit uterus	1	22000	5.8×10^{-3}		518
Ca	Chick and rat duodenum	1	28000	26		551
L-leucine	E. coli	1	36000	1	1	495
	Human r.b.c.				1800	501
Cystine	E. coli		28000	0.3	1	500
Arginine	E. coli		30000	1.5		500
Glutamine	E. coli		29000	0.15	0.06	500
Histidine	S. typhimurium	1	25000	1.5	1.5	552
Phosphate	E. coli	1	42000	0.8		553
Glutamic acid	E. coli	1		6.7		1234
Phenylalanine	Comamonas spp.	1	24000	0.1		1235

As shown in Table 6-12 quite a few binding proteins from several transport systems have been characterized. There is a considerable body of evidence which suggests involvement of these proteins, most probably along with other components, in the facilitated transport of various substrates. The following points of similarity between binding proteins and transport systems may be noted:

1. The K_t for transport and K_{dis} for binding of the substrate by the binding protein are usually within an order of magnitude of each other, although K_t is generally greater than K_d.

2. Binding protein can only be isolated from cells which show a functional transport system. Repressed cells do not contain either the transport system or the binding protein.

3. Binding protein and transport systems are lost from cells containing an intact transport system to similar extents upon osmotic shock. The shocked cells do not contain any BP, nor do they show facilitated transport.

4. Both can be simultaneously lost by mutation and regained on reversion.

5. Reversible inhibitors such as substrate analogs and protein reagents that block the binding of the substrate irreversibly also block transport into cells.

6. Osmotically shocked cells which have lost a major fraction of their transport activity restore their transport activity by the addition of the concentrated crude

fluid released by osmotic shock; increases are noted both in rate and extent of transport.[495,500]

Before proceeding with the discussion of the role played by binding protein in transport process it may be pertinent to elaborate on some of the features of binding proteins as a group. Binding proteins have an average molecular weight of 30000, and are usually globular in structure with hydrophilic exterior. They show little affinity for lipids or lipid solvents, and are fairly soluble in water and aqueous electroltyic solutions. They show little tendency for aggregation, whether free or bound to their substrates. They form stoichiometric complexes with their respective substrates, usually 1:1; the dissociation constant (K_d) of the complex is usually in a fairly narrow range for different binding proteins from the same organism. Binding of substrates usually shows a fairly flat pH–rate profile, thus implying the involvement of dissociable groups in binding is minimal. The binding proteins do not catalyze any of the more common metabolic reactions, although after binding of the substrate conformational change may be detected.[1124] In fact, the galactose-binding protein of *E. coli* has been shown to exist in two conformational states, only one of which undergoes a conformational change upon binding of substrate.[1298]

Certain differences in the behavior of binding protein compared to substrate interaction in transport process may also be noted. The temperature coefficient for binding is generally smaller than for the transfer of a substrate. Similarly, a binding protein does not induce transport of the substrate across BLM. Most of the transport systems show lipid specificity as shown by both genetic mutation and by direct analysis.[554,555,1119–1122] The lipids do not seem to have any effect on substrate binding by isolated binding proteins.

These physicochemical observations on binding proteins in general suggest that even though binding protein can specifically bind to a substrate, they may not be the carrier through the lipidic environment. Such an assertion is further substantiated by a few scattered observations. For example, a sulfate BP while part of the transport system can react with diazotized aminonaphthalene disulfonate, a protein reagent that cannot penetrate the lipid barrier. Similarly, the antibody to the sulfate BP inactivated the purified protein but did not inactivate this protein on bacteria. Also, treatment of the leucine binding protein with antiserum caused inhibition of leucine binding, but treatment of whole cells with antiserum caused no inhibition of transport.

These observations in general tend to suggest that the binding protein may act as carrier only in concert with other factors—lipids and/or proteins. This may account for a consistently small but significant difference observed in the binding of substrate and inhibitors by BP and by the transport system. It is consistent with the observed genetic mutation involving a component other than the binding protein; such mutants show specificity for substrate binding but are transport negative. Similarly, the effect of hormones (discussed in the next section) on facilitated transport system also implies that hormones interact with some component of the transport system without affecting its K_m for transport.

All these observations taken together suggest, rather strongly, that the binding protein is not the sole component of a transport system, and that binding of the substrate is not a simple expression of transport. Binding proteins may be associated with a lipoprotein complex such that binding of the BP to the complex is loose and polar in character. The translocation step may involve conformational change in the transport system. An important feature of this model is that the transport system without the binding protein may be shared by several binding proteins specific for different substrates.[1123] This requires that the shape and size of the binding protein/substrate complex from a variety of substrates and corresponding BPs be almost the same. The situation is complicated further by the fact that not all the transport systems from which a binding protein has been isolated show a simple passive carrier-mediated transport process. In fact, most of the bacterial transport systems are coupled to some yet uncharacterized energy source (see later). These systems may however be uncoupled in certain cases, such that the resulting system shows most of the features of passive carrier-mediated transport system. The study of binding proteins from some well-established carrier transport systems, such as red blood cell, is complicated by the fact that these proteins show considerably smaller affinity for the substrate.

Effect of Hormones on Facilitated Transport

It was Claude Bernard who recognized in 1879 that there are for any tissue or organism two environments, an external environment and an internal environment (*milieu interieur*). The constancy of the internal environment was stressed to be of particular importance, since external variations can be compensated for or brought into equilibrium in the internal environment. In higher organisms, highly complex mechanisms exist which regulate and stabilize such factors as temperature, blood pressure, the partial pressure of oxygen and carbon dioxide in the blood, and the concentration of H^+, Na^+, K^+, Ca^{2+} and glucose in tissue fluids. Hormones* play an important role in homeostasis, that is, in stabilizing the *internal* environment of the body.

Every hormone acts basically by modifying and modulating some aspect of cellular metabolism and they seem to have characteristic functional properties:[557] (a) They modify existing metabolic processes. (b) They are secreted in response to specific secretory stimuli. (c) They can be transported in the bloodstream, usually bound to a carrier. (d) They are continuously inactivated.

---

*"Hormone" means excitor; however, there are hormones that function as inhibitors, and many of the hormones that facilitate the activities of one group of tissues may retard those of another. The term, therefore, is now quite generally used in a wider sense to incorporate all humorally dispersed chemical substances whose functions are communication between various organs of the body, and the regulation or coordination of organic processes.

(e) They have a high degree of specificity, both with respect to target and the effect elicited at the target.

In general, molecular mechanisms for hormone action remain largely obscure. They are, however, generally conceived in metabolic terms as regulating devices closely linked to enzymatic and transport activity. Thus, it is not surprising that attention has been focused on regulatory aspects of facilitated transfer by hormones in animal tissues, since increased secretion of hormones or their administration is associated with growth or hypertrophy of tissues known to be target organs. Moreover, since hormones may affect biosynthetic mechanisms directly, questions may be raised concerning the dependence of protein synthesis on the transmembrane movement of metabolites and the size of their intracellular pools.[558]

It may be argued that one of the best means to control the metabolic activity would be through control of access of reactants to the system utilizing them. In fact, various hormones are known to modify the permeability characteristics of biomembranes (Table 6-13). Hormonal control seems to occur in actively me-

TABLE 6-13 MAJOR EFFECTS OF HORMONES ON TRANSPORT OF VARIOUS METABOLITES[a]

Hormone	Metabolite Affected	Target Issue
Aldosterone	Na	Kidney, muscle, toad bladder and skin, blood cells
Androgens	Na, K, amino acids, sugars	Kidney, skeletal muscle, uterus, *levator ani* muscle
Estrogens	Water, Na, K, sugars, amino acids, iodide	Uterus smooth muscle, thyroid
Glucocorticoids	Na, S_2O_3, P_i, SO_4, glucose, aminoacids, ascorbic acid	Kidney, liver
Insulin	K, P_i, sugar amino acids, water	Muscle, heart, liver, adrenals, mitochondria
ACTH	P_i, I^-, urate, sugar, amino acids, water	Kidney, adrenals, thyroid, mitochondria
Thyroxin	Water	Mitochondria
Neurohypophyseal hormones	Sugar, water	Adipose tissue, bladder, mitochondria
Parathromone	Mg, P_i, K, Ca	Mitochondria, bone cells, kidney
Prostaglandins	K, fatty acids	Human uterus and various other smooth muscles
Acetyl choline and other neurotransmitters	Na, K, Cl	Central nervous system
5-Hydroxy tryptamine	Na	Kidney, frog skin
Cyclic AMP	Sugar, water	Almost all tissues
Vasopressin	Water, Na	Kidney, bladder, and skin
Auxins	Water	Plants

[a]The transport of metabolite affected at the target tissue is correct only in a limited sense, that is not all the metabolites are affected at all the tissues.

tabolizing tissue: kidney, liver, and brain. Although the final effect of hormone action at the tissue level is reflected in altered permeability characteristics, the general picture, at both lower and higher levels of organization, is far more complicated. At the cellular and subcellular level, it is generally difficult to pinpoint whether the primary effect of a hormone is in regulation of protein synthesis, or due to its action directly on the membrane, or whether it is the consequence of an occurrence at another level of anabolic and catabolic reactions. The situation is best illustrated by considering the action of insulin on glucose metabolism in general and glucose transport in particular.

Administration of insulin into an intact organism or to isolated tissue produces a variety of effects (Table 6-14). Within our present state of understanding, these various effects of insulin cannot be explained by a single mechanism or site of action. The action of insulin on protein metabolism is regarded as an effect of the hormone at the level of the gene. This view is based on evidence that insulin produces marked changes in the amount of RNA in the cell, an effect which is blocked by actinomycin. The effect of insulin on glucose uptake in muscle or fat cells is regarded as the result of an activation of the cell membrane transport system for glucose, an effect which can be seen at low temperatures and one which is not blocked by actinomycin. Because insulin influences several routes for glucose utilization by cells, e.g., the synthesis of fat and the synthesis of glycogen, it has been argued that the hormone must act on some initial stage of utilization of glucose:

This stage could either be facilitated transport of glucose (I), or phosphorylation of glucose after its entry into the cell (II). In fact as shown in Table 6-15 both of these systems are known to operate. In muscle and adipose tissue the transport system has a K_t for glucose approximating the physiological range of blood glucose concentrations (3–10 mM). The intracellular glucose concentration is low. The K_m for glucose of hexokinase is low (0.1 mM or less) and the enzyme is inhibited by glucose-6-phosphate. The liver cells in contrast are freely permeable to glucose, the intracellular glucose concentration is high, and the glucokinase is present with a K_m for glucose of 10–20 mM. In summary, glucose utilization is controlled primarily by transport in the first system and by phosphorylation in the second.[559,560]

The effect of insulin on facilitated transport could either be due to an increased number of specific sugar carrier sites on the cell membrane,[561,562] or it

TABLE 6-14 ACUTE EFFECTS OF INSULIN ON METABOLISM[559]

Tissue	Action of Insulin	Site and Mechanism
Muscle, adipose tissue, liver	1. Action at molecular level	Not known
Muscle, adipose tissue	2. Glucose uptake enhanced	Membrane transport stimulated
	3. Glycolysis enhanced	Activation of phosphofructokinase secondary to 2
	4. Glycogen synthesis enhanced	Activation of UDP glucose glycogen glucosyl transferase
	5. Protein synthesis enhanced	Activation of ribosome; accelerated membrane transport of amino acids
	6. Esterification of fatty acids enhanced	? Secondary to 3; ? other direct mechanism
Muscle	7. Ribonucleic acid synthesis accelerated	
Adipose tissue	8. Lipolysis inhibited	Inhibition triglyceride lipase
	9. Lipogenesis augmented	? Secondary to 2, 3, 6, and 8
Liver	10. Glucose output restrained	Not known; ? secondary to 6, 8
	11. Protein synthesis enhanced	As in 5

TABLE 6-15 EFFECT OF INSULIN ON METABOLISM OF GLUCOSE[a]

Characteristics of a System of Type I	Characteristics of a System of Type II
Transport rate limiting	Hexokinase rate limiting
Present in muscle, adipose tissue, pancreas aciner cells (of mouse and other organisms)	Liver and may be islet tissue of mouse
K for glucose transport (10 mM)	
K_m for kinase activity (0.1 mM for glucose)	K_m for kinase activity (10–20 mM); almost freely permeable to glucose
Inhibited by phloridzin	Not inhibited by phloridzin
Inhibited by glucose-6-phosphate	Not inhibited by glucose-6-phosphate
Not inhibited by mannoheptulose	Inhibited by mannoheptulose

[a]See Refs. 559 and 560 for details.

could reflect an increase in the transport efficiency of those already present.[563] One way in which transport efficiency could be changed is by a reduction of the affinity of the carrier molecule for sugar. Thus in the first case V_{max} is altered whereas in the second case K_t is altered by the action of the hormone.* It may

---

*It may seem paradoxical that a reduction in the affinity of the carrier for a substrate could increase its rate of transport. However, one of the ways of looking at this is as follows: If a carrier has infinite affinity for glucose, then little or no glucose could be released into the cell, however low the concentration of glucose in the cell. Conversely, a carrier with a low affinity for glucose would also result in minimal transport since little glucose would be accepted at the outer surface of the cell membrane. Obviously, there is an optimal affinity of the carrier for the substrate molecule to effect a maximal transport rate.

be noted that the kinetic data alone can not establish a mechanism; thus, for example, in this case the values of K_t and V_{max} may be composite constants for the sequential processes of glucose transport and phosphorylation.

Thus, while insulin increases K_t for glucose transport in rat adipose tissue,[561,562] it reduces this parameter in perfused rat heart. In contrast, insulin stimulates the penetration of sugar into frog sartorius muscles without causing any significant change either in apparent K_t for 3-O-methyl glucose or in K_i for phlorizin; only V_{max} for sugar increases as much as tenfold.[563] It is interesting to note that trypsin, like insulin, also increases sugar uptake in frog sartorius muscle. When the rate of entry of sugar is studied at different temperatures in control muscles and in muscles that have first been exposed to either insulin or trysin under standardized conditions, the ratio of the rate of entry of sugar in stimulated muscles to the rate in control muscles falls as the temperature is raised. On this basis it has been suggested that insulin and trypsin may alter the structure of the membrane in a manner that permits sugar carriers to move rapidly within the membrane. It was also noted that in frog sartorius muscle and rat adipose tissue, the increased rate of sugar transport is not specific to insulin only, but polyene antibiotics and enzymes like phospholipase C, papain, ficin, and a protease from *Streptomyces griseus* also enhance the rate of sugar transport. For a review see Ref. 1232. Effects analogous to those of insulin on glucose transport in muscle are also produced by electrical stimulation. The maximal effect of the stimulation is uninfluenced by insulin but the submaximal effect may be increased by the presence of hormone. Electrical stimulation acts immediately, even at low temperatures, whereas the action of insulin requires a lag period.[564] Since these effects, as does anoxia, occur even in the absence of insulin, insulin cannot be a part of the glucose carrier. It is possible that the glucose transport system may interact with insulin by modification of lipoprotein structures or by some yet unknown mechanism. In this connection, it may be noted that the effect of insulin is prevented by preexposure of cells to maleimide, which presumably blocks thiol groups. It is also pertinent to recall that insulin has no effect on conductance across BLM. Thus, in view of the experimental results just described it may be reasonable to assume that *one of the effects* of insulin is to increase glucose transport across a variety of membranes. The results are also consistent with the idea that the target site of insulin action may be the *transport system as a whole,* although insulin does not seem to bind to the *binding protein.*

From the foregoing discussion it is obvious that one of the characteristics of the hormonal effect is the multiplicity of effects at higher levels of organization. Moreover, a single hormone may affect the transport of more than one solute at the same time and simultaneously act on a variety of tissue. Perhaps, in each case, one of the changes has overriding physiological importance and significance, but clearly the hormones frequently influence general aspects of the behavior of the membrane and not merely a single specific transport mechanism. A general discussion of these hormone characteristics may be found elsewhere.[558,565] It may be emphasized here by way of recapitulation that if the hormone acts only by altering the behavior of the carrier system, it is difficult to explain either the ability of the hormone to accelerate carrier-mediated permeant entry of the stimulation of transport step without affecting K_m or K_t, unless there is yet another component in a transport system. Some of the hormonal effects on transport may be secondary; that is to say, such effects could arise due to altered protein synthesis, or to an altered rate of metabolism, or to changes in the driving forces, or to changes in the shape, size, or surface density of the carrier. Thus for example, the primary interaction of hormone may be on a site consisting of receptor, regulator, and catalytic units, and may thus depend on an intact membrane structure for complete response.[566] The diversity in the ability of hormones to stimulate permeation of a variety of solutes also suggests that some general characteristic of membrane structure and function, as yet unidentified, is affected by the hormone.

Present Status

While working with the carrier diffusion model a reasonable description of substrate transport across various cellular membranes may be obtained. However, some small discrepancies have been consistently observed. LeFevre[567] observed that the affinity of the glucose transport system represented by the apparent dissociation constant of the carrier for the sugar ranges over a factor of 400 (K_m = 13.6 mM to 4.3 M) for the following sugars (in the order of declining affinity): D-mannose, D-galactose, D-xylose, L-arabinose, D-ribose, D-arabinose. On the other hand, V_{max} for transport is the same. Since the maximum velocity is independent of the substrate, it seems that either the frictional coefficient for the translocation of the carrier-substrate complex is virtually independent of the substrate, or the translocation of the complex is not the rate-limiting step. In contrast, K_m values for leucine, L-phenyl alanine, L-methionine, and L-valine for the L-leucine transport system of human erythrocytes are 1.8, 4.3, 5.2, and 7.0 mM, respectively; but, unlike for the glucose transport system, the V_{max} values depend upon the substrate undergoing translocation.[501,568] The L-leucine transport system has a rather broad specificity although a three-dimensional shape relationship between the substrate and the carrier is clearly present. The bulk of the substrate side chains appears to be one of the specificity determining factors.

The main objection against a simple carrier model for facilitated transport is its inability to explain the rate of movement of a sugar across the human erythrocyte as a function of the amount of a second sugar present on the trans side of the membrane. The loss of sugar from cells containing a high (saturating) concentration of labelled glucose was measured under two different conditions:

first, transport into a sugar-free solution (net transport), and, second, transport into solutions containing nonradioactive glucose at the same concentration (exchange). The results show that exchange transport is 1.8 to 2.5 times that of the net transport.[569,570] According to the simple carrier model which we have discussed thus far, there should not be any difference in these rates. As an explanation, it has been suggested that the carrier does not return at the same rate when complexed, but moves more rapidly in this form. Doubts about the validity of such assumptions arose, however, from the following observation.[503] Cells were loaded with nonradioactive sugars at saturating concentrations and then transferred to solutions containing a low concentration of a radioactive sugar, either mannose or galactose. The increase in radioactivity in the cells was then determined as a function of time and was found to rise rapidly to a value much in excess of the final equilibrium value and then decline to this value at infinite time, as shown in Fig. 6-19.[571] Miller then undertook a relatively rigorous mathematical analysis of the simple carrier model to explain this phenomenon.[572] A simple carrier diffusion mechanism where diffusion through the membrane phase is rate limiting, and two other mechanisms derived from it, that is, when (a) free carrier moves more slowly than the complexed carrier, and (b) a slow diffusion step occurs at each interface of the membrane, do not seem to provide an adequate quantitative description of the experimental data.[573] These conclusions have been essentially confirmed by more explicit mathematical analysis using Miller's data.[574] Thus, the situation is as yet unresolved.

Even before these results were available, it had been suggested that the simple carrier diffusion model is probably an oversimplification. It can, for example, be shown that models not involving mobile carriers can also predict a similar set of phenomena.[575] In the pore model, assuming that the region between two protein molecules is very polar and that passage through the pore is the rate-limiting step, the kinetics for transport are the same as those for a carrier which diffuses.

In principle, the model which we have considered thus far implies a physical separation of entry and exit and may be taken to be equivalent to models which consider an alternate state of the barrier rather than alternate states of the carrier. Some of the kinetically indistinguishable models are: (a) the allosteric model, involving transitions of a binding site between two alternate shapes;[575] (b) the H-bond model, involving transition of a combining site between two alternate hydrogen bonded states:[575] and (c) the gate or fixed site model, for which the transport process may involve an alteration of the barrier.[576] These models become especially important in view of the small thickness of the membranes and the large molecular dimensions of the transport systems as suggested by various physicochemical studies. It is also assumed in the model which we have adopted that diffusion across the membrane is one single step. The number of steps involved in these processes can be infinite and, in principle at least, the

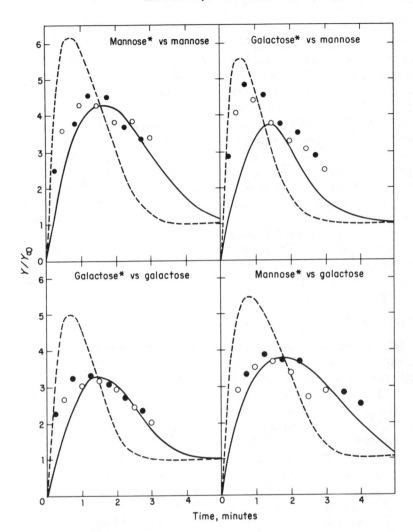

Fig. 6-19. Countertransport of sugar pairs. Y/Y_∞ is the ratio of the radioactivity of cells at the time indicated, relative to that at infinite time. The radioactive sugar (marked with an asterisk) is present externally at a concentration of 12 mM and the nonradioactive sugar present inside the cell at an initial concentration of 120 mM. The lines are theoretical curves.[503]

kinetic data can be elaborated to pinpoint a few of these steps. A theoretical analysis of twenty models of steady-state flux across a lattice model membrane may be found elsewhere.[577] For various other models with possible relevance for a carrier mediated transport system see Ref. 578.

The present status of carrier-mediated facilitated transport systems in bio-membranes may be recapitulated as follows. It is abundantly clear that transport of specific substrates across biomembranes is facilitated by specific sites. The simplest model which can accommodate this data to a first approximation requires a binding site which is mobile with respect to the osmotic barrier; that is to say, the binding site makes itself available to the substrate at both the interfaces. Quantitative elaborations on this theme have met with only limited success, although semiquantitative treatments have rather consistently supported the idea of a mobile site as carrier. The binding site of the substrate is most probably located on a protein; however, the possibility of lipids as carriers has not been ruled out completely.[579] Probably the most significant gap in our understanding stems from the lack of knowledge of the *immediate environment* in the membrane phase in which the binding site undergoes translational, rotational, or conformational changes. Thus, a transport system may be a multicomponent assembly.[579a] In fact, the "internal transfer model" consisting of a tetrameric transport system can account for several important aspects of glucose translocation in red cell.[1300]

Energized Transport

Typically, most living organisms survive in an environment which has a low and fluctuating concentration of nutrients and various other metabolites essential for their upkeep. Under such conditions, regulation of the rate and extent of various intracellular biochemical processes may be seriously affected by any change in the external concentration of these various solutes which must cross the membrane surrounding the cell. Thus, even under most optimal conditions of facilitated transport, the internal concentration of metabolites would fluctuate severely. However, the fact that most organisms survive under a variety of conditions strongly suggests that some *homeostatic* mechanism(s) operates to regulate the intracellular concentration of various solutes. It would not be unreasonable to assume that membranes play a very important, if not a crucial, role in such regulation. The genetics, physiology, and biochemistry of membrane transport processes all attest to this conviction. The object of this discussion is to describe some of the salient features and physicochemical events underlying coupling of two fundamental processes: the *translocation of solute* across the osmotic barrier *whereby energy is expended* for accumulation of a solute against its concentration gradient.

A considerable amount of work has been devoted to characterize processes in which translocation of a solute or its accumulation is accompanied by expenditure of energy. The following systems are best characterized (Fig. 6-20).

1. *Facilitated transport aided by a permeant or counter transport.* While discussing facilitated transport we noted that an apparent flux against a concentra-

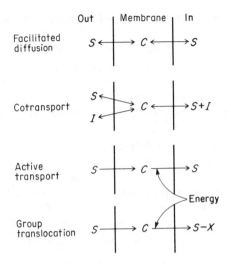

Fig. 6-20. A schematic representation of various facilitated transport systems. The carrier (*C*) located in the membrane phase combines with the substrate (*S* and *I*) for translocation. See text for details.

tion gradient may be observed as a special case of an interaction between simultaneous movement via a carrier transport system of two different substrates or of the same species in a differently labelled form. Thus, the movement of substrate (say S) against its concentration gradient (i.e., *uphill*) is energized by the simultaneous movement of a second substrate (say S') down its concentration gradient. This phenomenon is commonly attributed to a more rapid movement of carrier/substrate complexes (CS or CS') across the membrane compared with that of free carrier; in other words the flux of S is *inhibited* (by competition for the carrier sites) by the movement of S' down its electrochemical gradient. Thus, the flux of S' is *coupled* to the flux of S by virtue of the characteristics of the system.

2. *Cotransport.* If the mobile carrier mediating transfer of a substrate (say S) can form a ternary complex (of the type C.S'.S''), and the ternary complex has much higher mobility than that of CS or CS'' or C alone, S can be driven against its own electrochemical gradient at the expense of the electrochemical gradient of S''. Several specific examples of this type of energized transport are known to operate across biomembranes, as described later in this section.

3. *Group translocation.* If the substrate (S) upon its entry into another compartment is converted into yet another substrate which has a much lower affinity for the carrier, there will be a continuous influx of it accumulated as a chemically distinct species (say S*). Conversion of S to S* may occur either after it is released from the carrier into the second compartment (*isotropic group translocation*), or during the translocation step itself in the membrane

(*anisotropic group translocation*).* Both of these types are known, and will be described later in this section.

4. *Active transport.* The term active transport has been used in the literature rather loosely to designate *any* transport process in which accumulation of a substrate takes place. There are instances where such an accumulation can be shown to result from specific adsorption of solute, or its metabolism as implicated in isotropic group translocation. We shall restrict the use of term active transport to describe more specific situations. The term is taken here to imply solute translocation against an electrochemical or osmotic gradient without inducing any chemical change on the solute. The energy is supplied at the translocation step by a coupled energy-supplying reaction. The best known example of this type is the active transport of alkali metal cations energized by hydrolysis of ATP as discussed at length in Chapter VII.

Taken all together, the energized transport systems show the following characteristic features in addition to those characterizing passive facilitated transport (cf p. 217):

1. Solute can be accumulated within a cell from a medium of much lower concentration. K_t is of the order of one micromole/liter.

2. The rate of exit of these various solutes is much smaller than their rate of entry.

3. Mutant cells can be found which have lost the ability to accumulate a particular species of solute, even though these cells show all the characteristics of the carrier mediated facilitated transport system. This means that the energy coupling system has been *disengaged* from transport system.

4. Most of these systems show the characteristics of a facilitated transport system, such as substrate specificity, competitive inhibition, inhibition by protein reagents, etc. (see p. 219).

5. Most of these systems can be inhibited by metabolic poisons, anoxia, lack of an energy substrate, lack of proper ionorganic ions, etc.

6. Some of these systems are completely localized in the membrane phase, while others owe. their characteristics to the membrane along with the environment in which it is located.

Most of these characteristics point to specific and mediated processes which are spatially oriented. Now we shall proceed to discuss some of the characteristics of specific energized transport systems.

Cotransport. As the term implies, the transport of one substrate accompanies that of another solute in the same direction, at least one of which flows down its electrochemical gradient. Since energized transport is meant for accumulation of substances, it is usually the second component which flows down its electro-

---

*(It follows from the basic postulate of group translocation that the affinity of the carrier for the substrate is different at two interfaces, that is, this alteration is achieved of course at the expense of energy, either altering the substrate or by altering the carrier.)

chemical gradient. The species which is generally present in higher concentration in the external physiological environment is sodium ion (see Table 8-1) and it is the sodium ionic gradient which is often utilized to do the necessary work for uphill transfer. Quite a few sodium-driven transport systems have been characterized (Table 6-16); however, the following discussion of cotransport is based mainly on the studies done on Na-dependent intestinal transport of sugars and amino acids; other systems appear to have more or less similar characteristics. For an excellent review see Ref. 580. The cotransport system as a whole shows the following characteristics:

1. Sodium ions are absolutely essential and cannot be replaced by other alkali metal cations such as K, Rb, Cs, and NH_4. Li can be substituted for Na in only a few instances. Tris, guanidinium, or choline ions have no effect.

2. The movement of substrate is accompanied by the translocation of sodium ions, and since sodium ions are present in much larger concentration than the

TABLE 6-16 COTRANSPORT SYSTEMS DEPENDENT UPON EXTERNAL Na^+ [580]

Permeant	Organism	Coupling Factor (S/Na)	Remarks
Sugars (hexose)	Small intestine (rabbit, rat, mouse, hamster, guinea pig, chicken, frog, toad, bullfrog, human, dog, tortoise), jejunum (rat, human), kidney (rabbit, rat, frog), choroid plexus (dog)	Unity	May be electrogenic (see text)
Amino acids	Small intestine (rat, rabbit, dog, bullfrog, goldfish), kidney, toad bladder, adipose tissue, bone and cartilage, brain, eye lens, liver, striated muscle, cells (KB carcinoma, Ehrlich ascites carcinoma, marine bacterium, pigeon erythrocyte, mouse fibroblast, rabbit reticulocytes, human leukocytes), nuclei, mitochondria	From zero to one	Some of these systems are electrogenic (see text)
Sulfate, uracil, P_i, bile salts, ascorbic acid, Ca	Small intestine		
Uric acid, myoinositol, p-aminohippurate, Ca	Kidney		
Ascorbic acid, creatine, thiamine, acetate	Brain		
Iodide	Thyroid gland		
Ascorbic acid	Adrenal gland		
Glucuronate	Marine bacterium		
Phosphate	Marine fungus		

substrate, the direction of net flux of sodium determines the direction of accumulation of substrate; they are the same. Under physiological conditions, the high K concentration inside the cell may also facilitate dissociation of the ternary complex.

3. The substrate is not altered chemically when translocation is complete. Generally speaking these systems show poor substrate specificity. Thus, for example, the sugar transport system of hamster small intestine can transport all actively transported hexoses, such as glucose, galactose, 3-O-methyl glucose, xylose, and certain glycosides.

4. At low concentrations of sodium ions, the electrochemical gradient of the substrate itself can drive the transport.

5. The sugar transport system is inhibited by phlorizin; it appears to be a competitive inhibitor.

These observations are quite consistent with a model postulating formation of a ternary complex of carrier with substrate and sodium ions.[581,1130] At the other interface of the membrane the sugar and ion dissociate freely from the carrier as in a facilitated transport system. The active transport system can extrude the sodium ions (see Chapter 7) and thus maintain the ionic gradient necessary for driving cotransport.

The model described above (Fig. 6-20) is obviously oversimplified. A substantial amount of information is, in fact, now available to suggest certain specific details. However, these seem to differ from one group of substrate to another. Thus, for example, in certain superficial aspects the sugar influx system in the rabbit ileum is similar to the system for neutral amino acids influx in the same tissue.[582-584] Thus, both processes can be described by Michaelis-Menten kinetics, both are dependent on the concentration of sodium ions in the medium, and both involve simultaneous entry of sodium and nonelectrolytes into the cell. Thus the flux of a solute (permeant) is a hyperbolic function of solute concentration at constant Na concentration. In many systems, but not all, solute flux at constant concentration is a hyperbolic function of Na concentration. The transport systems can thus be described in terms of maximal flux rate (V_{max}) and "apparent Michaelis-Menten Constant" (K_t). Referring back to the interaction between Na and substrate influxes across the membrane, as indicated in the model in Fig. 6-20, one may elaborate further as follows:

Mucosal solution	Membrane			Cell
Na +	C	$\rightleftharpoons$	C	+ Na
	$\updownarrow$		$\updownarrow$	
S +	CNa	$\rightleftharpoons$	CNa	+ S
	$\updownarrow$		$\updownarrow$	
	SCNa	$\rightleftharpoons$	SCNa	

According to this model the substrate combines with a membrane component (C) to form a binary complex (CS). This binary complex may either be translocated across the membrane or combine with Na to form a ternary complex (CSNa) that is then translocated across the membrane. The rate at which C, CS, and CSNa are translocated are assumed to be first order, equal for all three forms, equal in both directions, and rate limiting so that the association–dissociation reaction at the membrane interfaces may be considered at equilibrium.

A comparison of various kinetic features of sugar and amino acid transport systems, with special reference to that observed in small intestine, reveals some differences as well. Of these the following may be noted:

1. The role of sodium ions in amino acid transport appears to be different than in the case of sugar transport. In the amino acid system, Na combines with a site that is already complexed with the substrate, leading to increased stabilization and consequently to an increased concentration of complexed carrier and enhanced influx. In such a system, removal of Na is equivalent to the addition of a competitive inhibitor; that is, its effect can be overcome by an increased concentration of amino acid. In the sugar transport system, removal of sodium is approximately equivalent to adding a noncompetitive inhibitor; the effect cannot be overcome by increasing sugar concentration. This implies that the role of sodium is to produce a complex suitable for translocation; in the absence of sodium this complex cannot be formed and hence influx does not occur.

2. Both systems exhibit Michaelis-Menten kinetics for influx of the permeant. All the sugars have approximately the same V_{max}; however, apparent values of K_m differ for different sugars.* In contrast, amino acid transport systems show the same Michaelis-Menten constant but different values of V_{max} for different amino acids.

3. There is a linear relationship between sugar and sodium influxes, whose quantitative aspects suggest that the ratio of the sugar/Na influx is approximately unity over a wide range of Na concentrations. The ratio of sodium to amino acid influx under similar conditions varies over a wide range.

4. Sugar influx appears to be more sensitive to sodium removal than does amino acid influx. Reduction of the Na concentration causes an increase in K_m for amino acids with little change in V_{max}; for sugars V_{max} decreases with only a small increase in K_m on reduction of Na concentration.

---

*There appears to be some species differences in the kinetics of sugar transport; for example, sugar transport into rings of hamster intestine is described by Michaelis-Menten kinetics, and reduction in Na concentration causes an increase in the binding constant (K_m) with no change in V_{max}. Completely opposite effects are observed for sugar transport in rabbit ileum as noted above[582] and in rabbit kidney slices.[585] In this respect it is of interest to note that the activity of intestinal sucrase (see later) from rat, hamster, and rabbit is stimulated by sodium.[586] Kinetic analysis has disclosed that Na decreases the K_m of rat and hamster sucrase with little or no change in V_{max}, but that Na increases the V_{max} of rabbit sucrase with little or no effect on K_m.

5. In the absence of Na, no evidence for saturation of the transport system could be obtained for sugar concentrations up to 40 mM; sugar influx is a linear function of sugar concentration. This suggests that the sugar (in this case 3-O-methyl glucose) influx under these conditions may be due to simple diffusion or that the K_m for influx is extremely high so that saturation cannot be detected at reasonable sugar concentrations. In the absence of sodium, amino acid influx appears to be primarily via a mediated process (downhill).

6. Phlorizin is a potent inhibitor of intestinal-sugar transport; however, in the absence of Na it has little effect on *residual* transport.

7. At least in one case (pigeon erythrocytes) resealed ghosts have been shown to have a substantially intact cotransport system for glycine.[587]

8. The coupling coefficient relating Na influx to substrate influx is unity for sugars and appears to be independent of Na and substrate concentration. In contrast, the coupling coefficient relating Na influx to amino acid influx is not fixed but depends strongly on the structure of the aminoacid, varying from about 0.2 for proline to 5 for cysteine, but seems to be almost independent of the concentrations of the two substrates.[1238]

Generally speaking, it appears that all these observations are consistent with the Na-coupled facilitated transport mechanism. Also, it appears that Na combines with a membrane site that is already complexed with amino acid or sugar, and may thus lead to increased stabilization and enhanced influx. Such a mechanism of transport should give rise to several interesting phenomena. For example, transport sites for two different substrates may be different but some cross inhibition of one by the other may be observed. This arises from the fact that amino acid and sugar transport, mediated by the same tissue, may produce inhibitory effects on each other's transport through localized alterations of intracellular sodium concentration. This localized increase in sodium concentration enhances the possibility of efflux of one of the substrates from the cell and hence, lowers the net transport. Such a mechanism has been invoked to explain the reduced rate of uptake of amino acids by intestinal mucosa of *Mustelus canis* (dog fish) in the presence of excess galactose.[588]

The concept of coupled transport (cotransport) as developed in the preceeding discussion has some important implications.[580] From kinetic considerations it is obvious that the accumulation of substrate depends both on the concentration of the substrate and that of the cations. This implies that, if sodium extrusion is inhibited, the transport of substrate would be inhibited as soon as the sodium distribution ratio approaches unity; however, exchange of substrate between the two compartments should still occur. This has been found to be so for the amino acid transport system in Ehrlich ascites tumor cells under the influence of metabolic inhibitors.[589] In fact cardiac glycosides, which block active transport of cations (p. 269) also block Na-coupled uptake of a variety of substrates in a variety of organisms.[580] These results are generally interpreted to reflect an indirect coupling between transport process and the exergonic metabolic reaction which replenishes sodium ionic gradient needed for Na-coupled transport system. However, a direct link between exergonic process (such as hydrolysis) is a possibility which has not been ruled out completely.[590]

Thus on the basis of the available evidence it may be concluded that asymmetric distribution of Na across the cell membrane does contribute to the energy required for the transport of other solutes against a chemical potential difference. This mechanism, however brings out several points of interest as far as the asymmetries of Na gradient and of the membrane are concerned; the potential difference across the membrane being the foremost.

Not very much attention has so far been paid to the characterization of a counterion which should accompany the ternary complex ($CSNa^+$) in order to maintain electroneutrality. In Ehrlich ascites tumor cells,[591] and marine Pseudomonad,[1237] transport of Na accompanies exit of K, and in halophile bacteria[592] Cl^- accompanies Na when the substrate is translocated. If a charge balance does not occur in the translocation step itself, an electrical potential will build up as cotransport progresses; this has been observed in rabbit ileum[593] and tortoise ileum[594] with various substrates. Similarly, the reconstituted pigeon erythrocytes actively extrude glycine when the Na concentrations on the two sides of the membrane are equal but the cell interior is electrically positive with respect to the medium. In these experiments the electrical potential difference was established by replacing extracellular Cl^- with the nonpenetrating anion toluene-2,4-disulfonate.[595] This evidence suggests that the translocation of the complex ($SCNa^+$) is influenced by the electrical field. Interpretation of results obtained from ileum is not easy.* Using microelectrodes, it can be shown that the increase in potential difference across epithelial cells during cotransport of substrates is due to the presence of an *electrogenic pump* (see Chapter 7 for details) at the serosal face of the epithelial cell.[594] Thus the change in potential difference across the whole tissue may not bear any relation to sodium flux during cotransport. It may, however, bear relevance to the observation that a changing electrical potential across the ileum does not alter the substrate flux. These conclusions are in contradiction to the observation where addition of substrate or Na^+ to the mucosal solution brings about a decrease in the electrical potential difference across the mucosal membrane, i.e., the cell interior becomes less negative.[597] These results suggest that the mechanisms responsible for the coupled influxes of Na and substrate across rat jejunum,[598] rabbit ileum,[597] and bullfrog small intestine[599] is electrogenic. This controversy has not been resolved as yet.

Progress toward biochemical characterization of cotransport systems has been

---

*Ileum is a complex tissue. The available evidence suggests that the brush borders of epithelial cells are responsible for cotransport mechanism. The brush borders may most easily be interpreted, in three dimensions, as being composed of a membrane covering a "skeleton" of rod-like elements embedded in an underlying meshwork.[596]

slow. Autoradiographic[600] and biochemical[596] studies suggest that the cotransport system resides in the brush border region of the cell; the basal membrane of the cell may be the site of both a diffusion barrier and a weak transport directed into the cell. In fact, Na-dependent binding of L-histidine has been attributed to a fraction of mucosal brushborders from hamster jejunum.[1126] Similarly analytical evidence has been presented suggesting that an enzyme *sucrase* (sucrose-glucose hydrolase) present in the brush border region may be a part of the sodium-dependent sugar transport system.*

Although the catalytic property of sucrase is not a part of the cotransport model which we have discussed thus far, there are a number of observations, summarized below, which do suggest a close relationship between sucrase and the sugar cotransport system:[601]

1. Both are located in the brush borders.

2. Both are activated by sodium.

3. Both are competitively inhibited by NH_4 and other alkali metal cations.

4. In the hamster, the sodium activation constant in the absence of sucrose and the one at infinite sucrose concentration are very similar for both systems. The ratio between the two Na-activation constants in the presence and absence of sucrose is large; therefore, Na-activation results in a marked decrease of the apparent K_m for sugar with little or no effect on V_{max}.

5. In the rabbit ileum, both the Na activation of sucrase and the Na activation of sugar transport are characterized by a large increase in V_{max} with little change in the apparent K_m values.

6. The enzyme is inhibited by most sugars which show Na-dependent absorption and it is unaffected by other sugars.

7. Antibodies directed against the sucrase-isomaltase complex (which do not inhibit sucrase activity) inhibit sugar uptake in vitro.

However, there are some significant differences in the behavior of these two systems, especially with regard to substrate and ion specificity and inhibition patterns. For such reasons the role of this protein in the transport system should be accepted only provisionally. Furthermore, the enzymes of glycolytic pathway, such as hexokinase, may be involved in these processes.[602]

Group Translocation. Facilitated diffusion isotropically coupled to a chemical reaction has been implicated in a variety of reactions.[611,620] Typically, these systems are inhibited by energy poisons and uncouplers of oxidative phosphorylation, under which circumstances the transport process behaves as a typical facilitated transport process. These processes are discussed further in the next section.

Vectorial coupling of a chemical reaction to a facilitated transport system

---

*It is interesting to note that specific sugar intolerance, a genetic defect, can be associated with the lack of disaccharidase activity in intestinal brush border.

(anisotropic group translocation) is of special interest. In such systems, there is specificity with regard to the direction (sidedness) of approach of substrate, of product exit, and of utilization of cofactors. The energy required for the translocation step is provided by the chemical reaction itself. We shall discuss this phenomenon with special reference to the phosphotransferase system for sugar transport, which operates in a variety of bacteria (Table 6-17). In fact, the possibility that phosphorylation is involved in sugar transport across bacterial membranes has been recognized for quite some time.[477] However, it was in 1964 that Kundig et al. reported[613] the isolation of a bacterial phosphotransferase system which catalyzes the transfer of phosphate from phosphoenol pyruvate (PEP) to various carbohydrates according to the following reaction sequence:

$$PEP + HPr \xrightarrow[Mg^{++}]{Enzyme\ I} Pyruvate + P \sim HPr$$

$$P \sim HPr + Sugar \xrightarrow[Mg^{++}]{Enzyme\ II} Sugar\text{-}6\text{-}P + HPr$$

$$PEP + Sugar \longrightarrow Sugar\text{-}6\text{-}P + Pyruvate$$

The phosphotransferase system consists of two enzymes (Enzymes I and II) and a heat-stable low-molecular-weight protein, HPr, which functions as a phosphate carrier in the overall reaction. Both HPr and Enzyme I are soluble and they are released by osmotic shock and other nonsevere methods of cell disruption. HPr has been purified to homogeneity, has two histidine residues, and has a molecular weight of 9500.[1239] It is phosphorylated by Enzyme I on a histidine residue; only PEP can be used as phosphate donor. Enzyme II is the membrane-bound component of the system since exhaustive washing of the membrane, or its rupture with either the French Press or by ultrasonic vibration, does not solubilize this component. It is responsible for specificity with respect to various sugars. Many Enzymes II are inducible while others are found in cells grown on glucose, and are considered to be constitutive. The solubilized and partially purified Enzyme II seems to consist of separate constitutive enzymes II for

TABLE 6-17 PHOSPHOTRANSFERASE-MEDIATED TRANSPORT SYSTEMS

Organism	Substrates	Ref.
Staphylococcus aureus	Galactosides and other sugars	603, 604, 605
Escherichia coli K-12	β-glucoside	606
Escherichia coli W2244	D-glucose and related sugars	607, 617
Escherichia coli W6	Proline	496
Aerobacter aerogenes	Fructose	608
Salmonella typhimurium	Several sugars	609, 610
Bacillus subtilis	α-Methyl glucoside	610
Clostridium thermoaceticum	Glucose and α-MG	610
Saccharomyces cereviciae	Sugars	610a

glucose, mannose, etc. Fractionation of Enzyme II from *E. coli* grown on glucose shows that three components are required for its activity: two proteins and a lipid fraction.[613] Omission of any of these fractions from incubation mixtures containing a phospho-HPr-generating system gives little or no sugar phosphate. The lipid has been characterized as phosphatidyl glycerol. Optimum activity is observed only when a particular sequence of mixing is followed.[1127]

The simplest scheme for explaining the translocation of sugar by the phosphotransferase system is shown in Fig. 6-20. The sugar on the outside is bound to its site on Enzyme II. A conformational change brings the sugar to the inside, still bound to Enzyme II. If the complex dissociates at a rapid rate, this process would explain facilitated carrier-mediated transport. Otherwise the sugar is phosphorylated and released simultaneously (?) inside.

The first evidence indicating that the PEP-phosphotransferase system might be involved in bacterial sugar transport was obtained using *E. coli* W2244 cells by an osmotic shock procedure. The shocked cells lost their ability to take up α-methyl glucoside and thiomethylglucoside (TMG); this capability could, however, be restored by addition of partially purified HPr to the reaction mixture.[607] It has also been noted that the shocked cells contain only a small fraction of their normal content of HPr, and that photooxidation or treatment with trypsin abolishes the ability of the partially purified HPr to correct the deficiency in the shocked cells. It may, however, be noted that only about half of the original activity was recovered under optimum conditions by addition of HPr. Genetic studies have shown that pleiotropic mutants of *Aerobacter aerogenes*,[614] *E. coli*,[615] *Salmonella typhimurium*,[603,609] and *Staphylococcus aureus*[616] which failed to grow on carbohydrates in general also lacked HPr and/or Enzyme I activities when tested in vitro. This defect results from a single mutation. Similarly, mutant has been described which fails to grow on mannitol, and, as expected, this deficiency is related to a lack of mannitol-specific Enzyme II activity. Experiments with cells grown on different sugars indicate that most enzymes II are inducible except for those which phosphorylate D-glucose and D-mannose and their analogs.

All sugars phosphorylated by the P-transferase system are of the D-configuration, and glycosides used as substrates are pyranosides. With the exception of fructose, in all the cases investigated thus far, the sugar is phosphorylated in position 6; glucose, mannose, the corresponding hexosamines and N-acetylhexosamines, galactose, and methyl-β-thiogalactoside have been studied.[607] Regarding fructose, it has been shown that there is a unique inducible P-transferase system which catalyzes the phosphorylation of fructose at the 1 position.[608] It has been shown experimentally that phosphorylated sugars do not diffuse across the membrane; this is confirmed by counterflow or exchange diffusion experiments with related sugars.

As shown in Table 6-17, group translocation by the phosphotransferase system occurs in a variety of lower organisms. Most of these systems behave in an identical fashion. The decided advantage of group translocation by the phospho-transferase system is that the energy used for transport and accumulation is conserved. Phosphorylation has the virtue that a single high-energy phosphate bond (see Chapter 7) is expended for both capture and activation of the substrate for subsequent metabolic reactions. In some bacteria, dephosphorylation of the phosphorylated sugar occurs before its metabolism.

Intensive study of phosphotransferase system has been possible using an osmotically intact preparation of a vesicular structure.[610] These sacs are prepared from normal cells having the desired transport properties. The cell wall is degraded or outgrown and the lysed preparation is subjected to differential centrifugation whereby plasma-membrane bound sacs varying from 0.5 to 1.5 μ in diameter are obtained. Studies with these systems have confirmed the conclusions arrived at from studies on intact cells, and have given some additional insight into the functioning of the system at the molecular level. Some of the conclusions arrived at are the following:[610,617]

1. Enzyme I and HPr are bound to the interior surface of the membrane, since in the external solution a high concentration of PEP is required for initiation of transport.

2. Neither in the intact cells nor in vesicles can PEP be replaced by any of a large variety of "high-energy phosphates" (see Chapter 7 for definition). PEP could, however, be substituted by 2-phosphoglycerate, which is about one-half to two-thirds as effective and the stimulation thus observed is completely abolished by NaF—an inhibitor of enolase, which catalyzes the conversion of 2-PG to PEP.

3. Mn, Zn, or Co can partially or completely replace Mg. Ca or Cu are highly inhibitory at concentrations where other cations are catalytic.

4. Using double labelling techniques, the sugar phosphate ester accumulated by the vesicles was shown to be phosphorylated by the PEP–phosphotransferase system during sugar transport, rather than after penetration into a pool of the free sugar inside the membrane.

5. Well over 90% of the sugar taken up by the vesicles is recovered as sugar phosphate, as is the case with nonmetabolizable sugars in intact cells.

6. Considerable stimulation of glycoside uptake occurs when HPr and Enzyme I are added in the presence of PEP ($> 5 \times 10^{-5}$ M). However, in the presence of NaF in the medium, no uptake of glycoside is observed even though phosphorylation of sugar takes place; all the phosphorylated sugar remains in the external medium. One of the possible explanations is that Enzyme II is capable of undergoing a conformational change in the presence of HPr and Enzyme I such that the complex formed between sugar and Enzyme II is oriented for *vectorial* phosphorylation.

7. With increasing temperature, the steady-state level for uptake of sugar

reaches a sharp maximum at 40°C (in about 30 min). On the other hand, the initial rate of phosphorylation, regardless of transport is maximal at 46°C. Thus, the P-transferase activity of the membrane increases exponentially with temperature up to 46°C, while the membrane becomes leaky above 40°C so as to decrease net uptake.

8. Variation of the osmolarity of the reaction mixture produces dissociation of transport from phosphorylation. A marked stimulation of transport is noted when buffer concentration is increased from 0.02 to 0.2 M potassium phosphate. Concentrations above 0.3 M result in progressive inhibition. Similar effects have been described with the proline uptake system in isolated membranes from *E. coli.*[496] In this case, maximum proline uptake is found at 0.5 M phosphate. One explanation for these observations is that, at low osmolarities, the membrane matrix is expanded, allowing leakage of transported substrates. At optimal phosphate concentrations, the matrix may become less permeable, allowing better retention of small molecules which have been transported. The experiments suggest that the ability of the membrane to retain sugar phosphate is experimentally dissociable from the ability of the membrane to carry out transport.

9. It is possible to cause changes in the barrier function of the membrane without affecting phosphorylation by altering the growth conditions of wild-type parent cells. Membranes prepared from *B. subtilis* grown on a minimal salt medium with glucose as the carbon source (*minimal membrane*) show a steady-state uptake temperature optimum at 27°C and an initial rate temperature optimum at 43°C. With cells grown on an enriched medium (*enriched membrane*), both the steady-state and initial rate temperature optima are at 37°C. It is noteworthy, however, that the rate of phosphorylation (regardless of transport) by these two membrane preparations is similar when assayed. The difference could, however, be traced to the fatty acid profiles of the phosphatides derived from these two membrane preparations. The phosphatides from the *minimal membrane* contain a higher proportion of shorter-chain fatty acids than those derived from enriched membranes. Similar changes in temperature characteristics induced by different growth medium for *E. coli* appears to induce a change in the proportion of phosphatides. It is possible, however, that the primary cause for these changes results from changes in the membrane proteins or in lipid–protein interactions.[618a]

10. The PEP–phosphotransferase system for transport of glucose in *E. coli* appears to be regulated by the immediate product of the transport reaction, glucose-6-phosphate, and also by glucose-1-phosphate and a variety of related hexose phosphates. The inhibitory sites for glucose-6-P and glucose-1-P are separate, distinct, and accessible from outside of the membrane, and the inhibitory effects of glucose-6-P and glucose-1-P on glucose uptake by the membranes vary independently. However, inhibition of glucose transport by glucose-1-P is antagonized by glucose-6-P, and vice versa. Glucose-1-P also inhibits the uptake of

other sugars, indicating that it may be involved in the regulation of the metabolism of sugars in general.[618]

The experimental findings just described leave little doubt concerning the presence of a phosphorylative transport system in a number of bacterial membranes (Table 6-17). However, at this stage little can be said about the molecular aspects of the system as a whole and the relationship between its parts. It seems pertinent to speculate, by way of recapitulation, on the rationale for the phosphotransferase system and its role in sugar transport. It appears that phosphorylation is not only the mechanism for trapping sugar once it enters the cell, but it may also be the energy-conserving mechanism in some cases. Furthermore the translocation step may be coupled to other intracellular reactions. The multicomponent character of the system as a whole offers possibilities by way of genetic regulation of transport processes.

Are There Other Energized Transport Systems?

Despite ingenious experimentation and sophisticated thinking, the answer to this question still rests heavily on parallel arguments or analogy. As summarized in Table 6-18, energy-yielding metabolic changes and perturbations have been considered to underlie a transport process when such changes occur in parallel with the transport process itself. Furthermore, metabolic energy has been considered necessary for the transport of a substance when the transport process has been influenced by blocking of the energy-yielding metabolic machinery. Such indirect approaches for assigning energy dependence for accumulative transport processes are at best circumstantial. By any of these experimental approaches it is not possible to assign the immediate or proximal source of energy. Depending on the kinetics of the processes involved the energy dependence of these processes may be traced at various levels of organization—both functional and morphological; for example, rate of respiration, rate of transport of other metabolites, rate of protein synthesis, etc. Also there are considerations of membrane uptake and renewal, of dynamic exchanges occurring on a somewhat larger scale, as, for example, in pinocytosis and phagocytosis. In fact, it should be emphasized here that transport activity is a kinetic link between the two faces of the plasma membrane. Obviously, *anything* happening on either side of the membrane (and on the membrane itself) will have an effect on transport processes. Otherwise, postulates such as the following[619] would be necessary to account for uphill transport: "a polyfunctional, mobile carrier system . . . . involved in the uphill transport of sugars, neutral and basic amino acids in the small intestine," and the membrane consists of a "mosaic of fixed specific membrane sites which acquire mobility as a result of deformations of the mobile membrane resulting in local transient engagements of the two protein surfaces, thus allowing bound substrates to be alternatively exposed to the extra- and intracellular fluids." Needless to say that this is a description of a cotransport system.

TABLE 6-18 FACILITATED UPHILL TRANSPORT PROCESSES LINKED TO METABOLISM AND CELLULAR ENERGY SOURCE

Permeant	Organism	Remarks	Ref.
Phosphate	Nitella	Light enhanced, requires ATP or cyclic photophosphorylation	629
	B. cereus	Arsenate and pyrophosphate share this system; E_{act} = 12 kcal; energy depletion has no immediate effect	630
Arsenate	Yeast	Inhibited by azide, uncouplers; may have P_i transport system too	631
Chloride	Nitella	Coupled to electron transport and photophosphorylation may be different from K system	632
	Atriplax spongiosa	Linked to light-dependent K system	633
	Intestinal mucosa	Metabolic inhibitors effective	635
Proton	Rat liver mito-chondria[a]	Respiration dependent	634
Nor-adrenaline	Brain (rat)	Na and K gradients are not the only energy source	636
Weak acids	Kidney (goldfish)	Metabolic inhibitors block uptake	637
Uridine	3T3 cells (cultured)	Blocked by metabolic inhibitors in 10–15 sec	638
Purines	Choroid plexus		639
Adenosine, inosine	E. coli	Inhibited by metabolic poisons	640
Arginine, lysine	Sugar cane cell suspension	Inhibited by metabolic inhibitors	641
Fe^{3+}	Rabbit r.b.c.	Inhibited by metabolic inhibitors	642

[a]May be a general characteristic of mitochondria and chloroplasts (see Chapter VII).

It must be obvious, however, that the facilitated transport processes as observed macroscopically have their molecular counterpart. The system as well as the phenomena need be characterized before much can be said about the molecular characteristics involved. Thus from a purely theoretical point of view energy coupling implicated in uphill transport processes could be due to any of the four processes suggested earlier (p. 236) and also due to certain others which are in principle closely related to these. Two aspects of this coupling may be emphasized. The energy process which drives the uphill transport may be either osmotic by itself or may be chemical; however, in either of these cases the work performed during the transport process is osmotic (see Chapter 7 for further discussion of these aspects). If the driving force for transport is osmotic, the medium surrounding the membrane should of necessity be asymmetric; that is, the medium on one side is at higher potential than the medium on the other side. In contrast, if the driving force for the transport is chemical, the membrane-bound transport system should be asymmetric; that is, the direction of approach of substrate, product, inhibitor, and cofactors will be anisotropic. Thus a distinction may be drawn in energy coupling to transport based on (a) asymmetry of the environment, and (b) asymmetry (anisotropy) of the membrane. These distinctions have their counterpart in the mechanisms involved, both at phenomenological and molecular levels. Thus, a cotransport system is an energized transport system of the type (a), whereas phosphotransferase and ATP-dependent transport systems belong to the type (b). Taken in a general sense quite a few *uphill* transport systems may be included in either of these categories:

1. Harold and co-workers[619] have implicated the presence of a proton-driven uphill transport system in *E. coli* and *Streptococcus faecalis*; the uphill transport of substrate is thus coupled to downhill transport of protons. The proton gradient is replenished by a *proton pump* utilizing the metabolic energy (see Chemiosmotic hypotheses on p. 289 for full significance). Thus one can account for the effect of uncouplers (proton conductors) and metabolic poisons on the transport process.

2. Quite a few bacterial transport systems[620] show capability for uncoupling their *energy source* or the energy source is uncoupled by a variety of reagents or by mutation.[1128] The uncoupled transport system however shows the characteristics of a passive facilitated transport system. Furthermore, the substrate does not undergo a chemical change during the transport. In such a situation the role of energy coupling is best rationalized in terms of altered affinity of the carrier for the substrate by the *energy source*.[611,620,1129] The two forms of the carrier may undergo interconversion by asymmetric metabolic reactions; thus one can evaluate the effect of metabolic poisons and competitive inhibitors. The model also predicts the presence of a steady-state accumulation ratio which is maximal for low external substrate concentration and which approaches 1 as the external substrate concentration rises. Furthermore, it should be possible to isolate mutants lacking the coupling mechanism, as, in fact, shown with *E. coli*.[621]

A corollary of this mechanism involves a *loosely coupled* energized transport system where the substrate undergoes chemical change after entering the cell and is thus *accumulated*.[622] In such group transfer processes the carrier system behaves essentially the same whether the energy source is functional or not. In contrast, in *tightly coupled* energized transport systems the affinity of the carrier is dependent on coupling. This may rationalize the observation that in *E. coli* energy is required even for the downhill transport of galactosides.[623] This would imply that the carrier is tightly coupled to the energy source and it cannot operate in the absence of an energy source.

The role of the energy coupling mechanism in the uphill transport may be elaborated further by both negative[624] and positive[625] feedback regulation as observed by loading the cells with the transport substrates. These processes thus imply that the maximum transport rate is limited by the energy coupling mechanism and not by the mobility of the substrate carrier complex. These observations have strong physiological and biochemical implications in the functioning of the energized transport system.[626]

3. Kaback and co-workers have recently published a compelling body of experimental evidence suggesting that sugar[627] and amino acid[628] transport is energized[1240] by oxidation of D-(-)lactate and α-glycerol phosphate in *E. coli* and *Staphylococcus aureus* respectively. Furthermore, the "carriers" may be intermediates in electron transport chain. The available evidence suggests that only the electron transport in the absence of oxidative phosphorylation is required for solute transport. Thus, the uptake is blocked by inhibitors of electron transport, uncouplers of oxidative phosphorylation, but not by arsenate and oligomycin (see Chapter 7 details on these aspects). Similarly, uptake requires oxygen but ATP or PEP do not stimulate uptake of the substrates. These results are consistent with a hypothesis involving oxidation-reduction of a membrane "carrier" protein with resultant conformational change. Alternatively, it is also possible that the proton gradient generated during the electron transfer from lactate to oxygen (see Chemiosmotic hypothesis, p. 299) is utilized from driving the uphill transport as discussed earlier.

All these observations and several others lead to a general conclusion that uphill transport of solutes is a general phenomena observed in a variety of organisms. The mechanism of energized transport has not been established even in one single case; however, the available experimental evidence suggests a wide range of similarities between the mechanisms involved in these various processes. Furthermore, it is perplexing to note that a large number of organisms show a wide range of similarities, and, in contrast, a single bacteria like *E. coli* (save for various strains) can synthesize and exhibit transport characteristics which appear to be different. The relationship between these various transport systems at genetic, structural, functional, and morphological levels is still beyond normal channels of speculation.

7/ Energy Transduction in Biological Systems

The distinction between animate and inanimate nature was for many centuries tied to the concept of energy. The living system requires considerable amounts of energy to survive and to multiply, to bring in substances from the environment, to move about, to contract and expand, and to do various forms of work. It is now well established that ultimately "life-as-we-know-it" depends on the sun's energy. In fact, the primary energy-conserving process in the biosphere is photosynthesis, which is essentially the synthesis of carbohydrates in higher plants according to the following reaction:

$$n\ CO_2 + n\ H_2O \xrightarrow{\text{light}} (CH_2O)_n + n\ O_2$$

680 kcal are conserved during the photosynthesis of a mole of glucose. The carbohydrates so produced are metabolized and produce other ready sources of energy, the most common being adenosine triphosphate (ATP). The overall process that accomplishes the synthesis of ATP in the respiring organisms is called *respiration*. As shown in Fig. 7-1, the energy stored in ATP is transformed into useful work—chemical or physical. The products of respiration, CO_2 and H_2O, are recycled. The overall process involves reactions, starting with those of photosynthesis and ending with those of respiration, linked to diverse processes such as synthesis of complex molecules, active transport of ions, etc. In this chapter we shall describe some of these energy transformations. By energy transformation or transduction we mean any process whereby the energy leaving a system differs in form from the energy entering the system. (A preliminary discussion of the conceptual framework for the present discussion may be found elsewhere.[643])

253

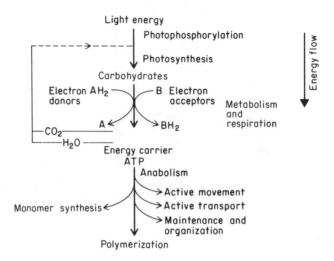

Fig. 7-1. Energy flow diagram showing conversion of light energy into chemical energy and its use in energy-consuming biological processes.

A variety of molecular means are available to the living system for the transformation of energy from one form to another. Thus, radiant energy can be converted into chemical energy, chemical energy into mechanical and radiant energy, radiant energy into kinetic energy, and kinetic energy into potential and osmotic energy under suitable conditions. All forms of energy are *completely* interconvertible, except one: thermal or kinetic energy. Thermal energy is the mechanical energy of the chaotic motion of atoms and molecules.

In a resting mass of gas or liquid, as many molecules move at a given time in one as in any other direction (save for the interface). The law of conservation of energy would not be violated during complete conservation of this energy into another form if the disordered motion were to occur in an ordered fashion. This conviction is expressed in a fundamental principle that says, in essence, that order does not arise spontaneously out of disorder; to create order out of disorder, a certain amount of energy must be expended. This is the second law of thermodynamics. The quantitative measure of disorder is called entropy, S, which is defined as (apart from a proportionality constant, k) the logarithm of the number (n) of different ways in which a particular state of an assembly of many particles can be achieved.

These factors assume particular relevance when considering the biological energy-transducing systems. The principal transducing devices of biological systems are the mitochondrion, chloroplast, sensory receptors, etc. These systems are composed of a highly ordered arrangement of molecules which are themselves synthesized as highly ordered combinations of precursor molecules. In no case do we have a complete understanding of why life processes require this degree of order; however, when it is lost, life is lost. We know that the organization of the cell is such that the availability of reactants, the disposition of products, and the occurrence of specific enzymes are all subject to precise control; any or all of these factors can produce marked changes in the actual free energy of any given reaction. Also the fact that products are constantly disposed of means that, for many processes, equilibrium is never reached, and, under these circumstances, reactions that would occur to an insignificant extent under test-tube conditions can be driven to essential completion.

Yet another characteristic of biological processes arises from the fact that they occur in association with membranes and related organized structures. The significance of this distinction is especially apparent for those reactions in which water is one of the participants: hydrolysis or hydration. Such reactions when carried out in aqueous solutions may show a change in free energy which is largely determined by the high concentration of water in the system. The same reaction in the presence of limiting amounts of water may have its equilibrium position shifted far in the opposite direction. This simple fact acquires an added significance when it is noted that a localized nonpolar environment forms a part of many functional biomembranes, and that certain fundamental biological processes actually have water as one of the participants.

Available evidence suggests that biological membranes can act as reversible energy transducers of four equivalent energy forms: electromagnetic, redox, group transfer, and ion gradient. Conversion of light (electromagnetic waves) into redox energy is one of the first steps in photosynthesis and a host of other photochemical reactions. The underlying process is best understood in terms of excitation of electrons which, in effect, amounts to lowering of the activation energy barrier for an electron-transfer reaction. In electron-transfer reactions the atom receiving electrons is said to be reduced while the donor is said to be oxidized. The energy differential in such redox reactions corresponds to a free energy change (ΔF)

$$\Delta F = -nF(E_a - E_d)$$

where n is the number of electrons transferred, F is the Faraday constant (23.068 kcal/volt), and E is the voltage-potential measured against a standard electrode for an isomolar solution of the oxidized and reduced forms of the acceptor (E_a) on the one hand, and of the donor (E_d) on the other hand. It is well known that some atoms are more *electronegative* than others; that is, some atoms have a greater attraction for electrons, a greater electron affinity, than others.* Oxygen is the best known of the electronegative atoms. When it accepts electrons from another atom the latter atom is oxidized, and oxygen is reduced. It is on this type of movement of electrons that most, if not all, of the activities of the living system are based. The molecules participating in these redox processes are characterized by an electron-transfer potential (Fig. 7-2), such that compounds of more negative potential tend to transfer electrons to those of more positive potential with an associated release of energy. Molecular oxygen is the ultimate acceptor for electrons, thereby establishing the energetic "zero point" for this type of reaction. It should be kept in mind however, that the actual ΔF for an oxidation-reduction reaction depends on the initial concentrations of electron donor and acceptor as in other chemical reactions.

The group-transfer mode of energy transduction is based on the same principles as those operating for electron-transfer reactions just described. The group-transfer mechanism of enzymatically catalyzed chemical reactions generally involves a spatially oriented *diffusion* of the group undergoing transfer from donor to acceptor relative to the anisotropic active center region of the enzyme. This has been one of the most useful concepts in quantitative biology. A simplified view of the concept is provided by the following phosphoryl group-transfer reaction:

$$R-O-PO_3H_2 + R'OH \rightleftharpoons ROH + R'-O-PO_3H_2$$

The most usual acceptor of phosphoryl groups from high-energy phosphate (the donor) is water, thus establishing the energetic "zero point" for this type of reaction. It may be noted here that the terms "low-energy" and "high-energy" bonds are poor terms in the sense that one is dealing here, not with the energy localized in a bond—a bond energy in the strict sense, but with the change in free energy that accompanies a group-transfer reaction, that is, with the differences in free energy between the reactants and their products.

In nucleotide phosphate hydrolysis, the ΔF for ionization of the inorganic phosphate product is part of the free energy change of the overall reaction. Since H^+ is a product of the reaction, the free energy of hydrolysis is very sensitive to pH and the nature of the ions. Since the bond and resonance changes as

---

*In the simplest case of the hydrogen atom, the energy is stored by the relative positioning of the nucleus (in this case a proton) and the orbital electron. The nucleus as such has no energy, nor has the electron. Thus if we were to measure the forces holding the electron relative to the nucleus, it would be equivalent to the energy needed to tear off the electron from the atom and to take it to infinity; that is, to a distance at which it does not interact with the nucleus. The energy needed for doing this work is called the ionization potential and is usually measured in electron volts. It is also implied that if the electron is dropped back to its empty orbital, an equivalent amount of energy will be released (the electron affinity). It may however be noted that the electron need not completely leave the sphere of influence of nucleus and thus only a part of the "ionization potential" will be absorbed or released depending upon the direction of electron (charge) transfer.

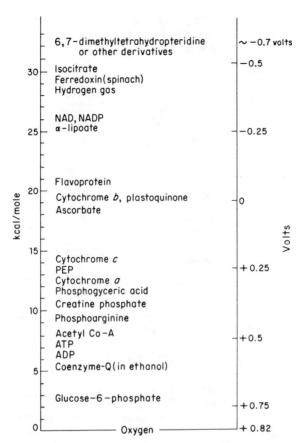

Fig. 7-2. The relative energy levels of some common phosphoryl and electron carriers. The reference line for phosphoryl carriers is the transfer of phosphoryl to water (hydrolysis), whereas the base line for electron carriers is the transfer of one electron to molecular oxygen.

well as heats of ionization may be expected to be roughly equivalent in hydrolysis of both high- and low-energy phosphates, one may ask why some phosphates have a significantly higher ΔF for hydrolysis than others (Table 7-1). The reasons for this are not precisely known, but it is thought to be due to a combination of factors which make the high-energy compounds unique.* Quantum mechanical calculations suggest that phosphates with relatively high free energies of hydrolysis show an unusual distribution of π-electrons, such that the molecule contains

---

*It has been pointed out that there is no unique phosphate bond in ATP and other energy-rich compounds, but rather the high free energy of hydrolysis is a result of a complex interplay of bond energies and heats and entropies of ionization, solvation, and complexation for both reactants and products.

TABLE 7-1 STANDARD FREE
ENERGY OF HYDROLYSIS OF
SOME ENERGY-RICH PHOS-
PHATES

Compound	$-\Delta F$ (cal/mole)
Phosphoenol pyruvate	12,800
1, 3-Diphosphoglycerate	11,800
Acetyl phosphate	10,100
Creatine phosphate	8,000
Aminoacyl AMP	7,000
ATP	7,000
Pyrophosphate	6,000
ADP	5,000
Glucose-1-phosphate	5,000
Fructose-6-phosphate	3,800
Glucose-6-phosphate	3,300
3-Phosphoglycerate	3,100
Glycerol-1-phosphate	2,300

chains of adjacent atoms all bearing a positive charge. In the ATP molecule, for example, a chain of six positively charged atoms occurs:

$$
\text{Adenosine-5}' \overset{+0.153}{-\!\!-} O \overset{+0.393}{-\!\!-} \underset{\underset{-0.89}{O}}{\overset{\overset{-0.809}{O}}{P}} \overset{+0.208}{=\!\!=} O \overset{+0.397}{=\!\!=} \underset{\underset{-0.805}{O}}{\overset{\overset{-0.805}{O}}{P}} \overset{+0.204}{=\!\!=} O \overset{0.306}{=\!\!=} \underset{\underset{-0.821}{O}}{\overset{\overset{-0.821}{O}}{P}} \overset{-0.821}{=\!\!=} O
$$

It has also been noted that most of the high-energy compounds are S or P derivatives; exception may be noted in acyl imidazoles, and in activated amino acids, in which the amino acid carboxyl group is joined in an ester linkage with the 2'- or 3'-hydroxyl of ribose in the terminal adenylic acid of transfer RNA. The uniqueness of high-energy compounds is thought to be derived from S and P atoms present.[644] These structural features ultimately result in resonance stabilization of the products of the group-transfer reaction by electrostatic repulsion of adjacent charged groups, by chelation with polyvalent ions such as Mg, and because of the energy of ionization, isomerization, tautomerism, and rearrangement of the product. All of these factors may contribute to the extra energy of high-energy phosphates as compared with the low-energy phosphates.

By far the widest range of high-energy compounds are derived from the phosphate group. Also important are three groups of energy-rich compounds of sulfur: acyl esters of thiols, mixed anhydrides of phosphoric and sulfuric acids, and sulfonium compounds. Others are uridine diphosphate glucose, acetyl CoA, S-adenosyl methionine, acyl imidazole, etc.

Although the idea of releasing energy by breaking a chemical bond is antithetical to fundamental principles of physical chemistry, the concept of a high-energy bond has stimulated a great deal of research in this area. However, two shortcomings of this generalized concept may be noted. (a) There is no special class of high-energy compounds and (b) the mechanism of energy transfer associated with transfer of a phosphoryl group is not the same in all phosphates, not even in all reactions of the same phosphate.

Energy in the form of ionic gradients has its origin both in the nature of the membrane and of the solution around it. Thus, as discussed in the last chapter, permselectivity of a membrane gives rise to diffusion potentials in asymmetric environments. In fact, the osmotic energy stored in ionic gradients can be utilized in a number of processes. The implications of involvement of ATP in these various processes is therefore crucial while considering the regulatory process. The following points are relevant in this connection: ATP is a universal stoichiometric coupling agent for metabolics; however its steady-state concentrations in most tissues is small, about $10 \ \mu M$. The turnover time of ATP in most of the living cells is 0.5 to 3 sec; however, the ATP level in the living cells is constant under a variety of metabolic conditions. Thus regulation of cellular activity is *perfect* and this has been attributed to the adenylate level termed as *energy charge*[*][645] which has been found to be constant for a variety of cells under quite different conditions. Such energy dependence for homeostasis has functional evolutionary characteristics.

The purpose of the discussion in this chapter is to elaborate various modes of transduction of one form of energy into another. For example, it is now well established that conversion of ADP to ATP, a group-transfer reaction, is one of the terminal stages in the electron-transfer reaction involving a reduced substrate and oxygen. Fundamentally, two different process are *coupled*, and the mode of coupling of such processes shall be the focal point of our discussion. Although quite different energy forms may be involved in these coupled processes, some of the salient features of coupled energy-transducing processes may be summarized as follows:

1. The overall coupled process proceeds with the conservation of free energy. Thus if one system goes from a lower potential to a higher, the coupled system must operate in the opposite direction, i.e., from a higher to a lower potential.

2. In most biological systems coupling appears to be stoichiometric.

3. The overall process is generally carried out by two or more separate but sequential and interdependent steps.

---

*Energy charge is defined as the ratio:

$$\frac{ATP + 1/2 \ ADP}{ATP + ADP + AMP}$$

4. Most of the coupled systems appear to operate reversibly.

5. It follows from the last point that most of these systems operate at efficiencies which approach 100%; for example, acid secretion in the stomach,[646] active sodium transport in frog skin,[647] and sartorius muscle contraction.[648]

6. Each component participating in the overall process seems to be oriented and compartmentalized to facilitate access of reactants and disposal of products. Thus, most of these systems are located in or on intricately designed lamellar structures which can provide for these features.

Before we proceed with a discussion of specific systems, it may be pertinent to elaborate on the last of these features.

Analysis of various structures mediating energy transfer, such as the mitochondrion, chloroplasts, and retinal rods, has brought biology to a level where microscopic structures merge with the atomic structures of molecules, and where the analysis of function merges with the principles of physics and chemistry. From such studies it has become increasingly evident that the coupling of suitable enzyme complexes is facilitated by lipids which are invariably present in some organized form (see Chapter 2). In fact, the organization seems to be one of the most significant features of the energy-transducing systems.* This would imply that the configurational and conformational identity of the macromolecules and of the supermolecular structures involved in the overall process is of prime importance. Thus, organization of these systems into highly ordered quasi-crystalline lamellar arrays may generate conditions which are favorable to physical energy-transfer mechanisms such as semiconduction, photoconduction, exciton migration, resonance transfer, and charge-transfer interactions. In this context parameters such as degree of freedom, quantum levels, translational and vibrational modes which define the energy levels in all molecular and atomic processes, assume an added dimension yet to be explored. It is in this context that controversies have arisen with regard to the nature of intermediates involved; in not a single instance has such a question been answered. It is however obvious that the organizational role of membranes in these various energy-transducing systems is not only to provide a larger reaction area but also to introduce a hydrophobic environment for various processes and to orient complex multicomponent proteins such that the possibility of long-range allosteric interactions may be realized.

In this chapter we shall discuss some of the better characterized biological

---

*Probably it is pertinent to note here that the nitrate-reducing bacteria show varying degrees of membrane organization depending upon the concentration of nitrate ion in growth medium. A considerably higher degree of organization of membranes is observed in starving bacteria; in contrast, bacteria grown with an adequate supply of nitrate show little organization of membranous structures.[649] For similar observations on the photosynthetic purple bacterium, *Rhodopseudomonas capsulata*, see Ref. 650.

energy-transducing and work-performing systems where membrane organization plays a dominant role.* These systems and their functions are:

1. (a) Active transport of cations where energy of a phosphate bond is utilized to accumulate ions against gradients.
   (b) Volume regulation and water metabolism is a process closely related to active transport of ions, and follows from it at a macroscopic level.
   (c) Phagocytosis and pinocytosis and related phenomenon involve substantial reorganization of the membrane (although only *transiently*) which may be energized by high-energy phosphates.
2. Electron transfer and oxidative phosphorylation where redox energy is converted to bond energy of ATP and into osmotic work.
3. Photosynthesis and photophosphorylation where electromagnetic energy is converted in bond energy of phosphates presumably via redox energy.
4. Receptor mechanisms where physical interactions can liberate large amounts of osmotic energy thereby giving rise to considerable amplification of the stimulus.
5. Primitive mechanisms of motility which involve conversion of chemical energy into mechanical energy (*mechanochemical coupling*).

It may be stated at the outset that of necessity the discussion is not exhaustive and is based on various reviews cited in the text.

ACTIVE TRANSPORT OF CATIONS**

A large variety of cells of both lower organisms and higher animals maintain a concentration gradient of sodium and potassium ions across their plasma membrane. Invariably, sodium is present in higher concentrations outside the cell whereas the concentration of potassium is higher inside. The functional reasons for accumulation of ions are more subtle than those for sugars and amino acids. The inorganic ions and their gradients serve a variety of functions, such as maintenance of resting potential, transmission of nerve impulse, driving cotransport, regulation of cell volume, secretion and/or activation of enzymes, hormones, and related substances, and protection against colloid osmotic pressure. More-

---

*Not all biological energy transducing systems are membrane bound. In fact, a specific protein, luciferase, isolated in pure form soluble in water can generate one quantum of light for each molecule of activated substrate (adenylo-luciferin) oxidized by peroxide.[651] Similarly, the "bioluminescent" protein, aequorin, undergoes a light emitting reaction on its interaction with calcium ions, leading irreversibly to a fluorescent product.[652] In both of these instances a membrane system is not required. It is, however, possible that the highly organized hydrophobic interior of proteins may provide the necessary asymmetric environment.

**For the definition of the term, see p. 238.

over, these ions provide the necessary polar environment for intracellular bio-chemical processes which are influenced to different extents by different cations. Of these various reasons volume regulation appears to be the most compelling reason which might have led to the complex ion-discriminating mechanisms present in most cells. With a membrane of limited structural strength, the colloid osmotic pressure of the macromolecules in the cytoplasm results in con-tinuous swelling. This problem has been met by two properties of the membrane: development of a specific cation-transport system which maintains a ratio of K/Na that is greater in the cells than in the medium, and a permeability that is greater for K than for Na. This is often called the "pump and leak" system. From the point of view of volume regulation evolution of the cell wall in lower organisms is well known; this occurs, however, at the cost of more subtle mem-brane functions such as cable properties (see Chapter 8).

The most remarkable feature of active transport of ions in biological systems is the specificity of the system; it can distinguish cations, especially sodium against potassium, with remarkable efficiency. Ionic active transport requires control of permeability by the membrane in such a way that it can regulate the entry and exit of the permeating species. This is an energy-requiring process which is accomplished by the coupling of transport with an exergonic chemical reaction. By the Curie-Prigogine principle of nonequilibrium thermodynamics, a vectorial flow of ions resulting from coupling of energy from a chemical reaction to the flow process can only be achieved in an asymmetric membrane system.[653] This is a consequence of the different tensorial order of the chemical and diffu-sional thermodynamic forces, which are macroscopically speaking, scalar and vectorial, respectively. Any membrane system which will transport ions against their electrochemical gradients using an exergonic chemical reaction as the driving force must, therefore, be asymmetric in the direction of *active flow*.

Such a process is essentially similar to the group-transfer reaction so common in enzymology. The part of the enzyme which catalyzes a group-transfer reaction does not, in reality, occur in the homogeneous aqueous solutions in which the enzyme-catalyzed reaction is ostensibly conducted. Rather, group transfer is a vectorially defined diffusion process, along the reaction coordinates of the tran-sition state complex, and it occurs in an anisotropic microscopic phase, otherwise called the "active center region" of the enzyme. In the overall reactions of such a system the real microscopic variables are negelcted or overlooked; however, such a simplification may not be possible for the membrane-bound "active sites" which are asymmetrically oriented in the membrane plane such that the reaction coordinates are perpendicular to this plane.

The phosphotransferase system for transport of sugars as described in the last chapter would, as expected, share some of the salient features of active transport systems. These are: the vectorial nature of substrate approach and product release, and macroscopic orientation of the chemical reaction in the membrane phase. Both are open systems in steady state, and can exchange energy and matter with the environment. However a distinction is drawn on the fact that in group translocation the transported species undergoes a chemical change and the driving force for such a change is the electrochemical field set up by the oriented chemical reaction in the membrane, as in a fuel or solar cell. In contrast, the species undergoing active transport do not undergo any chemical change and the energy for the translocation of the solute is derived from the chemical work done on the carrier itself.

As pointed out earlier (p. 236), despite ingenious experimentation and sophisticated thinking the identification of energized transport systems still rests heavily on arguments by parallel or analogy. We shall not delve here into elaboration of these aspects; we shall discuss only the cation active transport system. There is no fixed criterion for characterization of such systems, and variations are known with regard to the immediate source of energy. Here again we shall restrict the following discussion to ATP-dependent cationic active transport systems.

Active cation transport systems utilizing ATP as the immediate source of energy are implicated in ion uptake mechanisms operating in kidney, intestine, erythrocyte, nerve, muscle, skin, bladder, nuclear envelope, and several other tissues and organelles from a variety of organisms including human, marine organisms, bacteria, and plant cells. In all these instances, and in various others, the volume fraction of the cell membrane devoted to active cation transport is estimated at 0.002 to 0.4%. However, the amount of ATP consumed may exceed 50% of total cellular production. There is a considerably body of experimental evidence suggesting that active transport occurs through discrete sites randomly distributed over the membrane surface; that is to say, morphologically these sites may not be distinguished except perhaps by electron microscopy and/or auto-radiography. The biochemical studies to be described later in this section attest to this assertion. Since the active cation transport system of red blood cells and nerves is probably best characterized and seen to be representative of a large number of "sodium- or cation-pumps" in tissues of higher animals, we shall restrict our discussion to such a system only. However, it is believed that other active transport systems are closely related but not necessarily identical.[663]

Cationic active transport system of nerves and red blood cells has several morphologic and functional features which may be summarized as follows:

1. The active transport system is located in the membrane and is oriented anisotropically perpendicular to the plane of the membrane; the approach of the reactants and permeants, and their discharge, is vectorial.

2. The transport system can differentiate between sodium and potassium ions and their affinity for the system is spatially oriented; that is to say, sodium ion has higher affinity at the inner interface whereas potassium has higher affinity at the outer interface.

3. Translocation of ions occurs against their electrochemical potential gradi-

ents, and the energy necessary for driving the uphill transport is provided by hydrolysis of Mg-ATP, supplied from the side having higher affinity for sodium ions; addition of Mg-ATP to the other side or its absence from the inside does not drive transport. In some cases, however, it has been found that the simultaneous presence of aspartate ions[654] or of arginine phosphate[655] is necessary. The role of arginine phosphate appears to be the in situ synthesis of ATP, which is the proximate energy source. It may be noted that aspartate anion is one of the main free amino acids of the invertebrate nerve, and its role in physiology of these tissues is not certain.

4. The energy supplied by the hydrolysis of ATP is stoichiometrically coupled to the transport of cations as follows:

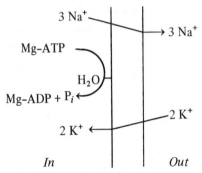

It may be noted that Na and K ions change the sides, and that the orthophosphate (P_i) and ADP liberated on the hydrolysis of ATP are found inside, that is, the side on which ATP and Na are present originally.

5. The functioning of the pump requires the simultaneous presence of sodium inside and of potassium outside. Thus the rate of influx of potassium and the rate of outflux of sodium are dependent upon the internal concentration of sodium and the external concentration of potassium ions.[656-661]

6. It has sometimes been suggested (and assumed) that the coupling of fluxes of sodium and potassium ions is so tight that the *pump* can be regarded as electrically neutral; that is, the outward transfer of each sodium ion is obligatorily balanced by the simultaneous inward movement of *one* potassium ion. This is in contradiction to the stoichiometry shown in the scheme just described. In fact, there is a rather large body of circumstantial evidence suggesting that in a large number of tissues and cells the pump is electrogenic, that is, for each mole of ATP hydrolyzed more sodium ions go out than potassium ions come in. Such electrogenic pumps have been characterized both by electrical measurements and tracer flux studies which can be correlated with and are found to be dependent upon a supply of metabolic energy. Rather similar ratios between the number of ions transported and the molecules of ATP hydrolyzed (Na/K of the order of 3/2 for each molecule of ATP hydrolyzed) are found in a wide variety of tissues, in-

cluding smooth muscle, snail neurons, red cells, nonmyelinated nerve, frog skin, frog and toad bladder, brain and kidney slices, squid axons, rat myometrium, crustacean stretch receptors, cardiac muscle, etc.

7. The stoichiometry of the overall reaction does not vary with the sum of the osmotic and electrical work involved in the process, nor does the potential difference across the membrane have any effect.[662,663] Thus, stoichiometry (three sodium and two potassium ions transported in mutually opposite directions for each molecule of ATP hydrolyzed or 1/3 mole of oxygen consumed) appears to be much the same whether pumping is downhill, on the level, slightly uphill, or steeply uphill. Neither does it depend on whether or not the cation is accompanied by a counterion (presumably through a leak) or whether (during short circuiting) it is moving alone.* The rate of transport however, seems to depend upon the concentration of the ion on the side from which it is transported.

8. The activity of the transport system is greatly inhibited by cardiac glycosides such as ouabain (strophanthin-g) only if the glycoside is present at the outer surface of the membrane. The inhibitory effect of these glycosides can be partly overcome by the addition of potassium in higher concentration to the medium outside.

9. The pump can operate even if all the cytoplasmic constituents of the cell, except for ATP and Na, are removed. In fact, studies done on resealed ghosts and perfused axons have given a considerable amount of information regarding membrane localization of the pump. If the erythrocyte ghosts are put into media rich in sodium but containing no potassium, the pump catalyzes a 1:1 exchange of sodium across the membrane, and hydrolysis of ATP does not seem to occur although presence of ATP is absolutely necessary. This exchange of internal and external sodium is completely inhibited by ouabain and partially by oligomycin.[659] This ouabain-sensitive Na exchange is progressively suppressed, synchronous with a progressive increase in the exchange of internal Na for external K, as external K concentration is raised from zero to 5 mM. The Na-K exchange represents active transport, and it is significant that the concentration of external K required for half maximal activation is 1-2 mM for Na-K exchange and the same value for abolition of Na-Na exchange. Thus the Na translocation step is not the same as the potassium translocation step.

10. Besides the uptake of K and extrusion of Na, the Na pump can catalyze other movements of both Na and K. A component of unidirectional K efflux from cells is sensitive to ouabain, even when the external solution is potassium-free, so that the efflux cannot be balanced by K entry. Similarly, when the intracellular concentration of P_i is increased there is an increased flux of K, and a reduction of internal P_i reduced efflux of K; P_i in the external medium has little

---

*An analogy could be a fixed gear bicycle in which the pedals rotate in the same relationship to the wheels whether the bicycle is moving forward on the level or uphill or moving backwards downhill.

or no effect. As noted earlier, even in high Na media containing as little as 5 mM K, there is no detectable ouabain-sensitive sodium flux. It follows that provided there is sufficient K in the external medium, the ouabain-sensitive K efflux should be thought of as part of internal and external K exchange taking place only when a sufficient amount of P_i is present in the internal medium. This further implies that this K efflux represents a reversal of the last step of the transport system.

11. A curious behavior of the pump is observed when red cell or crab nerve is immersed in solution lacking both sodium and potassium ions. Under these conditions, there is an ouabain-sensitive efflux of sodium ions, and in the crab nerve at least, this efflux is associated with a rise in internal phosphate as though ATP were being hydrolyzed.[662]

12. When the Na pump is inhibited, some of the unidirectional movements of Na and K occur by diffusion and these are thought of as leakage of ions determined solely by the driving force in the form of the concentration gradient. Thus, besides the exchanges of ions described above which depend upon operation of sodium pump, a second kind of pump has been described in red cells;[664] it is insensitive to cardiac glycosides but is dependent on the presence of Na in the external medium.

The observations summarized thus far can be best accommodated in the scheme shown in Fig. 7-3. It describes the operation of the pump in two steps: First the active site of the pump (see below for details) binds ATP and the complex brings about translocation of sodium. If sodium is exchanged for potassium, potassium is translocated. Thus, in summary, the Na pump, in addition to operating as coupled Na and K translocator, also mediates the following partial reactions: (a) Exchange of internal Na for external Na which is *inhibited* (not in the usual sense of the term) by ouabain and external K. (b) Exchange diffusion of internal and external K is increased by higher concentrations of internal P_i and inhibited by ouabain. (c) Hydrolysis of ATP accompanies the exchange of internal Na for external K.

The following characteristics of the pump may also be noted:

The maximum rate of outward movement of sodium ions in the squid axon is

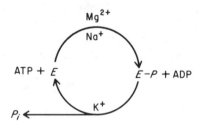

Fig. 7-3. A hypothetical scheme representing the operation of a cation pump.

$60\mu\mu$M-cm^{-2}-sec^{-1}, and the permeability coefficient for sodium ion is 4.6 $\times$ 10^{-7} cm-sec^{-1}. The pump displays saturation-type kinetics, and the maximum velocity is much reduced at low temperature ($Q_{10} \sim$ 2-3). The pH-rate profile is bell shaped with an optimum pH of 7.3.

The energy expenditure for the sodium pump in the isolated squid axon and the human erythrocyte is a minimum of 20% of the total energy produced by each cell. Although there is no general agreement as to the efficiency of the pump; however, it is significantly higher than 50%—possibly 95-100%.

In the red cell at least, there is experimental evidence that the transport system is reversible to a certain extent, as demonstrated by the incorporation of radioactive orthophosphate into ATP under large ionic gradients of both sodium (high outside) and potassium (high inside). Whatever the distribution of ions across the membrane and irrespective of the presence of ouabain, there is always a constant level of labelling of ATP. However, when ghosts rich in K and poor in Na are incubated in K-free Na-ringer, additional labelling was observed which was inhibited by ouabain or high external KCl. It seems that a labelling of ATP occurs at about 2% of the rate of ATP splitting.[665] Comparable evidence for synthesis of ATP have been obtained in perfused giant axons of *Loligo*.[666] In this case, however, there is evidence of involvement of an oxidative phosphorylation mechanism in the net ATP synthesis as determined by a luciferin-luciferase reaction. Further experiments have suggested that downhill movements of both Na and K are needed for labelling of ATP by P$_i$ in all these cases. It, therefore, appears that there is a connection between downhill ion movements and ATP synthesis.

Electrogenic Ion Transport Systems

As mentioned earlier quite a few energized transport systems translocate ions asymmetrically. In fact, it has been shown by the short-circuiting technique[667] that: (a) Under steady-state conditions there is a larger inward than outward flux of Na$^+$, thus originating a net flux of ions. (b) The charge moved per unit time by the net flux of Na$^+$ is equal to the electric current necessary to abolish the electrical potential developed under open-circuit conditions.* (c) The active transport of ions under these various conditions is also the sole source of the electric asymmetry. (d) In order to maintain this net transport of Na$^+$ (or other actively transported ion), an epithelial membrane consumes an extra amount of oxygen. Thus for a wide variety of tissues it can be shown that a microscopic and macroscopic electroneutrality is not maintained in the ion translocation in active

---

*Although in some cases (frog skin, urinary bladder, etc.), the net flux of Na is equivalent to short-circuit current (SCC), this is not necessarily the case with all other membranes. The magnitude of the SCC is given by the sum of the net transfer of all ions across the membrane, expressed as charge moved per unit time; thus, for example *passive leak* of Cl, K, etc., may reduce SCC to almost zero.

TABLE 7-2 ELECTROGENIC ION FLUXES IN BIOLOGICAL MEMBRANES[a]

Tissue	Ion	Open circuit potential (mV)	Inhibitor activators	Ref.
Frog skin	Na	90	Ouabain (I)	667
Toad bladder	Na	80-130	Ouabain (I) Aldosterone (A)	670
Skeletal muscle (frog)	Na	20	Ouabain (I)	671
Cardiac muscle (cat)	Na	5-25	Ouabain (I) DNP, azide (I)	672
Neurones (snail; leech; *Aplysia*, molluscan)	Na	< 30	Ouabain, PCMB (I)	673
Stretch receptor (crustacean)	Na	< 25-30	Li, DNP (I)	674
Nerve fiber	Na	< 30	Ouabain (I)	675
Myometrium (rat)	Na	46	Ouabain (I)	676
Chorio-allantoic membrane	Ca > Sr	40		677a
Turtle bladder	Cl, HCO_3	< 30		677b
Ciliary body	Cl	< 2		677c
Acetabularia mediterranea	Cl	80	Light (A)	1246
Plants	Cl, Na	> 100mV		1242

[a]Ion flux is from mucosal to serosal side for epithelial tissues; thus mucosal side is negative with respect to serosal side.

transport (Table 7-2). These electrogenic ion transport systems have several characteristic features. Some of these arise from the characteristics of the pump itself, such as sensitivity to ouabain, sensitivity to energy poisions and metabolic inhibitors, and dependence of kinetics on ionic environment. In contrast, in quite a few cases, the pump activity can be modified by hormones, particularly aldosterone. It is generally believed that the effect of aldosterone is at the level of protein synthesis. Also, the analysis of kinetic and biochemical data is complicated by the fact that the epithelial tissues consist of several layers of cells, and in some cases the transport of ion itself by the active process may not necessarily be the rate-determining step for the pump; furthermore, the diffusion barrier for all the ions involved in the overall process may not be the same. Thus, an analysis of the experimental results usually rests on the assumption of compartments in the system.

Most of these electrogenic systems can at best be treated as black box. The more detailed analysis of the nature of the frog skin potential led to a proposal *two-membrane model* of the epithelium.[668] It is assumed that the outward-facing epithelial boundary is permeable to chloride and other small anions and furthermore selectively but passively permeable to sodium. The inward-facing boundary, on the other hand, is readily permeable to K as well as Cl, but only slightly permeable to Na. This ion leaves the cellular compartment only by way of an inward-directed sodium pump. Such a system would account for the net transport of NaCl in the open-circuit skin, for the Na current in the short-circuited skin, and also for the maintenance of the epithelial cells in a state of osmotic and ionic balance.[669,1131]

In order to satisfy the requirement of the model, the pump might be an electrogenic pump, a Na/K exchange pump, or a sodium chloride pump. The fact that the pump requires potassium in the inside bathing solution seems to speak for the idea of the Na/K exchange system, although there is strong evidence that it is not a one-to-one exchange in the translocation step. See Refs. 1243 and 1244 for reviews.

The Molecular Basis of Active Transport

The properties of the sodium–potassium coupled pump have been attributed to an asymmetrically located enzyme in the membrane plane. The important work on the membrane fragments of various organisms containing the transport system, originating largely from the far-sighted studies undertaken by Skou,[678] has demonstrated conspicuously that although the intermediate complexes, including the alkali metal ions, must move up the electrochemical gradient during transport, there need not be a direct connection between the primary bond exchange involved in ATP hydrolysis and the translocation of alkali metal ions themselves. Now convincing evidence has accumulated that the hydrolysis of ATP and the simultaneous translocation of cations is mediated by a lipoprotein complex which is generally termed *transport* ATPase or Na+K activated ATPase, or ouabain-sensitive ATPase. Before discussing the mechanism of operation of the "pump," it seems worthwhile to discuss some of the salient features of the transport ATPase.

The Na+K-ATPase activity (EC 3.6.1.4 ATP phosphohydrolase) has been shown to be present by histochemical and biochemical methods in many different tissues from a number of organisms, including red cell, brain, kidney, liver, nerve, thyroid, retina, gills, electric organs, skin, ciliary body, etc. In most of the cases the distribution of enzymic activity parallels the intensity of active cation transport. For example, the potassium-rich red cells of the guinea pig and human have more activity than the potassium-poor red cells of the cat. Similar differences have been noted in the red blood cells of high-K and low-K varieties of sheep. Generally speaking, nervous and secretory tissues contain higher Na+K-ATPase activity than the inactive and noncellular tissues. It is also relevant to note that active accumulation of cation as well as the Na+K-ATPase activity in an organism is changed if the concentration of these solutes in the growth medium is changed. Thus, brine shrimp reared in more saline (NaCl) solution show higher ATPase activity than controls reared in low-salt solution; similarly, rats kept on a low-potassium diet show much higher ATPase activity.

The Na+K-activated ATPase is a lipoprotein complex. It has not been isolated thus far in completely soluble form, although some detergent *solubilized* preparations have been reported not to sediment at 140,000 g in two hours. Typically, the isolated protein shows specific activity of about 1–100 units (μM P_i mg^{-1} protein-hr^{-1}), although preparations with activity as high as 1300 units have been described.[679] The molecular weight of the complex isolated from various sources has been reported in the range $3-7 \times 10^5$. Other enzymic constants may be summarized as follows:

K_m (ATP) 10^{-2} to 10^{-4} $M \cdot l^{-1}$
K_m (Na) $\quad 10^{-2}$ to 10^{-3} $M \cdot l^{-1}$
K_m (K) $\quad 10^{-3}$ to 10^{-4} $M \cdot l^{-1}$
K_m (ouabain) $\quad 10^{-7}$-10^{-6} $M \cdot l^{-1}$: activation energy for binding is 39.5 kcal/ mole at $0°C$ and 7.8 kcal/mole at $37°C$
Q_{10} $\quad$ 2-2.7 for ATPase activity
pH optimum $\quad$ 7.3-7.4
Turnover number $\quad$ 2000-17000 (min^{-1})
Number of sites per mg of protein $\quad 10^9$-10^{12}
Number of sites per cell $\quad 10^3$-10^6
Area occupied by one molecule of the complex $\quad$ 5500 $Å^2$

The properties of the Na+K-ATPase from various sources show similarities which were noted in the transport systems from which they presumably originate. In fact, a number of biochemical and kinetic characteristics attest to it; some of these are summarized below (for Reviews see Ref. 1132):

1. Simultaneous presence of Na, K, and Mg-ATP is necessary in the medium for the ATP-phosphohydrolase activity. K can be replaced by other monovalent cations; half-maximal activation being in the order:

$$Tl > K > Rb > Cs > NH_4$$

Similarly Na can only be replaced by Li. This substitution pattern is the same for transport system and ATPase activity.

2. The ATP/ADP ratio also seems to be important and it changes significantly from one organism to another; ADP possibly acts as a cofactor or modifier of the pump.

3. For half-maximal activation, the concentration of K ions needed in the presence of sodium is 3 mM and 2.1 mM, respectively, for the enzyme and the intact transport system of human erythrocytes. The concentration of ammonium for half-maximal activation in the presence of sodium ions is 8 mM for the ATPase, and 7-16 mM for the intact transport system. Similarly, the concentration of sodium required for half-maximal activation in the presence of potassium is 24 mM and 20 mM for the same enzyme and the transport systems, respectively. Thus ion dependency of transport in the intact system is similar to the ion dependency of the ATPase activity in fragmented membranes.

4. The relation between rate of hydrolysis of ATP and the ratio Mg/ATP suggests that the optimal rates for ATP hydrolysis could be observed at different Mg/ATP ratio depending on the ATP concentration. This suggests an interaction of both reactants with intermediary steps of ATP hydrolysis.

5. The kinetics of Na and K activation of the ATPase from rat brain is of the Michaelis-Menten type; but it is second-order with respect to sodium and first

order with respect to potassium ion concentration, indicating that activation involves two sodium and one potassium ions.

6. One gram of a rabbit kidney preparation binds 0.6 μM of P_i from ATP in the presence of sodium ions alone, but only 0.074 μM of P_i is bound when potassium ions are also present.[680] Similar results have been obtained from several other enzymic preparations. Thus at an early stage of ATPase reaction cycle a phosphorylated intermediate, which is an acyl phosphate,[681] is formed. Electrophoresis of phosphorylated enzyme after dissolution in phenol–acetic acid–urea indicates that phosphate is bound to only one protein band out of 21 bands thus resolved. The same phosphorylated band is obtained when the enzyme is treated with P_i in the presence of potassium ions. Digestion of the phosphorylated enzyme with pepsin has yielded (about 2%) a peptide on which phosphate is bound to a glutamyl residue.[681]

7. ATP, ITP, and GTP are substrates for the transport enzyme and are hydrolyzed by the same substrate acceptor group. The ratio of the reaction rates is ATP:ITP:GTP = 27:2:1.[682] The relative magnitudes are, however, different for enzymes from different sources. An almost identical activation pattern of nucleoside triphosphate hydrolysis by sodium ions in the presence of saturating concentrations of potassium is found with all substrates (K_m = 10 mM NaCl). In contrast, the optimal stimulation by K at saturating concentrations of sodium is variable; hydrolysis of ATP is increased ten times, that of ITP four times, and that of GTP 1.2 times.

8. The enzymatic as well as the pump activity is inhibited by cardiac glycosides[683] and a variety of other compounds. These include, ouabain, hellebrigenin, ADP, IDP, P_i, aminonucleosides, furosemide, ethacrinic acid, thiol blocking agents, urea, organophosphates, phlorizin, phloretin, bergenin, marsalyl, oligomycin, Ca–ATP complex, and a variety of proteases and lipases. The concentration of ouabain for half-maximal inhibition of the enzyme is 10^{-7} M and for the intact transport system is 3–7 $\times$ 10^{-7} M under physiological conditions. It should be noted that there is not only an antagonism between Na and K for the inhibition by ouabain, but also between K and Mg. The enzyme can be labelled by analogs of cardiac glycosides such as 3-bromo acetate of hellebrigenin and strophanthin-G.[684] K_i for various inhibitors differs over several orders of magnitude from one organism to another.

9. The presence of lipids is absolutely necessary for the enzymatic activity. In fact, the preparations showing Na+K-ATPase activity are generally particulate, and any attempt to solubilize these preparations has invariably resulted in their inactivation. Treatment of microsomes with strong salt solutions, or low concentrations of detergents significantly enhance the Na+K-ATPase activity, provided the enzyme is still associated with an insoluble microsomal particle. However, as soon as the microsomal preparation is solubilized by detergent treatment, the Na+K-ATPase activity is lost. This suggests that the forces which are responsible for retaining the enzymatic activity in the insoluble microsomal particle

may also be responsible for holding the enzyme in a *catalytically active conformation(s)*. Alternatively, the Na+K-ATPase must consist of several protein or lipoprotein subunits which must act cooperatively in order to display phosphohydrolase activity. If solubilization of microsomal particles causes dissociation of units and a loss in conformational stability, the catalytic activity shall be lost. Enough experimental evidence has accumulated suggesting that phosphatidyl serine is absolutely necessary for the activation of the enzyme once solubilized by detergent treatment.[685] Similar activation could also be achieved by *n*-decyl and *n*-dodecyl phosphates in low concentrations.[686]

10. Active transport of cations can be induced in BLM with a suitable preparation of Na+K-activated ATPase.[687,1254] The reconstituted pump is electrogenic and the short-circuit current is dependent on the temperature, pH, relative and absolute concentrations of Na, K, and Mg ions, sidedness of ATP, Mg, and ouabain, and their concentrations. Although the evidence for incorporation of ATPase into the membrane is circumstantial, an absolute requirement of traces of didoecyl phosphate in the membrane-forming solution, correlation of pump-inducing preparation with ATPase activity, and the properties of the pump, all suggest that an active transport system has been incorporated. The reverse correlation, that is of the ATPase activity with the induction of pump is not observed; several enzymatically active preparations do not induce pump activity in BLM.

The properties of the membrane fragments showing Na+K-activated ATP phosphohydrolase activity reveal a large number of important features relevant to membrane behavior. Membranes sheared at low pressures and under relatively mild conditions are probably cleaved at structurally weak loci. Thus functionally intact units may be obtained. Fractionation and separation of fragments thus obtained is particularly difficult since they have exposed hydrophobic exteriors and, hence, a strong tendency to aggregate. These considerations bear particular relevance while considering reconstitution of functional modified BLM where the functional proteins from biological systems have been incorporated.[687]

All the evidence summarized thus far in this section shows that the Na+K-activated ATPase fulfils a number of requirements of a system for a coupled transport of Na and K across the cell membrane. Of these the following may be reemphasized: the transport system is located on the membrane; it has an affinity for Na higher than that for K; it has an enzyme system that can catalyze the hydrolysis of ATP at a rate dependent on the concentration of sodium and potassium; it is found (and can be induced) in all cells in which an active coupled transport of Na and K ions occurs; the inhibitors of one system also act as inhibitors for the other. This makes it reasonable to assume that the enzyme system is involved in the transport of cations. This raises the question whether it is part of the transport system or is the transport system itself. Although the evidence is in favor of the second possibility there still are some unanswered questions. To be a transport system, the enzyme complex should be able to transfer the energy from the hydrolysis of ATP into a movement of carrier-bound cations; the carrier sites must not only have specific affinities for the cations in question but they must also be able to undergo cyclic changes in affinity to accomplish transport. An understanding of these aspects is related to other aspects of ion translocation phenomenology; the nature of ion binding, of the translocation step, and the nature of coupling of translocation with the energy-supplying step. The translocation step involves movement of cations and if the ion-carrier complex is charged, the effect of applied external potential should be noticeable. In fact, external potential does not seem to have any effect on active transport across squid nerve at least. The results may however be rationalized by postulating a high activation energy step in the translocation step.

Until now, most of the pertinent data has focussed on the relationship of transport to the activity of the enzyme system. We are only beginning to obtain data that can elucidate the mechanisms involved. Analytical evidence suggests that lipids and proteins are both required for the catalytic activity. The results in general seem to suggest that the transport system is a multicomponent lipoprotein complex. It is certainly not a farfetched idea that in the transport process the liberation of P_i from ATP cannot be a one-step hydrolysis, but that between feeding of ATP into the system and the appearance of P_i (both inside the cell) some intermediate product must be formed which allows the displacement of Na and K ions. A guide to our understanding might be the basic fact that the directional effect of the pump mechanism on the cations has its counterpart in an asymmetry of ATP splitting device. Various ATP phosphohydrolase preparations catalyze, in addition to the Na+K-dependent hydrolysis, two other reactions which are probably related to the transport activity. One of these is transfer of the terminal phosphate of ATP to a microsomal component.[688] The extent of this phosphorylation increases with increasing sodium concentration and is decreased by potassium and other monovalent cations. The second reaction which is demonstrable only when ATPase activity is reduced by decreasing the Mg concentration or by allowing the microsomes to react with certain inhibitors, consists of a sodium-dependent ATP–ADP transphosphorylation. These partial reactions are best reconciled in terms of the scheme shown in Fig. 7-4. The supporting evidence follows.

Impressive support for the existence of the phosphorylated intermediate in two distinct forms comes from the following observations. There is a requirement of low Mg concentration for the ADP–ATP exchange reaction, whereas phosphorylation occurs in the presence of a higher concentration of Mg. Another indication of differences comes from the consequences of treatment of the ATPase with N-ethylmaleimide (NEM) which abolishes the ATPase activity but enhances the exchange reaction. Moreover, the amount of the phosphorylated intermediate in the NEM-treated enzyme is not reduced in the presence of K,

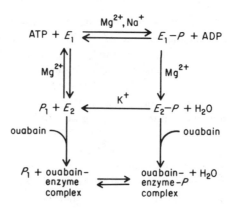

Fig. 7-4. A cyclic scheme representing partial reactions of Na + K-activated ATPase.[689]

which suggests that treatment with NEM inhibits conversion of E_1-P to E_2-P. Oligomycin has an effect similar to that of the treatment with NEM.

There is fairly substantial evidence that ouabain interacts with the E_2-P form of the enzyme and not with the E_1-P form. Furthermore, treatment of the de-phosphoform of ATPase with ouabain in the presence of Mg and at low ionic strength slowly makes it resistant to phosphorylation by labelled ATP. Addition of P_i accelerates this inhibitory action. Also, $^{32}P_i$ by itself rapidly phos-phorylates the ouabain-treated enzyme. Evidence that combination of ouabain with the enzyme after phosphorylation with ATP^{32} yields the same peptide as phosphorylation by $^{32}P_i$ after treatment with ouabain is provided by electro-phoretic comparison of the peptic fragments obtained by digestion of the enzyme after these reactions.[690]

The binding site of ATP is not very well characterized. It is interesting to note that diisopropylfluorophosphate (DFP) inhibits the enzyme irreversibly. This in-hibition is further aided by K and cardenolides, and is antagonized by low con-centrations of ATP but not other nucleotides. Thus the presence of a serine residue is indicated at the ATP binding site or at an allosteric one, which is not involved in the normal reaction cycle.

The role of ions in the overall process is fairly well characterized by tracer flux data and electrical measurements. However, the translocation steps are charac-terized as the ion-activation effects on the ATP-splitting reaction as shown by membrane fragments. Such ion-activating effects are well characterized in several enzymic systems.[691] Generally speaking, monovalent ions may exert their role by maintaining a specific protein conformation necessary for optimum catalytic efficiency. As expected, the ion-binding specificity is governed by the con-siderations described in Chapter 6. Furthermore, enzymes activated by mono-valent cations show a characteristic pattern of reaction. The intermediates for one of the substrates in monovalent cation-activated enzyme-catalyzed reactions have the structure

$$R—\overset{\displaystyle \|}{\underset{\displaystyle X}{C}}—Y—R \quad \text{or} \quad R—\overset{\displaystyle X^-}{\underset{\displaystyle |}{\overset{\displaystyle \|}{C}}}—Y—R$$

where X is O, N, or C and where Y is O or N. Obviously, involvement of glutamate residues as phosphorylated intermediates is consistent with this scheme. The kinetic data has yielded additional information. K^+ activates the turnover of enzyme hyperbolically at low concentrations, but with increasing concentrations of Na the activation curve becomes sigmoid. Similarly, at low concentrations of K, the Na activation curve is hyperbolic but at high concentrations of K it becomes sigmoid. These and additional data on the effect of monovalent cations on various partial reactions and ouabain binding have been interpreted to suggest a cooperative interaction between the Na binding sites.[692] The role of K in the dephosphorylation step may be related to this process.

At this point we have some knowledge of the effect of ions and substrates on the enzymic reactions of the intermediates, of the active site for the phosphohydrolase reactions, of the stoichiometry, and the kinetics and mechanism of ATP hydrolysis. However, very little can be said with regard to the mode of coupling of ion translocation to ATP splitting. There are no definite guidelines to approach this problem. It is conceivable that the enzyme can discriminate between Na and K, and change affinity for these ions by using subtle steric effects and/or allosteric changes. In fact, there is evidence from a large variety of sources (kinetic and biochemical) suggesting that the conformation of ATPase is different in the presence of Mg-Na-ATP and Mg-K-ATP.[692-696] Thus it is likely that conformational changes exerted by ATP and monovalent cations in the membrane-bound enzyme constitute the basis of the *rotation* of the active center, thus bringing about the vectorial translocation of cations.

The present state of knowledge for the cationic active transport system is thus at an optimum level for model building. Of course, there is no dearth of these models, although none of these is complete and satisfactory. In the light of the scheme shown in Fig. 7-4, various models may be evaluated. Whittham[697] suggests that potassium catalyzes the hydrolysis of phosphorylated intermediate as an allosteric affector. As a feedback of the exoenergetic hydrolysis step, two potassium ions are ejected in the opposite direction. Jardezky[698] has provided a model which uses allosteric opening and closing of a binding site which oscillates between K- and Na-directed sites. Opit and Charnock[699] favor a rotating model with binding sites changing specificity due to field-strength phenomena. The energy for rotation is provided by ATP hydrolysis. The model proposed by Lowe[700] is more or less a statement of the scheme shown in Fig. 7-4 in relation to hypothetical membrane-bound components located asymmetrically and operating vectorially. The model proposed by Middleton[701] emphasizes ion binding specificity of the enzyme-ATP complex by virtue of size-restricted sites such as may be present during various partial reactions. Ouabain is presumed to bind covalently to E_2-P intermediate. The model examined by Hill[702] is based on a set of hypothetical units which exist in two states and which interact cooperatively. Although the model is not specific for Na+K-ATPase, however, its general character offers considerable latitude for further elaboration.

Obviously, such exercises in model building have to go a long way. Some specific proposals with regard to specific questions may help to design new experiments. Thus, for example, it remains to be elaborated whether cations and ATP have the same or different binding sites; whether Na and K have the same or different binding sites; whether the system as a whole allows ions to pass in a single file or forms transient channels, or operates on the carrier diffusion type of mechanism.* There is disparity in the number of monovalent cations that are transported in opposite directions across the membrane. This raises the questions regarding both macroscopic and microscopic electroneutrality. The nature of the transition state, the nature of energy-transferring step, the efficiency of energy transfer, and reversibility of various steps involved in the overall process (microscopic reversibility), indeed, pose some fundamental questions to which we do not have any clue whatsoever (save for the extrapolations from classical thermodynamics which may not be relevant in the present context.)

METABOLISM OF SALT AND WATER

Conservation of water is a problem of the utmost importance for living cells. It is upon water that the chemical processes of life depend. Some of the physiological and pathological states associated with impairment of concentration or dilution control are given in the Table 7-3. In all organisms, however, the metabolism of water is closely connected with the metabolism** of salts; that is to say, when there is transfer of solute from one compartment to another across a semipermeable membrane, there is also a transfer of the solvent such as to restore osmotic balance. An important aspect of this metabolism is development

---

*It is unlikely that a freely moving carrier is involved in translocation step for either of the cations. It is difficult to conceive asymmetry in such a system unless some other membrane-bound component is involved in regulating the carrier. As the protein moiety of the transport enzyme has a molecular weight of at least 250,000, and has a strong tendency for association with specific lipids, the translocation would imply some sort of restriction with respect to the orientation of lipids in the membrane hyperstructure.

**For the present purpose "metabolism" may be considered as synonymous with "transfer" across biological membranes.

TABLE 7-3 PHYSIOLOGICAL AND PATHO-
LOGICAL STATES ASSOCIATED WITH
IMPAIRMENTS OF CONCENTRATION
OR DILUTION

Concentration-impaired	
Adrenalism	Hydration
Alcoholism	Hypermia
Medullary amyloidosis	Kalemia, hypo
Hypercalcemia	Chronic nephritis
Diabetes insipidus	Renal failure
Diuresis	Sicklemia

Dilution-impaired	
Adrenalism	Sicklemia and anemia
Hypopituitarism	Hypothyroidism
Hypervasopressini. n	Sodium retention

of regulatory mechanisms permitting maintenance of a constant chemical composition. At the organism level it is a complex process involving interaction between various systems. For the sake of illustration the principal areas of interaction between the water, salt, and potassium in the human body may be summarized as follows:

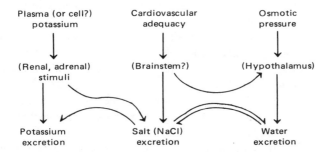

At the cell level, a major element of this regulatory system appears to be the control of the entry or exit of water and salt across the boundary membrane by an active or passive process. Transfer of water across the plasma membrane is a passive process, although under certain circumstances water can move against, or in the absence of, an osmotic gradient.[703] A possible explanation of water transfer against or in the absence of an osmotic gradient is that ion transport across the membrane is able to "drag along" water molecules. This may also be brought about by a "co-diffusion" effect in which solute transported in a water-filled channel would be assumed to exert a frictional drag against the walls and push a small column of water molecules along. Yet another explanation may be

TABLE 7-4 SOME CONDITIONS ASSOCIATED WITH PARALLEL CHANGES IN SODIUM AND POTASSIUM EXCRETION IN THE HUMAN

Increase in excretion

Living at high altitude	Ingestion of angitensin
Fluid or salt loading	Ingestion of mercurial diuretics
Heat failure	Ingestion of pituitary principles
Manic-depressive illness	Schizophrenia

Decrease in excretion

Abdominal compression	Alcohol diuresis
Anesthesia	Blood sequestration
Exercise	Water loading
Ingestion of serotonin	Schizophrenia

based on a local osmosis or *double membrane effect.*[*][704] Both of these models suggest that the uphill transport of water is only "active" in appearance.

Thus the "pump and leak" system can regulate the activity of water in cells by regulating the total ion content of the cell. As is in fact observed experimentally, reduction in the rate of active transport, for example, by chilling at 4°C, causes swelling of various tissues when buffered in sodium chloride. In contrast, incubation of these tissues at 37°C under appropriate metabolic conditions prevents this swelling.[707] Similarly, active transport of cations has been shown to be the primary rate-limiting event in the secretion of aqueous humor in the rabbit, in the formation of cerebrospinal fluid, pancreatic secretion, and other related processes. Thus any malfunction in the cationic active transport system should show up in major systemic disorders; in fact, quite a few of these are known (Table 7-4). Not all of these are the result of malfunction of ATPase, but they are all associated with disturbance in ion and/or water balance.[708] In some cases a direct role of the cationic active transport system can be demonstrated by inhibitors of active transport. Thus vegetative and synthetic cathartics such as cascara, podophyllum, ricinoleic acid, and phenolphthalein all block Na+K-activated ATPase.[709] It is believed that inhibition of active transport of cations (and consequent sugar cotransport) results in water retention, in an increased

---

*The thermodynamic analysis of this model was carried out as follows.[705] Three compartments in series are separated by two barriers. The first external barrier is assumed to be semipermeable and to be the site of active transport of solutes. The second barrier is nonselective and merely retards the diffusion of solutes without playing any osmotic role. The active transport across the external barrier will sustain a steady state in which the intermediate compartment is slightly hypertonic and water will therefore also flow through the external barrier as a result of "local" osmosis. Thus hydrostatic pressure will build up in this intermediate compartment, and will drive the water across the internal barrier into the internal compartment. Thus the system is able to accomplish "uphill transport of water." For another model see Ref. 706.

bulk in the lumen of the intestine, and a reflex stimulation of peristalsis.[710] The effect of hormones on ion/water balance is much more difficult to interpret. For example aldosterone is an hormone which stimulates sodium reabsorption in kidneys and (experimentally, but more conveniently) in toad bladder. It seems to work by stimulating protein synthesis through direct gene activation. It is believed that the protein or proteins synthesized as a result of aldosterone activation are transport proteins.

The volume regulatory processes have a rather strong influence upon the growth of the cell. As they are in equilibrium with their immediate environment, any change in size preceding cell division must occur as a consequence of a change in the osmotic gradient created primarily by the accumulation of electrolytes. It is obvious that as the cells divide and grow, as the quantity of macromolecules inside the cell increases and the vacuoles expand, there must be a concomitant uptake of appropriate ions both to maintain the correct milieu for the proper functioning of the cellular processes and to maintain or even increase the hydrostatic pressure which keeps the cells turgid. Whatever be the controlling mechanism at higher levels of organization (hormonal or otherwise), which maintains the orderly pattern of division, growth, expansion, and differentiation, ion accumulation is a necessary consequence of the expanding membrane surfaces at the cellular level.

Regulation of cell volume also has some evolutionary implications. Insofar as the living cell consists of macromolecules in close spatial relation to one another, one of the most fundamental requirements for the persistence of the system is the avoidance of the dispersion of macromolecules by diffusion. The formation of a membrane that is relatively impermeable to dissolved solutes was apparently the first step toward solution of this problem. This could however be adaptable to changing internal and external environment if a tough wall that could resist large pressure (osmotic and hydrostatic) differences was evolved, as in plant cells. For animals, nature seems to have chosen a more plastic course by evolving a weak and flexible membrane which is impermeable to ions, but the active transport system for these ions renders the membranes functionally permeable to the same ions. The membrane with its asymmetric environment acts as an ideal osmometer, as a storage battery, and a signal generator. The evolution of walled cells, by removing the need of osmotic regulation, allows cells to migrate into media of low tonicity but at the cost of more subtle (and maybe higher?) functions.

In summary, it is clear from the preceding discussion that constancy of body fluid, including volume as well as composition, is a particular instance of stabilization of the integrated aspects of body function such as blood pressure, metabolic rate, total body composition and, ultimately, growth and form itself. Inherent in this is the concept of steady-state control or energy-dependent *homeostasis* at the cellular and molecular levels.

PINOCYTOSIS, PHAGOCYTOSIS, AND RELATED PHENOMENA*

One of the most striking and dynamic properties of the plasma membrane is its ability to capture external materials by pinching them off into membrane-lined packets or vacuoles; these vacuoles may then be transferred rapidly to the deeper parts of the cytoplasm, where they undergo fusion, fragmentation, or appropriate kinds of processing. Such surface vacuolization or endocytosis has been observed in many kinds of cells including neurons, muscle fibers, kidney tubule cells, hair root cells, intestinal epithelium, gall bladder epithelium, liver cells, adipose tissue, osteoclasts, fibroblasts, insect oocytes, pollen mother cells of certain plants, amoeboid protozoa, etc. (see Ref. 711 for reviews). The most detailed studies of the phenomenon have been carried out with amoeboid protozoans and leucocytes whose capacity to engulf food particles was well known even in the last century.

When an amoeba approaches small food particles its membrane flows around and traps them into small vacuoles (Fig. 7-5) which are transported or carried across the membrane in several steps. The average number of pinocytosis channels varies predictably with temperature, pH, and the nutrient condition of the amoeba. When amoebae are transferred to a standard medium they begin to form channels within a few minutes, reaching a maximum number in about 15 min and tapering off to zero after 30 min. Subsequently, the complete pinocytosis cycle cannot be induced again until after the end of a digestive phase of 3-4 hr when streaming locomotion is resumed.[711 b]

A detailed EM study has recently appeared of vesicles in contact with the plasmalemma, in endothelial cells:[713] two types of contact are observed, one of which these authors associate with "discharge" and the other with "loading" of vesicles. In one type of contact a gamut of appearances is observed from tangential contact to extensive fusion of membranes, giving a vesicle with contents separated from the exterior by a thin diaphragm, which presumably ruptures. In the second type of contact the vesicles possess a neck or stalk, without a diaphragm, through which its lumen communicates with the exterior. This neck is assumed to elongate progressively and finally to become "pinched off"; in some cases a vesicle is apparently still attached to the plasmalemma by a solid strand of dense material.

In both amoebae and tissue cells, the size of individual vacuoles varies widely, partly due to rapid coalescence of small vacuoles into larger ones and partly to a reverse process in which smaller vacuoles are seen to bud off from larger ones. Quantitative studies suggest that the total volume of fluid taken up in a single

---

*The term pinocytosis is used to describe the process of excretion and engulfing small droplets of the outside medium through the membrane. In phagocytosis aggregates or particles are transported by a similar mechanism.

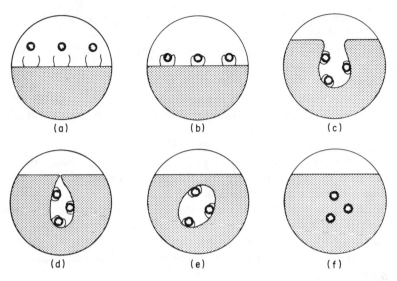

Fig. 7-5. Schematic representation of membrane flow during pinocytosis in amoebae. Pinocytosis begins with cessation of streaming followed by detachment from the substratum, wrinkling of the cell surface, and paralysis of contractile vacuole activity. Next (a), short hyaline pseudopodia form at various points of the surface giving the amoebae a rosette appearance (b). Finally at the tip of each hyaline pseudopod, a narrow, actively undulating channel is formed (c), from the innermost end of which small vacuoles begin to be separated (d, e, and f) into the cytoplasm.[712]

pinocytosis cycle is 1-10% of the total volume of the amoeba, and that the area of new membrane which must be synthesized is about $3 \times 10^4 \mu^2$, corresponding to about 6% of the total surface area of a streaming amoeba. In experiments during which the pinocytosis cycle is interrupted and resumed, the cells behave as if there is an absolute limit in the amount of endocytosis that can be carried out in a given time by a single cell. This limit is believed to correspond to the limit in precursor material available for the synthesis of new plasma membrane.

Several compounds such as proteins, basic dyes, and charged molecules induce formation of pinocytotic vesicles. For proteins the activity is closely related to the isoelectric point such that induction occurs primarily on the acid side of the isoelectric point. Similarly, other substances which possess a positive charge also induce pinocytosis; however, substances like glucose taken up during pinocytosis are not charged.[714] Thus considerable amount of histochemical, electron-microscopic, and immunological evidence suggests that the surface charges of the membrane, which are negative, are involved in the binding of positively charged inducer molecules.

That endocytosis is a process which requires expenditure of metabolic energy seems likely from the active and often violent undulations exhibited by the plasma membrane as observed in time-lapse films. Support for such an assertion may be derived from the following observations:

1. Mitochondria are conspicuously oriented near the active surface, and can sometimes be seen to exhibit remarkable saltatory or linear tension movements in the direction perpendicular to the plasma membrane.

2. The pinocytotic activity may be associated with contractile ATPase activity.

3. The rate of activity in tissue culture cells is increased by addition of ATP to the medium.

4. In amoebae ATP acts as an inducer of pinocytosis.

5. In polymorphonuclear leucocytes of the guinea pig, rabbit, and swine, phagocytosis is accompanied by the increased utilization of glycogen or of exogenous glucose and by increased lactate production.

6. Endocytosis is inhibited by respiratory and metabolic poisons, for example, iodoacetate and fluoride. On the other hand, the alveolar macrophage depends largely on aerobic metabolism for the ingestion of solid objects. Inhibition of electron transfer or uncoupling of oxidative phosphorylation also inhibits phagocytosis.

Thus it seems likely that endocytosis is an energy-dependent mechanism for the selective but gross transport of material across the plasma membrane.* Such a process has some far-reaching implications. For example, active transport of Ca, as observed across the embryonic chick chorio-allantoic membrane has been attributed to "a very specialized and selective form of pinocytosis or endocytosis."[716] Such a transport process has some distinct merits, e.g., by compartmentalizing substrate in transit, the internal *machinery* of the cell may be protected from excessive concentrations of Ca which may, for example, uncouple oxidative phosphorylation (see later) or markedly inhibit pyruvate kinase, pyruvate carboxylate, or some other metabolic enzyme. Thus by this mode of transport, Ca may be compartmentalized during transport before gaining access to the embryonic circulation. If it is in fact so, endocytosis can account for at least some of the "active transport" phenomena detected at the levels of tissue physiology.

ELECTRON TRANSPORT AND OXIDATIVE PHOSPHORYLATION IN MITOCHONDRIA**

As pointed out earlier, the usual source of energy utilized for the synthesis of ATP or some other energy-conserving process is the energy liberated during

---

*An important aspect of phagocytosis is implicated in the bactericidal action of leucocytes. Biochemically, this function has been assigned to the peroxidase of the leucocyte, that is myeloperoxidase. Genetic lack of this enzyme results in defects associated with (a) diminished leucocyte bactericidal and fungicidal activity and with (b) granulomatous disease, where the lysosome fails to rupture.[715]

**Greek, *Mitos* is for filament; *chondros* for granule.

transfer of electrons from a donor to an acceptor. The usual electron acceptor in higher organisms is oxygen, which accepts the electrons derived from a metabolic cycle characteristic of the organism. The overall process can be conveniently divided into three sections (see Ref. 1133 for reviews).

1. The rearrangement and oxidation of carbohydrates to produce reduced coenzymes which function as electron donors. We shall not consider various aspects of this process since they have little relevance to membrane phenomenology. For an excellent account see Ref. 717.

2. Oxidation of reduced coenzymes by molecular oxygen via the components of the respiratory chain designated as the electron-transport chain.

3. The energy released during oxidation-reduction reactions of the electron-transport chain is utilized for the synthesis of ATP or some other related energy-conserving process.

Thus the overall reaction of oxidative phosphorylation may be formulated as:

$$NADH + H^+ + \tfrac{1}{2} O_2 + 3X + 3I \rightleftharpoons NAD^+ + H_2O + 3 X \sim I$$
$$3 X \sim I + 3 ADP + 3 P_i \rightleftharpoons 3X + 3I + 3 ATP$$
$$\overline{NADH + H^+ + \tfrac{1}{2} O_2 + 3 ADP + 3P_i \rightleftharpoons NAD^+ + H_2O + 3 ATP}$$

It may be noted that X and I are hypothetical entities and nothing is known about their nature (however, see below).

Various steps involved in the overall reaction may be considered as separate events in space and time, coupled to one another by a *transducing* device intrinsic to the electron-transfer chain. The unifying feature of the mitochondrial energy-transducing process as observed in diverse organisms arises from the fact that the electrons from a donor of suitable potential pass to an acceptor and their flow can be coupled stoichiometrically to the synthesis of ATP. The mechanism by which electrons pass on from the original donor to the final acceptor involves a series of steps mediated by a sequence of electron donors and acceptors of intermediate potentials. The overall process as mentioned above thus carries little information as it is.

Mitochondria have been recognized as the center of cellular respiration.[718] Mitochondria are in essence constituted of a series of multienzyme complexes characteristic to various metabolic reaction pathways. The complexes are closely associated with sophisticated membrane systems which together organize and control the overall process of energy transduction. Mitochondria have two membrane systems (Fig. 7-6): a smooth outer membrane enveloping the mitochondrion and separated by a space from the much involuted inner membrane* (see Ref. 1134 for detailed description of cytoarchitecture). The folds on the inner

---

*Definitions of the mitochondrial inner and outer membranes, somewhat different than what is implied here have been proposed and employed by Green and associates. In their latest model a considerable overlap and "interlock" of two membrane systems is implied.[45]

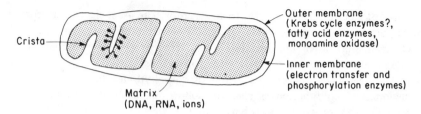

Fig. 7-6. A schematic representation of a mitochondrial membrane system.

membrane give rise to structures known as the cristae, which appear to be composed of smooth membrane if observed under the electron microscope after positive staining. If, however, they are negatively stained with phosphotungstate they appear to be covered with spherical subunits. The inner membrane encloses an apparently structureless matrix.

While discussing the properties of mitochondria it is interesting to recall that one of the theories of the origin of the mitochondrion postulates that the *ancestors* of mitochondria are bacteria which infected large host cells, and in the course of evolution ultimately developed a symbiotic metabolic relationship with them.[719] This might also account for the occurrence of similar energy conservation mechanisms in the membranes of the mitochondrion and bacterium. Furthermore, characteristic chemical differences between the inner and the outer mitochondrial membranes (see below) may correspond to vestiges of the original bacterial cell membrane and wall, respectively. Cardiolipin, for example, is the characteristic lipid of the inner mitochondrial membrane; a very similar class of lipids, the glycerol phosphatides is widely distributed in bacterial membranes.[720] Similar lines of evidence are now developing which suggest that the chloroplast of higher plant cells may also have arisen in evolutionary history from parasitizing chlorophyll-containing bacteria. It is obvious that unravelling of the energy-transduction mechanism from various organisms will have strong implications not only for cell biology, but also for evolution and related studies.

The inner and outer membranes can be separated by relatively mild treatments, such as swelling with hypotonic orthophosphate solution, treatment with low concentrations of digitonin or some other detergent, or by sonication. For isolation of separated membranes, advantage is taken of significant differences in their density; isopycnic centrifugation on a sucrose density gradient is generally preferred.[721] The properties of inner and outer mitochondrial membranes are compared in Table 7-5. The enzyme systems located in these membranes, in the intermembrane space, and in the matrix are summarized in Table 7-6.

The inner and outer membranes of mitochondria differ significantly both in morphological and functional aspects. The electron-transfer chain, and a set of systems that underlie the capability for coupled phosphorylation of ADP, energized swelling, energized transhydrogenation, active transport of various sub-

TABLE 7-5 PROPERTIES OF INNER AND OUTER MITOCHONDRIAL MEMBRANES

Property	Inner Membrane	Outer Membrane
Response to osmotic changes	Readily infolds and refolds	No reversible response; undergoes distension and rupture
Density	1.21	1.13
Shape	Folded	Distended
Thickness (Å)	50-70	50-70
Phospholipid/protein (w/w)	0.27	0.82
Protein content (percentage of total mitochondrial protein)	21	4
Phospholipid (mg/mg protein)	0.301	0.878
Cholesterol (μg/mg protein)	5.06	30.1
Cholesterol (μg/mg phospholipid)	16.8	34.3
Phospholipid conposition (rat liver mitochondria)		
Phosphatidyl choline	41	49
Phosphatidyl ethanolamine	35	31
Phosphatidyl inositol + phosphatidyl serine	2	17
Cardiolipin	21	3
Phosphatidyl glycerol	1	—
Ubiquinone	Present	Absent
Acetone extraction of 90% phospholipids	Double-layered structure retained	Membrane destroyed
Substructuring	Subunits may be attached	No significant structures
Enzyme distribution data	See Table 7-6; site of respiratory chain	See Table 7-6
Permeability characteristics		
Passive	Very limited except for low-molecular-weight substances	Permeable to substances up to 10000 mol. wt.
Energized and facilitated transport	Specific translocators for P_i malate, succinate, citrate, isocitrate, glutamate, long-chain fatty acids, divalent cations, etc.	Little evidence for energized or even facilitated transport
Surfaces	Outer surface smooth; inner surface covered with regularly spaced projecting subunits	Inner surface smooth; occasional projections on outer surface

strates, reversal of electron transport (reductive phosphorylation), and probably activation of protein synthesis seem to be associated with the inner mitochondrial membrane. The convoluted structure of the inner membrane to form cristae may have its functional significance in the fact that such an organization increases the total surface of the membrane without changing the volume of the organelle. The most important characteristic of these membranes arises from the anisotropy of orientation of various functional molecules on the membrane. This is probably best illustrated by the elegant experiments of Mitchell, Moyle, and Lee.[722] Isolated mitochondrial membrane fragments round off into vesicles, so these workers were able to demonstrate proton extrusion by the vesicles when the outside of the membrane was to the outside of the vesicle, and the proton uptake when the outside of the membrane was to the inside of the vesicle.

TABLE 7-6 LOCALIZATION OF ENZYMES IN LIVER MITOCHONDRIA[a] [721]

Outer Membrane	Intermembrane Space	Inner Membrane	Matrix
"Rotenone-insensitive" NADH-cytochrome c reductase	Adenylate kinase	Respiratory chain [cytochromes b, c_1, c, a, a_3, succinate dehydrogenase; succinate-cytochrome c reductase; succinate oxidase; "rotenone-sensitive" NADH-cytochrome c reductase	Malate dehydrogenase
NADH-cytochrome b_5 reductase	Nucleoside diphosphokinase		Isocitrate dehydrogenase
cytochrome b_5	Nucleoside monophosphokinase		Isocitrate dehydrogenase
Monoamine oxidase	Xylitol dehydrogenase	NADH oxidase; choline-cytochrome c reductase, cytochrome c oxidase; respiratory chain-linked phosphorylation	Glutamate dehydrogenase
Kynurenine hydroxylase			α-Ketoglutarate dehydrogenase lipoyl dehydrogenase
ATP-dependent fatty acyl-CoA synthetase			Citrate synthetase
Glycerolphosphate-acyl transferase		β-Hydroxybutyrate dehydrogenase	Aconitase
Lysophosphatidate-acyl transferase			Fumarase
Lysolecithin-acyl transferase		Ferrochelatase	Pyruvate carboxylase
Cholinephosphotransferase		δ-Aminolevulinic acid synthetase?	Phosphopyruvate carboxylase
Phosphatidate phosphatase			Aspartate aminotransferase
Phospholipase A_1		Carnitine palmityl-transferase	Ornithine-carboamoyl transferase
Nucleoside diphosphokinase		Fatty acid oxidation system?	Fatty acyl-CoA synthetase(s)
Fatty-acid elongation system		Fatty acid elongation system	Fatty acid oxidation systems?
Xylitol dehydrogenase		Xylitol dehydrogenase	β-hydroxybutyryl-CoA dehydrogenase
			Xylitol dehydrogenase

[a]The data in this table are based on results obtained in fractionation studies involving separation of the outer and inner mitochondrial membranes. Further information concerning the intramitochondrial localization of enzymes, originating from studies with intact mitochondria, is discussed in the text. For definition of mitochondrial compartments see Fig. 7-6.

The Respiratory Chain

In the early 1930s the work of Keilin and Warburg led to the general concept that certain biological oxidations were mediated by electron carriers and other redox catalysts arranged in a chain. The location and positioning of these components as assigned by spectrophotometric techniques developed by Chance and co-workers is shown in Fig. 7-7. The site of action of various inhibitors is also shown in this figure.

Such an examination of a respiratory chain by means of specific inhibitors and substrates rests largely upon the generality[724] that components of the respiratory chain on the oxygen side of the point of action of the added reagent become more reduced in the case of substrate, and more oxidized in the case of an inhibitor. If the inhibitor is added to a system in which the substrate and oxygen are already present, the components on the oxygen side of the site of action of the inhibitor will become more oxidized and those on the substrate side of the site will become more reduced. The point in the respiratory chain at which the effect of an inhibitor upon the steady-state shifts from an oxidation to a reduction, or vice versa, is called the *crossover point*. This crossover point locates the site of action of the inhibitor. To measure steady-state shifts, advantage is taken of changes in absorbance characteristic of various components of the chain.

All the components constituting the respiratory chain, with the exception of cytochrome c and ubiquinone (Q_{10}), are firmly bound to the inner mitochondrial membrane. Studies on sonicated mitochondrial membrane fragments point to three characteristic properties which seem to afford the structural basis for membrane function:[725] (a) Only 40% of the cytochromes c_1, c, a, and a_3 are accessible to charged anion, ferricyanide, or to cytochrome peroxidase enzyme substrate complex. This is evidence for the specific location of respiratory chains on both sides of the membrane permeability barrier, with the phosphorylating chains being inaccessible to the external oxidants. (b) The inability of the cytochromes which are accessible to ferricyanide to oxidize their complementary "internal" partners seems to be an indication of the impermeability of the membrane to electron flow at the c_1, c, a, and a_3 level. (c) The selective responsiveness of the added cytochrome c to controlled electron flow suggests that in the sonicated particles only the molecules that are bound tightly to the internal space can actively participate in the energy-coupling process.

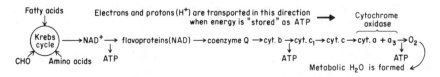

Fig. 7-7. The electron transport chain of mitochondria.

These three considerations indicate that the electron carriers involved in energy-coupling require a specific arrangement relative to the permeability barrier in a special region of the membrane. This condition makes it very difficult to isolate each component and then purify and reconstitute the complete system in the classical biochemical manner. There are other components of the inner mitochondrial membrane, not shown in Fig. 7-7, which also contribute to the overall function. For example, experimental evidence suggests that not all the lipid present in these membranes is necessary for morphological integrity, as seen in electron micrographs of acetone-extracted preparations. However, the electron-transport activity is lost on extraction of lipids. The inactivated membrane can be functionally reactivated by the addition of mitochondrial lipids or other unsaturated phospholipids. In general there appears to be no specificity for individual phospholipids in restoring electron-transfer activity in acetone-extracted preparations which still contain about 10% lipid. However, the degree of unsaturation of fatty acids may have some significance. The fatty acid residues of the phospholipid of mitochondria contain on the average 1.7 double bonds per fatty acid residue. Cardiolipin seems to be related to some specific function. Preparations relieved of 98–99% of their lipid can only be reactivated after addition of a suitable amount of cardiolipin along with other lipids. A specific role for cardiolipin is also indicated during the transition of the yeast cell from anaerobic to aerobic metabolism. As yeast mitochondria become fully functional, their total lipid content does not change in any significant way, while their cardiolipin content increases in parallel with their respiratory capacity and cytochrome content.

Still another indication of involvement of other components arises from the fact that some of the components have different properties after isolation; for example, isolated pyridine dehydrogenase is not sensitive to amytal, and isolated flavoprotein does not interact with cytochrome *b*. It could be that not all carriers have been identified or that during their removal from the membrane and purification, the physical state of the carrier is changed. Recognition of this possibility has promoted efforts to isolate the components of the electron-transfer chain in the form of segments. Green and co-workers have shown[445] that, under sufficiently vigorous conditions, it is possible to isolate four complexes, each capable of catalyzing one or more of the component reactions of the overall process. As shown in Fig. 7-7, these individual complexes consist of two or more components of the electron-transport chain and a considerable amount of lipid material which is essential for the activity of the complex. When these complexes are mixed in suitable stoichiometric ratios in concentrated solution and then diluted they yield more complex arrays which in turn are capable of catalyzing two or more steps of the electron-transfer reaction. These results indicate selective interaction of a given complex with the next, and each complex seems to have its components in a configuration which allows a recon-struction of the complete and active chain. More recently, Yamashita and Racker[726] have almost reconstituted the electron-transfer chain between succi-nate and molecular oxygen from the soluble components. But they had to add a particulate preparation of cytochrome b and it seems that this particulate frac-tion provides the structural material on which the individual components are organized.

Oxidative Phosphorylation

A considerable amount of energy is liberated in the overall process of electron transfer from reduced substrates to oxygen*, which is utilized for the synthesis of ATP from ADP and P_i in a coupled process. This transduction involves a system additional to, but intimately associated with, the complexes of the electron-transfer chain. Thus the rate of phosphorylation of ADP is determined by the concentration of reduced substrate, of oxygen, of ADP and P_i and of certain agents which can uncouple the electron-transport chain from the phos-phorylation mechanism. These observations imply that besides being dependent upon the cellular demand for energy the mitochondrion adjusts its respiratory rate to the availability of substrates and consequently to the metabolism. This compulsory coupling is intrinsically related to structural organization; it is not known whether some genetic or biochemical feedback loop operates in the system.

About 9 kcal of energy must be available in order to form a mole of ATP. Therefore one should expect to find the energy taps located (a) between NADH and its dehydrogenase, (b) between cytochrome b and cytochrome c, and (c) be-tween cytochrome a and oxygen. Thus as expected theoretically, the number of moles of ATP formed relative to the gram-atoms of oxygen consumed, termed as P/O ratio, is found to be an integral value for different substrates undergoing one-step oxidation. The true respiratory chain-linked phosphorylation in the oxidation of NaDH- or of NAD dehydrogenase-dependent substrates yield P/O ratio of 3, while that of flavoprotein-dependent dehydrogenase such as suc-

---

*The energy liberated during oxidation of NADH by atmospheric oxygen is easily calcu-lated as:

$$\Delta F = -2(23.068 \text{ kcal/volt}) \times [0.82 - (-0.32)] \text{ volts} = -52 \text{ kcal/mole}$$

All this energy need not necessarily be used for ATP synthesis alone; there are multiple ways in which the coupling mechanism and the energized state can be utilized by mito-chondria. In addition to ATP synthesis it can energize translocation of ions, volume change, reverse electron flow. Energy is conserved reversibly in the synthesis of ATP, but is dissi-pated in other processes.

TABLE 7-7 DIFFERENT CLASSES OF RESPIRATORY INHIBITORS AND UNCOUPLERS

	Class Compound	Specificity Affected	Concentration
A	HCN, HN_3, CO, H_2S	Inhibit cytochrome oxidase	$10^{-4}M$ (CN)
B	Antimycin	Blocks unknown factor between cyto-chrome-*b* and cytochrome-*c*	Stoichiometric with cyto-chromes
	British antilewisite	Blocks distal to the antimycin-sensi-tive factor	10 μM
	HQNO[a]	Same as antimycin	1 μM
C	Rotenone	NADH-flavine linked site	Stoichiometric
	Barbiturates (amytal)	same as rotenone	1 mM
D	Fat-soluble chelators (thenoyltrifluoro-acetone)	Peptide-FAD $\rightarrow$ Q and/or cytochrome *b*	0.1 mM
E	Oligomycin, aurovertin, and rutamycin	Block phosphorylation	1 μM
F	Proton donors (see Table 6-5)	Stimulate respiration when rate is limited by phosphorylation	
G	Arsenite, Ca, Cd	Inhibit respiration for keto glutarate and pyruvate only by blocking dihydro-lipoyl dehydrogenase	
H	Valinomycin type of cation conductors (see Table 6-6)	Slow and partial release of inhibition of respiration	1 μM
J	Nigericin type of cation conductors (see Table 6-6)	Exchange K for H^+; more or less same as valinomycin type	1 μM
K	Malonate	Inhibit succinic dehydrogenase	
L	Atractylate	Blocks phosphorylation probably by in-hibiting transfer of adenine nucleotides	1 μM
M	Oleate and other long-chain amphipaths	Interfere with energy coupling and with transport phenomena	

[a]HQNO = 2-alkyl-4-hydroxyquinoline-*N*-oxide.

cinate yield a value of 2.* Since all steps beyond the initial flavoproteins are similar for all substrates, one phosphorylation site must be localized within the complex I, The other two sites are localized in complex III and IV, respectively (Fig. 7-7).

Reversed electron transfer in the respiratory chain can be induced and mea-

---

*It is of interest to examine the energy relationship of these coupled reactions. Since phos-phorylation of ADP is the reverse of ATP hydrolysis, three phosphorylations therefore represent a conservation of at least 21 kcal of the 52 kcal theoretically available, that is, about 40%. Although this is much higher than the efficiency of a heat engine (less than 20%), the remaining 31 kcal may not necessarily be lost as entropy, but may be conserved by the mitochondrion in undefined ways, e.g., through intracellular control of presursor concentrations, ion transport, and other processes which are still poorly understood.

sured under suitable anaerobic conditions, by addition of ATP. The rates of oxidation of various cytochromes and other partial reactions of the electron-transfer chain are only about one-tenth of those with oxygen present, and equal rates are found for the oxidation of other components. Thus, the rate-controlling step in the overall process of oxidative phosphorylation appears to be determined primarily by the rate of interaction of phosphorylation of ADP.[727]

In fact, much of our knowledge concerning the actual mechanism of oxidative phosphorylation has come from the use of uncouplers and inhibitors. A large variety of chemical reagents and physical factors (see Table 7-7 in conjunction with Figs. 7-7 and 7-8) interfere with various steps in the overall process. Assignment of the site of interaction of inhibitors of the electron-transfer chain, as pointed out in the last section, is facilitated by the *crossover theorem*. It may also be noted that a certain degree of structural integrity of the complex appears to be required for their interaction since solubilized phospholipid-free enzymes do not exhibit the characteristic inhibition patterns. Various experimental ob-

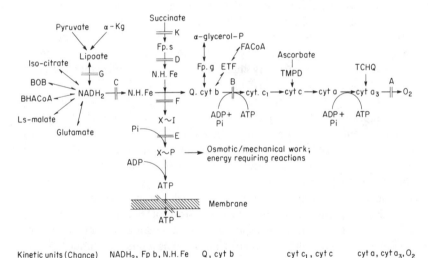

Kinetic units (Chance)	NADH$_2$, Fp b, N.H. Fe		Q, cyt b	cyt c$_1$, cyt c	cyt a, cyt a$_3$, O$_2$
(Half-time: red–ox.)	(50–500 m secs)		(30–80 m secs)	(2–5 m secs)	(0.5 m secs)
Reduced state	53%	20%	16%	6%	<4%

Fig. 7-8. Linkage of respiratory chain and phosphorylation in mitochondria. A–G are the sites of action of respiratory inhibitors (Table 7-1). The half time for the change from reduced to oxidized state of the components of the chain makes it possible to divide the chain into kinetic units. Key: FpD, NADH dehydrogenase; Fp · S, succinic dehydrogenase; N · H · Fe, non-haem iron; Q, ubiquinone; BOB, D-β-hydroxybutyrate; BHA, L-β-Hydroxyacyl-; fp$_F$, fatty acyl dehydrogenase; ETF, electron-transferring flavoprotein; TMPD, tetramethyl-*p*-phenylene diamine; TCHQ, tetrachlorohydroquinone.

servations, in fact, seem to suggest that the inhibition sites A, B, C (Fig. 7-8) may be identical with the three energy-transfer sites; thus for example, the partial release by ADP and P_i of inhibition of respiration largely removes the buildup of reducing equivalents of carrier as they become oxidized. Similarly, certain compounds which can inhibit various stages in the phosphorylation sequence (type E and L in Fig. 7-8) also inhibit the electron-transfer chain.[728] Some of these inhibitors seem to act at only one of the phosphorylation sites; the oxybarbiturates such as amobarbital and alkyl guanidine appear to be specific for site 1, and phenethylbiguanidine for site 2. The guanidines also compete with magnesium, which might imply that the site of action of this cation is before the phosphorylation step.[729]

The inhibited respiratory chain can be reactivated by a variety of organic compounds which usually contain a readily dissociable hydrogen atom and which have elaborate delocalized π-electron orbitals (Type F in Table 7-7). If the dissociable proton is substituted by some other group, the uncoupler molecule cannot act as proton carrier and hence cannot be an uncoupler.[730] At low concentration the uncoupler uncouples the electron transfer in functional mitochondria, even in the absence of ADP and P_i, from the energy conservation mechanism normally leading to ATP synthesis or ion translocation. Both phosphorylation and respiration are inhibited at higher concentrations of the uncouplers. However, this secondary inhibition appears to show different kinetic responses for different uncouplers and could be due to a direct action on some protein component of respiratory chain.[1249] As the uncouplers increase the rate of electron flow, it may be inferred that the control of electron flow exerted by the phosphorylation system has been released.[334] This does not necessarily imply that the inhibitors and uncouplers act at the same site. In the electron-transfer chain reconstituted from the complex II, III, and IV and cytochrome c, the uncouplers only inhibit respiration at concentrations much higher than those at which they can act as uncouplers. Thus the site of action of the uncouplers may not be the electron-transport chain.

The uncouplers prevent the formation of high-energy intermediates or states formed during normal electron transport. Most of the uncouplers are about equally effective at all three phosphorylation sites. Some uncouplers, however, are specific. For example, dicoumarol releases the inhibition induced by DBI, but not the inhibition induced by octylguanidine. Similarly, 5-chloro-3-*tert*-butyl-2 -chloro-4'-nitro-salicylanilide combines stoichiometrically with cytochrome oxidase and uncouples only at the cytochrome oxidase site.[731]

Uncouplers act prior to participation of P_i in the oxidative phosphorylation sequence. The uncoupling action is also exhibited in energy-linked reductions without participation of P_i (see Boyer in Ref. 732).

Uncouplers of type H do not produce extensive release of either type of inhibition, thus suggesting a somewhat different mode of action.[728] It has been found that the addition of valinomycin (type H) to rotenone-poisoned (inhibitor type C) mitochondria leads to a rapid loss of K and water, a countermovement (uptake) of protons, a rise in endogenous ATP, and an equivalent disappearance of ADP[733] The valinomycin induces ATP formation at the expense of a downhill movement of potassium ions. It is interesting to note that another class of antibiotics (type J) reverses these effects of valinomycin.[734]

The role of uncouplers in the regulation of oxidative phosphorylation and electron transport may also have some regulatory functions at the tissue level. Fatty acids (type M) uncouple oxidative phosphorylation by physical interaction with the *respiring* mitochondrial membrane. It would appear that fatty acids released by lipolysis would ensure maximum respiratory rate and direct transformation of energy released by oxidation to heat, *thermogenesis*.[746]

The chemical structures of various uncouplers of oxidative phosphorylation differ considerably. The only common characteristic which the uncouplers of the type F, G, H, J, and M (Table 7-7) possess is their ability to transfer protons and/or other monovalent cations through a hydrophobic phase such as a BLM interior (see Chapter 6 for an extensive discussion of this aspect). The relevance of this characteristic with regard to uncoupler activity of these agents for mitochondrial oxidative phosphorylation is not clear. However, its significance can be weighed with regard to various hypotheses forwarded to explain the nature of intermediate steps involved in conservation of energy at ATP.

We have now reached the point where understanding of the oxidative phosphorylation mechanism drops off precipitously. There are at least three separate phosphorylation sequences, but in no case is the identity of the intermediates known, as the reader must have guessed from the use of terms "X" and "Y" in Fig. 7-8. The experimental data can be recapitulated with the steps shown in this scheme as a minimum mechanism for oxidative phosphorylation. Any detailed assignment evidently requires an identification of the hypothetical entities, X, Y, I, etc. One of the reasons for this apparent failure is that the classical biochemical approach to solubilize, isolate, and purify the component enzymes, and then to reconstitute a functional system, has met with only limited success. The most notable attempt to achieve this is that of Racker and his associates (see Refs. 735–737 and unpublished data). Their approach consists of resolution and reconstitution of protein and phospholipid components, and study of the topography of the reconstituted membrane. Thus, both structural and functional aspects are emphasized. They have been successful in removing protein components which, if added back to nonphosphorylating *submitochondrial particles*,* have the effect of *re-coupling* phosphorylation to electron transport.

---

*SMP consist of vesicles of 2–20μ diameter formed from mitochondrial membrane. Topographically these vesicles are however inverted, that is, the membrane interface which was *inside* in mitochondria is now *outside*.

A total of four coupling factors* have been isolated. The coupling factor F_1, a soluble ATPase, is cold labile and insensitive to oligomycin, and it differs in these respects from the membrane-bound enzyme, which is cold stable and virtually completely inhibited by oligomycin. It has a molecular weight of about 250,000 and under the electron microscope appears as spheres approximately 85 Å in diameter. It probably constitutes the largest component of the inner membrane subunit.

An elegant series of experiments show that when the F_1 coupling factor is removed, the inner membrane subunits are stripped off the cristae, and that when the phosphorylating system is reconstituted they once again are attached to the membrane. Thus the subunits on the inner membrane appear to be the site for oxidative phosphorylation. Moreover, removal of F_1 from the membrane alters its stability and oligomycin sensitivity. A factor F_0 was also isolated which restores oligomycin sensitivity and cold lability.[736] F_0 still contains phospholipids and the entire electron-transfer chain. Further purification of F_0 yields a colorless fraction, CF_0, which is virtually devoid of cytochromes and phospholipids but retains ability to bind soluble F_1. In this case, however, binding of F_1 is accompanied by inhibition of ATPase activity. Upon addition of phospholipids, the ATPase activity is unmasked and the preparation is now cold-stable and sensitive to oligomycin. The negatively stained preparations of CF_0, when viewed in EM are rather amorphous. In contrast, the reconstituted oligomycin-sensitive ATPase complex appears as vesicles lined with spheres 80-90 Å in diameter. Although these vesicles are practically free of electron-transport carriers, their morphology closely resembles that of respiring submitochondrial particles.[737] While the combined biochemical and morphological evidence thus identifies CF_0 as a key component of the mitochondrial energy-transfer chain and the inner mitochondrial membrane, the detailed chemical composition of CF_0 is at present largely unknown. It has however been suggested recently that most of these components—the coupling factors (now about six in number) and some of the catalytic proteins (succinic dehydrogenase, etc.) of the electron-transport chain are largely localized on the inner face of the inner mitochondrial membrane; whereas, cytochrome c and c_1 appear to be localized at the outer interface (between the inner and the outer mitochondrial membrane). These proteins are interspersed by a *hydrophobic core* constituted of hydrophobic proteins, lipids, and cytochrome c.

Racker and coworkers have recently reconstituted *site* III into vesicular membrane-bound structures which catalyze (a) $^{32}P_i$-ATP exchange[1301] and (b) oxidative phosphorylation coupled to ion transport and respiratory control with

---

*The non-commital term "factor" was chosen because these components may be structural components of the membrane participating in the proper organization of the catalysts. rather than being catalyst themselves. It is possible, moreover, that the function of some of the coupling factors may be neither catalytic nor structural, but regulatory.

reduced PMS as substrate.[1302] The procedure for reconstitution essentially consists of fractionation of submitochondrial fragments in the presence of 2% cholate. The delipidated fragments mixed with sonicated liposomes of soyabean lecithin are dialyzed against 10% methanol to remove cholate. The coupling factors could be added before or during dialysis. Vesicles formed with phospholipid and cytochrome oxidase alone, without hydrophobic proteins, catalyze proton translocation during oxidation of reduced 1.4-naphthoquinone-2-sulfonate and PMS in the presence of valinomycin and K^+. Uncouplers eliminate the effect. These studies show that many aspects of energy coupling in mitochondria are present in purified cytochrome oxidase when it is properly incorporated into a phospholipid membrane, and when polarity is created by having cytochrome c on only one side of the membrane.

Thus due to a general lack of direct means of analysis of study of oxidative phosphorylation one has to resort to speculative theoretical analysis. Such an approach becomes particularly desirable if one takes into account the possibility that some of the steps involved in phosphorylation may not be biochemical in the conventional sense (involvement of gradients, etc.; see below). The task of coming up with a single theory capable of accounting for all the characteristics of oxidative phosphorylation and related phenomenon is great indeed. It must account for a number of experimental observations: action of uncouplers and inhibitors; role of cations, protons, and ion pumps; mode of coupling; relationship between ion transport and amount of ATP synthesized which is stoichiometric; and some predictions regarding the nature of the components involved in these processes or steps. Ideally, one would like to know the relationship between the means of oxidative and photophosphorylation, the relationship between energized transport systems in the plasma membrane and oxidative phosphorylation or photosynthesis, and the regulatory role of mitochondria in the physiology of the cell. The possible implications of these processes reach deeply into the question of the supramolecular organization of membrane structure, of the enzyme and electron carriers, and of the basic features of energy transduction in coupled molecular processes. None of the available hypotheses seem to account for all the observed facts, even less so for implications. However, there are some hypothetical mechanisms around which current experimentation is polarizing. A discussion of the salient points in these formulations follows (also see Ref. 738 for an excellent discussion).

One of the hypotheses for the oxidative phosphorylation mechanism is the more traditional and orthodox *chemical coupling hypothesis.*[739]

$$A_{red} + B_{ox} + I \rightleftharpoons A_{ox} \sim I + B_{red}$$
$$A_{ox} \sim I + ADP + P_i \rightleftharpoons A_{ox} + I + ADP$$

It proposes that high-energy intermediates are generated during electron transport and that these may be used either to form ATP or to transport ions by stoichiometrically coupled enzymatic reactions sharing a common intermediate.[740] It is suggested that the passage of a pair of electrons from each of three specific carriers in the respiratory chain to the next chain is coupled to the formation of an energy-rich bond, presumably an anhydride linkage between one of the two electron carriers and a third entity—an unidentified coupling factor (see Fig. 7-8). This high-energy intermediate ($C \sim I$ and/or $X \sim I$), directly or indirectly, can react with phosphate to form a phosphorylated high-energy intermediate ($X \sim P$ and/or $Y \sim P$) that can donate its phosphate to ADP. It is postulated further that such high-energy intermediates could be utilized for three different modalities of energy coupling, that is ATP formation, ion translocation, and mechanochemical coupling; and probably also in energy-dependent pyridine nucleotide transhydrogenation in mitochondrial membrane. This intermediate may be mitochondrial F_1 or chloroplast CF_1.[1251] This hypothesis can account for various features of oxidative phosphorylation, such as acceptor control of respiration and the action of inhibitors and uncouplers. One of the most serious shortcomings of the hypothesis is the fact that there is no experimental evidence for the generation of a biochemical high-energy intermediate in spite of several attempts. It is however possible that the high-energy intermediate is normally in a hydrophobic environment in the membrane and would become unduly labile if removed into an aqueous medium. Some experimentally feasible model systems have been proposed,[741,742] for example, oxidation of glutathione can be coupled to phosphorylation of AMP.[742]

In summary, chemical hypothesis for oxidative phosphorylation is consistent with a large amount of experimental information, but none of this indicates whether the high-energy intermediate is formed, at all, and if it is then at what stage. Also there is considerable difficulty in accounting for observed stoichiometry of ion translocation and high-energy intermediate consumed with chemical and thermodynamic clarity.

Chemiosmotic Hypothesis: Now we pass our discussion on to the chemiosmotic hypothesis for coupling oxidative phosphorylation to electron transport. It postulates that:[738,743]

1. There are enzyme complexes forming functional units which are an integral part of the membrane structure,

2. ATP is synthesized by reversing the activity of the enzyme ATPase which is asymmetrically located in the hydrophobic environment of the coupling membrane.

3. Only a small amount of water is available in the membrane phase; therefore, the activity of water is manifested in the kinetic step as follows:

$$\frac{[ATP]}{[ADP]} = \frac{[H_3PO_4]}{K[H_2O]} \cdot \frac{[H_0^+]^2}{[H_i^+]^2}$$

$$\log \frac{[ATP]}{[ADP][H_3PO_4]} = 2\left(pH_0 - pH_i + \frac{\Delta E}{Z}\right) - \log_{10} k(H_2O)$$

where square brackets represent the activities of the individual components. The activity of water will depend on the relative tendency of *that* water to move into the aqueous phase. The assumption is that ADP, P_i, and ATP all take part in the reaction from the same side of the membrane.

4. the membrane has highly specific permeability in direction and space for H^+ and OH^- ions, and thus allows for asymmetric escape of these ions.

5. The function of the respiratory chain is to produce a sufficient and effective concentration gradient of H^+ and OH^- ions in the active center region by a vectorial arrangement of the electron transfer components of the chain in the plane of the membrane. The polarity of the membrane is such that in mitochondria the protons traverse from the inside to the outside during oxidation; in chloroplasts from the outside to the inside.

Thus according to the chemiosmotic hypothesis, equilibrium is reached when the work done in transferring two reducing equivalents through a loop of the respiratory chain becomes equal to the work done in translocating two protons across the membrane; that is, when the redox potential span along the loop becomes equal to the *proton motive force*. A schematic representation of this is given in Fig. 7-9. Electron transport is visualized to generate a pH gradient across the membrane; this gradient can bring about either ion translocation by exchange diffusion or the gradient may drive the active transport pump backwards, that is, to synthesize ATP from P_i and ADP.* The H^+ and OH^- ions liberated during the synthesis of ATP are liberated asymmetrically and thus the synthesis takes place without reversing back to the ATPase activity. Hence respiration per se in such an organized membrane is able to generate a very strong gradient of protons without the intervention of ATP or other energy-rich phosphate bonds. Calculations show that proton motive force of about 230-270 mV generated by the respiratory chain under equilibrium conditions is suitable for maintenance of an ATP/ADP ratio of 1.

The chemiosmotic hypothesis envisages generation of "active" orthophosphate (a positive "phosphorylium" ion) as driven by the respiratory proton gradient. The observation that ^{18}O-labelled phosphate loses its label to water in the presence of intact mitochondria that is undergoing oxidative phosphorylation[744] is of great interest in this context. The rate of this [18]O exchange between P_i and water can proceed as much as ten times more rapidly than the ^{32}P exchange between phosphate and the terminal phosphoryl group of ATP.[745] Many of these observations could be explained by assuming that orthophosphate donates an OH^- group to the accumulated hydrogen ions; the electrophilic phosphorylium ion then reacts with various nucleophilic constituents, ADP, protein-bound histidine, etc., forming $\sim P$. When the $\sim P$ bonds are subsequently hydrolysed (say by ATPase), the released phosphorylium ion will incorporate an

---

*In fact incubation of guinea pig red blood cells in a high-Na K-free medium led to an increase in the concentration of ATP which could be wholly or partly prevented by ouabain.[665] Also see p. 263.

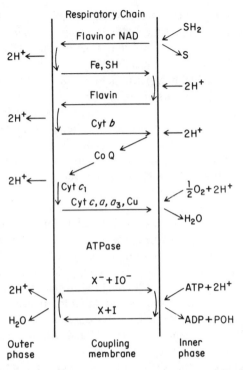

Fig. 7-9. Proton translocating respiratory chain and proton translocating ATPase of the chemiosmotic hypothesis. The respiratory chain system is illustrated for an NAD-linked substrate. Succinate taps into the chain at the second loop.[743]

OH which, if it stems from outside of the membrane, would contain ^{18}O if the experiment employs $^{18}OH_2$.

One of the most successful predictions of the chemiosmotic hypothesis has been the effect of uncouplers. It is suggested that these agents "short-circuit" the ionic gradients, and thus the high-energy intermediate in the osmotic sense is lost. It has now been shown that the unusually high affinity of the enzyme succinate oxidase in reconstituted mitochondria for the uncoupler may only be apparent if the function of the uncoupler is to provide protons. It must, however, be noted that such evidence cannot be regarded as an absolute proof for the validity of the chemiosmotic hypothesis. For example, yet another explanation of the uncoupling effect of the proton carriers could be that it competes for a nucleophilic intermediate in the coupling mechanism with the electrophilic energy carrier.[333,334] The proton might also cause the acid hydrolysis of the primary high-energy compound.

The Mitchell hypothesis can also explain a number of phenomena associated with the electron transport chain, charge separation by the membrane, nature of high-energy intermediate, establishment of electrical potential differences and possibilities for an ion exchange.[1250] The model predicts furthermore that the electrochemical potential difference of hydrogen ion necessary to keep the ATP/ ADP ratio at unity would be a pH differential of 3.5 units (acid inside) or a membrane potential of 210 mV, or a combination of both. In fact, changes in pH can be shown to occur across chloroplast lamellae and are associated with net ATP synthesis.[747] Similarly, a pH shift between inside and outside the lamellae could be used to make them luminescent (presumably reversal of photophosphorylation) so that energy due to separation of charge or any of the intermediates can reverse the reactions right back to the photochemical step at chlorophyll (see scheme in Fig. 7-11).

The chemiosmotic coupling hypothesis has yet to provide a complete description of the coupling mechanism. While considering the evidence supporting this concept of a proton pumping arrangement of the electron carriers, the same difficulty arises that prevents acceptance of the chemical coupling mechanism. There is simply insufficient detailed evidence available regarding the chemical identity of the electron carriers, their sequence, and their mechanism of interaction to pass a conclusive judgment as to whether all the postulated proton exchanges actually take place, and if so, whether they take place vectorially in the sense of the "sidedness" demanded.

When it was first proposed, it was claimed that the chemiosmotic theory has done away with the hypothetical intermediates $X - I$ and $X \sim P$ necessary for the chemical coupling hypothesis; this is still true, but it does not mean that these intermediates do not exist. In its more recent form, the model derived from the chemiosmotic hypothesis suggests that electron transport can be coupled to dehydration reactions and it is possible that the initial dehydration is from X-OH and IOH (uncombined forms of X and I) to form $X - I$, which then is a partial reaction characteristic of the "ATPase" complex. Thus it becomes apparent that the so-called chemical coupling and the chemiosmotic coupling are not necessarily very different. The chemiosmotic theory results in a proposed mechanism of $X - I$ formation that was very vaguely described in the chemical coupling hypothesis.[748] There is however one distinguishing feature of the chemiosmotic theory. It requires a vesicular structure of the membrane which allows for a clear separation of the two sides of the membrane; the chemical hypothesis does not. A demonstration of oxidative phosphorylation in a vesicle-free system would eliminate the major feature of the chemiosmotic theory.

Conformational Coupling Hypothesis: The conformational coupling hypothesis is based on the postulation that the *primary* site of energy conservation is conformational change in the molecular species present at the junction of the electron transport chain and the phosphorylation sequence. This is in accord with the reversible changes in light scattering, shrinkage, swelling, viscosity of the suspensions, reversible fluxes of cations, electrophoretic mobility, and other related properties of mitochondria and chloroplast suspension. Also there is evidence that changes in dielectric properties of the membrane precede chemical changes in the energy conservation system of mitochondria, as shown by anilinonaphthalenesulfonic acid fluorescence as a function of butacaine concentration.[749] All these changes can be blocked by suitable inhibitors and uncouplers.

These results have in general given a way to the idea that the material inside the chloroplast and mitochondrion undergoes conformational modification which can be triggered and energized by electron transport, and may be closely linked with H^+ and/or K^+ influx and accumulation. This can be best compared with other hypotheses of oxidative phosphorylation by a close examination of schemes shown in Fig. 7-10. Based on this idea, two rather general proposals have been made as to the nature of the molecules undergoing conformational change as a primary step in energy transduction.

Chemical Intermediate

Respiratory chain $\rightleftharpoons$ X ~ I $\rightleftharpoons$ ATP synthesis
$$\updownarrow \text{Proton pump}$$
H^+ translocation
$$\rightleftharpoons$$
Cation translocation

or

Respiratory chain $\rightleftharpoons$ X ~ I $\rightleftharpoons$ ATP synthesis
$$\updownarrow \text{Cation pump}$$
Cation translocation
$$\rightleftharpoons$$
H^+ translocation

Chemiosmotic Coupling

Respiratory chain $\rightleftharpoons$ H^+ translocation $\rightleftharpoons$ ATP synthesis
$$\rightleftharpoons$$
Cation translocation

Conformational Coupling

Respiratory chain $\rightleftharpoons$ Energized state $\rightleftharpoons$ ATP synthesis
$$\updownarrow$$
Work performance

Fig. 7-10. Three schemes which have been suggested for the interrelationships between respiratory chain activity, ATP synthesis, proton translocation, and cation translocation in mitochondria. See text for details.

Green and co-workers[750-753,45] have suggested that the inner mitochondrial membrane is constituted of an array of repeating units. The various configurational states of the membrane is determined both by the conformation of the repeating units (*nonenergized, energized,* and *energized-twisted*) and by the mode (configuration) of the cristae (*orthodox, aggregated* or *comminuted*). Thus the inner mitochondrial membrane can assume any of the following nine configurational states:[751]

Postulated functional status of repeating unit	Mode of Organization of the Cristae					
	Orthodox*		Aggregated		Comminuted	
Nonenergized (NE)	Classical	NE_{or}	Sheetlike	NE_{agg}	Not characterized	
Energized (E)	Vesicular	E_{or}	Energized	E_{agg}	Swollen	E_{comm}
Energized-twisted (ET)	Zigzag	ET_{or}	Zigzag	ET_{agg}	Snakelike	ET_{comm}

*The orthodox mode is found in mitochondria *in situ,* the aggregated and comminuted modes in isolated mitochondria.

Of these, eight have actually been recognized in electron micrographs.

It has been suggested that the transition from one conformational state of the repeating unit into other can be reversibly induced by either electron transfer or by the hydrolysis of ATP. However, discharge of the energized state for work performance (for example, translocation of ions, swelling, or energized transhydrogenation) is essentially irreversible. Thus certain reagents such as ATP will generate the energized state which may be discharged reversibly (by ADP, Ca, K or by NAD, NADP, or by inhibitors of electron transport), or irreversibly by uncouplers. Indeed, in the presence of these reagents the state of the mitochondrion appears to be changed in electron micrographs. Similar configurational changes have been observed in chloroplasts, sarcotubular vesicles, and erythrocyte ghosts.[751]

It is obvious that the energized state is described in terms of geometric, functional, and kinetic parameters. However, rationalization of this energized state in molecular terms such as specific "conformational perturbation of polypeptide chain," etc., is yet to be realized. It has however been suggested that *cooperative phase transition* of subunits may bring about an asymmetric distribution of surface charge which may lead to active transport of ions by redistribution.[753]

The second proposal is from Chance and co-workers,[754] who have been specific while proposing a conformation coupling mode of energy transduction. The primary event of coupled electron transfer at phosphorylation site II is identified with a modification in one of the two chemically distinct forms of cytochrome b, designated as the energy-transducing cytochrome b_T. This modification is expressed through a change in the redox midpoint potential and by an increase in its reaction half-time with cytochrome c_1. Cytochrome b_T has an absorption maximum at 564 nm, an energization time less than 200 msec, and *thermodynamic capability*—a potential span of 280 mV. The short-wave form functions exclusively in electron-transfer reactions. The energy conservation by cytochrome b_T may be described as a change in ligand-interaction energy. This could be the change in orbital energy of binding of heme iron which would alter the redox environment. Thus two secondary events of coupled electron flow may be identified with a charge separation across the lipid structure of the permeability barrier and/or a change in water structure. Both events result in an increased anilino-naphthalene sulfonic acid response (dielectric-dependent fluorescence changes). Also transitions from the energized to the deenergized state involve an absorbancy change in the soret band of cytochrome b upon the addition of dicoumarol to the intact mitochondria, and the reaction is complete within one second. These data seem to indicate that a portion of cytochrome b is under the control of an energy-coupling mechanism.

The "Scoreboard"

The discussion of respiratory chain and oxidative phosphorylation as mediated by mitochondria leaves certain gaps in our understanding of the basic physiochemical principles operating these processes; this becomes particularly evident while considering the steps involved in coupling of electron-transfer to phosphorylation. A number of theories, of necessity of speculative character, have been forwarded. Their status is probably best summaried in the following table.*

Phenomenon	Chemical	Chemiosmotic (latest form)	Conformational
Role of membrane	None	Active	Passive or none
Ion transport (anisotropy not always observed)	+	−	+
Action of uncouplers	±	++	−
Isolation of high-energy intermediate	−	±	+
$^{32}P_i$-ATP exchange	+	+	−
ADP-ATP exchange	+	±	?
$H_2^{18}O$ exchanges	+	±	−

The present status is thus at best in a state of flux. A large number of bacterial energy-transducing systems have varying degrees of structural and organizational complexity. Also, substrate level phosphorylation of ADP can occur without participation of membranous structures. Thus, a more general approach involving bacterial and other primitive energy-transducing systems may provide a clue to resolve the present state of affairs. Some comparative aspects of electron transport are summarized in Table 7-8. In the next section we shall discuss some of the features of energy transducing system involved in photosynthesis.

---

*(+) suggests that the hypothesis offers a consistent explanation; (−) means generally unfavorable and inconsistent explanation of limited validity; ? means no obvious explanation.

TABLE 7-8 COMPARATIVE ASPECTS OF ELECTRON TRANSPORT CONTROL IN MITOCHONDRIA AND SUBCELLULAR PHOTOSYNTHESIS SYSTEMS

Levels of Control	Mitochondria	Bacterial Chromatophore	Chloroplast
Energy source	Potential of e^- donor-acceptor	Photochemical—both intensity and wavelength (370–900 nm)	Photochemical—both intensity and wavelength (370–750 nm)
Energy utilization	Changes in redox and high-energy states	Trapping of quanta, charge separation, generation of high-energy states	
Energy span	1.15V	0.5–1.3V	1.25V
Energy taps (ATP)	3	1	2
Terminal acceptor	O_2	O_2, NO_3, SO_4, chlorophyll etc.	NAD or chlorophyll
Terminal donor	NADH	H_2S, S_2O_3, Fe^{2+}, light etc.	Light
Acceptor control	Respiratory—reverse electron transport	Reverse ET	Photosynthetic control
Effect of coupling factors and ion effects	Significant and rate controlling	Significant and rate controlling	Significant and rate controlling

PHOTOSYNTHESIS

During photosynthesis, electromagnetic energy is received, converted, transferred, and stored by cells. Thus, light energy is used to convert carbon dioxide, water, and relatively small amounts of salts into carbohydrates, amino acids, and other organic molecules. Essentially, photosynthesis is *redox* reaction between an oxidant (CO_2) with a potential of about $-0.4V$, and a reductant (H_2O) with a potential of about $+0.8V$, in which four electrons (or four hydrogen atoms) are transferred "uphill" against a potential gradient of about 1.2 V. In green plants the overall process proceeds in two steps (see Ref. 755a, b and 1252 for reviews):

1. The light reaction which converts absorbed quanta of light into chemical potential. Essentially, the process involves photolysis of water or other hydrogen donors ($H_2X = H_2S$, H_2O, NO_2, SO_3, S_2O_3, H_2, etc.) according to the following light-induced reactions:

$$H_2X \xrightarrow{\text{light}} H_2^* + X$$
$$BO + 2 H_2^* \longrightarrow BH_2 + H_2O \, (BO = CO_2, Fe^{3+})$$

Although it is true that water is normally decomposed in photolysis in plants, the concept of photolysis as the fundamental event is no longer strictly acceptable. The generally acceptable model for photosynthetic electron transport in plants involves two photochemical reactions that act in series and that occur in two different photochemical systems. Each system contains a reaction center at

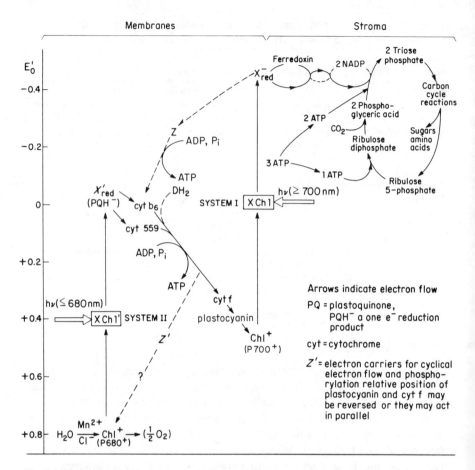

Fig. 7-11. A scheme of photosynthesis showing that the phtosynthetic unit, light reactions, possible linkage between two separate light reactions, electron transport from water to ferredoxin, and ATP formation take place in the membrane, whereas the carbon dioxide fixation reactions occur within the stoma. Ferredoxin is able to reduce pyridine nucleotides using electrons from excited chlorophyll, or alternatively from molecular hydrogen (in the presence of a dehydrogenase). The flow can also be reversed to release molecular hydrogen from protons in solution.[756]

which an oxidant and a reductant are produced (Fig. 7-11). In green plants, red light absorbed by photochemical system II (PS II) sensitizes a reaction that results in oxidation of water coupled to the formation of a weak reductant. Far red light absorbed by photochemical system I (PS I) sensitizes a reaction that yields a weak oxidant and a strong reductant that eventually reduces NADP. PS I and PS II are linked in series by several electron carriers, and the reductant produced by PS II is oxidized by the oxidant produced by PS I. There is at least one site for ATP formation during the course of electron transport from water to NADP, and there is evidence that this site lies between the two photochemical systems. It is obvious that once electron transfer has been mediated by the primary *photoreceptor*, the chlorophyll, electron transfer occurs by a sequence similar though not identical with the respiratory chain. There are striking similarities between the respiratory chain of mitochondria and the electron-transport chain of photochemical systems. Pyridine nucleotides, quinones, flavoproteins, and cytochromes participate in both systems and catalyze electron and proton transfers that are coupled to a device which generates ATP. Some inhibitors and uncouplers are effective in both and electron flow is controlled by the availability of ADP. Also a factor designated CF_1, which is remarkably similar to F_1 from mitochondria, has been isolated from spinach chloroplasts; however the protein exhibits specificity with regard to immunological and functional reactivity.[757]

2. The dark reactions fix carbon dioxide and regenerate CO_2 acceptor from the first products of photosynthesis according to the following reaction:

$$CO_2 + H_2O \xrightarrow{h\nu} CH_2O + O_2$$

to form carbohydrates. This step does not require light and it takes place at sites which are present at a concentration of about one (or less) per 100 molecules of chlorophyll. At very low light intensities the rate of photosynthesis is limited by a photochemical reaction, but at high light intensity the *dark* reaction is rate limiting. The temperature coefficient (Q_{10}) of the light step in photosynthesis is close to 1.0; the dark steps have Q_{10} of 2.0.

Many plants are structurally organized to receive the greatest amount of light possible. The large surface/volume ratio of the leaf, coupled with the large intercellular surface area and lamellar chloroplast structure, all facilitate maximum light absorption.[1135] The site of energy conservation in green plants is located in chloroplasts which are made up of a series of parallel membranes arranged as flat sheets or plates, and known as lamellae, distributed through a matrix of stroma which is surrounded by two outer limiting membranes (Fig. 7-12). In vivo experiments have suggested that chloroplasts contain not only all the chlorophyll and other pigments which contribute absorbed light energy for the photochemistry, but also apparently the whole of the enzyme complement of metabolic and biosynthetic pathways. Inside a cell, chlorophyll shows the ability to reorient relative to the direction of the light, either by swinging from profile to full face or by changing shape. Each chloroplast undergoes these changes independently and they are completely reversible. Structures similar to the chloroplast, like chromatophore in bacteria and plastids in algae, are involved in energy conservations in these organisms.

The role of a large number of pigment molecules associated with chloroplasts

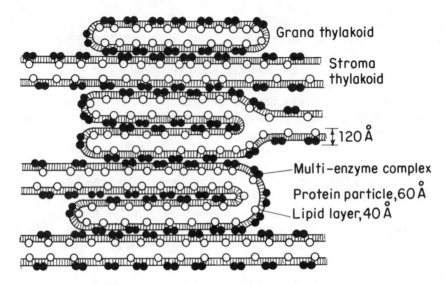

Fig. 7-12. An interpretation of granum lamella structure taking into account electron microscopic data.[758]

is not known. However, it seems likely that the light energy is transferred directly to the chlorophyll molecule which functions as the unique entry to the electron-transport mechanism of photosynthesis. The ratio of chlorophyll molecules to electron-transfer chain components is large (about 250:1). This implies that chlorophyll molecules cooperate to harvest the light energy. The energy absorbed anywhere in an aggregate of chlorophyll molecules is transferred to a single reaction site. Various modes of such energy transfer have been suggested, such as diffusion of substances, diffusion of electric charges including electron and holes, or resonance transfer, that is, migration of quanta of electronic excited energy either as a singlet or triplet.[755a] However, the rapid onset of photochemical events, and the efficient transfer of energy even at $1°K$ rules out mechanisms based on transfer or diffusion of atoms, radicals, or molecules. It is possible, therefore, that the electrons and holes may be transferred between different reaction centers. Thus, the chlorophyll molecules are considered to be arranged in a two-dimensional lattice and to behave somewhat like a semiconductor. This would imply that neighboring molecules are so strongly coupled that their orbitals fuse. Migration of the exciton, described as an electron and a positively charged hole migrating together through the lattice, may then be so rapid that the exciton cannot be associated with a single molecule at any one time. The possibility of chloroplasts being such a solid-state system has been supported by Arnold and Sherwood,[759] who found that if dried chloroplasts were illuminated at room temperature and then heated at temperatures rising to 140°C in darkness, they emitted light. This would occur in a semiconductor if some of the excited electrons were trapped in faults in the crystal lattice and were later released, by absorption of the infrared quanta, to drop back into the hole.

These ideas have been pursued further in the model shown in Fig. 7-13. In this scheme chlorophyll is visualized as being sandwiched between two layers, one containing an electron donor, possibly a cytochrome, and the other containing an electron acceptor, such as plastoquinone. This layer of chlorophyll is supposed to provide uniform horizontal conductivity in the lamellae. A quantum absorbed by chlorophyll will migrate by resonance transfer to a suitable site near the quinone where the electron transfer to quinone takes place. This is followed by migration of the electron vacancy or hole along the array of chlorophyll molecules until it becomes adjacent to the cytochromes, where an electron is transferred from the cytochrome to chlorophyll. Thus, whatever the final detail of

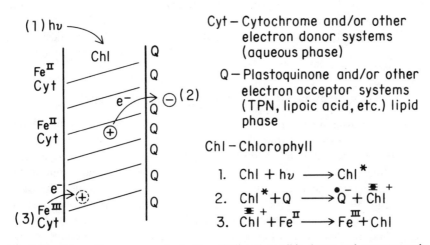

Cyt — Cytochrome and/or other electron donor systems (aqueous phase)

Q — Plastoquinone and/or other electron acceptor systems (TPN, lipoic acid, etc.) lipid phase

Chl — Chlorophyll

1. $Chl + h\nu \longrightarrow Chl^{*}$
2. $Chl^{*} + Q \longrightarrow {}^{\bullet}Q^{-} + \overline{\overline{Chl}}^{+}$
3. $\overline{\overline{Chl}}^{+} + Fe^{II} \longrightarrow Fe^{III} + Chl$

Fig. 7-13. Schematic arrangement of chlorophyll and possible donor and acceptor molecules in the chloroplast. The system in the chloroplast might structurally bear some resemblance to the mode shown here, the chlorophyll having associated with it on the one side the electron acceptor plastoquinone, in a lipid environment, and on the other side electron donor material such as the cytochromes, in an aqueous environment. Following the absorption of quantum in chlorophyll, it will migrate by resonance transfer to the quinone, where electron transfer to the quinone will take place. The resultant vacancy can migrate by hole diffusion, that is, electron transfer from normal chlorophyll into the vacant orbital of the neighboring chlorophyll positive ion. This process most nearly resembles the properties of a semiconductor and it permits the oxidant (chlorophyll positive ion) to separate from the reductant (electrons in the quinone orbitals) by a very nearly temperature-independent process. The oxidant then capture an electron from a suitable reducing agent, such as ferrocytochrome, thus producing ferricytochrome and regenerating normal chlorophyll.[760]

molecular organization and the underlying mechanism prove to be, a charge-separation step would be necessary in any mechanism describing electron transfer from chlorophyll to an acceptor.

Quite a few assumptions have been made in this model. These have little experimental support. Photoconduction is tacit in this mechanism of electron transfer.* It suggests that the first oxidation and reduction reactions occur on the opposite sides of a lamellar structure; thus the high-energy products are physically prevented from back-reacting with each other. The nature of acceptor and donor is at best tentative; considerable amount of experimental evidence, in fact, seem to point to some other primary electron acceptors and donors. Thus even though there is no evidence for the wholesale conversion of primary excitation into separated electrons and holes in the major pigment aggregates, various phenomena including light emission, light-induced spectral changes, electron resonance signals, and semiconduction properties of dried chloroplasts, and chromatophore preparations have in general been interpreted to suggest the presence of free electrons and holes in the chloroplasts.[761-763] The possibility that delocalized quanta of singlet excitation carry energy through a chlorophyll ensemble and to the photochemical reaction centers also appears to be acceptable. However, in summary, this attractive model to describe the primary photochemical act in photosynthesis is supported by experimental evidence which can, at best, be described as flimsy.

SENSORY RECEPTION AND ENERGY TRANSDUCTION

Sensory receptors are highly specialized functional units which can be defined at various levels of organization. They communicate with the nervous system by translating the stimulus into a train of nerve impulses and thus have an integrative function. As all sensory receptors eventually code their input into a neural pulse frequency, it is relevant to know the extent to which the transducer properties of a receptor are inherent in the form of their accessory structures and not in the neural membrane itself. The relevant properties of neural membranes are described in the next chapter; in the following section we shall discuss the characteristics of receptor processes without going into the philosophical question concerning how perceptions agree with physical reality.

In general, the receptor processes in their specialization lay the basis of sensory reception. A basic postulate concerning receptor function is that each receptor is adapted for detecting a particular kind of *stimulus* (or *energy?*). Thus we

---

*BLM formed from lipids containing chlorophyll do show the photovoltaic effect and photoconduction.[761] The mean distance between the porphyrin rings in the membrane surface is calculated to be 71 Å at chlorophyll/lecithin ratio of 0.1.[762] These results are consistent with the possibility of existence of mobile electrons and holes which can participate in intermolecular energy transfer.

speak of photoreceptors, chemoreceptors, mechanoreceptors, baroreceptors, audioreceptors, and those equipped for response to gravitational or angular acceleration. Selectivity and the range of sensitivity of a receptor determines what kinds of external stimuli are signalled to the C.N.S. (central nervous system). This is, however, subject to certain qualifications. Even the most specialized receptors may be excited by more than one kind of stimulus, although the threshold intensity of the stimulus may vary significantly. Receptor selectivity is thus relative and takes the form of a relatively low threshold to one kind of energy without excluding responsiveness to intense stimuli of other kinds. It may also be noted that there are significant differences in threshold behavior of various receptors. Some receptors have a sharp threshold, some respond only to stimulus changes in one direction, and some are primarily responsive to the rate of change in the stimulus amplitude. In any event, the amplification between subthreshold stimulus and the propagating all-or-nothing type of response is of the order of a million.[763]

The peripheral terminations of most afferent nerve fibers on which receptors are presumably located consist fundamentally of thin unmyelinated prolongations of the parent axon; the morphological configuration and shape of the terminals and their relations with nonneural structures are highly variable. However, correlation between gross morphological structure and functional specificity of receptors is not always possible, and it is generally agreed upon that differential sensitivity of receptors probably depends on more subtle structural differences than are revealed by light microscopy. Electron microscopic examination of receptors suggests that, in general, most receptor cells possess flagella or cilia. The latter can be regarded as the direct receiver of information or stimuli. The flagella or cilia of receptor cells have a number of special characteristics; one of these is their capacity for automatic, uninterrupted movement.[764] Apparently, this movement serves to search out specific stimuli in the surrounding environment, to correct or grade stimuli within the threshold range by means of autoregulation; i.e., to provide optimum conditions for reception. The site of initial interaction of stimulus with the receptor is thought to be located either on the membrane of the flagella or in other organelles. This action results in a number of molecular changes which form the basis for initiation of excitation in receptor cells.

In operation, receptor excitation characteristics contain static and time-dependent terms. The critical initial event at the site of stimulation is *depolarization* of the receptor neuronal membrane, as suggested by initiation of a train of action potentials in the axon (see Chapter 8 for detailed discussion of action potential and related phenomena). The magnitude of the initial depolarization (receptor and/or generator potential) is directly related to the intensity of the stimulus; if the latter is sufficiently intense, depolarization may reach a saturating value, generally manifested by the frequency of outgoing signals which cannot be increased any further. Of course, the subthreshold intensity of a stimulus does not initiate the train of action potentials even though local depolarization of receptor membrane occurs. Thus the sequence of the events may be written as:

$$\text{Specific stimulus} \longrightarrow \begin{array}{c} \text{Localized} \\ \text{perturbations} \end{array} \longrightarrow \begin{array}{c} \text{All-or-none} \\ \text{response} \end{array}$$

| Receptor-stimulus interaction | Receptor and/or generator potential | Action potential |

Thus the initial depolarization, generally termed receptor potential, is a local, nonpropagated depolarization in response to sub- and suprathreshold stimulus. In contrast, action potential is an all-or-none response which propagates along the axis of the axon (See Table 7-9 for other features).

Thus as a result of an interplay of these two processes, that is receptor and action potential, the receptor performs as a transducer; that is, as a structure which transforms one kind of energy (stimulus) into another (the nerve impulse). In this transformation of a local graded event to conducted, all-or-nothing signals amplification occurs not only of the amplitude (which is independent of suprathreshold stimulus) but also by frequency modulation; that is, the frequency of spike increases exponentially with the intensity of stimulus and vice versa up to a limiting value. (See Refs. 764-768 for various aspects of receptor mechanism.)

TABLE 7-9 FEATURES OF RECEPTOR POTENTIAL AND ACTION POTENTIAL

Receptor Potential	Action potential *
Characteristic of receptor neuron	Characteristic of all "excitable" tissues
Graded and stationary; their amplitude and duration depend upon stimulus parameters	Essentially invariant amplitude and conducted nondecrementally along the entire nerve axon
Probably ionic in character	Ionic in character
Show temporal summation, not necessarily linear	Show spatial and temporal summation, not linear
Show absolute and relative refractoriness, but not always	Always shows absolute and relative refractoriness
Both the peak amplitude and the rate of rise of RP increase with stimulus intensity	Only the rate of rise changes with stimulus intensity
In certain receptors RP is generated in on- and-off fashion even though the stimulus is maintained	Always generated as a train of spikes of equal amplitude; only the frequency changes as a function of stimulus (hyperbolic dependence)
The underlying process is not voltage dependent, hence not regenerative and rectifying	Show both regenerativeness and rectification
The magnitude of RP during sustained stimulation decreases (adaptation)	The frequency of spike-discharge *accommodates* to a sustained stimulation
Cannot be blocked by various agents which block action potential	Can be blocked by various agents such as tetrodotoxin, anesthetics, etc.

*The reader not familiar with characteristics of action potential may first familiarize himself with these characteristics as described in Chapter 8.

Large gaps exist in our knowledge of the excitatory train of events which occur in between these terminal events. The general picture may however be substantiated by a large number of scattered observations resulting from ingenious experimentation. One of the most significant set of results is derived from the study of photoreceptors (Fig. 7-14) of the eye and we shall discuss this system at some length.

On illumination of the retina with a flash of light, there is a silent period lasting at least 1.5 msec before the characteristic sensory impulse is detected. However, an *early receptor potential* (ERP) with very short latency (less than 10 μsec) has been identified in the mammalian retina illuminated with intense light flashes which precede the generation of the spike (Fig. 7-15). The amplification between the amplitudes of ERP and spike is of the order of one million. The ERP appears to have an amplitude which varies linearly with the amount of rhodopsin (the retinal receptor pigment) bleached. The action spectrum of rhodopsin (λ_{max} 500 nm) can be correlated with the amplitude of ERP. ERP persists even when cells are depolarized. It resists anoxia, remains unchanged at lower temperatures (-30 to $+48°$), and has a very small (less than 10 μsec) latency.[763] All these properties compared to the properties of the *spike* suggest a completely different origin of ERP. It is becoming increasingly clear that the ERP has its origin in the action of light on the *visual pigments* themselves. Thus ERP may be identified with the initial depolarizing event which initiates the train of spikes. However, ERP is not ionic in origin, even though the spike is.

Rhodopsin is the prime constituent of the membrane that composes the rod outer segments, and the receptor protein appears to be completely (and *perfectly*) oriented,[769,770] although it is free to rotate about an axis parallel to the rod axis.[1138] As in all photoreceptors* the initial event is absorption of light of a specific wavelength by a specific protein. Visible light absorption in rhodopsin is attributed to bound retinal, a carotenoid. Pigments from different animal groups differ in minor details of chemical structure; however, a protein-carotenoid molecular complex** appears to be invariably involved, and thus many of the specific details in the detection of light can be generalized.[1253] In fact a rhodopsin-like retinal-protein pigment has been characterized in the purple membrane of *Halobacterium halobium*.[1306]

Light absorbed by the rhodopsin molecule in the rod outer segments isomerizes rhodopsin to metarhodopsin which may bring about subsequent changes (Fig. 7-16). In all probability the number of quanta required for the photolysis

---

*Others being those involved in photosynthesis, phototaxis, phototrophy, and photoperiodism. For a discussion on these aspects see Refs. 771 and 1136.

**A pigment-protein complex from the calf retina has been isolated and characterized. It is a glycoprotein (mol. wt. 27000), and has a low content of charged residues and is relatively apolar. The protein devoid of pigment exists in a less compact, probably less ordered inactive state more sensitive to denaturation and proteolysis. This condition is reversed on addition of the pigment (retinol) and the complex shows increased hydrodynamic radius.[772]

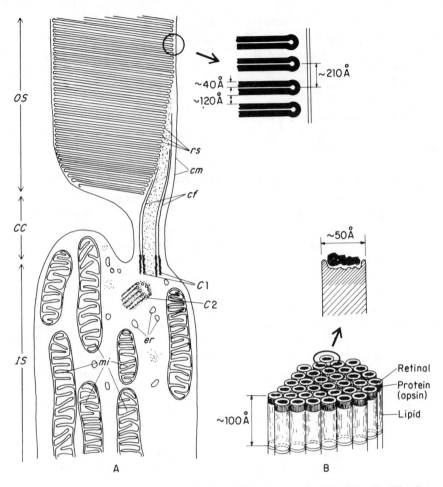

Fig. 7-14. (a) Schematic representation of a rod (light-receptor) cell in the retina. The discs at the top of the cell are folded and refolded to provide layers of membranes, each of which contains light-sensitive pigments on its surface, as shown in (b). 11-*cis*-retinal (in darkness) and its uncoupling to the all-*trans*-retinal in the light is shown in Fig. 7-15b. Structural data on retinal rod outer segments may be summarized as follows:

Animal	Diameter μ	Thickness of Dense Layers (Å)	Number of Dense Layers per Photoreceptor	Number of Rhodopsin Molecules per Photoreceptor	Cross-sectional Area of Rhodopsin (Å²)	Diameter of Rhodopsin Molecule (Å)	Calculated mol wt of Pigment-protein
Frog	5.0	150	1000	3.8×10^9	2620	51	60,000
Cattle	1.0	200	180	4.2×10^6	2500	50	40,000

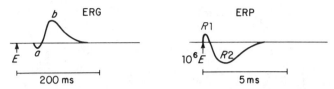

Fig. 7-15. The electroretinogram (ERG) and early receptor potential (ERP) in a dark-adapted vertebrate eye. Each of these responses is a biphasic fluctuation of potential involving cornea-positive (upward) and cornea-negative (downward) components. Unlike the ERG, the ERP has a small latency (less than 10 μsec). For both types of response to be comparable in amplitude, the flash that stimulates the ERP must be of the order of 1 million times more intense. Some fundamental differences between ERP and ERG must be noted. Intense illumination of structures not containing visual pigments, like pigment epithelium, skin, plant leaves, and iris, leads to the production of potential with appearance and properties very similar to the ERP. Hence, ERP is common to all structures containing light-absorbing pigments whether rhodopsin, melanin, or chlorophyll. Anoxia abolishes ERG within 40 min, but not ERP. Fixation with formaldehyde and gluteraldehyde abolishes ERG, but only modifies ERP. ERP is resistant to changes in ionic medium and temperature, whereas the ERG is very sensitive to such changes. These results suggest that ERP, in contrast to the ERG, cannot be the result of ion fluxes across the cell membrane.

of a molecule of rhodopsin is one (the *quantum yield*). Most of the intermediates in this sequence of transformations have been identified.[773-777] Since the decomposition of metarhodopsin II at room temperature is a slow process of the order of minutes and visual excitation takes only about a msec, the intermediates beyond metarhodopsin II can be left out of consideration when one is looking for the step immediately causing excitation. The transition of rhodopsin to metarhodopsin I takes place within 4 μsec at 0°. This then suggests that probably at the transition of metarhodopsin I to metarhodopsin II, there occurs a significant change (the nature of which is completely unknown) which by itself or coupled with some enzymatic process such as adenyl cyclase[1137] locally depolarizes the nerve membrane to produce a spike. To this end some direct experimental evidence has become available recently. It suggests a chemical or conformational changes in the opsin molecule following the photolytic step.[769] Also it has been shown that addition of all-*trans*- or 11-*cis*-retinaldehyde to a suspension of isolated dark adapted bovine rod outer segments causes an influx of sodium upon illumination.[778] It has been suggested that retinal which is attached to phosphatidyl ethanolamine as Schiff's base is transferred to the protein* (again bound as a Schiff's base presumably to a lysine residue) during transition of metarhodopsin I to metarhodopsin II. This transition is a first-order process with an enthalpy of 31 kcal and an entropy change of 60 kcal[774] (see Table 7-10 for other data).

---

*The protein opsin has a molecular weight of 28000 to which retinal is bound through a terminal lysine residue on about ten sites. This protein also has a strong affinity for lipids in general.

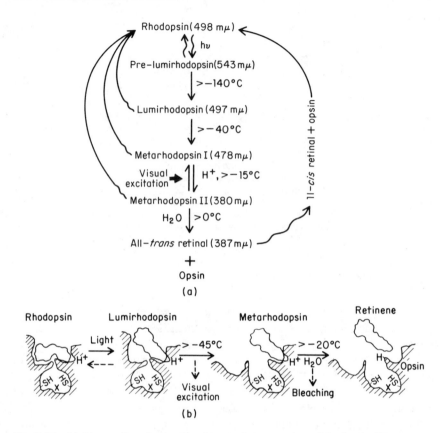

Fig. 7-16. (a) Intermediate in the bleaching and regeneration of rhodopsin. Wavy arrows represent photoreactions, straight arrows thermal (dark) reactions. Rhodopsin can be converted directly to metarhodopsin II by incubation with DCMB or urea in the dark followed by illumination with white light. The interrelationships of pararhodopsin with the final products of bleaching is still uncertain. (b) A schematic representation of steps involved in the action of light on rhodopsin. The absorption of light by a molecule of rhodopsin isomerizes its chromophore, 11-*cis* retinene, to the all-*trans* configuration, yielding as first product lumirhodopsin. The consequent destruction of fit between retinene and opsin labilizes the protein, which rearranges to a new configuration (metarhodopsin). This process exposes new reactive groups on opsin (including two sulfhydryl groups and one H^+-binding group) and may be responsible for triggering visual excitation. Vertebrate metarhodopsins are unstable, hydrolyzing to opsin and all-*trans*-retinene, the process that corresponds to bleaching.[773] See Table 7-10 for thermodynamic data.

The rod is strongly dichroic in respect to rhodopsin absorption, the polarization being essentially perpendicular to the axis. However no dichroism is induced after partial bleaching by exposure to a flash of plane polarized light. However, partial bleaching of retina treated with gluteraldehyde (a bifunctional reagent that cross-links adjoining rhodopsin molecules) leaves the rods markedly dichroic to end-on

TABLE 7-10 THERMODYNAMIC PARAMETERS
GOVERNING THE KINETICS OF THERMAL
DECAY OF PHOTOLYTIC INTERMEDIATES[a]

Intermediate	ΔF^* (kcal/mole)	ΔH^* (kcal/mole)	ΔS^* (e.u)
Prelumirhodopsin	—	11	8
Lumirhodopsin	18	4.5	-51
Metarhodopsin I	14	39	+82
Metarhodopsin II	21	7.5	-50

[a] ΔF^* = free energy of activation, ΔH^* = enthalpy of activation, ΔS^* = entropy of activation (in entropy units). These data are shown merely to indicate the order of magnitude and the sign. From Ref. 774.

view, the dichroic ratio increasing with the extent of bleaching.[1304] Formaldehyde engenders no such effect. In the short time scale of these experiments the bleaching of the rhodopsin proceeds only as far as isomerization of 11-cis– retinene to the all trans- state, but the pigment has not yet been released from the protein. The decay rate of the dichroism on a time-scale of microsecond follows a simple exponential law, and is retarded when the viscosity of the medium is increased by allowing sucrose or glycerol to diffuse into retina.[1305] From the rotational relaxation the viscosity of the environment of rhodopsin is calculated as 2 poise, which is in the range of thin oils or interior of micelles. The rod membrane, having practically no cholesterol is probably more mobile than most.

In spite of such vast amounts of experimental data available on visual receptors it is not possible as of yet to describe the molecular event involved in the receptor function. In fact, the significance of various events, particularly ERP, is not known with complete certainity. ERP have been observed in a variety of pigmented tissues and the underlying mechanism appears to be similar but not identical. More serious difficulty arises when one considers the mode of operation of chemoreceptors. The stimulating molecules have been shown to interact with hypothetical receptors which (for their cytoarchitecture see Ref. 1139) show an extremely high degree of specificity, comparable to enzyme-substrate interactions,[779-781] and receptor proteins (in fact binding proteins) have been isolated in some cases.[782] The stimulating molecules presumably interact reversibly and do not undergo any chemical change. Thus, the interaction energy must be small—of the order of a few kT. Thus, most of the energy must be derived from the receptor itself. This would imply that the stimulating substrates function as regulators or effectors, which increase the number of charges carried across the membrane in the equilibrium state. This is consistent with the observations on model systems involving carotenoids in an organized state (film or solid) where the efficacy of olfactor molecules has been correlated with their ability to increase conductance of these films.[783] It is surprising how such a high degree of substrate specificity is achieved for these interactions without incorporation of proteins. It is however quite possible that these molecules increase membrane conductance as is observed in the early stages of lysis (see p. 209, also see Ref. 458b). On the other hand, evidence has been forwarded suggesting a role for mucopolysaccharides in the receptor process.[784] However, the most promising suggestion arises from the correlation of receptor functions with an electrogenic pump in receptor cells.[767] It has been suggested that depolarization of the nerve may be mediated and modulated by the interaction of an electrogenic pump with the stimulus, i.e., the pump is coupled to the receptor at the molecular level. The nature of the evidence is, however, circumstantial.

In conclusion, electrophysiological data alone cannot closely define the primary process of transduction that follows the interaction of the *stimulus* with the receptor; neither the chemical kinetics of the photochemical reaction, nor the associated structural changes in the photoreceptor membrane are as yet understood. By far the most elusive property of the sensory receptor cell is the functional nature of the membrane itself and the factors affecting it. Little is known concerning the structural differences between electrically excitable and inexcitable membranes. The resolution obtainable by electrophysiological techniques is not nearly as fine as could be wished for, and *electrical excitability* is, in the strictest sense, a generic term embracing various types of responses which may have varied molecular origins. There is now good evidence to indicate that inherent differences in the excitability of the membrane are likely to occur from one part of the cell to the next. Thus, the site of impulse initiation cannot be obvious from the geometry alone.

Perception involves impact of information initiated simultaneously or in rapid succession at many sensory channels. Thus, one of the most intriguing phenomena in perception is study of sensory interaction, the modification of the input from one sensory channel by input from another channel. It is seldom clear even at higher levels of organization whether the effect occurs at the receptor level or at some point in the nervous system. Thus the importance of the receptor function can hardly be overemphasized.

PRIMITIVE MECHANISMS OF MOTILITY

Primitive mechanisms of motility include cell movement, cytoplasmic streaming, and related phenomena. Many types of specific cell movements are known. These include oriented contraction of muscle cells, beating of cilia and flagella, a wide variety of protoplasmic streaming phenomena including different types of pseudopodal or amoeboid movement, the specific positioning of organelles within cells, saltatory movements of some organelles, a rapid axial rotation of the nucleus which occurs in some cells, and various highly ordered movements of chromosomes and the cell surface which are involved in cell division. In no instance has the mechanism of movement been elaborated at the molecular level. Nevertheless, sufficient data is available to show that a marked resemblance exists between the more primitive mechanisms of motility and those of muscle contraction. Certainly, the most recurrent theme is the requirement of ATP as the direct energy source.[785-787] In this section we shall discuss some of the salient features of cytoplasmic streaming with special reference to amoeboid movement, mainly to illustrate the phenomena.

Some spontaneous movement of organelles and other inclusions occurs universally in cells and in fact may be regarded as characteristic of the living state. This motion is nonrandom and hence must be distinguished from Brownian motion. These protoplasmic movements embody a multitude of different forms, many of which seem to be specific to one or another of the cell type; for example, a cell may have a high degree of orientation or may display net movement of the cell, or may not show any of these characteristics. In simplified form the phenomenon may be described as follows:

When a cell is examined under the highest power of the light microscope, granules in the cytoplasm can be seen to be in continuous motion (*movement*). Part of this motion could be due to Brownian motion; however, much of it cannot be explained in this way. Time-lapse films of animal cells in culture also show streaming within the cell and many protozoa exhibit large-scale streaming during locomotion. The rate of cytoplasmic flow can be very high, reaching a velocity of 60 microns sec^{-1} in mature internodal cells. The maximum movement occurs adjacent to the cell wall. Besides cytoplasmic movement other subcellular particles can be seen rotating about their own axes. In amoeboid movement there is net locomotion by the cell; this implies deformations of the cell surface to form temporary cytoplasmic extensions of pseudopods. Within this broad definition a startling range of variations occurs, including the broad, lobose pseudopodia of *Amoeba proteus*; the extremely long, fine cytoplasmic processes of foraminifera, hornlike pseudopodia with or without helical twists, ruffled pseudopodia entirely free of cytoplasmic granules, and many other kinds. Such protoplasmic activity may also be observed in a great variety of animal cells. However, it is not certain whether "amoeboid movement" is a single phenomenon with numerous variations in detail with superficial similarities.

In general, it is agreed that the driving force for streaming originates at the contact point between the endoplasm and the cell membrane with its associated actoplasm.[785,788-790] However, the studies pertaining to the mechanism of these phenomena are in a very controversial stage.[785] One of the major difficulties in carrying out such studies is that, to date, no definite structures may be directly associated with such functions. Furthermore, it is not clear whether the motive force is attributable to changes in the membrane and/or underlying cytoplasm. The following observations, though quite vague, may provide some insight into the complexity of the system:

1. Typically the ATP requirement is correlated with the presence of one or more proteins which exhibit catalytic ATPase activity and which often show

actomyosin-like viscosity changes in the presence or absence of ATP and cations such as Na, K, Mg, and Ca. Although it is by no means clear that contractility per se is involved in all mechanisms of cell motility, the possibility that it is has not been excluded. It may be noted, however, that the contractility may be a general property for any long fiber formed from macromolecules like actomyosin.

2. Many cell movements are stopped when organisms are subjected to high hydrostatic pressure, which is associated with sol-gel transformations.

3. Studies with fluorescent antibody directed toward the surface coat suggest that there is extremely slow turnover rate of the membrane in the direction of movement.

4. Movement requires local areas of attachment between the plasma lemma and both the substratum over which the cell moves and the underlying plasma-gel. Also it seems that plasma–sol is converted into plasma–gel at the advancing end of the cell, and plasma–gel into plasma–sol at its rear end. However, the mechanochemical principles involved in these processes are unknown.

5. Contraction can be brought about at the periphery of the cell by local application of polylysine or other suitable polyelectrolytes to an advancing pseudopodium.[231]

6. The motive force for streaming is quickly abolished by inhibitors of anaerobic glycolysis, but inhibitors of oxidative phosphorylation have little effect. This suggests that mitochondrial ATP is unavailable for the streaming mechanism; therefore, the mechanism must be located in the plasma where reactions catalyzed by soluble glycolytic enzymes are the primary ATP source.

7. The velocity of streaming endoplasm is virtually constant at all levels (that is, not parabolic); on the other hand, a very steep velocity gradient exists between the outermost layer of moving endoplasm and the innermost layer of stationary ectoplasm gel. It appears that the endoplasm does not so much flow as glide en masse over the inner surface of the endoplasm. It implies that such cyclosis is brought about by some type of oriented shearing force applied along the ectoplasm-endoplasm interface. Such a streaming mechanism based on lateral shear forces is very reminiscent of the sliding filament model of striated muscle contraction.

8. By using intracellular electrodes on *Amoeba proteus*, it has been found that the membrane potential at the tip of an advancing pseudopod is about -30 mV, whereas that at the rear of the cell is about -70 mV. This implies a potential gradient[791] in the cytoplasm of about 1 V-cm^{-1}. Externally applied currents also cause distinct effects on amoeboid movements. A weak continued alternating current (ac) delivered through an amoeba in an alkaline medium causes lateral contractions with pseudopodal extensions in both directions perpendicular to the current flow. The amoeba thus flows into one of the two pseudopods and moves away. In stronger continuous ac the pseudopods continue to lengthen until the contracted areas turn yellow and disintegrate. A particularly attractive

outcome of these observations is that they provide a mechanism whereby environmental stimuli could modify amoeboid movement. For example, mechanical stimuli could modify such movement by altering membrane permeability and the bioelectric potential. In fact, changes in membrane permeability can be detected.[792] It has been found that the addition of lysozyme, ribonuclease, NaCl, and/or KCl to the culture media of *Chaos chaos* leads to a large drop (up to fifty-fold) in electrical resistance of the membrane and an increase in permeability. The amoeba cease to move when the membrane permeability is high. When the resistance is restored to control values, the organism begins to move freely again. The decrease in resistance is governed both in magnitude and duration by the Ca concentration in the medium.

From these observations alone it is very difficult to pin down the role of the surface membrane in amoebic locomotion. Some of the physical effects such as ion exchange, adhesion to substrate, response, feeding, and pinocytosis are functions usually associated with the membrane as a barrier to and an intermediary with the environment. The electrical potentials measured at the membrane suggest that they are most likely the result of activities of the locomotor mechanism rather than the cause of its activity. Thus a sol–gel transformation in a pool of actomyosinoid proteins energized by ATP hydrolysis gives rise to their movement. The system appears to function in a series of irreversible cyclic events in which polymeric actinoid fibrils are formed by bonding of actin monomers with ATP or GTP via a Mg-activated ATPase. These then interact and bond with myosinoid fibrils or aggregates and ATP. If Ca is available the myosin ATPase is triggered and ATP is split. The energy thus released shifts the chemical bonds between the actinoid and myosinoid proteins, and thus causes them to move in relationship to one another.[785]

Such streaming and locomotion of vacuoles is of importance in a number of aspects of the behavior of the whole organism. Contacts with fibers control movements parallel to the orientation of these structures, and contacts between neighboring cells also affect their movements. A contact prevents further movement of the cell in its original direction. If only a few contacts are formed, the cell may move in a new direction and break its original contact after considerable tension has been produced. But, if many contacts have been formed, largescale movement of the cell ceases, though the tissue can still move if the movement of the individual cells is concerted. Thus changes in intracellular adhesiveness with time can account for a large fraction of all morphologic movements. Many of the properties of cancerous cells indicate the loss of contact control of movements.[789] It is obvious that processes such as contact interaction, contact inhibition, peripolesis, cellular aggregation and segregation, chemotaxis, and chemokinesis underlie the behavior of the organism as a whole; such phenomena in concert with cell movement may have their origin in some such behavior of the membrane.[237] In conclusion, it may be noted that the macroscopic behavior of the organism is a consequence of the factors as described above, and any degree of control in these processes may underlie considerable macroscopic changes. In any event the whole question of the mechanics of the translation of orientational information through contact with the substratum is at the moment as vague as it is important.

Besides these specific cellular processes there are reactions in animals which may collectively be called behavioral responses. In the simplest form, for example, certain organisms use light stimuli as signals for navigation, to search for food, etc., which may be collectively termed *photomotion*. More specific definitions of behavior are included in this category: photokinesis, a light induced change in motility; phototaxis, the directed orientation of an entire organism to light of different intensity and wavelengths; phototropism, the orientation of one part of an organism toward light, as in the bending and twisting of plants to present leaf surfaces to the light.

Thus the translation of an internal effect into a surface action raises questions similar to those involved in the origin of nervous impulses in animal photoreceptors. The problem of motility, while complex in itself, thus involves the more general problems of light reception, excitation, contractility, and probably neural transmission. If there is a connection between the absorbed light and the motility of the cells, as suggested by the phenomena mentioned above, there must exist some mechanism yet unknown for transmission of information from the receptor to the effector.

8/ Excitability: Time-Dependent Phenomena in Membranes

At higher levels of organization of organisms the time-dependent electrical changes, though highly specialized but generally observed, assume special significance. The central nervous system may be considered the governing organ of *animative* activities. With the help of two peripheral extensions, the receptor and effector systems, it regulates the visceral and behavioral functions of the organism. It was suggested in the last chapter that the receptors and effectors cannot communicate by themselves. Transmission, decoding, and storing of receptor information, and the programming initiation, and feedback regulation of organic activities are dependent on the central nervous system with its conducting, relaying, and coordinating mechanisms.

Operationally, the axons are the communication lines of the nervous system, and transmit electrical impulses from one part of the nervous system to another. Axons are sometimes compared to electrical cables or transmission lines; however axons also have built-in amplifying and relay systems. Thus, depending upon the fiber diameter, the conduction velocity changes—lower for thinner fibers, and higher for thick fibers. A large number of other properties of the nerve fibers vary with the speed of conduction; for instance, the duration of the impulse, its rate of rise, its size, the duration of the inexcitable or refractory period following each impulse, the threshold of excitation, and the sensitivity of discharge to pressure on the nerve and to asphyxia; in short, an array of properties connected with impulse conduction, all of which need not vary in an exactly parallel manner. Similarly, several nerve types differ among each other in more than one significant way. Thus, sensory nerves not only have a different conduction velocity when compared with motor nerves, but they also show a lower threshold of excitation, and they put up less resistance to impulse generation (less *accommodation*) than motor nerves. These differences have far-reaching consequences for the proper functioning of the central nervous system. Without going into details it may be stressed that a central question in the investigation of the nervous system concerns the propagation of the nerve impulse. See Ref. 802 and 1255 for details.

The key to the axon's propagation of an impulse lies in the characteristic properties of its membrane. A unique feature of excitable cells is the brief transition in membrane properties which takes place when the transmembrane potential is reduced rapidly below (on an absolute scale) a threshold value. When the threshold is surpassed, a complex potential oscillation takes place. In normal neural activity this is brought about by an appropriate superposition of inputs. The membrane potential oscillations are generally known as *spikes* and result from changes in the permeability characteristics of the membrane (see later). Such changes can be recorded with an electrical circuit other than the stimulating circuit.

During suprathreshold stimulation, after the lapse of a small period of time (*latency period*), the inside of the nerve swings momentarily positive, giving rise to a transient action potential with a magnitude of about 100 mV under the normal conditions. Under certain conditions, such responses across the membrane can be initiated by a sudden reduction in temperature, a sudden reduction in the concentration of Ca^{2+}, or a sudden increase in the concentration of K^+. It may be noted in all these cases a small *depolarization* of the resting membrane is involved (see later).

The time course of a typical *spike* is shown in Fig. 8-1. The parameters which characterize a spike are the resting potential (diffusion potential in steady state), threshold for activation, peak value for transmembrane potential (*overshoot*), negative peak value following overshoot (*undershoot*), duration of spike, rise time of initial phase, and propagation velocity. Before going any further with the changes consequent upon generation of a spike or a train of spikes we shall elaborate on these parameters.

The minimal intensity of the current flowing through a nerve which produces excitation is a measure of the threshold. The threshold conditions for initiation of a spike are fairly sensitive though indirect measures of some of the electrical properties of excitable cells as implied in the effect of temperature, Ca, and K changes on threshold. The threshold current is not a fixed value; it depends on the prior history of the cells, the nature of the media, and the rate of application of depolarizing pulses. As the slope of a linearly rising stimulus is reduced, not only is the time to excitation increased, but the absolute stimulus level that must be reached before excitation will occur also increases. This is known as the phenomenon of *accommodation* of stimulus. Thus, there is a minimal strength of stimulus (as a square pulse) required to evoke a propagating spike, and a

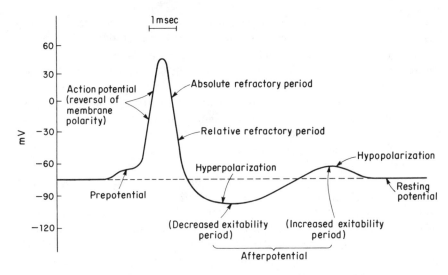

Fig. 8-1. Some of the characteristics of the different phases of an action potential recorded intracellularly from an axon.

suprathreshold stimulus, no matter how strong, evokes* a stereotyped response; that is, there is no change in the height of the spike. Only the frequency of spikes changes as a function of suprathreshold stimulus. The response of a suprathreshold stimulus is fixed in size, shape, duration, and conduction speed all over the length of the fiber, no matter how far it is from the stimulating point. This is termed *all-or-none* or *all-or-nothing* behavior. Threshold and all-or-none behavior appear to be different aspects of the same regenerative mechanism.

The value of the membrane potential at which an action potential (or spike) is just initiated by a depolarizing current is called the *threshold voltage*; a current that barely depolarizes the membrane to threshold is similarly called a threshold stimulus; however, these two quantities may not be simply related by Ohm's law (see later). The reciprocal of threshold is defined as excitability. If an applied current is slightly subthreshold, the depolarization persists for some time after the termination of the current; the size of a subsequent threshold stimulus is reduced during this early period, and the excitability is increased. However, the time course of this increased excitability is such that even when current flow continues at a constant level, excitability drops to a steady intermediate value. This decline from peak excitability during constant current flow is a manifesta-

---

*Owing to the spontaneous play in excitation process, a stimulus does not invariably evoke an impulse until its strength is slightly greater than threshold. In fact, a threshold stimulus is generally defined as one which produces impulses on 50 per cent of occasions.

tion of the phenomenon of accommodation as described earlier. Following cessation of current flow, excitability at the stimulating cathode declines below the resting level and recovers only slowly. This is known as post-cathodal depression. The blockage of impulse generation by depolarizing current occurs only in the region of the stimulating cathode. Thus impulses will propagate on either side of the blocked region, but not through it. This method of blocking impulse conduction is called cathodal block. Similarly, a sufficient hyperpolarization of the membrane can also block conduction in a nerve. Blockage also occurs if the membrane potential is made so large that local current flow from the hyperpolarized region into an approaching active region is insufficient to depolarize the hyperpolarized region to the threshold. This phenomenon is called anodal block or (better) hyperpolarization block. Similarly, a sudden withdrawal of the hyperpolarizing pulse causes excitation or initiation of a train of spikes; this is known as anodal break excitation. Various other phenomena associated with threshold behavior of nerves and the propagation of nerve impulses in a single nerve fiber are absolute and relative refractoriness, spatial and temporal summation of stimuli, enhanced and depressed phases following subthreshold stimulation, and the mutual destruction of colliding impulses travelling in opposite directions. As we have noted throughout our discussion of the threshold phenomena in propagating nerves, it is only a depolarizing current which is excitatory. This is again confirmed by the observation that if the two stimulating electrodes are separated by a distance from each other the impulse always arises at the cathode. When the stimulus is applied through electrodes separated by a larger distance the sequence of events is the same as observed for closely placed electrodes, except that the *latency* (time interval between application of the shock and the beginning of the spike) is larger for larger separations. Systematic investigation reveals that this latency is directly proportional to the distance between the stimulating and recording electrodes, and to the intensity of depolarizing pulses.

In elongated excitable cells, like nerves, an impulse once initiated by a stimulus is propagated rapidly from the stimulus site (or the cathode if both electrodes are separated) to adjacent regions of the membrane and thus spreads as a wave over the membrane of the entire cell. Such self-propagation can be phenomenologically described as follows. As the axon is depolarized during stimulation, the local region within the axon becomes positive with respect to the next adjacent section which still has a normal resting potential which is negative. Consequently, a current flows from the positive to the negative region, completing the circuit by returning to the positive region through the conducting solution outside the axon. The current arriving at the region of the normal resting potential triggers the generation of a spike like that in the region it has just left. In this manner the impulse is regenerated from point to point along the axon, and travels from one end of the axon to the other.

As pointed out earlier, the passage of each impulse in the fiber is always followed by the refractory period—a silent period during which the nerve membrane is unable to carry a second transient change in potential. The maximum frequency at which a nerve fiber can transmit impulses depends upon the time length of the absolute refractory period—a minimum duration between two impulses, or the time taken to recover from the excited state. Thus the limiting frequency for spikes will be determined by the relationships shown in Fig. 8-2. The upper limit is usually about $2/t$ impulses per second, where t is the absolute refractory period in seconds. Thus for $t = 10^{-3}$ sec the maximum number of impulses that can travel along the fiber is about 500 impulses per second. The normal working range in the human body is about 5 to 100 impulses/sec. A value as high as 1600/sec has been noted for the discharge of the electric organ of *Gymnotid* and other electric fish.[793]*

The velocity of conduction of an impulse (train of spikes) depends upon the diameter of the fiber and is usually in the range of 1 to 100 m/sec. Empirically, it can be given approximately as $V = 2.50\ D$ m/sec, where D is the diameter of the fiber in microns. This relationship depends on the nature of the nerve. However, the whole surface of a fiber may not take part in the conducting impulse. Myelinated nerve fibers, for example, are enclosed in part by the insulating myelin sheath. Only the uncovered parts of such nerves, known as the nodes, participate in the conduction of the nerve impulse in a slatatory fashion from node to node.

The passage of an impulse across an axon is a propagated event and it is very difficult to study the changes involved in such a process since *time* and *space* are variables. To study the time-dependent phenomena across a piece of axon, an axial metal electrode (usually platinum or silver) is placed within the axon for short-circuiting longitudinal conduction. Improved survival of the axon results when the end ligatures and the axial electrode punctures of the axon are covered with flowing isotonic sucrose.[794] Such a short-circuited axon is defined as being space clamped, that is, the membrane potential is independent of the space coordinate along the axis of the axon, and the potential changes simultaneously over the whole length of the axon, synchronously.

---

*The electric organs of these various fish give rise to a powerful electric discharge. Although a variety of discharge pulse shapes are produced, the basic operation always involves a common mechanism: two faces whose membranes develop different potentials to generate external currents. A striking histological feature of most electric organs (electroplax) is the considerable folding of the nonnervous membrane. They are arranged in series such that there is an addition of every single potential difference generated by the nervous face of each electroplax, thus giving rise to powerful electric discharges, which may sometimes be as high as 500 volts from 5000 cells potentials. The nervous system of the electric eel is hooked up to each top membrane of a pile of cells in such a way as to short them out by a nerve impulse and thereby deliver the resulting high voltages through its electric organ.

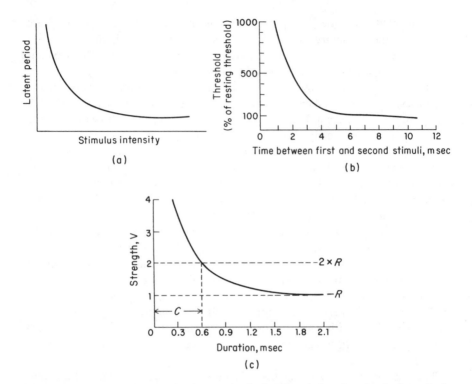

Fig. 8-2. (a) A curve showing the decrease in the latent period as the stimulus intensity is increased. (b) A curve showing the decrease in the refractory period as the stimulus intensity is increased. (c) A curve (the solid line) showing strength–duration relationship. In this example, after approximately 1.8 msec (duration of stimulus) further increases in duration do not result in a lowering of the strength of the stimulation required. This is the rheobase (R). Chronaxie (C) is determined (arrow) as that duration along the strength–duration curve at a strength twice the rheobase.

In a space-clamped axon only one spike or action potential is generated after application of a square pulse of sufficient magnitude and duration. In quite a few ways space clamped action potential is similar to a propagating spike; it shows a similar sequence of potential changes, refractory period (both, relative and absolute), anode break excitation, overshoot and undershoot, duration of spike, characteristic rise time, and other related phenomena. An approximate relationship between some of the variables giving rise to these phenomena is given in Fig. 8-2.

It may be noted that threshold characteristics of the membranes under space-clamped conditions are not comparable to those of a propagating spike and thus many of the related phenomena would not be exhibited in a space-clamped membrane. It has been shown that the all-or-none or sharp transition phenomenon characteristic of the propagating impulse degenerates into the potential response which is a smoothly continuous function of the stimulating current.[795]

Studies with space-clamped axons as such and application of various other techniques have given rise to the experimental evidence suggesting that the origin of potential changes is in ionic gradients across the membrane. Thus, by changing the ionic concentration in the external medium it can be shown that the resting potential is determined primarily by the concentration of potassium ions; in contrast, the height of the action potential is determined by the gradient of sodium ions. Since the concentration gradients of these two ions operate in mutually opposite directions the transient swing in potential during the course of the action potential is approximately determined by the difference between the potassium and sodium equilibrium potentials. These observations raise a number of questions which cannot be answered by the simple space clamp technique. In fact, the period following introduction of the space clamp technique has witnessed the introduction of a large number of additional techniques. Before discussing results obtained from these techniques it may be pertinent to discuss the techniques and materials employed to obtain the data. Readers familiar with basic techniques of *axonology* may skip over this section.

* * *

For the study of excitable membranes the nerve fiber of squid has been particularly favorable, and it is generally taken as a model for the electrophysiological properties of nerves in general. The axons of squid nerve, commonly referred to as "giant axons" are about 0.1-0.5 mm in diameter, and have allowed a unique control of the experimental variables which include potential difference, ionic concentration, and composition of aqueous phases. Many other invertebrate species also have axons with diameters of 100 microns or more*; they are *Planorbis, Sepioteuthis, Cambarus, Dosidicus, Carcinus, Maia, Astacus, Doryteuthis, Myxicola, Lumbricus, Carcinus, Homarus, Procambarus,* and *Loligo* spp. The giant plant cells *Valonia, Nitella,* and *Halicystis* have also been used. More realistic material, such as muscle, purkinje fibre, brain cortex, electroplax, node of Ranvier, skin, and bladder from various species of higher animals have also been used, but with limited success since these tissues are difficult to handle and the results are sometimes much too complicated to interpret.

---

*Why some animals and not others have evolved giant axons is not fully understood, but giant axons seem to be involved in escape mechanisms. The giant axon of squid is part of the mechanism by which the animal flees its enemies in water. In order to dart away rapidly the squid uses a jet propulsion system which calls for the synchronous contraction of muscle located throughout its body mantle, and therefore all those muscles must receive the message from the brain simultaneously. Thus the farther the muscle is from the brain, the larger is the diameter of the axon leading to it. Indeed, experiments have shown that the thicker the axon is, the faster it conducts the impulse. Obviously, the squid could not escape the electrophysiologists!

From the point of view of technique, the most significant advance was the introduction of the *voltage-clamp technique*.[796,797] Under normal conditions during the course of the action potential it is not possible to study the effect of potentials intermediate between the resting potential and the peak of the action potential. Such studies are possible by clamping the voltage and following the changes in the flow of current (for theory see Ref. 1141). The basis of the voltage-clamp technique is to produce a sudden displacement of the membrane potential from its resting value through a pair of electrodes and to hold the potential across the membrane at this new level by means of a feedback amplifier (Fig. 8-3). The current that flows through a definite area of the membrane (maintained by space clamp) under the influence of this applied voltage is measured with a separate pair of electrodes and a separate amplifier. The voltage-clamp technique has the important effect of eliminating the contribution of membrane capacitance to the electrical phenomena. Through such experiments it has been established that a sudden, sustained change in membrane potential, whether an increase or a decrease, shows a graded response (see later), in contrast to action potential whose magnitude changes very little with the stimulating current.

Another experimental advance has been the introduction of the perfusion technique.[798-800] The cytoplasmic fluid of the axon can be squeezed out by a series of firm strokes with a roller. The giant axon is then perfused with solutions of varying ionic strength and composition. Once perfused, the axon can be tied off at both ends and handled like a normal axon. It is more convenient,

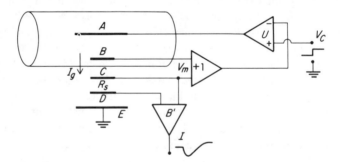

Fig. 8-3. A simplified schematic representation of squid axon voltage clamp. Electrodes A and E are current-injecting electrodes. Electrodes B and C record membrane potential V_m. Electrodes C and D determine the membrane current I_g by recording the voltage drop produced across the external saline resistance, R_s. The operational amplifier of high gain U is connected in a negative feedback loop so that the membrane voltage, V_m, follows that of the control voltage, V_c. Amplifiers B and $+1$ are ordinary differential amplifiers. At current electrode A, $V_A = U(V_c - V_m) = V_m + a$, where a represents potential drop at the surface of electrodes A and E, and the IR voltage drops across the membrane and external saline. Therefore, $V_m = [U/(1 + U)] V_c - [1/(1 + U)]a$, for very large amplifier gain. For more detailed treatment see Refs. 796 and 797. Figure and explanation courtesy Dr. A. Strickholm.

however, to mount it in such a way that a microelectrode can be inserted into the axon. In an alternative method of perfusion,[799] a length of about 1 cm in the middle of the axon is perfused with the solutions passing through pipets inserted at both ends of the axon. Although this method is less likely to damage axons, it has the disadvantage that appreciable quantities of the original axoplasm remain in the perfused region.

So long as the appropriate salt concentration, pH, and osmotic conditions are maintained inside and outside the axon, electrical excitability and other characteristic features of the membrane, including active transport, can be maintained. From perfusion experiments, the only requirement for excitability appears to be that the solution used for perfusion inside be rich in potassium, and poor in sodium ions; the nature of anions is not very critical except for the stability of the axon. To produce an action potential and related phenomena it is necessary that the perfusion fluid is primed with a sufficiently high concentration of potassium ions. A variety of probes and markers can also be introduced into the perfusion fluids and thus their effect on the membrane phenomenology can be assessed.

Although perfused axons show similar, if not identical, behavior to native axons,[799] it is possible that there are several, as yet unidentified, components of the axoplasm which fulfill an essential role in the maintenance of the transmembrane potential and propagation of the nerve impulse. All that can be said therefore is that the perfused axons are still fully functional and can conduct up to about one-half million impulses even after they have been washed by a flow of artificial solution amounting to 100 times the volume of original axoplasmic fluid. Also, electron micrographs of perfused axons indicate that the structure of the axon is not seriously affected by perfusion and about 95% of axoplasm is removed during perfusion.[798] Thus while considering the integrated properties of nerve cells it must be kept in mind that the cell interior is highly organized, containing various organelles with specific and general functions. Nevertheless, it is convenient and meaningful to regard the cell fluid as a single aqueous phase when discussing ion transport across cell membranes.

* * *

Equipped with these techniques neurophysiologists have produced vast amounts of experimental data reviewed elsewhere[801-807,1255,1307] and theorized from time to time. Unhappily, there are some significant differences of opinion (and sometimes of facts) as to the interpretation of the data. The rest of this chapter first deals with some of the experimental data and then the theoretical attempts to consolidate it. For simplicity and brevity of exposition the experimental results are treated in two sections: (a) the steady-state characteristics of the excitable membranes, and (b) the time-dependent characteristics. The overall discussion of biomembranes pertains to the observations made on squid axon (unless stated otherwise), however it is believed to be part of some general pattern as stressed while discussing physical and molecular theories of excitation. It may also be noted that no attempt has been made to elaborate the differences observed in the excitability properties of various organisms.

EXPERIMENTAL RESULTS

Steady-State Characteristics of Excitable Membranes

Resting Potential. It was mentioned in Chapter 7 that the composition of the fluids inside and outside any living cell are considerably different, although they are isotonic. Thus ionic gradients exist across a living cell membrane (Table 8-1). A consequence of the existence of ionic gradients in these biological systems is the development of an electrodiffusion potential and dissipation of the concentration differences which are maintained by active transport.

The magnitude of the diffusion potential is of the order of several tens of millivolts. Through various electrophysiological techniques it has been found that the inside of a resting fiber is 50–85 mV negative to the external solution.

TABLE 8-1 IONIC COMPOSITION[a] OF IONS IN NATURAL WATERS, SAP FROM
CENTRAL VACUOLES OF PLANT CELLS, BODY FLUIDS, AND
INTRACELLULAR FLUIDS OF SEVERAL ORGANISMS[b]

	Na	K	Ca	Mg	Cl	Organic Anions	Diffusion Potential (mV)
Seawater	417	9.1	9.4	50	483	1	—
	(100)	(2.2)	(2.3)	(12)	(120)		
Squid blood	410	18	9.4	51	490	2	—
	(100)	(4.4)	(2.3)	(12)	(120)		
Squid axoplasm	54	415	0.5	10	40–150	300	−70 to 90
	(100)	(800)	(1)	(20)	(100–300)	(600)	
Cat Plasma	178	5	3	1	128	2–3	—
Cat muscle	28	151	1	15	18	50–100	−100
Valonia (in seawater)	90	500	2	Trace	597	Trace	−50 to 90
Pond water	1.3	0.019	1.08	0.9	1.08	—	—
Hydrodictylon in pond water	4	76	2	—	55	Trace	−30 to 50
Frog muscle	10	130	—	—	3.0	—	−90
Ascite tumor	54	119	—	—	61	—	−16
Schwann cell (squid)	312	220	—	—	167	—	−40

[a]Due to complications arising from structural and chemical compartmentation of cell, the intracellular concentrations are mostly only approximations, and the term "apparent concentration" may be more appropriate.
[b]Upper numbers are expressed in mM; lower numbers are relative concentrations, expressed in terms of 100 units of sodium ions. See Ref. 811 for data on several excitable and inexcitable tissues.

This potential difference is generally called *resting potential*. The value of resting potential is a function of the electrolyte concentration of any given medium and axoplasm. No potential difference has been observed within the cytoplasm. The magnitude and sign of the resting potential can be best accounted for by the assumption that the contribution of potassium ions is major, and that, to a first approximation, the contribution of chloride and sodium ions can be neglected. There are some serious deviations from ideality in the permeability characteristics of most biological membranes. Some of the reasons for these deviations are: biionic and multiionic potentials may be appreciable, divalent cations may regulate the passage of other ions, there may be some dependence of ionic fluxes on the concentration of other ions, metabolic state, and the contribution of the electrogenic pump. In some cases at least some of these components have been found to be significant.[808]

Also the value of the resting potential is a function of a variety of external factors. Treatment of the axon with heavy metals, perforation of the membrane with a sharp instrument, or high intracellular hydrostatic or osmotic pressure, or large thermal gradients result in the reduction, sometimes stepwise,[809] of the resting potential.[810] The concentration of divalent cations, especially Ca^{2+}, is critical. At normal Ca^{2+} and K^+ concentration disturbances caused by temperature, low-intensity current, stretching the axon lengthwise, pH, or osmotic pressure changes have little effect on the resting potential. In the case of low Ca^{2+} in the external medium, a continuous and graded reduction in resting potential may be observed under certain conditions. These include decreasing pH (0.2 to 2 pH units), stretching (5–20%), application of an outward passing current (0.1–10 milli-amp-cm^{-2}), reducing osmotic pressure (5–15%), or by varying temperature (above 40°C). All these factors appear to be interdependent; however, the mechanism is not clear.

Electrical Properties of Excitable Membranes in Steady State. The behavior of a resting axon membrane on the application of an electrical pulse is strongly dependent upon several factors. These include the polarity and time course of the applied current, the magnitude of the resting potential, and the ionic composition of the internal and the external media. Under normal physiological conditions, the magnitude and the time course of the potential developed across the axon membrane in response to the passage of small current is more or less independent of the direction of current flow, and the system shows ohmic behavior. However, it should be noted that the application of small-amplitude, long-duration current pulses to a space-clamped squid axon may cause the membrane voltage to oscillate before setting to a steady-state value.[811a] At higher voltages, the conductance is non-ohmic as shown in Fig. 8-4. In fact, the electrical conductivity is highly anisotropic and is time dependent. This phenomenon is however a special manifestation of more general characteristics of

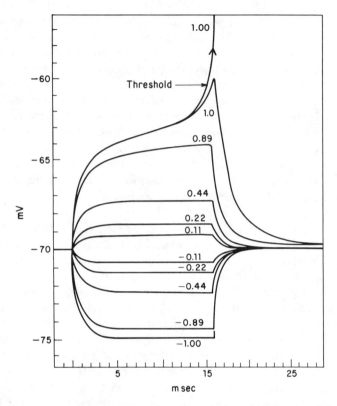

Fig. 8-4. Response of axon to rectangular waves of current of different intensity. The numbers on each record give the strength of current relative to threshold.

excitable membranes. These characteristics are best manifested as a function of the electrical field and of the ionic environment of the excitable membrane. The interplay of voltage-dependent conductance characteristics give rise to more complex phenomena, including the action potential. In the last analysis the action potential, and related phenomena, are found to be the consequence of the ionic fluxes (conductance) which may, however, be completely described in terms of an initial condition, a time constant, and a steady state; each of these is a function only of the membrane potential. In its simplest form the voltage dependence of conductance (permeability) of excitable membranes is probably best manifested in their current–voltage, conductance–voltage, voltage–time, and current–time relations in a symmetrical ionic environment.

For the sake of brevity and continuity of overall discussion we shall first discuss in this section the behavior of modified BLM. As pointed out earlier (Table 6-13) the adsorption of several substances such as alamethicin, monazomycin, and EIM on BLM results in a lowering of the resistance. The behavior

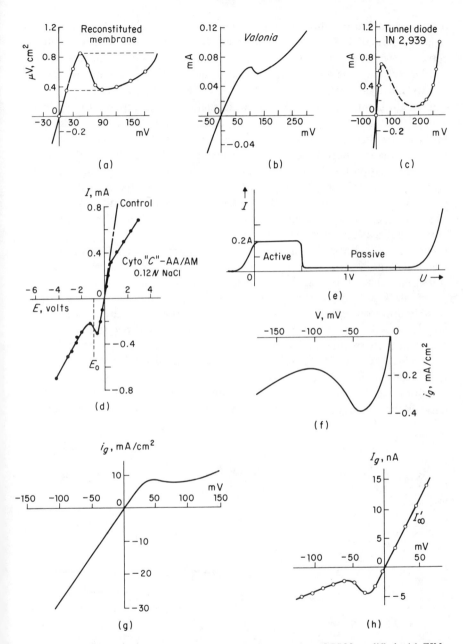

Fig. 8-5. Steady-state current–voltage curves of (a) reconstituted BLM modified with EIM; (b) *Valonia* cell; (c) tunnel diode; (d) polyampholite membrane; (e) passive iron wire; (f) squid axon in isoosmotic KCl; (g) eel electroplax; and (h) frog node.

of these modified membranes could also be altered by external factors, such as electrical potential. This is best illustrated by the steady-state current–voltage curves shown in Fig. 8-5. It is obvious that the conductance of these modified BLM is a strong function of both the magnitude and direction of applied current and its consequent potential. For example, BLM modified with EIM show a region of differential negative resistance. In such cases, in response to increasing voltage, the current first increases in accordance with the ohmic relationship, but a further increase in voltage decreases the current flowing across the membrane. Increasing the voltage still further results in a rise in the current, usually ohmic in relationship with a different conductance from the conductance observed at lower voltages. The region of the curve in which the current decreases as the voltage is increased is called a *negative resistance* region. More precisely this is a region of negative differential resistance since it is only the slope of the current–voltage curve that changes in sign.

The differential negative characteristic of a current–voltage relationship is also manifested in the transient response of the membrane to an applied suprathreshold value of rectangular constant current or its non-ohmic counterpart potential (Fig. 8-6). In such cases, the resistance of the membrane shifts reversibly from one stable value to another till it attains a steady value. For example, in EIM-modified BLM, the transition from one resistive state to another occurs at a sharp threshold value which may, depending on conditions, range from 15 to 50 mV, above which the resistance decreases five- to tenfold to a new steady value which persists until the applied current is turned off. The transition can be initiated in some cases by potentials of both signs if the *I–V* curve is symmetrical. The threshold is generally lower (in the case of EIM-doped

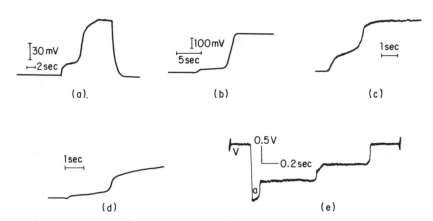

Fig. 8-6. Potential recordings of responses to applied constant currents: (a) reconstituted membrane: BLM modified with EIM; (b) tunneldiode; (c) frog nerve in isomolar KCl; (d) *Valonia*; and (e) polymerized lipid membrane.

BLM) when the cathode is located on the side containing EIM. The resistance change is *regenerative*, that is, the value of the resistance cannot be held steady anywhere between the two limiting values. The rate constants of the regenerative transitions can be varied experimentally from 10 to 100 per second by increasing the concentration of EIM. The regenerativeness also implies that a short pulse of the same sign applied on a subthreshold pulse can initiate the transition if it is applied on another subthreshold pulse, i.e., the effect is additive but not necessarily linear. The transition from a low-resistance to a high-resistance state takes place with a latency which varies inversely with the value of the applied suprathreshold voltage. The reverse transition, that is the decay of the pulse, has a sigmoid time course, provided the applied potential remains above zero. If it is pulsed again before fully returning to the resting resistance value, a second transition occurs with shorter latency. For any given modifying agent the ratio of the two resistance levels in the steady-state $I-V$ curve, the threshold potential for rectification, and the time constants for regenerative transitions are strongly dependent upon the lipid type used to form the BLM.[111,123] This observation once again suggests specific interactions between the modifying agent and the BLM lipids.

A very strong analogy can be seen in the steady-state properties of BLM modified with EIM (and other related agents) and the corresponding properties displayed by various model and biomembranes. Rectification of an electrical pulse is shown by Valonia cells,[812] frog skin,[813] frog node of Ranvier,[814] electroplax,[815] lobster[816] and squid[817] axons in isotonic potassium chloride solutions, that is, when the potassium concentration of the two sides of the membranes is the same. Model membranes such as oxidatively polymerized lipid membranes,[61] polyampholite sandwiched membranes,[818] porous silica membranes,[819] all show similar characteristics (see Figs. 8-5 and 8-6). Although the mode of conduction is not the same, a semiconductor tunnel diode also exhibits similar characteristics. The potential-dependent regulation of ion permeability has also been observed in mitochondria.[820]

The nonlinearity of the $I-V$ curve in most of these systems is also reflected in time–current, time–potential, conductance–voltage, and other related phenomena. Intrinsically, such a behavior is due to a specific instability of the system toward the suprathreshold stimulus. The molecular mechanism underlying such a behavior has not been resolved yet. In solid-state devices rectification and related phenomena have been attributed to quantum mechanical tunneling or to the typical behavior of electrons in conduction and valence bands.[821] The possibility of electron tunneling (quantum mechanical) can be ruled out in the systems with which we are concerned, since the charge carriers in these membranes are ions and not electrons, and the charge carriers move through pores (or some related system) and not through electronic conduction bands. Other related possibilities such as the increase in the number of current-carrying ions or excitation of ions to higher energy states (less mobile states) appear to be less probable. Yet another, physically equivalent situation occurs in the acceleration of charge carriers to such a high energy that when they collide with the atoms and molecules in the system they produce more free electrons or charge carriers. Once this happens and if the process is maintained the voltage necessary to produce a given current declines, and the I–V curve becomes nonlinear. However, such a situation in the systems with which we are concerned would lead to dielectric breakdown, and the field strengths encountered in these systems are not large enough to set in processes leading to such a phenomenon. In fact, at potential differences between 150–300 mV irreversible changes in a number of model and biomembranes including the squid axon,[822] lobster giant axon,[823] *Chara* and *Nitella* spp.,[824] and BLM occur (see Chapter 4).

Thus, one remaining possibility, involving modification of the *charge carrier* by an electric field may be given serious consideration. It has been proposed that an electric field (external) can alter the average distribution of charged sites on a polyelectrolyte macromolecule whose conformation may be affected allosterically by such field-induced charge perturbations. In fact, such a situation may arise when a macromolecule is membrane bound and it can reorient itself asymmetrically in the electric field across the membrane. Thus, any perturbation in the electric field would affect the relative stability of its two (or more) different configurations. Such an assertion is based on the general thermodynamic principle that in an equilibrium between two forms or states of a system or subsystem imposition of an electric field will cause a greater decrease in free energy in the more polarizable of the two states. Although we shall not go into the detailed implications of this assumption, it should suffice to indicate here that the electric field of a resting membrane will tend to keep the polar groups oriented (on one interface) in such a way that the positive groups are tucked into the membrane. If the field orienting these groups is perturbed such groups will tend to swing free. If one assumes that in the resting state, around −70 mV or so, the conformation of the molecules forming the conducting path is blocked due to the field across the membrane, and when such a system is perturbed by depolarization, it will tend to reorient to a conformation in which the conducting path is not blocked. With this assumption of two states of conducting pathway it is possible to predict the rectifying and nonlinear characteristics of the systems with which we are concerned. Although a justification for these assumptions will be presented later in this chapter, we shall proceed to elaborate the model inherent in the assumptions we have just described.

The model may be stated as follows:

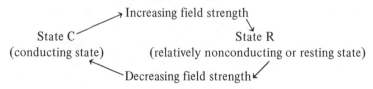

Increasing field strength

State C
(conducting state)

State R
(relatively nonconducting or resting state)

Decreasing field strength

The population distribution in these states may be obtained from the following relationship:

$$\frac{[R]}{[C]} = \exp\left[- a(V - V_h)\right] \tag{8-1}$$

and
$$R + C = 1$$

where a is a constant, V_h is the potential at which the conducting pathways are equally distributed between the states R and C; and V is the applied potential. Thus, on the application of an electric field across the membrane, the current flowing through the membrane will increase linearly with an increase in voltage provided that the voltage is small; this is because the ions are accelerated in the lower energy conduction state C. If the electric field is large enough the conduction sites will tend to go into the higher-energy but less conducting state R. Thus as the electric field (voltage) across the membrane is increased further, more and more conduction paths are in the state R. Thus, there will be a region (depending upon the value of V_h) on the voltage axis in the I–V curve where the current may decrease when the voltage is increased; this is because sites in state C are being continuously and *rapidly* (see later, however) converted to state R. Fig. 8-7 shows a set of typical I-V curves based on the preceding model. It may be noted that the resting state need not necessarily be relatively non-conducting (corresponding curves are not shown in this figure).

Thus one essential requirement for differential negative conductance characteristic of the I–V curve is that the system should be capable of exhibiting two field-dependent conductance states which are separated by a small energy gap. Muller and Rudin[825] have forwarded a proposal to this effect and the consequent treatment is in excellent accord with the experimental results. In fact, direct evidence indicating effect of electric field on the population distribution of conducting channels induced by EIM has also been obtained (p. 206). As pointed out earlier (Chapter 6), at low concentrations of EIM in the medium the conductance of BLM is limited to a few discrete levels. From the study of these fluctuations under the influence of an applied external voltage, it has been concluded that the EIM-doped BLM contains ion-conducting channels capable of undergoing transitions between two states of different conductance. The difference in current passing through "open" and "closed" (corresponding to states C and R, respectively) states is directly proportional to the applied membrane potential, that is, the conductance of each channel is ohmic. Furthermore, the fraction of the total number of channels that is open varies from unity to zero as a function of potential, according to a relationship of the type predicted by equation 8-1.

Although discrete conductance fluctuations have not yet been observed for excitable biomembranes, there is evidence that these membranes do contain discrete conductance channels.[826–827]

The steady-state characteristics of excitable membranes in general show differential negative conductance in current–voltage curves. This and ancillary

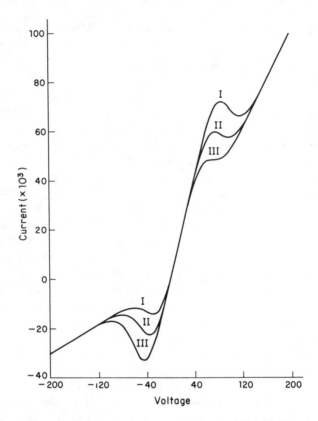

Fig. 8-7. Calculated plots of steady-state current–voltage relationship for three arbitrarily chosen values of V_H: −25 and +90 for curve I; −38 and +78 for curve II; −53 and +63 for curve III (see equation 8-1; $a = 0.0809 \, mV^{-1}$).

phenomena associated with these membranes can be best accommodated in a model based on discrete conduction pathways which can exist in two distinct states, the population distribution of which is voltage dependent. The consequences of this model are elaborated later in this chapter. Other models assuming two steady states[828] or other criteria[1140] have also been suggested.

Time-dependent Phenomena in Excitable Membranes

Biomembranes. As emphasized earlier several experimental techniques were evolved in order to describe the molecular changes accompanying the progress of the action potential. The problem has by no means been solved. However, the available data has given considerable insight into processes operating and underlying these phenomena. In this section are described some of the characteristics of excitable membranes, mostly from squid axon, under conditions which are close to physiological or which bear strongly on the behavior of the *physiological* axon. No attempt will be made to review all the available data. The intention is to illustrate the behavioral pattern of axons and nerves which bears strongly on the underlying mechanism of the generation of the action potential. The phenomenology characterizing action potential and related time-dependent phenomena in squid axon (and other related systems) may be summarized as follows:

1. The strength of an abruptly applied and terminated current (square wave) required to initiate an impulse under space-clamp conditions depends on the duration of the current flow. For an action potential to be initiated, the membrane must be depolarized for a length of time which is a function of the applied current. The capacity charging time for various membranes depends on the speed of the clamp and is of the order of 25 μsec. Thus the stimulating pulse should be of duration longer than this time, in fact much longer than this. The shape of the strength-duration curve is hyperbolic (Fig. 8-2c) and is very nearly the same for all tissues, although the time and current scales vary.

2. The action potential appears to be almost insensitive to large (from 25 to 6 volume % of glycerol) increases of osmotic pressure. The squid axon is also relatively insensitive to hydrostatic pressure changes if the axoplasm is not removed. Only the axonal membrane and ionic gradients seem to be responsible for generation of action potential following a stimulus. As shown by tracer flux studies, the ionic movement occurs down the electrochemical gradients; thus the potential energy stored in ionic gradients is utilized to drive the overall process.

3. The processes related to the action potential are not directly coupled to any metabolic source of energy except for the replenishment of ionic gradients. In the presence of metabolic and "pump" inhibitors, an axon is capable of generating a few hundred thousand impulses simply by drawing on its accumulated ion store without recharging it. Similarly, an axonal membrane is still excitable at lower temperatures when all the metabolic processes have ceased.[829]

4. Temperature has very little effect upon resting potential until about 25°C, above which there is a slight decrease in resting potential as the temperature increases (however see Ref. 1143). Temperature variation, however, has a significant effect on the height of the action potential and the frequency of the repetitive firing. Using an isolated single *Electrophorus electricus* L. fiber, the temperature coefficient (Q_{10}) and the energy of activation of the electrical activity have been evaluated over a wide range of temperature.[807] The duration of action potential and the latency period decrease with a rise of temperature, whereas the amplitude of the spike is little affected. When the logarithm of the reciprocal of the half height width of the action potential is plotted against the reciprocal of the temperature, a straight line is obtained (Arrhenius plot). The energy of activation has been calculated as 21 kcal/mole and the temperature coefficient (Q_{10}) in the neighborhood of 3.6. The Q_{10}'s of the latency period and of the post-synaptic potentials are very close to 2.6 and the energy of activation is around 16 kcal/mole for these processes. Similarly, in the squid axon,

the rate of rise of the action potential ($Q_{10} \sim 2.0$), its rate of decline ($Q_{10} \sim 3.2$), and the rate at which the ionic currents change with time ($Q_{10} \sim 3\text{-}3.5$) are all markedly temperature sensitive.[830-833] The absolute magnitude of the current attained at any voltage probably varies with temperature ($Q_{10} \sim 1.3$), but less markedly with time. All these observations point emphatically to the presence of a high-energy *barrier*—almost 20 kcal/mole or 0.8 eV—in the generation of action potential (however, see the discussion later in this chapter).

5. There is heat production concomitant with the passage of the action potential. The early rapid phase (60 msec) of heat production averages 1-10 μcals/g of nerve per impulse at $0°C$. In the second phase, lasting several seconds, more heat is generated.[834] Immediately after the first phase is a period of negative heat production during which about 80% of the heat produced in the first phase is reabsorbed. The positive and negative phases, both occur during the course of the action potential and are associated with the electrical depolarization of the membrane.[835]

6. The change in ionic conductance in the excitable membrane depends very strongly on the direction and magnitude of the applied current (Fig. 8-4). The hyperpolarizing currents, even of considerable magnitude, do not seem to affect the membrane in any significant way except for a slight decrease in conductance. In contrast, depolarizing current of suitable magnitude can initiate a series of changes after the lapse of some time (latency period). Also the depolarization shows threshold phenomena and for suprathreshold depolarizations the response is nonlinear and does not have any counterpart in hyperpolarization region. This may be interpreted to suggest that the changes resulting from the effect of applied potential are due to anisotropy of orientation of the conduction mechanism.

7. The transference numbers of monovalent cations through the axon membrane is much greater than that for anions and divalent cations.[836] In the resting state the membrane is moderately permeable to potassium and chloride ions, but is relatively impermeable to sodium and the internal anions. In contrast, the membrane excited with a suprathreshold depolarizing pulse becomes momentarily permeable to sodium and then the potassium permeability is restored. The passage of a single impulse is accompanied by a loss of $3.0\text{-}4.0 \times 10^{-12}$ mole/cm^2 of potassium and a gain of $3.5\text{-}3.8 \times 10^{-12}$ mole/cm^2 of sodium at $18.5°C$. But the experimental values are significantly high at high temperatures and low at low temperatures. The total quantity of sodium which enters the fiber is more than sufficient to provide the necessary electric current for the nerve signal.*

---

*The reason being that, although the fluxes are high, they last a small time, and chemically speaking, the amount of ions necessary to charge the membrane is small. In fact, the minimum net influx of sodium required during activity is simply the amount of charge necessary to change the voltage across the membrane capacitor roughly from potassium equilibrium potential to sodium equilibrium potential. A similar net efflux of potassium suffices to recharge the membrane.

These results also suggest strongly that the change of sodium and potassium permeabilities are out of phase. The results are of great importance since the isotope flux measurements are made by a completely separate technique, in no way relying on the internal rules of the complicated electrical system.

8. Careful and quantitative analysis of the permeability changes occurring during the course of the action potential suggest that following depolarization under voltage-clamp conditions, there is a momentary surge of outward current due to discharge of the membrane capacity. This surge of current lasts only a few microseconds and it is seen under all experimental conditions. This discharge is followed by an inward current which declines after about one millisecond and gives way to a prolonged outward current which declines only when the potential has returned to its resting value (Fig. 8-8). A small depolarization usually does not produce any measurable sodium current, and superimposition of yet another large depolarizing pulse on this depolarization after a lapse of about 20 msec reduces the I_{Na} caused by the superimposing pulse. If the initial conditioning step potential was a hyperpolarizing one, I_{Na} elicited by superimposed depolarizing pulse is increased. The effect again depends on the duration of hyperpolarization.

If sodium ions in the external medium are replaced by organic cations such as choline, the early inward current is abolished, but the maintained outward current persists unchanged. By subtracting the current in the absence of sodium ions from that in their presence, it is possible to divide the observed current into sodium ion-dependent and non-sodium ion-dependent components. The magnitude of the sodium ion-dependent component is governed by the steepness of the electrochemical gradient down which the sodium ions are moving, and it becomes zero when there is no electrochemical gradient for sodium ion. Similar reasoning indicates that the non-sodium ion-dependent component of the current is carried by potassium ions. Thus the total current flowing across the membrane following a clamped depolarization has both the sodium- and potassium-dependent components (Fig. 8-8). The corresponding changes during the course of an action potential are shown in Fig. 8-9. Some of the features of the components of the ionic current under voltage clamp conditions may be summarized as follows:

(a) Following depolarization, the sodium and potassium components rise along a sigmoid course; however, the inward (sodium) current increases five to ten times more rapidly than the outward (potassium) current.

(b) Repolarization of the axon to the resting potential shuts off both currents along a simple exponential course. Thus repolarization of the membrane during the transient inward current does not necessarily give rise to outward potassium current before returning to the resting high-resistance state.

(c) The inward (sodium) current is only transient. Under a depolarizing voltage clamp, with a short initial delay it reaches a peak in less than one millisecond and then falls off exponentially to a final equilibrium value that may be only

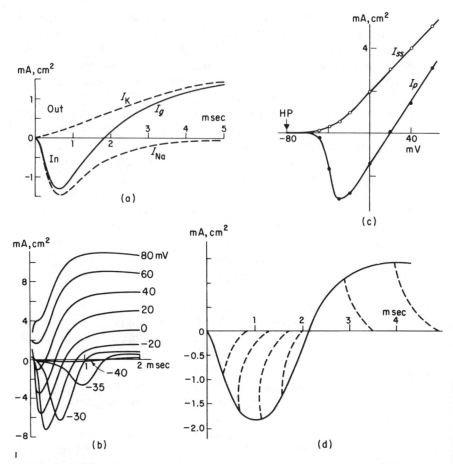

Fig. 8-8. (a) The time course of ionic current (I_g) after a voltage clamp at 56 mV above the resting potential. This is analyzed into the early transient inward sodium current I_{Na} and the later outward potassium current I_K. (b) Typical voltage-clamp curves for various clamp potentials as labelled. (c) Peak early transient current (I_p) and steady-state late current (I_{ss}) as function of the clamp potential. (d) After withdrawal of clamped voltage at various times, the current, whether fast or slow, decays (broken lines) to zero with a first-order rate constant dependent upon the final membrane potential.

slightly above the initial level. The total period of inward current is greatly reduced by cutting short the period of depolarization. The magnitude of the sodium current elicited by a step depolarization to a fixed potential is a function of the membrane potential in the period preceding the step, and is large if the fiber is hyperpolarized in this period, and small if it is depolarized.

(d) The rate at which sodium inactivation develops depends on the membrane potential, and is more rapid for large depolarizations than for small ones.

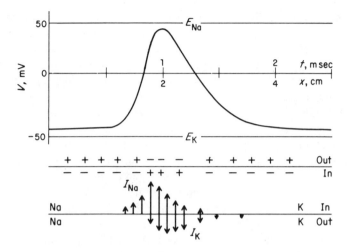

Fig. 8-9. Description of an impulse in an axon propagating to the left as a function of time at one point or distance x at one point. At rest, left, the potential is near potassium equilibrium potential. The increase of potential as the impulse approaches allows an inward sodium current which further increases membrane potential until sodium equilibrium potential is reached; now the potassium current again takes over to complete the cycle, and the membrane potential returns to the resting potential.

(e) The outward potassium current increases with the magnitude of the depolarization and the equilibrium value is maintained as long as depolarization persists; upon repolarization it closes along an exponential time course. A slow inactivation for potassium steady current has been described recently.[837] Also following the injection of pentyl trimethylammonium iodide, potassium permeability shows inactivation characteristics that are qualitatively indistinguishable from the sodium inactivation.[838]

(f) Equilibrium potentials for delayed conductance may be different than resting potentials. For example, in frog sartorius muscle the equilibrium potential for delayed conductance rise is more positive (about -80 mV) than the resting potential (-92 mV) and the potassium equilibrium potential (about -100 mV). Thus, during the period immediately following the spike the delayed conductance is still open; in the squid axon this gives rise to the undershoot (Fig. 8-1), and in frog muscle to the negative afterpotential. In the frog muscle the delayed conductance *channel* behaves as if it were less selective than the resting membrane.

Besides these voltage-dependent components of ionic currents, there is always a voltage-independent linear "leakage" component. Various methods have been suggested to evaluate the magnitude of this component. It is usually taken to be equivalent to the conductance value at large hyperpolarizing voltages or alternatively equal to the conductance corresponding to the initial surge of current at sodium equilibrium potential. Thus, the initial burst of current at relatively positive voltages is due to the leakage component. Some care must, however, be exercised with such assertions. Recently it has been shown that this initial 'blip' in the 25-75 μsec range also contains a small but significant voltage-dependent component.[839] The nature of this component (that is, of the charge carriers) and its significance has not been established as yet, though, it is believed to be due to Ca^{2+} and/or H^+ ions. In fact, extra Ca^{2+} influx has been observed during voltage-clamp pulses in perfused squid axon.[1142]

Thus the nature of voltage-clamp curves depends on several factors such as the membrane potential before depolarization, the magnitude of depolarization, the concentration and the nature of monovalent and divalent cations present, both internally and externally, of pH, and temperature. The depolarization brought about under physiological conditions by the flow of local circuit currents sets in three major potential-dependent processes in the membrane: a rapid rise in sodium permeability, a less rapid process which blocks or inactivates the increase in potassium ionic permeability. As the temperature increases the inward current decreases slightly, and the amplitude of the outward current changes significantly. The voltage-clamp data also shows that the ionic current is dependent on previous membrane potentials, but not on the past history of membrane currents per se. Thus, the membrane potential not only influences the driving force for the movement of ions, but more significantly, it also uniquely determines the rate coefficient of conductance (and hence current) change.* The peak conductance for sodium and potassium, that is, for inward and outward currents increase sigmoidally with the strength of depolarization provided the resting potential is held at - 100 mV or more negative. In both cases the conductances reach a maximum value in the vicinity of zero absolute potential. Also, the membrane loses its capacity for increasing its permeability to sodium when its resting potential is maintained at progressively positive levels by replacement of internal potassium ions or by depolarizing pulses.

9. Neither sodium nor potassium ions are absolutely necessary to elicit transient conductance·change. When the squid axon is perfused with a solution containing only potassium but no sodium, following a depolarizing step to potentials

---

*Apparently contradictory behavior is, however, observed when the perfusate is diluted with isotonic sucrose solution. In this case action potentials can be obtained even though the resting potential is reduced to zero.[840] This would imply that the sodium inactivation mechanism, which is voltage dependent, shifts in the direction of positive transmembrane potential. Such a shift is a consequence of reduction in ionic strength rather than a decrease in potassium concentration.[800] A tentative explanation of this phenomenon is based on the hypothesis that both the fixed charge layer and the ionic strength should be considered as determining the potential difference across the membrane, as opposed to potential difference between bulk solutions. The shift in threshold and inactivation curve can then be explained by the supposition that the activation and inactivation parameters are related to the membrane difference of potential rather than the total difference in potential.

greater than about 60 mV (inside positive), one still observes a transient outward bump of current.[800,806] It has been shown that the current is carried by potassium ions, but it has the kinetics of the channel through which sodium ordinarily passes. Since in the present situation, with sodium outside and potassium inside the axon, a finite sodium potential does not exist, the apparent biionic equilibrium potential would be given by the equation:

$$V_{eq} = \frac{RT}{F} \ln \frac{P_{Na} [Na]_o}{P_K [K]_i}$$

Thus from the observed value of equilibrium potential, the permeability ratio P_{Na}/P_K was computed as about 12. This ratio remained essentially unchanged on replacement of chloride by sulfate, or by changing the pH of the bathing medium from 7.3 to 6.4. When ammonium ions are substituted for external sodium the value of the fast transient conductance is about a third of that normally observed. But when internal potassium is replaced by ammonium the slow steady-state conductance is about half of that normally observed. The time course of the *turn on* of the early current is unchanged in the presence of ammonium ions. The ammonium ion action potentials are only obtained by anodal break excitation because ammonium is also a depolarizing agent and hence a nerve with ammonium ions in the external solution is always inactivated in the resting state.[841] The relative ease with which various ions enter the axon during the rising phase of the action potential is:[1257]

$$Na \simeq Li > NH_2OH > NH_2NH_2 > NH_4 > \text{guanidine} > \text{formamide} >$$

Na ≃ Li		NH₂OH	NH₂NH₂	NH₄	guanidine	formamide
1	1	0.9	0.6	0.16	0.12	0.11

$$K > Rb > Cs > \text{choline}$$

K	Rb	Cs	choline
.07	.025	.016	.013

In contrast, the specificity sequence for the resting potential is:

$$K > Rb > Cs >> Na \simeq Li$$

Axons perfused internally with phosphate salts of Li, Rb, Cs, Na, or substituted ammonium ions can maintain excitability and show an action potential if the axon is bathed externally with divalent cations such as Ca, Sr, or Ba, despite the completely unphysiological nature of the internal and external ions.[842] No simple explanation has been forwarded to account for this observation; however, it appears that the effect of ions is different on the two sides of the membrane, and the selectivity sequences for transient and steady states are also quite different. In fact, K, Cs, or Rb ions are external inhibitors of the initial conductance; however, when perfused internally they do not block the outward slow current.[800,843] All these observations, therefore, suggest that the ion-conducting systems are asymmetrically disposed and different alkali metal cations have different actions on various phases associated with the development of the action potential.

10. In general a change in the univalent/divalent cation ratio affects the excitability threshold and the effect of pH[1258] as may be observed under many different conditions: (a) When spontaneous repetitive discharge occurs in alkaline media, addition of divalent cations to the external medium tends to restore normal excitability. (b) When depolarizing univalent cations are applied externally, addition of divalent cations to the external medium tends to restore normal excitability. (c) A decrease in the calcium concentration in the external medium increases excitability. In the complete absence of calcium or other divalent cations the squid axon membrane becomes rather *linear* and electrically inexcitable. The effect of Ca^{2+} deprivation on lobster axon includes a reduction of spike amplitude, the development of the undershoot, and an increase in the refractory period. Thus, the presence of divalent cations,* such as Sr, Ba, Mg, Ca, etc., appears to be absolutely necessary; however, the effect of these cations is not the same on all the excitable tissues, nor can they be replaced completely by one another. The amounts of divalent cations needed for excitability seem to depend upon the concentration of sodium ions and pH of the external medium; other alkali metal cations do not seem to have any effect. The role of Ca and resting potential on sodium conductance suggests that the depolarization of the resting potential changes the effective binding constant of the membrane for polyvalent cations. It has been observed that an e fold change in Ca concentration results in a 5 mV shift of the voltage–conductance curve on the voltage axis (Fig. 8-10), and a similar shift of all Hodgkin-Huxley parameters on the membrane potential axis[845] as described later. Increased Ca^{2+} concentration (>0.1 mM) in the internal medium abolishes excitability altogether. It may also be noted that although, under physiological conditions divalent cations seem to affect permeability and excitability, they do not contribute significantly to the fast and slow components of ionic currents. Most of these effects of divalent cations could be accounted for by a screening rather than a binding phenomenon.[1259]

11. Excessively high or low internal pH decreases the resting potential which is followed by the loss of excitability. However, no significant effect of external pH (3.5-10.5) on resting potential is observed until after excitability is abolished; the membrane resistance in the resting state remains at its normal level under these conditions. High calcium concentration protects against the effect of low pH, thus suggesting that an oxygen ligand is likely to be the binding site

---

*The alkaline earth metal ions (Ba and Sr), Mn, Fe, Ni, Zn, Cd, and lanthanides (La and the other seventeen elements) could all substitute externally for Ca in lobster axon.[844] All the eighteen lanthanides at 10-20 mM give rise to a considerably lengthened spike with an absolutely slow falling phase, high threshold, large overshoot, normal or slightly subnormal rate of rise.

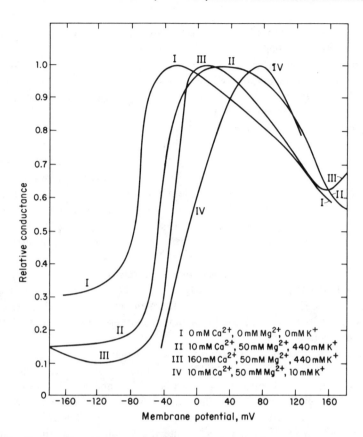

Fig. 8-10. Change of squid axon conductance with impressed potential (net) for various divalent ion concentrations in the external medium.[846]

($pK_a \sim 6$) for calcium. Carboxyl or phosphate groups are implicated in this binding.[844]

12. Oxidizing agents (such as peroxides, etc.) are detrimental to the nerve fiber and cause lengthening of the spike; in contrast, reducing agents cause a shortening of the spike and are generally *beneficial* for the axon. The action potential can be blocked reversibly by the application of a number of thiol blocking agents such as NEM applied internally or externally;[847] however, there does not seem to be any gross change in the integrity of the membrane, as evidenced by the largely unchanged resting potential. The action potential can generally be restored by reversing the effect of oxidizing agents by reduced thiols such as cysteine, mercaptoethanol, etc. A block similar to that induced by oxidizing agents could also be induced by internal or external application of fluoresceine–mercury acetate, a reagent that cannot penetrate the hydrophobic core of the membrane. The heavy metal ions Pd, Pt, Pb, Au, Ag, Hg, and Cd are

all toxic to lobster axon;[844] these metals are known to bind preferentially to sulfur ligands, and their blocking effect can be removed by thiols. These observations, therefore, suggest that a thiol group is involved in the functioning of these axons during excitation. Since all these changes refer to protein-bound thiols, the results are usually interpreted as indicating some configurational rearrangement in the membrane proteins.

13. Perfusion of the internal side of axonal preparations with proteases abollishes the action potential within 60 min. The resting potential decreases more slowly and comes to zero after more than 2 hr.[848,1144] The membrane resistance in the resting state is decreased, and both the sodium and potassium permeabilities increase correspondingly; the current–voltage curve also becomes linear. Proteases do not have any effect on the external surface of the membrane. Similar effects have been observed by perfusion with antibodies directed against the axoplasmic proteins.[849] Both of these observations imply participation of membrane-bound proteins in excitation processes. Phospholipids also seem to be involved, since application of phospholipases produces loss of excitability and of resting potential almost simultaneously. The effect of phospholipases may be lytic as the resistance and capacitance of the membrane almost falls to zero. It is interesting to note that the bound Ca from lobster nerve is released under the influence of ribonuclease.[850]

14. From the foregoing discussion it is evident that the surface layer of axons is composed of proteins and phospholipids. Alteration of these aggregated structures and their constituents by various biochemical and physical means would be expected to profoundly affect its excitability. Indeed, various phases of the ionic conductance can be influenced by a variety of agents added to the external medium (Table 8-2). Their effect is to raise the threshold, slow the rate of rise, and decrease the amplitude of the action potential without affecting the resting membrane potential. These effects can be studied and explained by a specific diminution of the ionic current under voltage-clamp conditions. Drugs like tetrodotoxin (TTX) and saxitoxin (Type A) block the early transient component of the action potential, that is, the rising phase of the sodium inward current.[851] Their action is at least partly reversible, and they are effective only when applied externally. As these agents are active at extremely low concentrations, $K_m = 3.31 \times 10^{-9}$ M,[852] it has been estimated as the upper limit, that the interaction of thirteen molecules of TTX per square micron of membrane surface is sufficient to block the early transient component.[851]

Relatively large concentrations of Ca and local anesthetics (Type B) such as procaine reduce the excitability by antagonizing the active increase in sodium conductance and, to a varying degree, by suppressing the secondary increase in potassium conductance.[859,860] Studies at various pH suggest that tertiary amine local anesthetics penetrate the nerve membrane in uncharged forms and block the action potential from inside the membrane in charged form (also see p. 210).

TABLE 8-2 PHARMACOLOGICAL AGENTS AFFECTING EXCITABILITY[a]

Agent	Blocks Inward Current	Blocks Outward Current	Blocks Decay of Peak Transient	Ref.
Tetrodotoxin	+++			851–853
Saxitoxin	+++			854
Procaine	+	+++		855–859
Dibucaine and other local anaesthetics	+	+++		853, 857–860
Benzyl alcohol	+	++		861
Hemicholinium-3		++		862
DDT			+++	853, 863, 864
Veratrine			+++	856, 865
Condylactis toxin			+++	866
Scorpion neurotoxin			+++	867
Allethrin				868
Internal	++			
External	++	++		
Aldehydes			+++	869
Tannic acid			+++	870
Tetraethyl ammonium and other tetraalkyl				871
ammonium ions		+++		871
Acetylchlinesterase				
inhibitors		+		872
Pronase		++		876
Fluorobenzene derivatives	+++			873
Acridine orange	++		+++	874
Benzene sulfonic acid (trinitro-)				875
Batrachotoxin	The resting permeability for Na^+ is increased.			1145, 1260
Histamine	Suppresses AP and RP			1262

[a]+++ Half-maximal inhibition at $\leq 1\ \mu M$; ++ half-maximal inhibition at $\sim 100\ \mu M$; + half-maximal inhibition at $\gtrsim 1\ mM$.

The blocking agents of Type A and B do not have any significant effect on the time course of various components of the action potential, steady-state conductance, and on the conductance of the leakage component.[853] In contrast, the time course of sodium conductance is significantly affected, probably by interference with the closure of the conductance mechanism (the falling phase of sodium current) by the agents of Type C such as DDT and varatrine. These agents (Type C) suppress delayed rectification. Their effect on action potential is such that the increased negative after potential is further augmented and prolonged, forming a plateau resembling action potentials of cardiac muscle upon removal of potassium from the bathing medium. Thus the agents of Type C stabilize the sodium conducting mechanism once it is developed, without much affecting the kinetics of activation of either sodium or potassium conductance. The time constant of the falling phase of the sodium conductance is prolonged about 4.4-fold, whereas the time to reach its maximum is prolonged 1.27-fold only; the steady potassium current is slowed only by a factor of 1.6. It should be noted that the sodium current is inactivated to about half of its initial value for depolarization, while the potassium current is not quite at its steady-state value. The prolonged sodium current in the presence of varatrine or DDT is blocked by tetrodotoxin.[865] For further discussion see Narahashi (p. 423) in Ref. 1307.

The blocking effect of tetraethylammonium ions in some species appears to be due to competition for the potassium permeability mechanism, that is the slow channel.[877] Similarly, the effect of pentyl-triethylammonium iodide has been shown to be on the potassium deactivation process.[871] Internal application of Cs blocks the delayed currents normally carried by potassium ions without much affecting the transient sodium current.[877]

Certain agents such as 2,4-dinitrofluorobenzene (FDNB) and 1,5-difluoro-2,4-dinitrobenzene (FFDNB) when externally applied suppress the action potential and the delayed potassium current, irreversibly. These agents react by displacement of fluorine atoms and formation of a covalent bond with free amino, histidyl, sulfhydryl, or tyrosyl groups of proteins; reaction with lipids is also possible. One of the related compounds, 2,4,6-trinitrophenol, suppresses only the early current with no effect on the delayed current. The effect of trinitrophenol is usually considered in terms of charge-transfer complex formation or as a proton conductor (Table 6-5).

The internal application of FFDNB, tannic acid,[870] glutaraldehyde,[869] Cs, NaF, etc., prolong the falling phase of the action potential enormously, sometimes up to 1000 msec or more. A similar, though not as specific, effect is exhibited by allethrin,[868] scorpion venom,[867] hemicholinium-3,[862] and condylactis toxin, CTX.[866] CTX is a large molecule (mol.wt. 10,000–15,000) and affects only the turn-off process in lobster axon and crayfish stretch receptor; it has no effect on squid giant axon. Although most of the inhibitors act rather indiscriminately, there may be some species specificity, which is indicative of a difference in the components responsible for the conduction mechanisms or their environment.

15. Some physical changes have been shown to be associated with the excitation activity of the axon. For example, it has been shown that the membrane shortens reversibly with excitation,[878] and there may be volume flows and pressure changes during an action potential.[1147] Similarly the light scattering and birefringence changes accompanying the action potential in two types of nonmyelinated nerve fiber have been measured.[879] Such birefringence changes could originate either from radially oriented molecules associated with the mem-

brane in an electric field caused by the mutual attraction of the ionic double layers.

Changes in IR-Spectra,[1308] fluorescence (tissue stained with 8-anilinonaptha-lene-1-sulfonic acid, ANS), turbidity, and birefringence have been associated with the nerve excitation.[880] The findings are consistent with the view that the process of excitation is accompanied by conformational changes in macromolecules in the nerve. However, it should be noted that there is an enormous increase in the rate of interdiffusion of cations across the nerve membrane during excitation; consequently, it is expected that the ionic composition in the immediate neighborhood of the membrane is rapidly and drastically altered during excitation. Such an alteration could also bring about the observed changes in the optical properties of the membrane. It should be noted that the dye ANS does not fluoresce unless its molecules are bound to the hydrophobic sites of the macromolecules; also the dye does not seem to penetrate the axonal membrane.

It is evident from the foregoing discussion that the characteristics of the excitable axonal membrane depend upon the time, previous history, potential, relative and absolute concentration of various ions, and the presence of pharmacological agents. Relatively simple changes in ionic conditions may cause changes in permeability for one ion or for all. This may result in an anomalous response, as observed with perfused axon.[842] The results thus suggest that the ionic selectivity is as much the intrinsic property of the membrane as it is adopted from the medium.

The ionic changes during the course of the action potential need not always be the same as those occurring in the squid axon membrane (Table 8-3). Action potentials produced by alternating cation–anion permeabilities are found in the algae *Chara*[885] and *Nitella.*[886] Voltage-controlled chloride permeability also contributes to action potentials in electroplax.[881] Similarly, summer frog muscles show only a transient inward sodium current.[892]

Also the quick restoration of the membrane potential is not a universal phenomenon. Thus, for example, during the electrical response of heart muscle the sodium conductance does not itself shut off completely and the restoring potassium conductance does not take place immediately. The result is the appearance of a prolonged plateau of depolarization in the heart action potential. Such a state may provide a sufficiently long maintained stimulus for the systolic contraction of the heart muscle fiber. In fact, important functional differences between the cardiac currents and the currents in squid axon are the following:[883] (a) There are at least three (may be four in the case of purkinje fiber) independent current-controlling systems whose kinetics resemble those of the squid potassium current in form, although not in detail. These cardiac currents are activated by different ranges of potential. (b) The activation reactions are very slow; thus the longest time constants are of the order of 2 sec at about $35°C$.

TABLE 8-3 PERMEABILITY CHARACTERISTICS OF SOME EXCITABLE
MEMBRANES AND RELATED SYSTEMS[a]

System (Ref)	Charge Carrying Species		Time Duration of AP	Depolarizing Force
	Resting	Transient		
Nerve, muscle	K	Mainly Na	1–5 msec	Electrical or chemical
Electroplax (881)	K	Cl	1–2 msec	Electrical
Heart cells (cultured) (882)	K	Na, Ca	1–5 msec	Electrical or chemical
Heart muscle (883)	K	Cl, Na (?)	Up to 1 sec	?
Tenebrio muscle (884)	K	Mg, K	Up to 100 msec	Electrical
Algae *Chara* (885) and Nitella (886)	Cation	Anions	—	Electrical
Esophageal cells of Ascaria (887)	K	K	100 msec	Electrical
Ganglion cell (bull frog) (888)	K	Anions	1–5 msec	Electrical or chemical
Giant nerve cell (Molluscan) (889)	K	Ca	10 msec	Electrical
Snail nurone (890)	K	Ca	2–10 msec	Electrical
Herbivorous insects (891)	K	All ions	10–100 msec	Electrical
Frog muscle (summer) (892)	—	Na	3–10 msec	Electrical
Phaseolus leaf (893)	—	—	Several sec	Electrical
Inhib. Post-synaptic memb. (lobster nerve) (894)	K	Anions	—	Electrical
Photoreceptor cells (898)	K	Na	0.5–5 msec	Electrical
Smooth muscle (1190)	K, Cl	Ca	0.5–3 msec	Electrical
Polyampholite membr. (818)	Cations	Anions	2–10 msec	Electrical
Oxidized lipid (61)	Cations	Cations	2–20 msec	Electrical
BLM modified (111,123,1149)	Cations	Anions	0.5–20 msec	Electrical
Passive iron wire (895)	Electrons	Electrons	10 msec	Electrical/chemical
Porous glass or silica (902)	Ionic	Ionic	Up to hours	Electrical, hydrostatic
Lipocollagen Membranes (1150)	Ionic	Ionic	Up to 200 msec	Electrical

[a]For other systems see Refs. 896, 897 and H. Grundfest (p. 477) in Ref. 1307.

(c) The temperature dependence of the kinetics is much greater ($Q_{10} \simeq 6$) compared to nerve (Q_{10} 2-3). (d) In contrast to the "channels" of the squid nerve membrane which behave as linear resistances once activation has occurred, the cardiac channels are grossly nonlinear. (e) Step changes in potential produce exponential (not sigmoidal) current changes in cardiac muscle.

While discussing electrogenesis of excitable biological systems, it is pertinent to note that the graded response (in contrast to the apparent *all-or-none*) has also been noted in several tissues such as kidney, salivary gland, Plasmodium, smooth muscle, certain skeletal muscle, muscle spindle, and dendrite of pyramid cell.[899] Thus, sharp threshold characteristics of propagating action potential as well as of space-clamped action potential may only be a special case of some general property of excitable membranes, and seem to have developed as a corollary to the establishment of central control of peripheral organs and the resulting need to transmit the impulse over a finite distance. The graded response may be primitive and fundamental, but such behavior of the membrane may be involved in the depolarizing and integrative mechanisms involved in sensory transduction and information transfer.

Time-Dependent Phenomena in Model Membranes

Transient oscillatory changes are not characteristic to certain biomembranes alone, and for that matter to membrane systems only.[1263] As described in the following section excitability is a rather general characteristic of both biological and artificial systems. In fact, in the last seventy years several model and analog systems have been proposed to account for the observed phenomenology of excitable biomembranes. Single and repetitive firing as a consequence of electrical or some other stimulus in various model systems is shown in Fig. 8-11. The observed correspondence in some cases may be just superficial, while in other cases the similarity of phenomena is believed to arise from identity of molecular architecture. Some of these model systems are considered below.

Action Potential in Modified BLM. When a BLM modified with EIM is titrated with basic proteins such as protamine or polyarginine, it displays time-dependent electrical changes depending upon the ratio of the basic protein to EIM, the composition of BLM-forming lipids, and the nature and concentration of ionic gradient around the BLM.[111]

Large protamine to EIM ratios (>10) under the conditions described above drive the system from a cationic to an anionic conductance level. The resting potential under these conditions is essentially determined by the anionic concentration gradient across the modified BLM. The basic features of the conduction mechanism appear to remain unchanged after the treatment with protamine except that the BLM becomes an anion conductor, and the sign of the restorative and regenerative transitions is consequently inverted. Protamine itself does not have any effect on BLM in the absence of EIM. This implies that protamine couples directly or indirectly with EIM. Other basic proteins such as histones and cholinesterase cannot substitute for protamine. EIM and protamine function almost equally well when they are placed either on the same or on opposite sides. This implies that only the salt gradient determines the polarity of the membrane.

Use of a small ratio of protamine to EIM, or change of ion type in the medium may result in instabilities of the resting potential levels—both anionic and cationic. On the application of an external electrical stimulus the resistance of the membrane decreases sharply and briefly and the potential shifts in the other direction, both of these changes being completed and reversed in a few milliseconds. This cycle of events is analogous to the action potential in nerve fibers. Under suitable conditions the system behaves as a monostable or unstable multivibrator, that is, the cycle of these transitions developed upon stimulation changes from one potential to another and back with a certain frequency. The successive cycles are separated by a time gap of passive state during which the system can only be stimulated by a suprathreshold potential. Such a *silent*

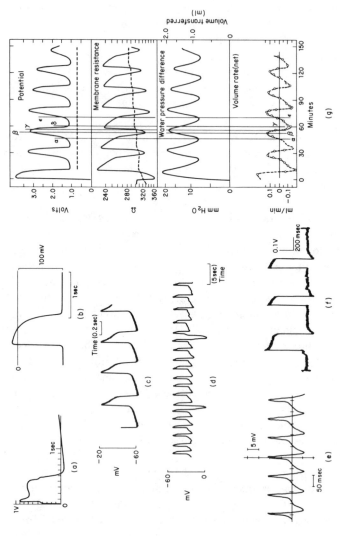

Fig. 8-11. (1) Action potential in passive iron wire. (b) Intracellular action potential of frog ventrical heart cell. (c) Rhythmic action potentials of BLM modified with EIM and protamine; spontaneous rhythmic firing in the absence of applied currents was obtained by using K_2HPO_4. (d) Same as (c) but under a steady depolarizing system. Both (c) and (d) are taken from Ref. 111. (e) Rhythmic action potential in methylacrylate/acrylic acid membrane.[818] (f) Rhythmic action potentials in oxidatively polymerized lipid films.[61] (g) Undamped oscillations obtained with the Teorell's membrane oscillator. The data is for silica membrane in a pressure gradient and electrochemical gradient. Open circles on the volume rate curve calculated from Teorell's equation. Vertical lines illustrate phase relationships. The dashed line in potential figure represents a correction of the zero voltage level.[902,903]

period is left behind each spike and is generally called a refractory period. Suprathreshold stimulating voltages do not change the magnitude (amplitude) of the action potential; only the frequency of the repetitive changes is altered. The frequency of firing is also determined by the temperature. Under suitable conditions the rhythmic state of the action potential can be maintained for extended periods of time without an applied stimulus.

Such an oscillating system acts as though only a part of the EIM-induced cationic conductance has been changed to an anionic conductance by titration with protamine. Additional protamine or EIM can lead to cationic or anionic states, respectively, which do not show instability. The change from a cationic to an anionic state is aided by a high ratio of sphingomyelin to lecithin, monovalent to divalent cations, and a high chloride ion concentration. Thus, all the components of the system, lipids, proteins, and electrolytes, influence the excitability of the BLM. Furthermore, the transient phenomena induced by EIM and protamine on BLM are reversibly blocked by physiological concentrations of cocaine (2%). Similarly various other compounds including acridine, yohimbine, phenothizine derivatives,[111] and pheromones such as farnesal, etc.,[900] also affect these transient processes.

Time-dependent phenomena similar to those induced by EIM can also be induced by alamethicin[123] and monazomycin.[414] These compounds develop a cationic resting potential, negative slope resistance, delayed rectification, bistable changes in the resting potential, and single and rhythmic action potentials in BLM. Best results with alamethicin are obtained with membranes formed from oxidized cholesterol in octane. It appears that six molecules of alamethicin aggregate to form the channel (Fig. 6-14). The rise time of an applied pulse increases with the sixth power of the alamethicin concentration. Also, the maximum change of the steady-state conductance, and the steady-state conductance itself, varies exponentially with the sixth power of the potential and the sixth power of both the alamethicin concentration and the ionic concentration in the medium. The membrane current and conductance undergo a sigmoid rise with time following an instantaneous displacement of the potential by the voltage-clamp method. The time constant of the response is voltage dependent. The maximal rate of potential change following the applied stimulus depends strongly on the lipid type, ionic strength, the temperature ($Q_{10} \sim 3-4$), and the presence of protamine. When these parameters are held constant the maximum rate is found to increase with the sixth power of the alamethicin concentration. All these observations indicate that the molecularity and the kinetic order of the reaction underlying the development of the conductance is six with respect to both alamethicin and ions even when extrapolated to zero ionic concentration. The hexamolecularity of the system is retained even when the membrane is stimulated. The significance of all these observations will be discussed further in this chapter.

Oxidatively Polymerized Lipid Membranes. The formation of these membranes was described in Chapter 3. The polymerized crosslinked membranes from a variety of lipids show electrical properties such as rectification, nonlinear I-V curve, and action potential.[61] They develop biionic potentials which indicate the order of cation permselectivity as:

$$Li < NH_4 < Na < K < Rb < Cs$$

| 1 | 14 | 43 | 55 | 62 | 69 |

This selectivity sequence is essentially the decreasing order of their hydrated ionic radii. The current voltage curve is nonlinear. The pH optimum for conductance is in the neighborhood of 4.75. It falls considerably on the acid side of this pH. The value of the conductance also depends upon the presence of foreign substances such as hexamethonium ion, acetylcholine, guanidinium ion, and ephedrine, which also develop biionic potentials of their own. The magnitudes of these biionic potentials are 12, 96, 120, and 138 mV, respectively, measured against an equal concentration of lithium chloride (0.1 M). Since the intercationic selectivity and the biionic potentials attain their maximum value in the same pH range (4.7-6.0), it has been suggested that the selectivity depends upon a limited amount of water present in the membrane which is sufficient to ionize some of the acidic groups contained in the membrane.

When an electrical pulse is applied to these crosslinked membranes they show rectification (Fig. 8-6), and under suitable conditions (temperature 20-25°C, pH 5-7) a pulse of 0.2-0.5 V initiates a repeated response and the system oscillates between two widely different values of potential. The frequency of oscillation is increased with the current strength up to a certain value. Above this value the conductance of the membrane remains at its peak value, that is, the restoration process for further response is blocked. The properties of these membranes do not seem to depend much upon a definite procedure for their initial oxidative polymerization, and a wide variety of unsaturated oils or synthetic lipids can be used for their formation.

Polyampholite Membranes. A very interesting type of excitable membrane system can be constructed by bringing in contact a polyanion and a polycation.[818,901] This can be achieved either by spreading a drop of a polyacid solution over a concentrated solution of polybase, or by depositing these polyelectrolytes electrophoretically onto a suitable neutral hydrophilic polymeric matrix (50-1200 Å pores) as a support. Simple polyanion membranes in solutions containing suitable concentrations of calcium also behave in similar fashion. It appears therefore that these polyampholite junctions are essentially anion- and cation-exclusion barriers; it is obvious that the surface facing the polyacid solution is positively charged, and the other interface of the membrane would be negatively charged. These membranes differ considerably in thickness (1000-

TABLE 8-4 POLYAMPHOLITE SANDWICH MEMBRANES WHICH
SHOW EXCITABILITY

Polyacid	Polybase	Support	Spike Potential (mV)
Yeast RNA	Ca	Polyethylene glycol	15–25
Acrylic acid/acrylamide copolymer (1:1)	Cytochrome	''	20
Methyl acrylate/acrylic acid copolymer (1:1)	Ca	''	200
Polyglutamic acid	Poly-L-lysine	''	20–40
Polyglutamic acid	Ca	''	100
Dextran sulfate	Poly-L-sarcosine	''	50
Dextran sulfate	Ba	''	—

5000 Å) and the pore size (50–3000 Å diameter). The membranes have been prepared from various polyelectrolytes such as polyglutamic acid, polylysine, egg albumin, etc. (Table 8-4).

When the polyampholite membranes are inserted between two sodium chloride solutions of equal concentrations and a constant potential is applied, the current flow exhibits transient electrical pulses; the magnitude (1–100 mV) and duration (1–20 msec) of these transitions resemble those of normal spikes in biomembranes. An analysis of the requirements for eliciting the electrical response in these membranes shows that there are three necessary factors:

1. There must be an anisotropy of chemical composition such that a cation-exclusion barrier is separated from an anion-exclusion barrier by a neutral polyampholite zone.

2. Electrolytes must be present on both sides of the membrane to act as the current-carrying species.

3. The flow of current and its behavior is dependent upon the direction of the applied pulse. Making the electrode on the polyacid side positive and the electrode on the polybase side negative, I-V curves can be obtained which show a region of differential negative resistance (Fig. 8-5). The direction of current flow is asymmetric and rectification can be obtained in the positive quadrant only. The I-V curves are linear if the polarity is reversed.

If the preceding conditions are fulfilled, current flow through the membrane results in a sequence of mechanical and electrical events as a function of the applied voltage. At a critical voltage the recording electrodes indicate an instability region in which the membrane becomes sensitive to mechanical vibrations. Further increase in the applied voltage results in a loss of the mechanical stability and the generation of "spikelike" activity. Higher voltages cause a decrease and finally a complete loss of these properties. Thus, with a suitable constant applied voltage (typically 0.7 V) the current shows a series of spikes. The threshold voltage depends upon the membrane thickness. The spikes are of 1-10 msec duration and the period of the sequence is of several milliseconds and depends upon the nature of the polyampholite and the intensity of the applied pulse. Spikes of 1-100 mV can be observed in which successive spikes sometimes have constant amplitude (Fig. 8-11). At higher voltage (1.3 V) the firing suddenly ceases and there is a constant current flow through the membrane.

The shape and the features of the spike are different from the ones we have discussed thus far for other excitable systems. There is a fast rise time in its development which is followed by decay and inversion of current flow. This reversed current flow usually lasts as long as 50 sec, that is, the refractory period is very long compared to other time constants.

The "Teorells Oscillator"[902,903]. In this well known oscillatory system a homogeneous thin porous silica or porcelain separates two compartments containing stirred solutions of different electrical conductance; in the simplest case two solutions of a salt at different concentrations. The pores in the membrane are wide enough to permit filtration of solution but narrow enough to support a path for diffusion that is free from disturbance.

An electric current of a constant strength (usually 5-40 mamp) is allowed to pass across the membrane. In the case of a negatively charged membrane (such as silica) the high conductance compartment (I) is the negative pole and conversely the low conductance compartment (II) is the positive pole. For positively charged membranes the direction of the steady-current should be reversed.

An electroosmotic flow of solution takes place when a current is applied such that fluid volume increases in compartment I and decreases in compartment II. If both, or at least one of the compartments is of a small, finite volume and open in the sense that free surface levels are established, it is obvious that the electroosmotic flow will induce a hydrostatic pressure difference (P) across the membrane. From the simple geometrical layout of the model it follows that at any instance the mean linear velocity of flow of the fluid across the membrane. will be related to the hydrostatic pressure difference across the membrane.

The fundamental concept in the oscillatory mechanism centers around the changes of the membrane resistance (electrical) as related to the velocity of fluid movement. This resistance is in turn dependent on the course of the "concentration profile" across the membrane. Thus it is possible to predict without elaborate calculations, that the steady-state membrane resistance (R_s) will be a function of electroosmotic flow rate (V), varying around an intermediate resting value of resistance R_0 at $V = 0$, with a low and a high limiting value at extreme flow velocities. Thus $R = f(V)$ is a sigmoidal function.

A change in flow velocity (V) cannot induce an instantaneous attainment of the corresponding steady-state value of R. On the contrary, the adjustment from one concentration profile to another requires time due to the relative slowness of the diffusion process. The time delay expressed as the rate of change of the instantaneous membrane resistance (dR/dt) occurring at any given V is a complicated function. However one can make an approximation by assuming that the rate of change of the membrane resistance is directly proportional to the "deviation from the steady-state."

The relation between the electric current density, i, the transmembrane potential, E, and the hydrostatic pressure, P, can be derived from the irreversible thermodynamics as $i = E/R + \ell P$, where ℓ signifies an electroosmotic permeability coefficient. For membranes with relatively wide pores and with not too small i, the contribution of the streaming current ℓP can be neglected and the validity of Ohm's law assumed.

In a system characterized by the preceding assumptions and when a critical value of a constant current from the external source flows, the following are observed: damped and/or undamped oscillations of transmembrane potential, of hydrostatic pressure difference, of membrane resistance, and of water flow across the membrane. Typical results are depicted in Fig. 8-11. The water level difference corresponding to the electroosmotic pressure and the hydrostatic pressure is also therefore of the relaxation type, like the action potential in the nerves. The pressure variation shows a phase lag in relation to the potential waves; the two effects thus appear to be coupled. It has some distinctive features when compared to an axon as an oscillator. Teorell's oscillator can be stimulated both by an electric current and by mechanical (hydrostatic) stimuli. For such reasons the Teorell's oscillator is an "electro-hydraulic excitability" analog, which in a great many cases gives an excellent formalism for describing and reproducing actual biological excitability phenomena: threshold properties, refractoriness, all-or-none responses, abolition, "frequency modulation," voltage-clamp data, etc.

THEORIES OF ACTION POTENTIAL

The phenomenon of "animal electricity" has intrigued observers probably ever since the Egyptians noticed the electric fish (recorded in 2600 BC on a tomb at Sakkara). The modern studies on excitation and propagation of impulses have their origin in Galvani's discovery of bioelectricity. More than a century after Galvani's discovery, Ostwald suggested a mechanism to explain cellular electrical potentials. Studying electrical potentials generated across semipermeable precipitation membranes Ostwald wrote: "at this time it is perhaps not too presumptuous to predict that not only the current in muscle and nerve, but also the mysterious action potentials of the electric fish, can be interpreted in terms of the characteristics of the semipermeable membrane." Indeed, as we shall elaborate in this section, our understanding of semipermeability (permselectivity) phenomena seems to support such a view.

Considerable progress has been made in the last fifty years in our understanding of *excitability* in terms of simple physical concepts. Some of the recent experimental developments in the field of "axonology" as discussed earlier in this chapter are concise: however, excellent reviews on the biological and physical concepts may be found elsewhere.[801-807] Although we shall not go into the historical developments of the subject, it may be noted, however, that the present concepts of membrane excitability seem to have developed in two major phases. The first phase originated toward the end of the last century with the definition of equilibrium diffusion potential by Nernst and Planck,[904] then was elaborated by Bernstein[905] to describe action potential in terms of ionic permeability changes in the nerve membrane, and finally the *actual* experimental demonstration of such changes was made by Hodgkin and Huxley in 1952.[830,831,906-908] The second phase probably started sometime around 1960 and in the decade to follow quite a few attempts have been made to account for the molecular and quantitative basis of the observed phenomenology developed during the first phase. For obvious reasons the general tendency in this phase is to treat the earlier theories and treatments with some skepticism. In the following pages we shall discuss some of the theories and the underlying concepts which have been put forward to account for the phenomenology associated with the action potential, both in model and biomembranes.

As early as 1900 Wilhelm Ostwald pointed to a simple reaction of a "passive iron wire," that is, an iron wire covered with a coating of its oxide as a possible model for nerve impulse conduction. When such a wire dipped in 67% nitric acid is touched with a zinc rod, a reaction sets in on the surface of the wire, showing the regions of maximum and minimum activity which move along the length of the wire in the form of a wave front. This iron wire analog and various other physical systems (see 896; see also Table 8-3), all show some of the features in common with the excitability characteristics of the nerve axon. Some of these features are threshold activation, all-or-none response, refractory period, periodicity, steep but sigmoid rising phase, spontaneous propagation, and accommodation phenomenon. The similarities between all these models may be coincidental, but it may be worthwhile to consider some of their common physicochemical features. As seen from the electrochemical standpoint all these models consist of two electrically conducting phases which are separated by a thin intermediary layer or membrane which is *unstable* to specific electrical or chemical stimuli. For the nerve, it is the thin protoplasmic membrane and for the passive iron wire it is the thin oxide layer (50 Å thickness). In principle, all the properties exhibited by such a system imply *force-dependent specific kinetic instability of the flow* (transport) system on which other properties such as conductance, osmotic pressure, etc., depend.[895,896] For example, when a pulse of 20×10^{-8} amp/cm^2 is applied to an oxidatively polymerized lipid film, the membrane attains a plateau potential which drops after a variable latency to a lower level, indicating an increase in membrane conductance. Frequently, however, repeated responses of spike occur during the pulse of the current and the conductance then oscillates between two widely different values. By increasing the magnitude of the applied pulse the frequency response increases, and finally the repetitive response stops and the conductance of the membrane remains at its peak value.[61] Similarly, the anionic and the *bistable* conductance state in the BLM can be stabilized by adjusting the amount of EIM and protamine in the medium.[111] The consequence of this metastability is of course reflected in the nonlinear dependence of *flow* on *forces*. The flow and force relationship of various oscillating systems have been characterized; the current voltage relationship is a relationship of this type. One of the earliest suggestions of this type is implied in the hypothesis put forward by Bernstein in 1902.[905] The two states of the membrane are considered in his hypothesis when he postulated: (a) That the living cell has an electrolytic interior and exterior. (b) In the resting state the cell membrane is permeable to K ions only, which gives rise to the diffusion potential. (c) During excitation the resting potential changes to a relatively lower value by permitting the passage of all ions.

At this time the existence of the membrane was merely an hypothesis. For some forty years Bernstein's hypothesis was widely accepted and has been the guideline for most of the subsequent research in nerve physiology. Later findings supported the view that the membrane capacity played an important but passive role, while the excitable properties were confined to membrane conductance only. The first disturbing observation was that the resting membrane of the muscle fiber was permeable to potassium as well as to chloride ions.[909] However, this finding does not materially affect Bernstein's predicted results on membrane potential since the Cl_o/Cl_i does not differ very much from K_i/K_o on a logarithmic scale. The most significant modification was to be brought forth as a result of the observation that the action potential reverses its sign and makes the inside positive by as much as 40 mV at its peak value. This lead to the reformulation of Bernstein's hypothesis into Hodgkin and Huxley's hypothesis (HH hypothesis).

The HH hypothesis[831] describes how a depolarizing potential change resulting from the flow of stimulating current causes an increase in the sodium conductance of the membrane, which by positive feedback tends to depolarize the membrane still further. This regenerative process of excitation is followed by a secondary decrease of sodium conductance and a simultaneous delayed increase of potassium conductance, both of which act together to make the membrane potential more negative than the resting potential, producing the undershoot. Thus for the squid nerve membrane, the current records (the voltage-clamp data) can be accounted for by postulating the existence of three voltage-dependent first-order reactions. The fastest of these (described by the variable m) controls the activation of the inward sodium current which depolarizes the membrane. The slower reactions (controlled by the variables h and n) control the inactivation of the sodium current and the activation of the outward potassium current, respectively. All these reactions occur within a few milliseconds following step changes in the independent variable, the membrane potential.

The HH theory assembled "within its grasp much of everything that had gone before and moulded it all into a self-contained, coherent structure."[801] The theory offers a quite successful phenomenological description of the excitability phenomenon as observed in nerves and muscle in terms of chemical and electrical potentials. The sodium and potassium flux is defined in terms of an electrical conductance and their displacement from equilibrium potentials. Thus the excitable system in the membrane may be considered as being constituted of two ion-selective current components, and a leakage current component which is not voltage dependent. They can be represented by an equivalent circuit diagram (Fig. 8-12) consisting of the bilayer electrical capacity (C), a sodium battery, a potassium battery, and a leakage battery, each having a voltage V and a series of internal resistances R (with appropriate subscripts) and all in parallel. The sodium and potassium batteries are assumed to be separate in space and time, and the resistance of the leakage battery does not change during activity. The sign and the value of the resting potential are best described by the Goldman constant field equation.[910] From these assumptions, the current for each individual ion (I_i) can be considered as a function of the membrane potential:

$$I_i = G_i(V_m - V_i)$$

where G is the conductance, V_m is the applied potential and V_i is the equilibrium for the ion i. Thus, the total current flowing through the membrane should be given by:[801,803]

$$I = I_c + \Sigma I_i + \frac{a}{2r\theta^2} \cdot \frac{d^2 V}{dt^2}$$

$$= C \cdot \frac{dV}{dt} + \Sigma g_i (V - V_i) + I_p$$

$$= C \cdot \frac{dV}{dt} + \tilde{g}kn^4 (V - V_k) + \tilde{g}_{Na} m^3 h (V - V_{Na}) + \tilde{g}_1 (V - V_1) + I_p$$

where I_c is the capacitative current and I_p is the sum of the pump and leak currents. After making certain assumptions, the total current can be regarded as being constituted of an inward early transient component and an outward

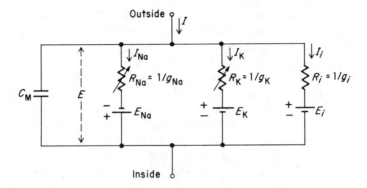

Fig. 8-12. Equivalent electrical circuit representing excitable membrane. Sodium conductance (g_{Na}) and potassium conductance (g_K) vary with time and membrane potential. The other components such as sodium equilibrium potential (E_{Na}), potassium equilibrium potential (E_K), leakage potential (E_l) and leakage conductance remain unchanged.

steady slow component. The apparent membrane capacity $(1 \ \mu\text{F-cm}^{-2})$ remains constant, and the pump current is relatively unimportant since it is much smaller during the action potential than any other current component. a is the radius of the axon, r is the resistivity of the axoplasm, and θ is the conduction velocity of the action potential. The terms n^4 and m^3h are potential–time dependent functions introduced to relate the time course of the ionic conductances g_k and g_{Na} with the maximum ionic conductances $\tilde{g}_k$ and $\tilde{g}_{Na}$. The functions $n, m,$ and h can be derived from the voltage-clamp data; these three dimensionless parameters vary between zero and one, and they describe the "on" and "off" process of the sodium conductance $(m$ and $h)$ and "on" of potassium conductance (n). The parameters m, n and h are assumed to obey first-order kinetics:

$$dn/dt = \alpha_n(1 - n) - \beta_n n$$

with forward and reverse rates α_n and β_n to give

$$n = n_\infty - (n_\infty - n_0) \cdot \exp(-t/\tau)$$

where n_0 and n_∞ are initial and final values and the time constant

$$\tau_n = 1/(\alpha_n + \beta_n)$$

The fast activation m behaves much the same as n; m^3h is the fourth-power analog to n^4. The separation of m and h is a log–log graphical routine to find τ_m/τ_n and $m_\infty^3 h_0$, and so are obtained m_∞ and τ_m. One of the most significant features of this treatment is that the rate constants α and β, the corresponding steady states $\alpha/(\alpha + \beta)$, and the time constant $1/(\alpha + \beta)$ are functions of the step voltage alone. Thus, the kinetics of n, m, and h, and consequently the time course of g_{Na} and g_k, are also functions of voltage alone for a step voltage input.

A physical significance for the theory may be found by assuming that a channel for potassium transport is formed when four charged particles move to a certain region of the membrane under the influence of the applied potential. Similarly, for the sodium channel it is assumed that three simultaneous events, each of probability m open the channel to sodium and that a single event of probability $1 - h$ blocks it. However, it should be noted that the choice of three constants and the forms of the equations may have been one of convenience or one based on some hope and belief that it might correspond to a physical mechanism; one of the most attractive is just described. Various models have been suggested to give the physical basis to the HH theory;[9,11] however, in the theory itself there is not a single indication as to how the membrane performs its functions.

A complete description of the action potential can be achieved from the set of equations just described. The differential equation to describe the time course of the action potential is obtained by equating the current passing through the axon as a result of the local circuits with that due to the applied pulse. Thus, if the action potential is elicited over a length of nerve, there is no net longitudinal current and no net radial current at any time after the stimulus is applied. Under these conditions $I = 0$ for any $t > 0$, the boundary conditions being the initial displacement of V. The equation has been solved by choosing arbitrary constants and the action potential curve can be computed. Thus, the theory provides a plausible description of the kinetics of the processes controlling the current flow, and of the behavior of the axon in subthreshold, threshold, and response regions. For digital computer solutions of HH-equations see Y. Palti (p. 183 and 215) in Ref. 1307.

A similar explanation has been put forward to account for the action potential phenomenon in BLM modified with EIM and protamine[111] or with protamine-alamethicin.[123] In both of these cases it has been postulated that the resting potential is essentially due to cation-conducting channels formed by EIM or alamethicin only; whereas the spike peak current is drawn through anion-conducting channels formed by a complex of protamine and EIM or alamethicin. The modification of BLM by protamine results in heterogeneous conductances developing two opposing resting potentials, one of which is voltage dependent. Because of the relative rates, magnitudes, and directions of two conductances change, the magnitudes and signs of two emfs are such that their interplay develops action potentials. Several other changes can also be predicted theoretically, and most of such changes have been observed experimentally.[111]

Despite great advance forward with the quantitation of axon behavior by HH theory, there have been dissatisfactions with this approach. It is evident from the foregoing discussion that the theory is empirical (although with a possible physical basis) and that at the very best the equations are curve-fitting expressions of the voltage–clamp data. Treatment of the data accorded by the equations is only approximate and of limited scope. There are definite and consistent differences between calculations and experiments, particularly in the threshold region. Also, the equations are so specific as to invite alterations and to require modifications from axon to axon and tissue to tissue. The potential range is restricted, and there is no hint as to what might happen as this range is exceeded.

Mechanistically, the treatment has no *necessary* connection with the mechanism of ion permeability. The separation of the sodium and potassium components of ion current indeed implies separate permeability mechanisms for these ions under normal conditions. Beyond this, the equation and the equivalent circuit are purely empirical expressions of the experimental facts. Indeed, the treatment in terms of equivalent circuit diagrams may sound simple, but it is far from being so. The circuit with two batteries of constant emf and two variable resistors (Fig. 8-12) is equivalent to a circuit with one battery of variable emf (equal to $(g_{Na}E_{Na} + g_kE_k)/(g_{Na} + g_k)$) and series resistor $1/(g_{Na} + g_k)$. Physically this would imply that all these observations could be accounted for by postulating a single channel which could undergo a transition from one state to another, and through which different ions could pass at different rates. These schemes may be represented as follows:

(a)   Impermeable $\longrightarrow$ Sodium permeable $\longrightarrow$
                                Impermeable $\longrightarrow$ Potassium permeable
(b)   Sodium channel (impermeable) $\longrightarrow$ Sodium channel (permeable)
       Potassium channel (impermeable) $\longrightarrow$ Potassium channel (permeable)

From the evidence available at present it is very difficult to decide in the favor of either of these possibilities. (For a theoretical discussion see Ref. 1051.) Hypothetically speaking, a compelling observation in favor of a two-channel hypothesis would be that in which both ionic conductances were simultaneously and fully "turned on," giving rise to a total membrane conductance of twice the normal maximum value for either of the ionic species.[912] As the time constants for the two processes are different it may not be possible to "open" both the channels without the use of some *modifying agent*. Also, a coupling between separate sodium and potassium channels may not be demonstrable easily.

It is implicitly assumed in the treatment that ions move by virtue of differences in concentration and electrical potential as in free solution; that is, the ionic concentrations at the interfaces are directly proportional to the concentrations in the adjoining aqueous solution. It is also implicit in the argument that the ionic permeabilities, which are proportional to the mobilities and the distribution coefficient of the ions, are unaffected by the ions or membrane potential, and that there are no interactions between ions moving in opposite directions or between ionic species. This is commonly referred to as the independence principle.* However, the general validity of this principle cannot be fully justified on theoretical and experimental grounds.[913]

It may also be noted that the ions involved in the so-called leakage conductance have not been conclusively identified. The role of Ca ions is also obscure, however, divalent cations seem to screen interfacial charges rather than bind on to specific sites.[1259]

As noted earlier, the *channels* do not appear to be absolutely permselective. The perfusion experiments in fact show that relatively simple changes in ionic conditions may cause changes in permeability for one ion or the other. For example removal of potassium from the external medium reduces the permeability for potassium in many cells, whereas an increase in external potassium concentration results in an increased permeability of this ion. Thus the permeability coefficients for ions may not be constant under different conditions; in fact, under appropriately controlled conditions various anomalous responses have been observed in perfused axons.[800,801,806] Some of these results can be accounted for in terms of altered equilibrium diffusion potentials as a function of ionic concentrations, which determines the conductance level in excitable membranes.

Another objection of somewhat general character against HH theory may be in the basic assumption regarding the applicability of the Nernst-Planck equation or the assumption for the constant field. The fundamental theory of diffusion potential as formulated by Nernst and Planck is based on the simplifying assumption of a homogeneous membrane of macroscopic and microscopic electroneutrality, and a membrane thickness that allows for continuity of the voltage gradient. The validity of these assumptions in the present context is somewhat questionable. Thus the possibility of methodological correspondence and the agreement between theory and data may be expected under limiting conditions only. In fact, relations similar to those derived from constant-field assumption have been derived from a completely different set of assumptions (see p. 155). Thus, a constant field approximation may be fairly valid. However, the contribution of interfacial potentials (V_i, V_0) to the total membrane

---

*Here a distinction should be made from independence of material flows. An example of this is the irreversible thermodynamic treatment of membrane phenomena in general, when it is assumed that all phenomenological cross coefficients are zero.

potential (V) may be somewhat difficult to evaluate since

$$V = V_0 + V_m + V_i$$

where V_m is the diffusion potential. So far V_m is taken equal to V; however, $V_0 = V_i = 0$ cannot be strictly met since potassium would have to have a negative distribution coefficient.[914] It has been shown that surface potentials are or can be of the same order as the transmembrane potential, and thus the value of V_m is apt to differ markedly from the observed value of V. It implies that relating the resting potential to potassium ionic gradients can be misleading. Unfortunately no suitable alternative seems to be available at present: however, a proposal for an excitation mechanism based on junction potentials has been made.[914a]

In view of these limitations of the HH theory, it is no surprise that a considerable amount of work, both theoretical and experimental, has been done to modify, rectify, and amplify the theory or to suggest alternative explanations. Some of these are restricted to the description of a single phenomenon and have been tested rather casually. For example, a system producing an action potential-like response or a uniform propagation may have no other recognizable similarity to a nerve fiber. For such reasons most of these alternative formulations are to be regarded as no more than an inadequate hypothesis until they can duplicate at least all or many and various accomplishments of ionic theory. Thus, a model or an analog which will reproduce the voltage-clamp data will equally well represent most of classical electrophysiology. With this goal in mind we now proceed to examine some of the alternative theories.

Physical Theories of Membrane Excitability

It is assumed that present knowledge of the properties of biomembranes is only of the total system, i.e., membrane plus surrounding solutions. The intrinsic macroscopic variables and the physicochemical attributes of the membranes themselves are largely unknown. Until these are known, one cannot apply to studies of macroscopic biological membrane phenomena a theory which makes, implicitly or explicitly, a particular, restricted set of assumptions about the intramembrane transport mechanism. For the present, therefore, the only class of theory which may be properly applied to biological membranes is one which is applicable to all membranes as a general principle of membrane behavior. Thus the physical theories of membrane behavior are generally based on first principles and four fundamental physical processes: *generation, recombination, drift,* and *diffusion.* We do not intend to go into the mathematical jargon of various theories forwarded to account for oscillatory behavior in general, and action potential in particular; however, the following general description of oscillatory systems may be of some help in understanding more complex theories.

A linear oscillatory system may be described by the following second-order differential equation:

$$M\ddot{X} + R\dot{X} + CX = 0$$

It states that the sum of the forces acting on the "oscillator" is zero. The first term being the inertial force, the second term is for resistive or frictional force, and the third is for restoring force. The mass, M, the resistance or friction coefficient, R, and the oscillator constant, C, are the proportionality factors relating the above forces to acceleration, velocity, and displacement, respectively. The above equations can be generated by the combination of a set of first-order differential equations describing two coupled processes, such as the following pair of equations:

$$dX/dt = aX + bY$$

$$dY/dt = cX + dY$$

The periodic solution of these equations arises when

$$C/M > (R/2M)^2$$

or

$$bc < 0 \quad \text{and} \quad |bc| > (a - d)^2/4$$

in the second equation. These relationships state that the oscillations result from a particular coupling between various forces such that whenever one is in its equilibrium position, the other tends to "drag" it out. Similarly, an equation may be written for the nonlinear systems:

$$M\ddot{X} + f(\dot{X}) + CX = 0$$

where the function $f(\dot{X})$ must behave in such a way that the damping force is in the direction of the velocity for small velocities, but in the opposite directions when the velocity reaches a critical value. When the nonlinearity of $f(X)$ is pronounced, relaxation oscillations occur. An unstable system would not oscillate, however, unless it contains at least two simultaneous relaxation processes corresponding to capacitance and inductance in the electric circuit.[915] It may be noted that due to the dissipative nature of excitation, a continuous input of energy is required for the maintenance of such oscillations.

One of the foremost approaches of this kind was used by Teorell to describe the phenomenology of Teorell's oscillator (see p. 358). In comparison with the HH model which operates on two forces, chemical and electrical potential, the Teorell's membrane oscillator is a three-force system, chemical, electrical and hydrostatic, where no specific permeability need be considered. The behavior of this system has been described by assuming a charged membrane, applicability of Ohm's law and of Fick's law for diffusion with convection, electroosmotic flow and microscopic electroneutrality in the membrane phase, and a resistance relaxation equation:

$$dR/dt = -K[R - R_s(V)]$$

where R is the membrane resistance and V is the volume flow rate. The last relation is empirical (however, see below) which serves to describe the data quite well. A set of six equations is derived from these assumptions to describe the state of the system, and the solutions are obtained by numerical and analog computer methods. In fact, solution of these six empirical equations for membrane potential E and flow rate V gives two differential equations of the form:

$$dE/dt = aV + bE$$

and

$$dV/dt = cV + dE$$

The solution of differential equations of this type, as noted earlier in this section, gives rise under certain conditions to integral curves showing an oscillatory time-dependent variation of E and V. Damped, undamped, or growing sinusoidal oscillations of E and V are obtained when $[(b - c)^2 + 4ac] < 0$: their character depends on whether the "damping factor" $(b + c)$ is < 0, equal to 0, or > 0, respectively.

The simplified picture of oscillations is visualized as water flow acting as a distortive factor on the concentration profiles, while the slow diffusion process acts as a restorative factor. These two processes chase one another continuously if sufficient energy is supplied for maintaining harmonic type of oscillations. Thus, the picture arising out of this treatment is one in which the oscillation is essentially the result of coupling between two flow processes: the volume flow through the membrane and the flow of salt within the membrane. Both processes are coupled to the constant external current from which the energy to sustain oscillations is derived. Thus, the system is an excellent analog for the physiological *mechanochemical transduction*.

The main drawback of the model is in the assumption of the existence of a steady current source. An alternative treatment[819] utilizes the Nernst-Planck equation for ion transport to explain the current-voltage characteristics of Teorell's oscillator and the limiting conditions in which the membrane assumes *bistability* and oscillates. Like Teorell's treatment[903] the time-dependent behavior is interpreted in terms of non-steady-state behavior of the membrane.

Several other attempts have been made to describe Teorell's oscillator. Fujita and Kobotake[916,917] assume (a) that the membrane is constituted of a set of parallel capillaries having a diameter that is large with respect to the thickness of the electric double layer at the capillary walls; (b) equations of flux from thermodynamics of irreversible processes; (c) steady motion of fluid in the capillary described by the Navier-Stokes equation; (d) steady-state behavior in the membrane with time-dependent boundary conditions; (e) charged capillary walls described by the Debye-Hückel approximation to Poisson's equation. The relationships derived from these assumptions suggest that the flip-flop current-voltage characteristic is a consequence of the dependence of the electroosmotic coefficient on salt concentration.[916] The theoretical treatment is in accordance with the experimental conditions; for example, the total mass of the system is held constant, constant electric current is passed through the membrane, and the pressure difference is allowed to vary in accordance with the transport of fluid. The analysis suggests typical relaxation oscillations of the mass flow as well as of the differences of the potential and pressure across the membrane.[917] The predictions are in accord with Teorell's experimental observations, and assumption (d) also eliminates the necessity of Teorell's resistance relaxation equation. However, the theory has many weaknesses, especially in assumptions (a) and (e).

The periodic behavior of the membrane for variations in solvent flow, electrical potential, concentration, and pressure relaxation can also be explained as a consequence of the hydrodynamic processes. In a multicomponent system the motion of each species can be described by the equations of motion.[918] These equations describing an isothermal quasihomogeneous system can be adapted to Teorell's type of oscillator.[919] Using perturbation methods, these general equations of the solute species, of motion of the membrane, and of the barycentric flow, all derived from first principles, show that the periodic behavior is a consequence of the effect of boundary conditions on hydrodynamic stability (a differential equation for the first-order perturbation in pressure gradients). The stability condition obtained from the above treatment is general in the sense that it allows either the onset of unstable turbulent flow, or damped motion, or purely oscillatory or increasing oscillatory motion, depending upon the values of various experimentally measurable quantities and constants. Under simplifying conditions these general equations of motion are reduced to those of Teorell's,[903] Frank's,[819] or Kubotake and Fujita's[916,917] as described earlier. The resistance relaxation equation of Teorell can be arrived at from first principles, thus eliminating the necessity of its assumption, by taking into account the gradient of chemical potential and the dependence of current flow on pressure. In this treatment, the nonlinearity is actually eliminated by various approximations; nevertheless, the equations may yield periodic solutions if acceleration terms are included. This possibility is a mathematical one; whether or not periodicity is predicted for the specific conditions in the experiment has yet to be ascertained. In this situation it is not possible as yet to determine whether the membrane steady-state assumption of Kubotake and Fujita corresponds to reality.[920]

An alternative treatment of Teorell's oscillator has been given in terms of the equations of motion of the membrane lattice containing fixed charges,[921] which is consistent with statistical theory of the transport of electrolytes. When coupled with the equation of fluid motion[919] the resulting relation gives an equation for simple harmonic oscillations.[922] It implies that the force present at any instant on the oscillating system of localized charges is proportional to the negative of the displacement from their equilibrium positions. Thus the force to which these systems are subjected has a periodic dependence on time.

An interesting proposal has been made to account for the behavior of the excitable membrane in analogy with semiconductors.[923] The membrane is assumed to have two interface potentials (V_o and V_i; cf p. 367) and the diffusion potential (V_m). The active current carriers are positively charged ions in the outer and inner solutions which may be regarded as a p-type material. Since the observed membrane potential is positive outside and negative inside, the outer junction is biased in the forward direction and the inner junction is biased in the reverse direction. Thus the axon is much like a p-n-p transistor, both in its physical arrangement and its biasing, and therefore, it is expected to follow the same physical laws. In fact, the predictions based on this model have been shown to be in general accord (only in a limited sense) with the experimental data on the nerve axon[923] and Lillie's iron wire.[924] However, the model as a whole (at least in the form as proposed first) gives little consideration to the neurophysiological observations, and the author has been hypercritical of neurophysiology in general. However, the similarities between the p-n-p transistor and the bimolecular lipid membrane (of the axon) have been the subject of models which are claimed to duplicate most of the bioelectric phenomena.[925,1264] A model electric circuit has been proposed using a single p-n-p transistor; this circuit is developed within the constraints of the HH theory and its equivalent circuit analog.[925] The model reproduces patterns (threshold, overshoot, etc.) and forms similar to the responses obtained from excitable biomembranes. Quite a few physical explanations can be forwarded to interpret these analogs (for example see Refs. 914a, 923, and 924), however none appears to be satisfactory.

Katchalsky has outlined the thermodynamics of irreversible processes as applied to membrane phenomena. The phenomenological equations can be obtained by assuming: (a) that the transport is not accompanied by a flow of solvent; (b) the driving forces are the differences of the electrochemical potentials of two ions across the membrane; and (c) the oscillating system is a multilayered structure. The phenomenological equations derived from these assumptions suggest a nonlinear dependence of flow on the driving forces. The nonlinearity is expressed in electrical terms as rectification or non-ohmic dependence of current on potential, in the shape of flip–flop kinetics, and in the nonlinear shape of the curve of log concentration gradient vs resting potential. Indeed, the observed experimental data support all these predictions.[926]

The events underlying the phenomenon may be described as follows. The application of the potential across the membrane presumably causes an accumulation of the salt in the intramolecular spaces. The local increase in the concentration raises the conductance, and hence the sharp rise in the amount of current passing through the membrane until the peak of the conductance is reached. The accumulation of salt has, however, an effect that counteracts it. The analogy may be found in the behavior of polyelectrolytes in salt solutions. It is known that polyelectrolyte molecules which are stretched at low ionic strength contract or collapse at high ionic strength.[927] Thus, when the peak of the spike is approached, breakdown of permselectivity takes place; water flushes out the accumulated intramembrane salt and the membrane returns to its initial state. Thus the potential dependent change in permeability is one of the intrinsic properties of the asymmetric composite series membranes.

Yet another irreversible thermodynamic analysis of the "active-state" of nerve membrane suggests[1265] that by "suitable transformations of the transport equations a correspondence with the theory of thermal explosions results." This phenomenological correspondence may thus allow for all-or-none phenomenon. However this analysis is essentially based on the assumption that the energy released during downhill ionic diffusion of ions is coupled to the propagating depolarization disturbance modified by charge density fluctuations in the membrane dielectric and the charge transfer network in the vicinity of the inner surface of the membrane. The validity of quite a few of these assumptions may be questionable, however, more importantly this theory has yet to come up with predictions of known experimental results.

All the studies described in this section suggest that the theoretical analysis is no simple task in this problem, and an exact analytical description of the membrane oscillator is probably too much to hope for. It seems that the chief problem encountered in the analytical description of the membrane oscillator is the concentration dependence of the local phenomenological coefficients and the effect of this on the integrability of the local equation. An added complication is that in order to approach the problem at all with the present techniques (and may be methods) one is forced to approximate time-dependent processes in the membrane by steady-state equations.

There are several other difficulties with such macroscopic systems when one considers the behavior of bimolecular lipid membrane, modified or naked. In a thin membrane the number of surface molecules is an appreciable fraction of the total population of molecules. Such a membrane cannot be considered a thermodynamic phase in the Gibbs sense. A second difficulty concerns the usual one-dimensional simplification. If the approaches based upon the average behavior of particles are to be applied, the membrane must be considered as a series of numerous planes each of which contains a large number of particles. Only in this way can the idea of a continuous density function of real particles be retained. The most important difficulty is that for a very thin membrane the physical meaning of diffusion coefficient and mobilities, which in the kinetic theory are related to mean free path, become unclear when the dimensions of the membrane are such that only a small number of scattering collisions can occur. Last but not the least, the weakness of these analyses lies in the "black box" adopted for their phenomenological description. Even from a purely phenomenological standpoint a central question in the understanding of membrane oscillations is whether the oscillations arise from coupling between two processes or from an instability in a single process. Furthermore, owing to the dependence of some of the coefficients on solute concentrations in the membrane, the distinction is not sharp. Through such coefficients all concentration changes in the membrane involve, in a sense, coupling of the solute flow to other flows.

The Molecular Basis of Excitability in Biomembranes and Related Systems

The general lack of progress in this direction since the work of Hodgkin and Huxley in 1952 is in striking contrast to the progress made in molecular biology during the same period. An obvious difference between the two fields is that experiments on membranes require an intact membrane system in contrast to microbiological experiments, most of which can be done in cell-free systems. Furthermore, excitability characteristics are kinetically fast on an anisotropic system which is extremely heterogeneous with respect to functional and molecular species. Since our knowledge regarding chemical changes occurring in excitable membranes (especially of biological origin) is still rudimentary, this section will be rather fragmentary.

In contrast to the assumptions made by the exponents of physical theories of membrane processes, the proponents of the so-called molecular theories have been rather bold in assigning specific functions to specific molecules or molecular aggregates, these functional species being of course hypothetical. It is now widely recognized that the *site* of excitatory processes is the surface of the cell; however, it is not *intrinsic* to the components or the organization of the *bimolecular lipid membrane*. In fact, the evidence summarized below suggests that excitation does not reside in the lamellar lipid structure, and may be associated with a small fraction of the total membrane area.[1052]

1. Black lipid membrane prepared from a variety of lipids does not show any excitability characteristics; however, such a behavior can be induced by a variety of non-lipid components.

2. During excitation the electrical capacity of nerve and model membranes remains unchanged, i.e., the bulk of the membrane retains its dielectric properties.

From this data alone it can be computed that the proportion of the surface area of the axon membrane which might be occupied by a gating mechanism would be approximately one part per million, even in the excited state. For some other tissues this figure may be approximately 100 ppm. Similar conclusions may be arrived at from inhibition studies; for example inhibition of transient currents by tetrodotoxin suggests that the number of sodium-conducting pathways per square micron may not be more than 13 for squid axon,[851] 75 for rabbit, 49 for crab, and 36 for lobster nerves.[1146] See also Ref. 1320.

3. Spin label studies with suitable probes suggest that the interior of the nerve membrane is such that the hydrophobic probe suffers a restriction of internal motion. However, the ESR spectrum of the lobster nerve does not change during action potentials.[928] Such a result may be due to the location of the probe at a site where no conformational changes occur during excitation. This would imply that only a relatively small fraction of nerve membrane undergoes molecular changes during excitation.

4. It is hard to imagine that the lipid molecules alone could show an extremely high degree of specificity as is implied in ionic selectivity, in the type of changes implied by the effect of voltage on conduction or permeability, and with regard to the type of effect which some inhibitors seem to have on excitability.

5. Protein reagents (such as NEM, IAA, DNFB, etc.; see Table 8-2) and some antibodies specifically modify or block excitability characteristics; however, the amount of reagent bound to the membrane is usually much smaller than the amount that could combine with all the protein in the membrane.

6. A significant amount of experimental evidence has accumulated suggesting that the excitable membrane may have individual, isolated conduction pathways which show a voltage-dependent on-and-off process.[663]

7. A small but finite amount of heat production has been associated with the generation of the action potential.[834] The early phase of heat production (of the order of 1–10 μcal/g/impulse) is immediately followed by negative heat production during which about 80% of the heat produced in the first phase is reabsorbed. These phases of heat production are of special interest since they either accompany or very rapidly follow various phases of the action potential. The positive heat is associated with the rising phase and is augmented by replacing sodium with barium in the bathing solution. The negative heat is augmented by replacing sodium with lithium. This is indicative of two roughly equivalent processes involved in the generation of the spike, and since the amount of heat is minute it may be involved in the process of membrane reorganization at only a few scattered molecular sites. It would be of interest to know if heat production is associated with the magnitude of depolarizing potential and consequently to induced ionic fluxes.

8. A behavioral deficiency produced by a single gene mutation in *Paramecium aurelia* has been traced to impaired electric excitability[1309] of the cell membrane. The mutant membrane does not exhibit the normal depolarization—activated increase in calcium conductance responsible for regenerative depolarization in the

wild type. Other electric properties characteristic of the wild-type membrane remain normal in the mutant.

9. A binding component from lobster axon membrane has been characterized.[1310] This phospholipoprotein binds nicotine to the same extent as the wet nerve does, and the binding is completely blocked by acetylcholine, procaine, bungarotoxin. It has been suggested that this "receptor" is located on the internal surface of the axon membrane, and is a component common to both the ionic (Na and K) gates.

Thus most of the attempts to describe the molecular changes during excitation start with the axiom that voltage-dependent changes occur at only a few discrete conduction sites. However, this is the beginning of more serious problems. For example, questions regarding the distribution and nature of an ion conduction site can at best be answered with speculation. Then there are the questions such as: Is it graded or completely on-or-off? Is each site completely independent of all others? Or do some cooperate and act together in sizable groups? In a steady state is the pattern of sites fixed? Is each unit of the population varied to give only a constant average current? Do sodium and potassium pass through the same channel or through different channels? Is there any coupling between sodium and potassium conduction pathways?

These questions cannot be answered in terms of the experimental data available on any single system. Extrapolation from one system to another is obviously left to the choice of the proponents of any model. The question of ionic selectivity is rather puzzling, however, the treatment given in Chapter 6 provides a rationalization for most of the assumptions in this regard. The model systems corresponding to most of the assumptions regarding the effect of potential on distribution of sites and their behavior are rather scanty (see below). Thus interpretation of excitability in molecular terms is largely based on extrapolations, which have not been substantiated by direct experiments; however, the criteria of success of a model in explaining known phenomenology and predicting new experiments may for the time being be taken as sufficient evidence for its feasibility.

One of the most characteristic features of excitable membranes is their all-or-none response in generating an action potential (a propagating spike in biomembranes, to be more correct). The usual, but in this case a misleading, way of interpreting this phenomenon is in terms of stimulus response relationship, which is usually postulated as steeply sigmoidal for such situations. As supporting evidence is cited the fact that the rising phase of sodium and potassium currents in voltage-clamp curves also has a sigmoid time course. Since sigmoidicity is treated as a characteristic of stimulus response relationship in cooperative processes, this would imply that the electrical or chemical perturbation of a few specific sites on the membrane causes a change in their conformation and that this change is *physically* transmitted to neighboring subunits in an apparently cooperative manner. Since the proteins are polyelectrolytes, it appears likely that conformational changes, even limited ones, will lead to a shift of charge, thereby inducing a series of processes and changes which might markedly affect ionic permeability.

Cooperativity is a general property of multiunit systems. The general thermodynamic principle involved here is that, in an equilibrium between two forms or states of a system or subsystem, imposition of an electric field or binding with a ligand or some other suitable stimulus will cause a greater decrease in free energy in the more polarizable of the two states, and hence will shift the equilibrium in a direction favoring that state. When the equilibrium is a cooperative one, a "phase transition" may occur. The allosteric model of ligand-protein cooperative interaction proposes two or more sets of equivalent binding sites which remain equivalent upon transformation of the protein from one state to another, but the microscopic dissociation constant characterizing each set of equivalent sites is different in the two states. As a consequence of interaction with the ligand, the equilibrium shifts between these states, increasing the fraction of the total protein in the state with higher affinity for that ligand and concomittantly altering the microscopic affinity for other ligand molecules. In such a system the multiplicity of processes or of valence usually gives rise to a high-energy step; this might either be reflected in the high temperature coefficient for the process or in abrupt changes as a consequence of a small change in the stimulus.

The allosteric model has been successfully applied to account for the kinetics of multicomponent regulatory enzymes.[929,930] The idea of cooperative changes underlying excitability has been considered to account for the action of acetyl choline on the post-synaptic membranes to elicit post-synaptic action potential. (For an excellent account of excitability associated with synapses see Ref. 937. However, for the present purpose synaptic and nerve excitability may be considered to be identical processes, except that the stimulus for synaptic action potential is provided by acetyl choline or some such chemical ligand in contrast to the depolarizing current in nerves.)

The model describing axon excitability as a cooperative process postulates the following significant aspects of membrane organization to account for the observed data:[930–936]

1. Membranes are made up by the association of macromolecular lipoprotein units generally referred to as "protomers." The protomer is the "primitive cell" of the lattice and does not necessarily possess in itself a particular property of symmetry.

2. The protomers can exist in several reversibly accessible conformational states subject to lattice constraints similar to the quaternary constraints involved in the organization of the quaternary structure of the oligomeric proteins. It also implies that the conformation of these units may or may not differ when

they are organized into the membrane structure or dispersed in solution; however as a rule they assume the lowest free energy conformation.

3. The modification of the conformation on the binding of a specific ligand is reflected in the rearrangement of the membrane organization, and presumably the other way around.

Using these constraints on membrane organization, one can postulate further that the two states of the hypothetical protomer are $P \rightleftharpoons D$, where P represents the state when the membrane is polarized and D represents the state when the membrane is depolarized. These correspond roughly to the states R (for P) and C (for D) postulated on p. 336 to account for the steady-state characteristics of the excitable membranes. The changes of membrane potential are determined by the fractions of protomers which undergo transition to state D. Antagonism between activators and inhibitors is then described in terms of differential stabilization of either the P or D conformation. The apparently cooperative response seen with activators is due to transitions in neighboring protomers, which could give rise to either a graded or an all-or-none response depending upon the free energy of interaction between protomers. Since a small energy input* suffices to trigger the action potential, the amplification characteristics are postulated to be of an all-or-none type with protomers having a small energy gap.[932-935]

One of the most serious drawbacks of the preceding model stems from the fact that it does not explicitly state the nature of transient kinetics. Also it is hard to find a rationalization for action potential elicited with a depolarizing current. Furthermore, the theory does not elaborate on the nature of ionic permeability changes, voltage-clamp data, and interionic selectivity. Also quantitative treatments do not seem to support conformational cooperativity within a channel.[1267]

Another series of attempts to elaborate molecular characteristics of excitable membranes have exploited the ion-binding characteristics of membranes. Referring back to the interaction of polyvalent cations with the membrane and their effects on ion flow, we noted that Ca can be expected to bind phosphate group of lipids or protein or both. Along with this feature we must also consider the ion specificity properties of the ion-transfer mechanism. From the study of a three-dimensional model of a phospholipid, especially in a closely packed situation with other lipid and proteins around it, one is tempted to postulate that configurational changes in proteins brought about by dipole rotation may have a strong influence on the local configuration of the anionic groups and thereby make it possible for them to have specific ion-binding

---

*In biological excitable systems it is noted that the applied potential of about 10-50 mV suffices to bring about excitation. The energy input for such a change per charged group is therefore $1 - 5 \times 10^{-2}$ eV or about 230-1150 cal per unit charge. This is an amount much lower than that required to change any chemical structure or to open even weak bonds.

properties which change as the configuration changes. Thus, the large field of resting potential can provide a configuration which can bind polyvalent cations, and this would provide a *lock,* as it were, for the ion gate. On this general pattern Tasaki and Kobatake[806] have proposed the so-called macromolecular theory of membrane potential. The theory makes the following major assumptions:

1. In the resting state the outer region of the axon membrane contains bound divalent cations, probably Ca, derived from the external medium. On stimulation some of the bound Ca is replaced by univalent cations derived from the internal medium (potassium ions in the normal nerve); thus the membrane acts as a cation exchanger.

2. The changes in the cation concentrations within the membrane, resulting either from the external concentration changes or from applied current flow, may change the state of the membrane to one in which it becomes more permeable to cations, probably as a consequence of an increase in the density of the fixed charges.

3. The ions are supposed to move through the same site.

4. Both the anion and cation selectivity sequences observed in squid axon are directly related to the effects of the ions concerned with the charged groups of proteins and lipids, presumably those located on the surface.

5. A macromolecular change occurs following the suprathreshold depolarization, which makes the membrane more permeable to all cations, but to a varying degree.

Thus the excitation process is considered to represent transitions between two stable states of the membrane. The potential variations associated with transition from the resting to the excited state primarily represent variation in the phase boundary potential at the outer interface. The theory or the assumptions do not explicitly elaborate the "macromolecular change" or the "cooperative interactions" involved in cationic permeability changes following excitation, even though the membrane is assumed to be a cooperative cation exchanger. The model as a whole has been successful only to a limited extent, that mostly on a qualitative basis.

Yet another closely related model based on a dilation–contraction equilibrium regulated by the divalent–monovalent ion-exchange properties of excitable membranes has been suggested[938] and quantitatively analyzed.[939] For theoretical purposes the membrane is considered as a two-dimensional cooperative cation exchanger constituted of a lattice of finite dimensions. In the resting state the lattice units of the cation exchanger bind divalent ions such as Ca. Upon depolarization or decrease of the Ca activity the resting state becomes unstable and the univalent cations are bound to the lattice units; this process may be considered as a two-dimensional phase transition. The ion movement accompanying the exchange gives rise to an inward excitation current. The kinetics of such a process have been described in analogy to the phenomenon of nucleation and nucleus growth. The predictions of the theory agree with the results of voltage-clamp experiments, but some significant differences may be noted. The equations and the magnitude of the data-fitting parameters suggest that the area per lattice unit may be 21 X 21 Å² in the squid axon. Similarly, the differences in the interaction energy of a lattice unit in two binding states, with the rest of lattice being in one of the two states, may be of the order of 2.5 kcal/mole. The physical mechanism of membrane processes and the predictability from quantitative treatment are still not clear; however, it is obvious that the two ion-exchange models and the allosteric kinetic treatment described earlier are rather complementary. These systems, if they bear any resemblance to reality, need to be elaborated so as to predict the complete pattern of behavior of excitable biomembranes.

A significant number of attempts have been made to account for excitability in terms of the solution properties of the electrolytes which so conspicuously participate in the overall process. For quite some time it has been known that at very high voltages the conductivity of an electrolyte is a function of the electric field strength.[940] For strong electrolytes the effect is rather small; 0.3% change at 180 kV/cm. This could be attributed to the relaxation of the ionic cloud: the atmosphere of the ion could not follow its extreme speed in the high electric field. The ionic cloud thus becomes distorted, allowing the (central) ion to move at higher speeds than normal at that concentration. This is called the first Wien effect. Weak electrolytes show a significantly greater conductivity increase at high field strengths—almost 6% increase for acetic acid at 120 kV/cm.[941] This effect with weak electrolytes is dependent upon concentration, and at high dilution saturation effects are observed; saturation corresponds to the conductivity at infinite dilution. This is known as field dissociation or the second Wien effect and is ascribed to the dissociating effect of large fields. Thus it can be shown that in liquid dielectrics ($\epsilon \simeq 10$) the dissociation constant of dissolved electrolytes may increase some 100-fold.

According to the theory of Onsager[942] the field dissociation effect arises because of the modification of the ionic atmosphere (first Wien effect) which as a secondary effect enhances the degree of dissociation of any ion combined in a pair. The effects are particularly significant with 2:2 electrolytes.[943] In the latter case the high field enhancement of conductance is due to increased dissociation of site-localized counterions, and the conductivity does not immediately return to a normal (low-field) value after application of a high-field pulse; the relaxation time is a function of valence of ions involved, their mobility, the viscosity and the dielectric constant of the solvent, and the absolute temperature.

A model for the generation of the action potential, based on the field dissociation effect has been proposed.[944] It suggests that *local alkalosis* due to the Wien effect on acid dissociation is responsible for initiation of the events leading to electrical depolarization via chemical and conformational changes. It is further postulated that the effect of calcium ions may either be in parallel to or be sequential to the action of protons. Little experimental data exists to preclude or confirm the theory, presumably because of difficulty of verification.

The coupling of the action of Ca ions and of protons has some merits. It has been known for some time that an asymmetry is observed with respect to the effect of divalent cations and protons. A qualitative suggestion regarding the mechanism of membrane excitability is based on the assumption that (abrupt) removal or reversal of the resting membrane potential must alter the electrochemical gradient for protons across the membrane.[945] This gradient of the order of one pH unit per 100 Å may be created momentarily which may seriously influence the membrane permeability characteristics. It is further assumed that the sodium-selective channels are normally blocked by calcium ions to negatively charged sites near the outer end of the channels. The calcium ions can be displaced competitively by hydrogen ions, opening the channels to sodium. According to this model, depolarization of an excitable membrane causes an outward flow of hydrogen ions across the membrane. The consequent transient increase in hydrogen ion concentration at the outer surface of the membrane displaces calcium and opens the sodium channels. Thus a reduction in pH of 1.5 to 2 pH units at the outer membrane surface is sufficient to displace nearly all the sodium channels. The experimental data however shows that it does not happen.[1268]

Jain, Marks, and Cordes[946] have elaborated the two-state hypothesis (see p. 336) to account for the kinetic and the steady-state aspects of the phenomenology of excitation. The details of the model are outlined in the following sets of postulates:

1. Conductance across the membrane is the consequence of ionic species flowing down their electrochemical gradients.

2. Ionic flux across the membrane occurs through a limited number of specific sites on the membrane; these sites function independently of one another.

3. Under steady-state conditions each of the sites exists in either of the two states, designated R (*rest*) and C (*conducting*). The relative occupancy of the two states depends on the strength of the electric field across the membrane as given by:

$$f_c = \frac{[C]}{[R] + [C]} = \frac{1}{1 + \exp\left[-a(V - V_h)\right]}$$

(See p. 337 for details.)

4. Ionic flux through individual sites in a particular state is linear in the driving forces and in the gradients of the chemical and electrical potential.

5. The specific conductance, g_R, of a site in state R is considered to be small in comparison with that, g_C, of a site in state C. Thus, as a result of any

change in membrane potential each site has a quantal behavior; however, the macroscopic change in membrane conductance is continuous.

6. After application of a more positive potential (*clamped*) to the membrane at its resting potential (for which most of the sites are in the state R), a certain fraction of the sites, given by equation (8-1a), are converted into state C according to the following pathway:

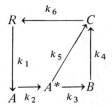

Here, R, A, A^*, B, and C represent distinct kinetic and functional states of conductance sites on the membrane; in contrast to states R and C, states A, A^*, and B are sparsely populated under steady-state conditions. In this kinetic scheme, three quantities are voltage dependent. The depolarizing potential applied to the membrane determines in accordance with equation (8-1a) the extent to which sites in state R are converted to state A, which eventually decays to state C by one of the two routes indicated in the scheme shown above. Also, the first-order transformations of R into A, and of C into R, are characterized by voltage-dependent first-order rate constants, k_1 and k_6, given by

$$k_1 = b \, \exp \left[\frac{a}{2} (V - V_h) \right]$$

$$k_6 = b \, \exp \left[-\frac{a}{2} (V - V_h) \right]$$

All interstate conversions are kinetically first order. The rate constants are not related to the rate of movement of ions, but to the rate at which structures change from one state to the other.

The postulates from (1) through (5) are supported by the evidence summarized on p. 337, although the nature of the evidence is circumstantial. The postulate (6) is even more speculative in its origin. In any event, with these postulates as the basis, the whole range of the behavior pattern of excitable membranes can be predicted. Thus for example, the current–voltage relationship in steady state and as derived from voltage-clamp data, the kinetics of inward and outward current under voltage-clamp conditions (Fig. 8-13), action potential and threshold stimulation, and quantitation of the effect of inhibitors have been described elsewhere[946] (see p. 348). Besides this, the data can be used to quantitate the whole range of steady-state and kinetic characteristics of the excitable membranes; for example, deviation of resting potential from the

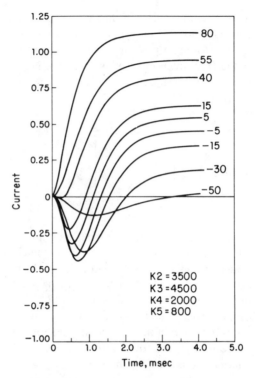

Fig. 8-13. Calculated plot of total ionic current vs time for several impressed voltages under voltage-clamp conditions. Rate constants correspond to those in the scheme. Absolute voltages are given at the end of each curve; the current scale is arbitrary. Values: $a = 0.0809$ mV^{-1}; $b = 2000$ sec^{-1}; $V_h = -38$ mV. The current scale is arbitrary and extends from -1 as a maximum for Na current to $+1$ as a maximum K ion current (achieved at the equilibrium Na potential of 55 mV). Compare Fig. 8-8.[946]

theoretically expected values, the effect of temperature on specific rate constants, and the effect of divalent cations and chemical reagents on specific intermediate states can all be related to one or another of six parameters and only one variable, that is, membrane potential. Furthermore, the whole experience of enzymology can be related to structural features participating in the voltage-dependent processes.

Thus by way of recapitulation, it is fair to say that the model systems, particularly modified BLM, have offered a great many constraints within which one may look for the mechanisms of the behavior patterns of excitable membranes, including axons. Obviously, excitability is an extrinsic property of the lamellar bimolecular lipid layer. Furthermore, voltage dependence of the membrane as a whole seems to arise from the effect of voltage on membrane-bound structures that seem to function independently of each other. Within these constraints, one can now proceed to interpret the data on the physiological axon.

9/ The Membrane Hyperstructure and Organization

Organization, which for the moment may be thought of as some set of structural relations, is the dominant feature of all biological systems and appears at all levels, from molecular to social. Moreover, insights into the forces responsible for this organization may be obtained from the statement of Chuang Tzu in the third century BC: "The harmonious cooperation of all being arose, not from the orders of a superior authority external to themselves, but from the fact that they were all parts in a hierarchy of wholes forming a cosmic pattern, and what they obeyed were the internal dictates of their own nature." Whatever may be the metaphysical consequences of such a dictum, it is now widely accepted that in molecular aggregates the organization is dictated by "information" present in the "units" of the system. How the components of membranes are organized and what gives stability to the basic structure is a matter of great concern and is dealt with in this chapter as a recapitulation of the functional aspects of membranes which we have developed in the preceding chapters.

From the mode of formation of BLM and from their behavior and properties (*intrinsic*), it is apparent that simple monomeric units aggregate to form more complex structures. A biological advantage of such self-assembling systems accrues from the fact that the structure of the aggregates can be completely specified by the genetic information required to direct a synthesis of component molecules. Regulation of size, shape, number, and conformation of monomeric units may still be necessary and be subject to control by external conditions. These factors may influence the assembly process either directly or by changing the conformation and the proportion of the assembling monomers.

In principle, it should be possible to predict the behavior both of the monomers and aggregate on the basis of differences in free energy of the molecules in their various states. These matters were dealt with in Chapters 2 and 3 (also see Ref. 947) while considering the aggregation of amphipaths generally. Even crude calculations of this type can provide useful information about the relative stability of various forms and the nature of the underlying molecular interactions. Thus, for example, in tobacco mosaic virus the free energy of addition of protein components (subunits) to the helical nucleoprotein structure is of the order of -7 kcal/mole. In BLM, the free energy of addition of phospholipid molecules to the bilayer is about -20 kcal/mole. Both these structures have stability toward thermal disruption; however, the morphology of the virus is stable whereas that of BLM is not. This may be ascribed to the fact that the intermolecular bonding in the virus has stereochemical specificity; in contrast, there seems to be little such specificity involved in the interaction of phospholipids. Such examples concerning design and stability of biological structures suggest the involvement of factors other than those evident in gross energetic considerations based on superficial assumptions. In the following pages we shall elaborate specific aspects of the interaction of various membrane components; this discussion is then followed by consideration of gross structural and organizational features related to biomembranes.

ORIENTATION OF POLAR GROUPS OF LIPIDS

The importance of the surface charge on membranes, as we emphasized earlier (Chapters 5 and 6, Ref. 1054), is reflected in various membrane functions, such as aggregation, ionic selectivity in transport processes, and accumulation of ions in the interfacial regions which may be important for the functioning of the membrane-bound enzymes. These aspects may also have a regulating effect on aspects of cell propulsion, and the contribution of proteins and polysaccharides to them may be significant. However, we shall restrict our discussion to those characteristics ascribable to the polar head groups of lipids.

Both experimental measurements and theoretical calculations suggest that phosphatidyl serine, phosphatidylethanolamine, and phosphatidylcholine, as well as sphingomyelin, have dipole moments near 15 Debye units. The pK_a values of the polar groups are such that, at neutral pH, the N_1 has a free positive charge, whereas the phosphate oxygen bears a full negative charge. Such dipoles undergo marked rotational changes when a sufficiently large electric field is applied or removed across them. The most significant points regarding the stereochemistry of the head group of phosphatidylcholine seems to be (a) that the choline residue exists in the *gauche* conformation rather than the extended zig-zag form and, (b) that the conformation of the glycerol residue can be *gauche-gauche*.[948]

Molecular models of the phospholipid head group indicate that the quaternary ammonium group can either be coplaner to and 5–6 Å below the phosphate groups, or at the other extreme, both these groups can be in one and the same plane—parallel to the bimolecular leaflet. From the former orientation a complex double layer would result giving rise to an electrokinetic potential. However, the experimentally observed electrokinetic potential is near zero; a result expected only if these groups are coplanar with the plane of the membrane. A similar arrangement has been suggested to occur in phospholipid monolayers on the basis of surface potential measurements.[949] Coulombic interaction seems to be the major force stabilizing this coplanar configuration of polar head groups.

Several models have been developed for quantitative treatment of electric fields associated with membrane interfaces. Quantitative analysis of charge distribution and dipole orientation in the plane of the membrane (BLM type) suggests that for the cholesterol, lecithin, and fatty acid type of molecules the point dipole model is adequate.[65] Similarly, for a simulated protein sandwiched lipid BLM (Dawson-Danielli model), the theoretical analysis suggests that the layer of dipoles which are allowed to rotate freely within the plane of the lipid bilayer align themselves within the field of adjacent charged macromolecules. Such alignment of dipoles creates comparatively large force fields. At a certain critical separation of charged molecules from the lipid-protein layer, small perturbations in the dipoles could cause the force fields to abruptly change sign. The attractive or repulsive forces on a charged molecule due to such sign reversal may be critical in certain regulatory, selection, and ordering processes. For various aspects of such *electrodynamic* analyses see Ref. 65.

ORGANIZATION OF WATER ASSOCIATED WITH MEMBRANES

The problem of the role of water in biological systems has become one of the most discussed in natural sciences. In fact properties of water have been called upon to an ever greater extent to explain otherwise incomprehensible biological phenomena. This state of affairs is due to the striking value of water for biosystems arising from the fact that (a) water forms the greater part of the biosystems and (b) water has anomalous properties, thus making it a unique fluid even from the purely physical point of view. The water content of living cells ranges from 70-90%, and water intimately associated with membranes may account for 30-40% of their total mass.

The anomalous properties of liquid water are a consequence of the structural peculiarities of individual water molecules. X-ray diffraction studies suggest that water molecules in both liquid water and ice are tetrahedrally coordinated. In the molecules of water, the three nuclei form an isosceles triangle with the two protons at the base of the H–O–H forming almost a tetrahedral angle. Of the ten electrons of the water molecule, two electrons are close to the oxygen nucleus whereas the other eight move in pairs over four elongate elliptic orbitals. The solitary electron pair moving along the orbits in the plane perpendicular to the

H-O-H plane accounts for the comparatively high electron density in the peripheral part of the water molecule.

Water molecule exhibits a variety of nonbonded interactions. One of the more simple ones is the Coulombic interaction between the electronic charges of two water molecules. In fact, in ice, the water molecules are arranged such that they touch at the different poles. However, this is the extreme example where the charges are completely delocalized from their orbitals. The more commonly observed situation is partial delocalization of the electrons on contact of the molecules of water with each other. Such a partial linkage between an electron donor and proton donor is usually called a hydrogen bond. Thus, there are two moles of H-bonds per mole of water molecules in ice. On melting of ice, 46% of these bonds are ruptured, and with the rise of temperature the number of ruptured bonds increases further. The energy of rupture of H-bonds is about 4 kcal/mole. Thermodynamic studies further suggest a cooperative character in the mode of formation and rupture of H-bonds. However, very little is definitely known about the structure of bulk water, though there has been considerable speculation.[950]

The associative character of water gives rise to a variety of phenomena, especially in the presence of extraneous ions, molecules, and other associative structures. One of the simplest situations of this type arises because of the ability of a proton to find itself in two possible positions in the structure of ice. The displacement mechanism shown below has been suggested for the anomalously high mobility of the proton in ice:

Such a displacement is known as Grotthus conduction. One of the most important features of such conduction is that the structure of the ordered molecules of water–ice possess the capacity for *distant action*.

In the presence of other ions, the electrostatic field of an ion will tend to orient the dipoles of the molecules of water near the ion radially and, at the same time, redistribution of the charges in the molecules of water themselves will strengthen their H-bonds with the neighbors, thereby stabilizing the ordered structure of ice in water. The ionic sequence increasing the degree of organization of water is:

$$\text{Cations:} \quad Mg > H > Ca > Na$$
$$\text{Anions:} \quad OH > F$$

Similarly, the sequence of ions decreasing the degree of organization of water is:

$$\text{Cations:} \quad K > Rb > Cs$$
$$\text{Anions:} \quad ClO_4 > Cr_2O_7 > I > Br > NO_3 > Cl$$

Unlike ions, nonpolar molecules interact with the molecules of water only through weak van der Waals forces. They frequently increase the degree of structural organization of water. There are several apolar molecules, however, which disrupt the normal structural organization of water. Compounds in the later category include tannic acid, tetraphenyl boron, guanidinium ion, hexafluorophosphate, Nesslaer's reagent, and hydrocarbon chains in the amphipathic molecules.

Hydrophobic interactions stabilizing ice-like structure in liquid water are numerous and their biophysical implications are enormous. Such interactions are generated by the structural features of the apolar molecules and macromolecules which come into contact with water.[1166] Thus the conformation of the macromolecule as a whole in aqueous solution must depend upon its influence on the structure of the surroundings. Study of such systems has given rise to terms such as bound water, ice-like structure, organized water; and to phenomena such as, screening effect, solvation effect, etc. Such terms lack accuracy and clarity, and reflect ignorance of the precise structures formed and of the consequences of their formation.

Various forms of loose binding of water have been characterized by physicochemical means. Some of them are more specific than others and include: hydration, incorporation into molecular chelates, trapped water in clathrates or cage structures, and hydrogen-bonded water bridging. Such structures often depend upon steric factors, such as bond angles, and the structural configurations of water dependent upon its physical state, such as the tetrahedral configuration in 'ice-like' structures. Interestingly, there are only a limited number of ways in which water molecules can fit into surrounding structures such as crevices and holes; even then, there is considerable difficulty in assuming the validity of structures involved in the formation of crevices, etc., as might be the case at the membrane interface. Therefore, at best, such approaches have only limited utility at the microscopic and molecular level.[65]

Thus it seems axiomatic that the structure of water within or at the surface of membranes depends upon the molecular organization of the membrane components. The ability to form ice-like structures depends upon the relative magnitudes of two effects: (a) the electric field, and (b) the random thermal energy of the water molecules. A third effect involving hydrophobic interactions, as discussed earlier, may be critical in determining the configuration of membrane components, but is difficult to evaluate. Thus, the limiting factor in determining the length of the water sheath covering BLM will be the field gradient arising from the presence of polar groups at the interfaces. The electric field is asymmetric on both sides of the membrane; it goes to zero at a midplane between the interfaces, but does not become zero on either side of the membrane at finite distances. In principle, the membrane field will, therefore, align the water dipoles along the lines of the electric field and help to form ice-like structures which should be associated with the polar groups. The importance of structured water stabilized by macromolecular surfaces has been established by a variety of experimental observations.[951] If it is assumed that the assembly of the polar ends of lipids have water-ordering properties comparable to those of the surfaces of cellulose acetate or polar glass. Ordered water up to a maximum sheet thickness of 22 Å (on each interface) could be intercalated with the interfaces of bimolecular lipid leaflets. Such structured water has peculiar properties: high viscosity and almost zero solubilizing capacity for salts. Considerable amount of information about water structure at the membrane interface can thus be adduced from reversible and irreversible changes associated with temperature, permeability, and transport characteristics, and surface properties including electrokinetic mobility. Evaluation of entropy factors may be particularly significant for quantitative studies. Unfortunately, very little can be said about the validity of these various aspects at this moment.

ORGANIZATION OF LIPID MOLECULES

Studies with molecular models suggest the strong possibility of van der Waals interactions between various lipid components of the membranes. The complex formed between certain phospholipids and cholesterol as established by monolayer studies (see Fig. 2-7) can be accounted for by manipulating molecular models. Such a complex could be formed with almost all types of phospholipids regardless of the nature of the nitrogenous base and the lengths and unsaturation of the tails. These complexes have the following features:[952]

1. Phospholipids can form a molecular complex $(1:1)$ with cholesterol (Fig. 9-1). Such complexes have generally similar morphological appearances. However, the phosphoglycerides and sphingosine complexes have neither the same end view contours nor are the positions of the polar groups superimposable.

2. If these complexes are arranged such that each unit is perpendicular to that of the neighboring unit, alignment of the nitrogens of choline results. Considerable symmetry is also observed in positioning of other groups.

3. The tail-to-tail arrangement of *di*molecular complex units accounts for the relative composition of lipids and for the phosphorus-to-phosphorus distance as determined by x-ray diffraction studies (51.5 Å). Provision can be made for the formation of pores, accommodation of varying chain lengths, water content, and water structure.

4. A uniformly complete coverage of the lipid surface by protein can be made by taking the globular structure of proteins into consideration.

Such studies have some bearing on real membrane structure; these structures account for the thickness of the membrane, the phosphorus-to-phosphorus distance, and the area of each component, especially of the hydrocarbon chains. Thus the hydrocarbon chains are oriented perpendicular to the plane of the membrane, whereas the polar groups have freedom of orientation with external

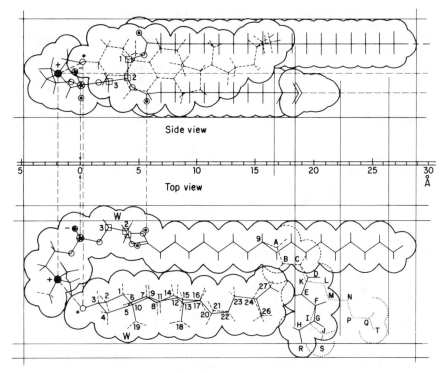

Side view

Top view

Fig. 9-1. The sphingomyelin–cholesterol complex. Similar complexes can be formed from other lipids, although the morphological details are different.[952]

fields depending on the molecular packing order. The density of packing is fixed between 20 and 30 Å^2 per chain and the average packing is 50 Å^2 per polar group; both values are in accord with x-ray diffraction data for model[953] and biomembranes.[954] As expected, and as indicated by monolayer studies, the area per chain is a function of the surface pressure of the monolayer constituting the bilayer, and the degree of unsaturation and branching of the hydrocarbon chains (Tables 9-1a and 9-1b). These values reveal that branched-chain fatty acids and unsaturated fatty acids occupy more area than their saturated straight-chain homologs. The projected cross-sectional areas of various fatty acids vary over a wide range, and this variation may exert considerable influence on the packing of the lipid molecules. From examination of molecular models, it can be seen that substitution of additional methyl groups in fatty acids, twisting of substituted chains, or introduction of a double bond in the hydrocarbon chains of phospholipid molecules constitute further obstacles to the compact organization of lipid films. This difference in the packing order may be the physical basis for permeability differences observed between membranes with varying amounts of

TABLE 9-1a EFFECTS OF CHAIN LENGTH AND BRANCHING ON
PACKING OF FATTY ACIDS IN MONOLAYERS

Compound	$t(°C)$	$Å^2$/Molecule	Surface Pressure (dyne/cm)
Myristic acid	25	37	5
Pentadecylic acid	25	37	5
Palmitic acid	25	24	5
Stearic acid	25	23.5	5
Arachidic acid	25	23	5
Oleic acid	15 to 20	48	5
Erucic acid	21	40	5
2-methyl octadecanoic acid	20 to 25	30	5
16-methyl heptadecanoic acid	20 to 25	32	5
Phytanic acid (3, 7, 11, 15-tetramethyl hexadecanoic acid)	20	36	24
Phytol (3, 7, 11, 15-tetramethyl hexadecene 2-ol-1)	20	34	24
Palmitic acid	25	20	24

TABLE 9-1b EFFECT OF CHAIN LENGTH, UNSATURATION AND CHOLESTEROL
ON PACKING OF LIPID MOLECULES IN MONOLAYERS

Lipid	Molecular Area $(Å^2)$	Area after Addition of Cholesterol
1, 2-Didecanoyl lecithin	79.5	79.5
1, 2-Dimyristoyl lecithin	72	61
1, 2-Distearoyl lecithin	45.5	45.5
1-Stearoyl-2-oleoyl lecithin	75	64
1, 2-Dioleoyl lecithin	83	77
1, 2-Dilinoleoyl lecithin	99	99
Egg lecithin	96	56

Cholesterol was added in a molar ratio of cholesterol : lecithin of 1 : 1. Surface pressure was
12 dyne/cm, except with egg lecithin, which was examined at 33 dyne/cm.

unsaturation and branching (Chapter 5). Furthermore, it may also be reflected
in the temperature characteristics of these membranes since the Kraft point
(transition temperature) is a function of unsaturation, branching, chain lengths,
and packing. The subtleties of the molecular association between lipid components (e.g., phospholipid and cholesterol) also imply considerable interchain
attraction as is in fact manifested in PMR,[955,1155] ESR,[1156] fluorescence depolarization,[1184] ^{13}C Fourier transform NMR,[956] x-ray diffraction,[953,954] partition,[1157] ORD[1316] and other studies. Thus, for example, the liposomes prepared from mixed lipids (cholesterol and lecithin) show a broadening of the
methylene signals, although the choline peak remains sharp. Such a differential
broadening suggests that the molecular motion of the polymethylene chains is re-

stricted but there is less effect on the molecular motion of the choline groups.[955] These physical methods have also given significant qualitative and semiquantitative information about the physical state of lipids, for example, bending of the hydrocarbon chains,[1158] rotational freedom of chains,[956,1186] the average orientation of probes attached at various points on the chain,[1158] inside–outside transition of lipids in bilayer,[1159] thermal phase transition,[1160-1163] binding of lipids to protein[1164] and drugs,[1315] cation-induced organizational changes,[1165] and relative orientation and positioning of lipid components in a given phase[1185] and state.[1297,1311]

From such studies the following salient conclusions may be drawn:

1. The hydrocarbon chains lie parallel and rather extended in a plane perpendicular to the plane of the membrane, although slight bending of the chains (60° to the plane of the bilayer) has been indicated by ESR studies.[1156,1158]

2. The hydrocarbon chains are engaged in rapid rotation along their long axis and toward the methyl end the chains are in liquid-like environment. The presence of double bonds on the chain or the presence of cholesterol seem to confer greater rigidity to the chains, thus reducing the random motion.

3. The polar boundary is diffuse rather than abrupt.

4. The ions—Ca and Mg, for example in physiological concentration—act by diminishing the charge repulsion between the phospholipid head groups, thus allowing the bilayers to contract, so as to constrain the motion of the hydrocarbon chains.

5. The carbon atoms along the chain have different degrees of freedom of rotational motion. The rotational correlation time (τ_c in psec) and relaxation time (T_1 in sec) for various carbon atoms in ^{13}C Fourier transform NMR are as follows:[956]

$$\tau_c$$

$$
\underset{2.1}{\overset{7}{CH_3}} - \underset{1.1}{\overset{24}{CH_2}}(CH_2)_{4-6}\underset{0.5}{\overset{90}{CH}} = \underset{0.26}{\overset{90}{CH}}(CH_2)_8\underset{1.1}{COO} - CH
$$

$$\overset{97}{CH_2} - R$$

$$CH_2 - O - \overset{O}{\underset{\underset{O}{\downarrow}}{\overset{\|}{P}}} - O - (\overset{87}{CH_2} - \overset{}{CH_2}) - \overset{36}{N(CH_3)_3}$$

$$T_1 \qquad 2.4 \qquad 0.27 \qquad 0.43$$

The methyl groups in the choline moiety and the terminal methyl groups are highly mobile. The rotational freedom decreases as one moves away from the terminal methyl groups. The methylene groups near the polar head appear to be highly restricted in their rotational motion.

6. Thus the middle of the bilayer is in effect a "liquid," but at the edges it behaves essentially as a layer of "solid" domains. Such a structure when perturbed by an incoming apolar group may bring about a "domain effect"—successive

clusters of chains swinging from one bent state to another.[1158] Such a mechanism has strong implications on a mechanism for transport (see p. 135) and depolarization (see p. 378) processes.

7. There may be difference in packing of the same phospholipid molecules in different structures formed by varying degree of dispersion, for example sonicated and unsonicated liposomes or intact cells and vesicles.[1297]

These conclusions in general may be elaborated to suggest the nature of lipid packing in biomembranes. Thus it seems probable that while the lipid bilayers of myelin and other cholesterol-containing membranes are highly condensed, other bilayers (?), such as those of mitochondria (low cholesterol and high unsaturation), are expanded with much looser molecular packing. Such observations also imply that the lowering of temperature, increase in chain length, or addition of cholesterol will all produce the same effect, that is, a change to a smaller area per polymethylene chain and a decrease in entropy associated with closer packing. The unsaturated and branched-chain fatty acids cannot be packed tightly because of steric hindrance and other geometrical factors; the position and stereochemistry of double bonds or branching is important in this respect. Thus chlorophyll, phytol, and other naturally occurring linear polyisoprenoids (carotenoids, vitamin A, coenzyme Q, etc.) inhibit formation of condensed bilayers* and may lyse them (see Chapter 6).

In biomembranes there is a small proportion of lipid apparently involved in direct interaction with proteins, although the apparent lipid/protein ratio of membrane preparations varies from about 4:1 for myelin to perhaps 6:1 for some mitochondria (Table 9-2). Furthermore, a large proportion of membrane-bound protein is held hydrophobically (see next section). This contradiction apparently arises since the protein molecules are about 100-fold larger than the lipid molecules and may project outside the lipid layer. Provided that the tertiary structure of proteins is compact, the proteins need immobilize only a fraction of their weight of lipid. This is in complete accord with the available data. These aspects are discussed further in the next section.

The effect of temperature on chain mobility in lipid membranes is of particular interest. Calorimetric and spectroscopic data suggest that at around 37°C most of the model and biomembranes undergo phase transition.[959,1160-1163] At higher temperatures the chains appear to be relaxed and in random orientation. Thus the area per methylene chain in the plane of the membrane would show

---

*In this situation a more loosely organized arrangement is preferred and several possibilities for arrangement present themselves. One of the most interesting speculations in this regard is made to explain the formation of gall stones. It has been observed that lecithin, cholesterol, and bile salts form bilayer membranes as well as micelles of fixed stoichiometric composition; consequently such structures cannot incorporate more than a fixed amount of cholesterol.[957] Thus any excess of cholesterol synthesized by the organism should accumulate in a crystalline form, such as a gall stone. A similar metabolic disorder for ingestion of phytol and related isoprenoids may result in Rufsum's disease.[958]

TABLE 9-2 COMPOSITION OF SOME TYPICAL PLASMA MEMBRANES

Type of Cell	Protein (%)	Lipid (%)
Ox brain Myelin	18–23	73–78
Human erythrocyte	53	47
Saccharomyces cereviciae NCYC 366	49	45
Pseudomonas aeruginosa	60	35
Saccharomyces cereviciae ETH 1022	37[a]	35
Bacillus megatorium	70	25
Micrococcus lysodeikticus	68	23
Rat muscle	65	15
Rat liver	85	10
Avian erythrocyte	89	4
Rod outer segment	40–50	20–40
Chlorophylls	35–55	18–37
Mitochondria (total membrane)	70	30
Mitochondria (inner membrane)	75	25
Sarcina lutea	57	23
Mycoplasma laidlawii	47–60	35–37
Bacillus spp.	58–75	20–28
Micrococcus lysodeikticus	65–68	23–26
Staphylococcus aureus	69–73	25–30

[a] Up to 27% mannan was found in the preparation.

considerable variation as a function of temperature. In fact, the area per polymethylene chain in BLM and related structures is not quite that of the area in
crystalline lipids or the molecular dimensions (25, 18, and 14–15 Å, respectively). However, the hydrocarbon chains may not be in a completely random
orientation. Complexation of membrane constituents to form a relatively condensed phase and the stability between the interfaces of high dielectric medium
suggest that the system as a whole must have some type of order determined by
various participating molecular species. Energetically, strong interactions favor a
crystalline phase which is highly ordered; in contrast, entropy favors the less
well ordered liquid state. The situation is further complicated by participation
of interfacial layers of water. Thus the order may mean fixed relations between
parts, but not necessarily a regular repeat pattern as in crystals. It also implies
that a long-range disorder in the surface and local crystalline domain may be
manifested as in liquid form, without imposing serious restrictions on membrane, permanent or transient. The long-range disorder may arise as a consequence of the increase in entropy which compensates for the increase in the
energy resulting from the departure from the optimum packing. This may result
in departure of 'domains' from the average membrane composition.

Some degree of specificity for tissue, cell, or cell organelle can sometimes be
deduced with respect to variations in chain length, number and position of
double bonds, and occurrence of branched-chain and cyclopropane structures.

Environmental and genetic factors of various types appear to influence the fatty acid pattern[960,1167,1278] and polar group distribution[961] in various organisms. See also Ref. 1195, 1279. These changes are of interest in evaluating the importance of the phospholipid composition for the membrane function, such as osmotic fragility,[1312] activity of membrane-bound enzymes[1314] (also see below) passive permeability (cf. Chapter V), and stability. This is to emphasize that one should not consider this flexibility of composition of phospholipids as an argument for a nonspecific function of fatty acid constituents in the membrane. In the preceding chapters we have emphasized time and again (and will discuss at some length later in this chapter) that some specificity is in fact observed in membrane function with respect to phospholipid composition. A simple-minded way of looking at this is in terms of intermolecular associations between lipids and proteins to give stabilized structures. In this regard, the degree of unsaturation, branching, etc., which has a destabilizing effect on the packing of lipid chains alone, may stabilize the condensed lipoprotein structure. In fact, experimental studies suggest that the fatty acid composition of phosphoglycerides is regulated in order to yield molecules with properties suitable for membrane foramtion, and that this regulation occurs not only at the level of fatty acid biosynthesis but also at the stage of phospholipid biosynthesis.*

The nonrandom distribution of hydrocarbon chains of different apolarity on two positions of many natural PL invites a quantitative description in terms of molecular species. With some reservations one could envisage that a control of the biosynthesis of constituent units would ultimately result in the control of the factors which govern properties of the membrane, such as lipid-protein interactions, spatial arrangement of lipoproteins, the degree of liquidity and flexibility of hydrocarbon chains, etc. It is probably too early to speculate on the molecular significance of these changes. However, a participation of lipids at different levels of membrane function already indicates that it is impossible that any single molecular arrangement accounts for such a multiplicity of functions. This discussion is taken up further while discussing lipoproteins later in this chapter.

---

*These assertions are probably best substantiated by the following studies. The PL from aerobically grown yeast cells contain a considerable quantity of monounsaturated fatty acids, which under anaerobic conditions are greatly compensated for by the appearance of fatty acids of medium chain length (C_{10}–C_{12}).[962] Moreover, the short-chain acids are found to be located only at the 2-position of glycerol, which is found to be predominantly occupied by monounsaturated chains in lecithin from aerobically grown yeast. In this respect it is of interest to note that a synthetic didecanoyl lecithin was found to be highly hemolytic to anaerobically grown yeast.[963] Various other lipids also have a similar lytic effect. With respect to the acquisition of the structural requirement of PL in serving as membrane constituents there is probably a good reason for a cell to avoid the production of significant quantities of phosphoglycerides with two hydrocarbon chains with medium chain length.

LIPID–PROTEIN INTERACTIONS

In the preceding chapters it was emphasized that proteins are usually associated with various functions of the membrane since BLM as such do not show any of these functions. The proportion of protein in biomembranes ranges from about 20 to almost 70% (Table 9-2). In searching for the unifying principles of membrane design, we must be aware of the wide variations in the composition of the diverse membranes with respect to their lipid and protein composition,[964] and particularly of the possibility of a large number of possible modes of lipid-protein interactions.

Early experimental attempts to clarify the nature of association of proteins with lipids in cellular membranes include determination of the ease with which various lipids could be extracted from (usually dried) tissues or from isolated membrane preparations. For example, washing human erythrocytes ten times in physiological saline at 37°C removes about 25% of their lipoproteins, 18% of their cholesterol, and 5% of their phospholipids.[965] Further washing produces little additional effect and these changes are accompanied by very little hemolysis. Similarly, a large variety of membrane-associated proteins can be loosened by osmotic shock.[966] After such treatments the membrane usually appears to be leaky, even when reconstituted, and some of the functional characteristics, such as ionic specificity, facilitated transport, active transport, and excitability are lost. As suggested by such studies, at least part of the protein is loosely bound to the membranes and can be separated by washing with saline or by osmotic shock; that is, by relatively *mild* treatment. The proteins released by such mild methods show a number of characteristics implying that they are localized on the membrane surface. These are:

1. Most of the enzymes and proteins interact with extracellular substrates.

2. Histochemical localization shows their positioning on the surface of the membrane.

3. Some of these enzymes and proteins react with corresponding antisera, and, in contrast, other membrane-bound proteins do not.

4. Surfaces of several membranes contain a system of fibrils. It is generally believed that these fibrils lend the membrane deformability and elasticity. Such an assertion is supported by morphological observations of the vesicular fragments that remain subsequent to low-ionic-strength dialysis and protein solubilization. When examined under the electron microscope, thin sections of these vesicular structures, which have lost their membrane fibrils, sometimes exhibit an open-ended broken membrane.

Thus it is believed that most of the proteins released by mild treatment may be bound to the membrane by electrostatic forces. Frequently, but not always, such proteins may be part of the membrane coat and thus not a part of the plasma membrane per se.[967]

Another significant part of membrane proteins is rather strongly bound to lipids *or* the lipid is strongly bound to membrane proteins, whichever way one chooses to describe it. Through extraction procedures involving organic solvents, the membrane components can be separated into loosely bound, weakly bound, and strongly bound lipid fractions (Fig. 9-2). This implies that several types of lipid–protein interactions exist in biomembranes. It is reasonable to assume that molecular expansion within the membrane during *solubilization* and/or *homogenization* occurs more readily under conditions in which the electrostatic interactions of the membrane constituents with the solvent tend to counteract the internal cohesive forces. The most important factors decreasing electrostatic interactions with the solvent are low ionic strength, low concentration of divalent cations, and high pH. Therefore it is not surprising that the size and morphology of the membrane fragments depend not only on the mechanical means used to disrupt the cell but also on the nature of the suspension medium.

Extraction of lipids and solubilization of membrane proteins by certain organic solvents,[1168] detergents,[1169] or by some other method[1170] which stabilize ionic interactions constitutes strong evidence for domination of hydrophobic and nonpolar bonding between most of these membrane components. Since the polarization forces governing such interactions fall off very rapidly with distance and since they are weak, such forces would be unimportant unless some protein interdigitates through the bimolecular lipid layer. As a rule, such interactions are characterized by the lack of chemical and major conformational changes; therefore, the driving force for hydrophobic interactions

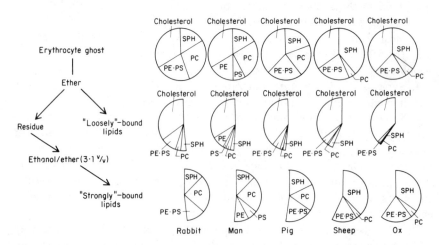

Fig. 9-2. Scheme for the extraction of loosely and strongly bound lipids from erythrocyte ghosts and the relative distribution over the corresponding fractions, presented for a series of five mammalian species. Abbreviations: SPH = sphingomyelin; PC = phosphatidylcholine; PS = phosphatidylserine; PE = ethanolamine containing phosphoglycerides.[982]

between proteins and lipids may be derived from several physicochemical factors: (a) Weak interactions between apolar residues. (b) Low dielectric constant of the hydrophobic environment. Hydrogen bonding and ionic interactions are strengthened by exclusion of water. (c) The ordering of water around exposed apolar groups causing the unfavorable loss of entropy.

Such factors were considered at some length in Chapter 2 and 3, and have been fully discussed elsewhere.[968-979] However, such interactions are not expected to be specific. In the following paragraphs we shall elaborate on some of these aspects.

Both the model lipid systems and biomembranes show lamellar structure in the electron micrographs. Furthermore, x-ray diffraction studies suggest that the distance between polymethylene chains is in the fairly narrow range of 4.5 to 4.8 Å. As was emphasized in Chapters 6 through 8 most of the functional characteristics are ascribable to membrane-bound proteins and that in these membranes nonpolar interactions are implicated in the organization of tertiary structures of catalytic proteins. Such interactions have several specific features so as to enhance the catalytic role of the bound proteins.[979,980] They may organize tertiary structures of catalytic proteins, may provide a nonpolar environment in which polar reactions may proceed more efficiently, may assist binding of apolar ligands, may bring various active sites in favorable tertiary and quaternary conformation, either directly[1171] or as allosteric effectors,[1172] and may assist in *steering* the orbitals of the substrate along the path which takes advantage of this directional preference.[981] Thus, it seems probable that the nonpolar chains of phospholipids play some functional besides the structural role. It must also be emphasized that lipid and ion-binding requirements for catalytic or functional proteins are a corollary of anisotropic orientation of these proteins in the membrane plane. All these aspects are particularly reflected in the absolute requirement of specific phospholipids for various membrane-associated proteins (Table 9-3). These studies do not constitute proof in themselves, but they are very suggestive of participation of lipids in some critical catalytic processes, as implied by the following observations:

1. Dissociation and perturbation of lipid-protein matrix by hydrolytic enzymes alters catalytic behavior.[982,983,1232]

2. Reconstitution of active lipoproteins by admixture of the protein with suitable lipids or lipid mixtures.

3. Extraction of lipids by solvents and consequent inactivation of the proteins.[983a]

4. Manipulation of physicochemical factors implicated in lipid–protein binding such as temperature, pH, salt concentration, solubility characteristics, changes in absorption spectra, and modification of specific protein groups by chemical treatment.

It may be pertinent to note here that the effects exhibited by lipids on these

TABLE 9-3 INTERACTION OF LIPIDS WITH FUNCTIONAL PROTEINS

Protein	Source	Lipid Needed	Ref.
Electron transfer chain	Mitochondria	Unsaturated lipid	984
NADH-cytochrome c reductase	Mitochondria	Lecithin + lysolecithin	985
β-Hydroxy butyric dehydrogenase	Mitochondria	Lecithin	986
Succinooxidase activity	Mitochondria	CoQ and purified PL	987
Malateoxidase activity	M. phlei	Phospholipids	988
Cytochrome c oxidase	Mitochondria	Cardiolipin	989
Oxidative phosphorylation	Mitochondria	Intact membrane structure	990
Na + K-ATPase	Red cells, etc.	Phosphatidyl serine, dialkyl phosphate	685, 686
ATPase activity	Beef heart mitochondria	Crude phospholipid to induce rutamycin sensitivity	991
Glucose-6-phosphatase	Rat liver	Lecithin + lysolecithin	992
Phosphatase activity	Rat kidney	Cetyltrimethylammonium bromide	993
ATPase activity	Muscle (rat, rabbit)	Lecithin, detergents	994
5'-nucleotidase	Rat liver	Sphingomyelin	995
Lipase Y(−)	Mucor javanicus	Phosphatidyl inositol	996
Protoheme ferrolyase	Erythrocyte (duck, chicken)	Crude lipids from egg yolk	997
Fatty acid desaturase	Liver (hen)	Crude lipids	998
	Liver (rat)	Crude lipids	999
	Yeast	Double bond in 9 position of lipid	1000
Cyclopropane synthatase	Clostridium butyricum	Anionic amphipaths	1001
Glucosyl diglyceride synthatase	Mycoplasma laidlawii	Fatty acid or detergent	1002
Mannosyl diglyceride synthatase	Micrococcus lysodeikticus	Anionic amphipaths	1003
Aminoacid-transfer RNA ligase	Mammalian tissues	Cholesteryl 14-methyl hexadecanoate	1005
Transport systems:			
Lactose and other related sugars	E. coli	Unsaturated fatty acids	1004
Lysine	Neurospora crassa	Lecithin	961
Ca	Muscle (rat, rabbit)	Lecithin and related lipids	994
Antigens	Red blood cells (cattle)	Lipid solvents inactivate	1006

For other enzymes see Ref. 1007. The lipids have also been implicated in prothrombic activation,[1008] growth of E. coli,[1009] bacterial respiration,[1010] structural integrity of an arbovirus[1011] and bacteriophage,[1287] interaction with proteins as membrane components,[1012] solubilization of cytochrome c,[1013] histamine release[1285] and possibly nerve excitation.[1014]

various proteins may be the consequence of various different types of changes; thus, nonspecific effects of lipids may cause changes in charge profiles of the macromolecule and thus the accessibility of the substrate. Both specific and nonspecific lipid activation of proteins may result from conformational alteration in the enzyme by direct interaction of lipid with protein.

As collected in Table 9-3, quite a few systems involve lipid–protein interactions of a varying degree of specificity and sophistication. A particularly striking example of the specificity of such interactions is provided by the Na+ K-activated ATPase. Treatments which solubilize these proteins result in the loss of ATPase activity, which may or may not be reversible. Thus, deoxycholate treatment abolishes the ATPase activity from a bovine cerebral cortex preparation, and the inactivated preparation may be reactivated by a variety of amphipaths[685] including dialkyl phosphates.[686] Among these, didodecyl phosphate is quite interesting; at low concentrations it increases enzymic activity but then becomes inhibitory at higher concentrations. Similar effects have been observed with reconstitution studies of the cationic pump (p. 272) on a black lipid membrane containing traces of didodecyl phosphate. Phospholipid specificity has also been noted in the reconstitution of UDP:galactose-lipopolysaccharide-α, 3-galactosyl transferase system in monolayers.[1288]

The participation of lipids in such catalytic and structural functions can hardly be overemphasized while considering various examples collected in Table 9-3. Of particular interest concerning the role of lipids in catalytic activity of proteins is the effect of various inhibitors. For example, rutamycin and dicyclohexylcarbodiimide do not act on the factor F_1 (see p.294) itself but indirectly, e.g., by preventing the activation of ATPase activity of these particles by phospholipids.[994] It is thus conceivable that inhibitors such as oligomycin, rutamycin, ouabain, etc., may act by stimulation of proteins by phospholipids. In converse, the effect of hormones and other regulators may be stimulatory on some similar system. It is apparent, moreover, that a requirement for a lipid for catalytic activity need not imply its action at the active site (see Ref. 979 and 1232 for some related aspects).

It has been suggested that the occurrence of hydrophobic interactions between lipids and proteins may impose considerable constraints on tertiary and quaternary structure of proteins, which may be reflected in their amino acid composition.[1015] In fact, it has been noted that the amino acid composition of the soluble and the membranous proteins differs significantly (Table 9-4). The soluble proteins contain more charged and hydrophilic amino acids than do the membranous proteins. Also, the ratio of hydrophilic to apolar amino acids is markedly higher in the soluble proteins. The characteristics of the constituent amino acids of membranous proteins are essential for their adaptation to the apolar environment (for example, see p. 203 for incorporation of macrocyclic compounds into BLM). The relatively apolar proteins can interact with lipids in

TABLE 9-4 AMINO ACID COMPOSITION OF SOLUBLE AND MEMBRANE LIPOPROTEINS[1015]

Protein	Species	Organ	Hydrophilic (Mole %)	Apolar (Mole %)	H/A Ratio	Total Charged (Mole %)	Small (Mole %)
Soluble lipoproteins							
High-density (HDL-3)	Human	Plasma	53.2	24.8	2.1	39.8	11.8
Low-density Sf 6	Human	Plasma	50.0	31.1	1.6	33.0	10.9
Low-density Sf 44	Human	Plasma	52.5	27.9	1.9	35.2	11.8
Chylomicron	Human	Plasma	50.3	24.3	2.1	35.6	13.6
Vitellenin	Chicken	Yolk	48.4	30.0	1.6	34.6	12.2
α-Vitellin	Chicken	Yolk	45.9	28.7	1.6	31.6	13.4
Fat globule, soluble	Cow	Milk	48.1	26.3	1.8	25.8	13.3
Chylomicron, soluble	Dog	Lymph	50.0	25.2	2.0	34.6	15.2
Mean			49.8	27.3	1.8	33.8	12.8
S.D.			2.4	2.5	0.2	4.0	1.4
S.E.M.			0.8	0.9	0.1	1.4	0.5
Membrane lipoproteins							
Chylomicron, insoluble	Dog	Lymph	46.9	26.2	1.8	25.6	15.9
Fat globule, insoluble	Cow	Milk	42.4	33.1	1.3	27.2	13.1
Nascent "chylomicron"	Rat	Intestinal mucosa	44.5	31.5	1.4	29.5	15.1
Structural	Cow	Mitochondrion	36.5	27.3	1.3	25.8	18.0
Proteolipid	Cow	Myelin	32.9	32.5	1.0	17.1	22.8
Heme-proteolipid	Cow	Microsomes	34.5	35.6	1.0	22.7	21.5
Membrane	Rat	Smooth endoplasmic reticulum	47.0	27.9	1.7	35.6	17.0
Membrane lipoprotein	Human	RBC	47.4	29.8	1.6	32.6	17.0
Membrane structural	Sheep	RBC	45.7	29.3	1.6	31.7	13.6
Membrane, insoluble	Cow	Mitochondrion	43.7	28.7	1.5	30.3	17.8
Structural	Spinach	Chloroplast	51.2	27.7	1.8	39.0	13.3
Fl-fraction	Beet	Chloroplast	39.6	29.1	1.4	28.9	18.1
Mean			42.7	29.9	1.4	28.8	16.9
S.D.			5.7	2.8	0.3	5.8	3.0
S.E.M.			1.6	0.8	0.1	1.7	0.9

many ways which may result in aggregation (quaternary structure) in which the polar groups are buried inside the aggregate. Probably such a process is involved in the strong propensity for aggregation of a lipoprotein isolated from human erythrocytes.[1016]

Such observations in general suggest that the proteins can, and in all probability do, interact with hydrophobic chains of lipids. The aggregate resulting from such an interaction would have a surface charge profile which is completely the reverse of that for globular proteins in aqueous solution. Thus, a lipid–protein complex formed by hydrophobic interactions can have both a hydrophobic and a hydrophilic exterior. Identification of the conformational states of these lipid–protein complexes, has met with only limited success in spite of some claims of success.[1017,1018] It has been shown that ORD and CD data is complicated by the presence of artifacts common to all optically active particulate systems.[1019] However, after elaborate series of corrections it may be "expected" that membranes may contain as much as 50% protein in α-helical conformation. Thus caution is necessary in ascribing functional significance to ORD and CD data. Similarly, the story of "structural proteins" and "core proteins" with very attractive properties as membrane components from various biomembranes also appears to be in trouble.[1020]

Before moving on to the discussion of mode of organization of lipid–protein complexes, it may be pertinent to note some unusual bonding modalities between lipid and proteins in various biological systems. One such interaction has been noted in the envelopes of *Halobacterium halobium*.[1021] Solubility properties, dissociation studies, and amino acid analysis indicate that, in the envelopes, the lipids are bound to at least two types of proteins, say I and II. The behavior of protein I suggests that the lipids are bound hydrophobically to an apolar part of the protein. Protein II, after extraction with chloroform–methanol, is soluble in $8M$ urea, dodecyl sulfate, water, and acetic acid, from which it can be recovered. However, from a cresol solution the protein complex precipitates on addition of chloroform–methanol, thus indicating that protein II is attached to the lipids by polar bonds. The usual tests for the presence of polar interactions between the two are negative. On the basis of the titration curves of free and complexed protein and various other observations a tentative structure has been proposed for this lipoprotein complex (Fig. 9-3). It consists of tetradentate intermolecular chelate between protein, magnesium, and the lipid head group together with ionic linkages between terminal phosphate and the arginyl group to maintain electroneutrality.

Yet another unusual type of lipid–protein interaction is that between thiol groups of proteins and lipids.[1022] Using brain microsomes as a model membrane system, it has been demonstrated that membrane permeability is increased and ion discrimination is decreased by two classes of compounds: thiol binding agents and peroxidizing agents such as ion plus ascorbic acid. As an ex-

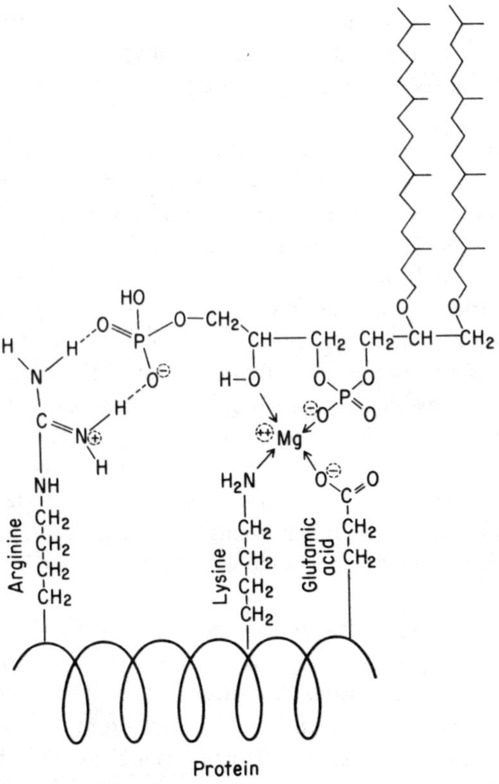

Fig. 9-3. Possible binding between lipid and protein in the envelope of Halobacterium halobium.[1021]

planation, it has been suggested that the formation of thiol–olefin linkages takes place and that, consequently, the membrane characteristics are altered. However, it is possible that the thiol-containing proteins may be involved in transport processes; the experimental data does not rule out this possibility.[1174]

Recently, a possible function of phospholipids in amino acid transport, and in protein and membrane biosynthesis has been suggested (see Tria and Barnabei in Ref. 969). A considerable amount of experimental evidence has been obtained to the effect that covalently bound lipid–protein complexes may be functional constituents of protein synthesizing membranes.[558,1023] The experimental evidence is circumstantial; however, it is a possibility that may find rationale in some situations which warrant involvement of membranes in group-transfer processes such as activation of amino acids for protein synthesis.

Before closing this discussion of lipoprotein complexes, it must be noted that not all the lipid of membranes is associated with proteins, and that there is considerable difference in the binding of different lipid molecules which form the membrane. This is further supported by exchange studies[1024] in which different lipoproteins exchange their lipid moieties with lipid in solution. Thus, individual serum lipoproteins containing ^{32}P-labelled phospholipids have been isolated which after incubation with nonradioactive serum lipoproteins of different structure exchange the lipids.[1025,1026] A similar lipid exchange has been noted in low- and high-density serum lipoproteins,[1027] in bovine serum albumin,[1028] in liver,[1029] brain[1313] and red blood cells,[1030] and from ER to mitochondrial membranes.[1175] Furthermore, these various serum lipoproteins can exchange their lipid with that of the liver cell or erythrocyte membrane. In all these cases, the exchange appears to be molecular and the rate of exchange is sometimes a function of the lipid already present in the lipoproteins. The time for half exchange is usually much smaller than the time for the turnover of phospholipid; thus, in adult rat brain the time of half-exchange for most of the ^{31}P-labelled phospholipids is approximately 30 min; in contrast, the turnover rate for phosphoinositides is 11 days, 17 days for phosphatidylethanolamine, and 40 days for sphingophosphatides.[1031] It is also pertinent to note here that a protein fraction isolated from rat liver microsomes greatly increases exchange of phospholipids between mitochondria and microsomes. This exchange between biomembrane fractions is temperature dependent, and the extent of exchange is a function of the amount of the protein present in the medium.[1032]

These results can be interpreted in several ways. One possible explanation of these phenomena is that these exchange protein fractions release a *carrier* phospholipid from membranes by proteolytic and/or detergent action. The possibility of weakening of lipid–lipid bonds, or the possibility of formation of collision complexes between two exchanging species or some of their components is more likely, although not supported as yet by conclusive experiments.

MEMBRANE MODELS

Subunit Structure and Organization

On the basis of the discussion of functional and morphological features of biomembranes as developed so far, it is clear that the membrane considered only with phospholipids as matchsticks joined tail-to-tail is not only inadequate but also inconsistent with a variety of experimental observations. The role of proteins is dominant when one considers functional aspects of biomembranes. Lipid requirements for catalytic and related functions and substrate specificity for transport processes place further constraints on membrane structure, a corollary of which is an anisotropic orientation of lipid–protein complexes in the structural organization. Thus, interdigitated lipid–protein structures may occur in specific regions of matchstick ensembles or BLM, or be more randomly distributed, or may be so densely distributed that they give an appearance of an hexagonal array. The concept of mosaic membrane was first suggested by Hober.[1188] Direct evidence for incorporation of proteins into the lipid matrix of biomembranes has been provided in electron microscopy by the freeze-etching technique.*[1033] The results suggest that particles are embedded within the membrane matrix but may protrude, giving the membrane surface a roughened appearance. The protusions are easily distinguished from adhering nonmembrane particles, as the latter can be removed by EDTA washing.

A variety of cells have been examined by the freeze-etching technique, and all were found to have a rather roughened appearance. The etched surfaces of yeast, red cells, chloroplasts, and neurons appear granular with particles dispersed more or less irregularly across the membrane. Not only cross-sectional and oblique views but both sides of the membranes can be visualized where the fracture line has cut across the membrane to expose the outer surface. This fracture line may be related to the differential chemical composition of the inner and outer membrane surfaces.[1033] The structural continuities have not been identified with any degree of certainty; however, in all probability they are enzyme assemblies, ion pumps, ion and molecular carriérs, and special ion gating mechanisms responsible for various specialized functions, and thus, expressions of specificity of membrane functions. This is further supported by the observation that fractured faces of pure lamellar phases and most of the plasma membranes, with the possible exception of mitochondrial membranes, have a planar structure in the interior of the membrane.[1034] Any globular substructure appears on or near either surface of the membrane.

Details in membranes from rat liver, rabbit liver, red blood cells, beef heart, onion stem, and spinach chloroplasts have been described as sheaths, fringes, globules, granules, pebbles, and bumps while interpreting the electron-microscopic data on negatively stained preparations.[1035] To those untrained in electron microscopy, it is difficult to decide how much evidence these images should command. This is certainly a matter that will provide interesting historical reading some years from now.

To support the idea of subunit structure in biomembranes, analytical evidence has also been relegated. As pointed out earlier, a large number of lipid–protein complexes have their constituents rather strongly bound together. Also, these complexes show a strong tendency to organize themselves spontaneously into vesicular structures which closely resemble, at least in form, the membrane

*The freeze-etching technique consists of (a) rapid freezing of the sample, (b) sectioning or fracturing at low temperature in vacuum, (c) coating the sectioned sample with a thin layer of platinum plus carbon, and examining the platinum–carbon imprint with the electron microscope. This sectioning procedure results in fragmentation along lines which reveal microanatomical structural organization. Other advantages of the technique include the lack of dehydration, chemical staining or high-temperature steps, and visualization of the functioning structure as inferred from the survival of yeast cells after platinum–carbon coating.

structure from which these lipoprotein complexes originate. As collected in Table 9-5 quite a few biological membrane systems have been dispersed by relatively mild treatments. The dispersed preparations (usually in electron micrographs) seem to consist of relatively stable lipoprotein complexes of estimated dimensions, usually $60 \times 60 \times 40$Å or so. When the dispersing agent is removed from these preparations they usually aggregate to reconstitute membrane-like structures. A small amount of divalent cations usually facilitate this process. To a first approximation, the results of disaggregation and reaggregation studies may be summarized as follows:

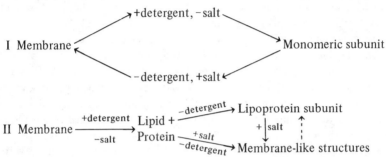

The first scheme assumes the integrity of the lipoprotein complex during aggregation and disaggregation. Several experiments contradict this assumption. Comparison of the original membranes of *M. laidlauii* with reaggregated products formed after removal of detergents show that, although they were similar chemi-

TABLE 9-5 LIPOPROTEIN SUBUNITS CHARACTERIZED IN BIOMEMBRANES

Membrane Source	Dispersion Method	Method of Identification	Dimensions of Subunit	Ref.
Mitochondria	Detergent SDS, DOC, sonication	EM	$90 \times 60 \times 50$ Å	1036
Myelin	Lysolecithin treatment	Electrophoresis, sedimentation; EM	$60 \times 40 \times 40$ Å	1037
Mycoplasma laidlawii	SDS	EM, sedimentation	70–150 Å diam.	1038–1040
Micrococcus lysodeikticus	SDS	Sedimentation		
Hydrogenomonas facilis	Mechanical	Disc electrophoresis[a]	Quasicrystalline solid	1041
Neurospora crassa	Mechanical	Disc electrophoresis	Quasicrystalline solid	1041
Retinal receptor disc membrane (frog)	Native	x ray (also EM)	40–50 Å diam.	1042
Dictyostelium discoideum	Detergent, urea, AcOH	Sedimentation; reconst.	70–100 Å diam.	1043
Pseudomonas aeruginosa	High–pH EDTA	Sedimentation	mol. wt. 10^5	1044
Mouse urinary bladder	Negatively stained	EM	50 Å diam.	1045
Axolemma	Negative staining	EM	40–50 Å diam.	1046
	Native	Freeze etching	40–50 Å diam.	1047
Mauther cell synapse (goldfish brain)	Negative staining	EM	95 Å diam.	1178

[a]Although disc electrophoresis shows heterogeneity, it is attributed to aggregation.

cally (composition, etc.) and morphologically (however, see Refs. 1038–1040), the former are sensitive to dispersion by urea and guanidine whereas the latter are disrupted preferentially by anionic detergents. Results of this kind imply that hydrogen bonding plays an important role in stabilizing the in vivo membrane, but that hydrophobic bonding is the dominant stabilizing factor in the re-aggregated product. Thus, irreversible changes may have modified the lipid–protein and/or protein–protein interactions during the disaggregation–reaggregation cycle; it can not be assumed that the lipoprotein subunit obtained upon dissolution of the membrane represents an in vivo subunit of the original membrane. Similarly, arguments may be forwarded to counteract claims of subunits in negatively stained preparations. See also Ref. 1169 and 1321.

Disaggregated subunits may be further dissociated into several other components (Table 9-6). These solubilized components can interact in a variety of ways to produce a number of different reassociated structures.[1187] Only under a limited set of conditions does the *majority* of the disaggregated material re-aggregate to form assemblies which closely resemble the original membrane; however, in not a single case has it been possible to reconstitute a functional membrane (however, see Chapter 7 for instances of limited success).

In summary, evidence of this kind indicates that only a simplistic view of various functional apparatus may be adopted in terms of vesicles and even less so for subunits and knobs. However, the functional membrane in these processes is not only expected to possess a mechanical adhesional role, but a more active one, as implied by their vectorial or bulk anisotropic aspects. Such an asymmetry has its origin in the marked and consistent asymmetry of distribution of particles between the two fractured faces of a membrane as observed in freeze-etched

TABLE 9-6 "STRUCTURAL PROTEINS" FROM BIOMEMBRANES[a]

Protein	Source	Molecular Weight	Remarks	Ref.
"Structural protein"	Mitochondria, bacteria, Rbc, myelin, etc.	22,000–28,000	Hydrophobic; have some role in restoring membrane function[b]	1020 1048
"Structural protein"	Mitochondria (*N. crassa*)	3500 ± 400	"Structural protein" appears be homo- or hetero-multimeric	1049
"Miniprotein"	Rbc, mitochondria, retinal rods	4000–6000	Suggested as the basic unit for membrane structure	1051
"Structural protein"	Mitochondria (Beef)	50,000–70,000		1052
Sarcotubular protein	Skeletal muscle	17,000	May be 80% of the total protein; exists as aggregate	1180

[a]For a review see Ref. 11790. For experimental results indicating heterogeneity of these proteins see Refs. 1050 and 1049. Thus, at best these "structural proteins" represent a class of proteins.
[b]May be inactivated enzymes (1050).

specimens; most of the membrane particles appear on the concave fracture faces, whereas fewer particles appear on convex faces.[1033] Although it is possible it is hard to rationalize that in a model implying identical subunits in its structure each element is asymmetrical and constituted of other weakly bound fragments. Furthermore asymmetry of the secondary and tertiary folding of the polypeptide chain leads to asymmetry of recognition site. For subunits which are not chemically identical, such equivalence of association between chemically different units, i.e., "pseudo-identity," is rather exceptional in other biological macromolecules. Furthermore, experimental evidence in favor of repeating subunit model is tenuous and is based on empirical correlations such as those briefly summarized earlier in this section. Also, phase transition studies, effect of uncouplers, permeability characteristics of biomembranes, and other physicochemical studies (for reviews see Refs. 37k, 1054–1056), all taken together seem to suggest that there are extended regions of BLM-type structures in biomembranes.

Thus the pattern emerging from such diverse studies and from a number of theoretical models[1053–1056,1195,1289] may be summarized as follows:

1. The biological membrane is a coherent and condensed hydrophobic phase separating hydrophilic phases. The membrane phase originates from the noncovalent assembly of a variety of lipids, proteins, and polysaccharides.

2. Membranes are infinite unlimited structures in two dimensions; in contrast, multicomponent enzymes and other proteins are made up of a finite and well-defined number of repeating units.

3. To a first approximation, in analogy with a three-dimensional phase, one may find in membranes two-dimensional sols, gels, crystals, clathrates, microscopical biological anatomies, and the like. Additional complexities are introduced if the local viscosities, elasticities, or changes of state that control displacement patterns in the plane of the membrane are sensitive to potential differences across the membrane.

4. The membrane proteins are bound hydrophobically to the lipids, such that they may interdigitate through the hydrophobic region of the lipid bilayer. These proteins may be packed in the form of subunit assemblies.

5. The whole membrane matrix may be constituted of distinctive (perhaps identical* or perhaps different) repeating units with respect to size, composition, function, and geometric shape. These units are packed in their assemblies such that their axes lie nearly normal to the membrane surface.

6. It is possible that the amino acid residues may show a certain degree of specificity with regard to interdigitation with lipid chains.

7. The subunit assembly may or may not give a long-range order to the membrane.

---

*It is topologically impossible to build a uniformly curved surface with, say, planar hexagons alone. This will not allow any flexibility in the sheet and at some point the deformation energy may exceed the bond energy, and the sheet will become unstable. Flexibility can however be granted by introducing pentagons.

8. The state of the aggregated complex and the degree of aggregation may be strongly dependent upon the presence of other components and the physical state of the system: pH, ionic strength, temperature, divalent ion concentration.

9. It seems likely that the organization of the lipoprotein microstructure will depend on the same factors of closeness of fit, multiple attachment and matching of polarity, etc., that are recognized to determine the subunit aggregation in proteins.

10. The intrinsic asymmetry of the membrane could arise either from asymmetry of the orientation of the subunits or from the asymmetric distribution of functional units on the interfaces of the membrane. The role of an asymmetric environment in this respect is yet little understood but cannot be neglected.

11. Cooperative allosteric interactions between various units in the membrane matrix (*cis* or lateral cooperativity) appears to be improbable as a general characteristic, though not impossible. Such cooperativity may however be involved in amplifying properties of receptor membranes. Allosteric transitions within the membrane subunit perpendicular to the plane of the membrane (*trans* or radial cooperativity) is probably a fairly widespread phenomenon in membranes.

12. Highly specific functions analogous to enzyme–substrate interaction, such as specific ion binding, etc., may be the function of specific units of the membrane, and may be governed by factors similar to those governing catalytic processes in proteins.

13. It follows that the membrane structural unit which carries at least one receptor site may exist in several conformations of differing affinity for ligands in reversible equilibrium, thus introducing a possibility of cooperative (alosteric) interactions.

14. The role of surface coat and related material in relation to structure and function of plasma membrane is largely unknown. However, it is quite likely that surface structures may regulate and modulate the properties of plasma membrane.

15. It must be noted that the terms lattice, units, cooperativity, and asymmetry have been used rather loosely here, and must be treated in a generalized context.

Thus a discussion of membrane phenomenology can be best recapitulated by an excerpt from *Little Gidding* (by T. S. Eliot):

> And the end of all our exploring
> Will be to arrive where we started;
> And to know the place for the first time.

Personally, I cannot help feeling that we are not there yet. However, the subject matter we have discussed is in a highly interesting state. The newer methods of biochemistry are starting to pay off, and the systematic attempts now in progress to extract and characterize the membrane components are beginning to impose some sort of pattern which can be recognized with and related to a diverse array of membrane function.

References

1. Oparin, A. I., *The Origin of Life on the Earth*, Academic Press, Inc., New York, 1957.
2. Chambers, R., and E. L. Chambers, *Explorations into the Nature of the Living Cell*, Harvard University Press, Cambridge, Mass., 1961.
3. Revel, J.-P., and S. Ito, in *The Specificity of Cell Surfaces*, Eds., B. D. Davis and L. Warren, p. 211, Prentice-Hall, Englewood Cliffs, N.J., 1967.
4. Keeton, W. T., *Biological Science*, W. W. Norton & Co. Inc., New York, 1967.
5. Ponder, E., in *The Cell*, Vol. 2, Eds. J. Brachet and A. E. Mirsky, p. 1, Academic Press, 1961.
6. Ling, G. N., *A Physical Theory of the Living State, the Association-Induction Hypothesis*, Blaisdell Publishing Co., New York, 1962.
7. Ling, G. N., *Intern. Rev. Cytology*, **26**, 1 (1969).
8. Troshin, A. S., *Problems of Cell Permeability*, Pergamon Press, New York, 1966.
9. Kurella, G. A., in *Membrane Transport and Metabolism*, Eds. Kleinzeller and A. Kotyk, Academic Press and Czech. Academy of Sciences, 1961.
10. Chambers, R., *J. Gen. Physiol.*, **5**, 189 (1922).
11. Overton E., *Vjschr. Naturforsch. Ges. Zurich*, **41**, 383 (1896).
12. Gorter, E., and F. Grendel, *J. Exptl. Med.*, **41**, 439 (1925).
13. Hill, J. G., A. Kuksis, and J. M. R. Beveridge, *J. Am. Oil Chemist's Soc.*, **41**, 393 (1964).
14. Schmitt, F. O., R. S. Bear, and E. Ponder, *J. Cell. Comp. Physiol.*, **9**, 89 (1936).
15. Mudd, S., and E. B. H. Mudd, *J. Gen. Physiol.*, **14**, 733 (1931).
16. Mudd, S., and E. B. H. Mudd, *J. Exp. Med.*, **43**, 127 (1926).

17. Dawson, J. A., and M. Balkin, *Biol. Bull.,* **56**, 80 (1929).
18. Chambers, R., *Biol. Bull.,* **69**, 331 (1935).
19. Hober, R., *Plugers Arch.,* **133**, 237 (1910).
20. Hober, R., *Plugers Arch.,* **148**, 189 (1912).
21. Frick, H., *J. Gen. Physiol.,* **9**, 137 (1925).
22. McClendon, J. F., *J. Biol. Chem.,* **69**, 733 (1926).
23. Cole, K. S., *Trans. Faraday Soc.,* **33**, 966 (1937).
24. Collander, R., and H. Barlund, *Acta Botanica Fennica,* **11**, 1 (1933).
25. Overton, E., *Vjschr. Naturforsch. Ges. Zurich,* **44**, 88 (1899).
26. Harvey, E. N., *Biol. Bull.,* **60**, 67 (1931).
27. Harvey, E. N., *Biol. Bull.,* **61**, 273 (1931).
28. Cole, K. S., *J. Gen. Physiol.,* **12**, 37 (1928).
29. Danielli, J. F., and E. N. Harvey, *J. Cell Comp. Physiol.,* **5**, 483 (1935).
30. Danielli, J. F., Cold Spring Harbor Symposia on Quantitative Biology, **6**, 190 (1938).
31. Danielli, J. F., and H. Davson, *J. Cell. Comp. Physiol.,* **5**, 495 (1935).
32. Robertson, J. D., *Progress in Biophysics,* 343 (1960).
33. Robertson, J. D., *Protoplasma,* **63**, 218 (1967).
34. Sjostrand, F. S., *Radiation Research* Supplement, **2**, 349 (1960)
35. Schmitt, F. O., R. S. Bear, and K. J. Palmer, *J. Cell. Comp. Physiol.,* **18**, 31 (1941).
36. Finean, J. B., *Engstrom-Finean Biological Ultrastructure,* Academic Press, 1967.
37. Finean, J. B., *Progress in Biophysics and Molecular Biology,* **16**, 143 (1966).
37a. Stein, W. D., and J. F. Danielli, *Disc. Faraday Soc.,* **21**, 238 (1956).
37b. Staehelin, L. A., *J. Ultrastruc. Res.,* **22**, 326 (1968).
37c. Finean, J. B., in *Biochemical Problems of Lipids,* p. 127, Butterworths, London (1956).
37d. Muhlethaler, K., H. Moor, and J. W. Szarkowski, *Planta,* **67**, 305 (1966).
37e. Gross, L., *J. Theoret. Biol.,* **15**, 298 (1967).
37f. Sjostrand, F. S., *J. Ultrastruc. Res.,* **9**, 340 (1963).
37g. Frey-Wyssling, A., *Macromolecules in Cell Structure,* Harvard University Press, 1956.
37h. Lenard, J., and S. J. Singer, *Proc. Natl. Acad. Sci.,* **56**, 1828 (1966); also Wallach, D. F. H., and P. H. Zahler, *ibid,* **56**, 1552 (1966).
37i. Vanderkooi, G., and D. E. Green, *Proc. Natl. Acad. Sci.,* **66**, 615 (1970); also see earlier papers cited therein and Ref. 45.
37j. Benson, A. A., *J. Amer. Oil Chemist's Soc.,* **43**, 265 (1966).
37k. Dewey, M. M., and L. Barr, *Current Topics in Membranes and Transport,* **1**, 1 (1970).
37m. Vandenheuvel, F. A., *J. Amer. Oil Chemist's Soc.,* **40**, 455 (1963); **43**, 258 (1966).
38. Korn, E. D., *Science,* **153**, 1491 (1966).
39. Moretz, R. C., C. K. Akers, and D. F. Parsons, *Biochim. Biophys. Acta.,* **193**, 1, 12 (1969).
40. Stoeckenius, W., J. H. Schulman, and L. M. Prince, *Kolloid-Z.,* **169**, 170 (1960).

41. Branton, D., *Proc. Natl. Acad. Sci. (U.S.)*, **55**, 1048 (1966). See also 1033.
42. Kavanau, J. L., *Structure and Function in Biological Membranes*, Vol. I, Holden-Day Inc., San Francisco (1965).
43. *Ibid.*, Vol. II (1965).
44. Lucy, J. A., and A. M. Glauert, *J. Mol. Biol.*, **8**, 727 (1964).
45. Green, D. E., and H. Baum, *Energy and the Mitochondrion*, Academic Press, New York (1970).
46. Cuthbert, A. W., *Pharmacol. Rev.*, **19**, 59 (1967).
47. Watkins, J. C., *J. Teoret. Biol.*, **9**, 37 (1965).
48. Dean, R. B., *Nature*, **144**, 32 (1939).
49. Saunders, L., *J. Pharm. Pharmacol.*, **15**, 155 (1963).
50. Gaines, G. L., *Insoluble Monolayers at Liquid-Gas Interfaces*, Interscience Publishers, New York (1966) p. 257.
51. Villegas, R., F. V. Barnola, and G. Camejo, *J. Gen. Physiol.*, **55**, 548 (1970).
52. Shah, D. O., and J. H. Schulman, *J. Lipid Res.*, **8**, 215, 227 (1967).
53. Shimojo, T., and T. Ohnishi, *J. Biochem.*, **61**, 89 (1967).
54. Joos, P., *Chem. Phys. Lipids.*, **4**, 162 (1970).
55. Tobias, J. M., D. P. Agin, and R. Pawlowski, *J. Gen. Physiol.*, **45**, 989 (1962).
56. Mysels, K. J., K. Shinoda, and S. Frankel, *Soap Films*, Pergamon Press, New York, (1959).
57. Overbeek, J. Th. G., *J. Phys. Chem.*, **64**, 1178 (1960).
58. Koelzer, P. P., and H. Steinnach, *Arzneim-Forsch.*, **17**, 507 (1967).
59. Rosenberg, M. D., *Protoplasma*, **63**, 168 (1967).
60. Tobias, J. M., D. P. Agin, and R. Pawlowski, *Circulation*, **26**, 1146 (1962). Also see Kobatake Y., A. Irimajiri, and N. Matsumoto, *Biophys. J.*, **10**, 728 (1970).
61. Monnier, A. M., *J. Gen. Physiol.*, **51**, 26s (1968).
62. Rothstein, A., in *Ion Exchangers in Organic and Biochemistry*, Eds. C. Calman and T. R. E. Kressman, Interscience, New York (1957).
63. Coleman, R., and J. B. Finean, *Protoplasma*, **63**, 172 (1967).
64. Gary, K. R., *Proc. Natl. Acad. Sci. (U.S.)*, **60**, 614 (1968).
65. Friedenberg, R. M., *The Electrostatics of Biological Cell Membranes*, North-Holland Publishing Co., Amsterdam (1967).
66. Planck, M., *Ann. Physik u. Chem.*, *Neue folge*, **39**, 161 (1890); **40**, 561 (1890). Also see Cole K. S., *Physiol. Rev.*, **45**, 340 (1965).
67. Kipling, J. J., *Chem. Ind.*, 1007 (1964).
68. Chapman, D., *Introduction to Lipids*, McGraw-Hill, London (1969).
69. Davies, J. T., and E. Rideal, *Interfacial Phenomena*, Academic Press, New York (1961). Also see *Chemistry and Physics of Interfaces*, American Chemical Society Publications (1965). D. A. Kadenhead, *J. Chem. Ed.*, **49**, 152 (1972).
70. Liquid Crystals: *Proc. International Conference on Liquid Crystals* held at Kent State University, Gordan and Breach, Science Publishers, New York (1967). Also see *Advances in Chemistry*, **63**, 157 (1967).
71. Osipow, L. I., *Surface Chemistry: Theory and Industrial Applications*, Reinhold Publishing Corporation, New York (1962). Bikerman, J. J., *Sur-*

face Chemistry: Theory and Applications, Academic Press Inc., New York (1958). Adamson, A. W., *Physical Chemistry of Surfaces*, Interscience Publishers, Inc., New York (1960). Harkins, W. D., *The Physical Chemistry of Surface Films*, Reinhold Publishing Co., New York (1952).

72. Winsor, P. A., *Chem. Rev.*, **68**, 1 (1968).
73. Defay, R., and I. Prigogine with collaboration of A. Bellemans, *Surface Tension and Adsorption*, John Wiley & Sons, Inc., New York (1954).
74. Chapman, D., *Ann. N.Y. Acad. Sci.*, **137**, 745 (1966).
75. Van Deenen, L. L. M., *Ann. N.Y. Acad. Sci.*, **137**, 717 (1967). See also 960.
76. Small, D. M., *Fed. Proc.*, **29**, 1320 (1970).
77. O'Brien, J. S., *J. Theoret. Biol.*, **15**, 307 (1967).
78. Tien, H. T., and E. A. Dawidowicz, *J. Colloid Interface Sci.*, **22**, 438 (1966).
79. Mukerjee, P., *Advances in Colloid and Interface Science*, **1**, 241 (1967).
80. Dervichian, D. G., *Progress in Biophys. and Mol. Biol.*, **14**, 263 (1964). D. Ghosh, R. L. Lyman, and J. Tinoco, *Chem. Phys. Lipids*, **7**, 173 (1971).
81. Sutula, C. L., and L. S. Bartell, *J. Phys. Chem.*, **66**, 1010 (1962).
82. Adam, N. K., *The Physics and Chemistry of Surfaces*, Oxford University Press, London (1941).
83. Luzzati V., F. Reiss-Husson, and E. Rivas, T. Gulik-Krzywicki, *Ann. N.Y. Acad. Sci.*, **137**, 409 (1966).
84. Luzzati, V., and F. Husson, *J. Cell. Biol.*, **12**, 207 (1962).
85 Segerman, E., in *Surface Chemistry: Proceedings of 2nd Scandinavian Symp. on Surface Activity*, Ed. V. Runnstrom-Reio, Munksgaard, Copenhagen, p. 157 (1965).
86. Luzzati, V., T. Gulik-Krzywicki, and A. Tardieu, *Nature*, **218**, 1031 (1968). See also (954).
87. Hooke, R., Communications to the Royal Society, March 28, 1672. See T. Birch, *History of the Royal Society*, A. Millard, London, Vol. III, p. 29 (1757).
88. Corkill, J. M., J. F. Goodman, D. R. Haisman, and S. P. Harrold, *Trans. Farad. Soc.*, **57**, 821 (1961).
88a. Fergason, J. L., *Scientific American*, **211**, 77 (1964, August). G. H: Heilmeir, *ibid.*, **223**, 100 (1970).
89. Bungenberg de Jong, H. G., and J. Bonner, *Protoplasma*, **24**, 198 (1935).
90. Teorell, T., *Nature*, **137**, 994 (1936).
91. Dean, R. B., H. J. Curtis, and K. S. Cole, *Science*, **91**, 50 (1940). Danielli, J. F., *J. Cell. Comp. Physiol*, **7**, 393 (1936).
92. Setala, K., O. Ayrapaa, M. Nylholm, L. Stjernvall, and Y. Aho, *Fourth International Conference on Electron Microscopy*, Vol. II, p. 181, Springer-Verlag, Berlin, (1960).
93. Langmuir, I., and D. F. Waugh, *J. Gen. Physiol.*, **21**, 745 (1938).
94. Van Deenen, L. L. M., V. M. T. Houtsmuller, G. H. de Haas, and E. Mulder, *J. Pharm. Pharmacol.*, **14**, 429 (1962).
95. Haydon, D. A., and J. Taylor, *J. Theoret. Biol.*, **4**, 281 (1963).
96. Mueller, P., D. O. Rudin, H. T. Tien, and W. C. Wescott, *Nature*, **194**, 979 (1962).

97. Mueller, P., D. O. Rudin, H. T. Tien, and W. C. Wescott, *Circulation*, **26**, 1167 (1962).

98. Mueller, P., D. O. Rudin, H. T. Tien, and W. C. Wescott, *J. Phys. Chem.*, **67**, 534 (1963).

99. Mueller, P., D. O. Rudin, H. T. Tien, and W. C. Wescott, in *Recent Progress in Surface Science*, Vol. I, p. 379 (1964).

100. Mysels, K. J., H. F. Huisman, and R. I. Razouk, *J. Phys. Chem.*, **70**, 1339 (1966).

101. Howard, R. E., and R. M. Burton, *J. Am. Oil Chemist's Soc.*, **45**, 202 (1968). Also see Mueller, P., and D. O. Rudin, in *Laboratory Techniques in Membrane Biophysics*, Eds. H. Passow, R. Stampfli, Springer-Verlag, Berlin, p. 140 (1969).

102. Hanai, T., and D. A. Haydon, *J. Theoret. Biol.*, **11**, 370 (1966).

103. Huang, C., and T. E. Thompson, *J. Mol. Biol.*, **15**, 539 (1966).

104. Hanai, T., D. A. Haydon, and W. R. Redwood, *Ann. N.Y. Acad. Sci.*, **137**, 731 (1966).

105. Hanai, T., D. A. Haydon, and J. Taylor, *J. Theoret. Biol.*, **9**, 433 (1965).

106. Tsofina, L. M., E. A. Liberman, and A. V. Babakov, *Nature*, **212**, 681 (1966).

107. Andreoli, T. E., J. A. Bangham, and D. C. Tosteson, *J. Gen. Physiol.*, **50**, 1729 (1967).

108. Miyamoto, V. K., and T. E. Thompson, *J. Colloid Interf. Sci.*, **25**, 16 (1967).

109. Hanai, T., D. A. Haydon, and J. Taylor, *J. Theoret. Biol.*, **9**, 422 (1965).

110. Hanai, T., D. A. Haydon, and J. Taylor, *Proc. Roy. Soc. (London)*, **281A**, 377 (1964).

111. Mueller, P., and D. O. Rudin, *J. Theoret. Biol.*, **18**, 222 (1968).

112. Mueller, P., and D. O. Rudin, *Nature*, **213**, 603 (1967).

113. Conley La Force, R., *Rev. Sci. Instrum.*, **38**, 1225 (1967).

114. Folch, J., M. Lees, and G. H. S. Stanley, *J. Biol. Chem.*, **226**, 497 (1959).

115. Sotnikov, P. S., and Y. I. Melnik, *Bifiyzika*, **13**, 222 (1968) (Eng. Trans.).

116. Pagano, R., and T. E. Thompson, *Biochim. Biophys. Acta.*, **144**, 666 (1967).

117. Simons, R., *J. Mol. Biol.*, **36**, 287 (1968). Also see Van den Berg, H. J., *J. Mol. Biol.*, **12**, 290 (1965).

118. Takagi, M., K. Azuma, and U. Kishimoto, *Ann. Report Biological Works, Fac. Sci. Osaka Univ.*, **13**, 107 (1965).

119. Sessa, G., G. Weissmann, J. H. Freer, and R. Hirschhorn, *Fed. Proc.*, **27**, 248 (abstr.) (1968).

120. Tien, H. T., and A. L. Diana, *Nature*, **215**, 1199 (1967).

121. Tien, H. T., and H. P. Ting, *J. Coll. Interface Sci.*, **27**, 702 (1968).

122. Tien, H. T., S. Carbone, and E. A. Dawidowicz, *Nature*, **212**, 718 (1966).

123. Mueller, P., and D. O. Rudin, *Nature*, **217**, 713 (1969).

124. Tien, H. T., and A. L. Diana, *J. Coll. Interface Sci.*, **24**, 287 (1967).

125. Lesslauer, W., A. J. Slotboom, N. M. Postema, G. H. De Haas, and L. L. M. Van Deenen, *Biochim. Biophys. Acta.*, **150**, 306 (1968).

126. Henn, F. A., and T. E. Thompson, *J. Mol. Biol.,* **31**, 227 (1968).

127. Liberman, E. A., E. N. Mokhova, V. A. Skulachev, and V. P. Topaly, *Biofizika,* **13**, 188 (1968).

128. Bielawski, J., T. E. Thompson, and A. L. Lehninger, *Biochem. Biophys. Res. Comm.,* **24**, 948 (1966).

129. Van Zutphen, H., L. L. M. Van Deenen, and S. C. Kinsky, *Biochem. Biophys. Res. Comm.,* **22**, 393 (1966).

130. Miyamoto, V. K., *D.C. Electrical Properties of Lipid Bilayer Membranes,* Ph.D. Thesis, The Johns Hopkins University, Baltimore (1966).

131. Finkelstein, A., and A. Cass, *Nature,* **216**, 717 (1967).

132. Tien, H. T., *Nature,* **219**, 272 (1968).

133. Huemoeller, W. A., and H. T. Tien, *J. Chem. Ed.,* **47**, 469 (1970).

134. Perrins, J., *Ann. Phys.,* **10**, 160 (1918).

135. Tien, H. T., *J. Physical Chem.,* **71**, 3395 (1967).

136. Vrij, A., and J. Th. G. Overbeek, *J. Am. Chem. Soc.,* **90**, 3074 (1968).

137. Babakov, A. V., L. N. Ermishkin, and E. A. Liberman, *Nature,* **210**, 953 (1966).

138. Ohki, S., and A. Goldup, *Nature,* **217**, 458 (1968).

139. Huang, C., L. Wheeldon, and T. E. Thompson, *J. Mol. Biol.,* **8**, 148 (1964).

140. Gingell, D., *J. Theoret. Biol.,* **17**, 451 (1967). Also see Ilani A., *J. Membrane Biol.,* **3**, 223 (1970). G. M. Schwab, *J. Coll. Interf. Sci.,* **34**, 337 (1970).

141. D'Agostino, C., Jr., and L. Smith, Jr., *Proc. Cybernetic Sci. Symp.* Los Angeles, p. 11 (1964), Spartan Books Inc., Washington, D.C. (1965).

142. Henn, F. A., G. L. Decker, J. W. Greenwalt, and T. E. Thompson, *J. Mol. Biol.,* **24**, 51 (1967).

143. Haydon, D. A., in *The Molecular Basis of Membrane Function,* Ed. D. C. Tosteson, Prentice-Hall, Inc., Englewood Cliffs, New Jersey (1969).

144. Cook, G. M. W., W. R. Redwood, A. R. Taylor, and D. A. Haydon, *Kolloid Zeitschrift,* **227**, 28 (1968).

145. Sessa, G., and G. Wassermann, *J. Lipid Res.,* **9**, 310 (1968).

146. Attwood, D., and L. Saunders, *Biochim. Biophys. Acta.,* **98**, 344 (1965).

147.(a) Chapman, D., D. J. Fluck, S. A. Penkett, and G. G. Shipley, *Biochim. Biophys. Acta.,* **163**, 255 (1968). (b) Huang, C., *Biochemistry,* **8**, 344 (1969). (c) Huang, C., J. P. Charlton, C. I. Shyr, and T. E. Thompson, *Biochemistry,* **9**, 3422 (1970).

148. Papahadjopoulos, D., and N. Miller, *Biochim. Biophys. Acta.,* **135**, 624 (1967).

149. Papahadjopoulos, D., and J. C. Watkins, *Biochim. Biophys. Acta.,* **135**, 639 (1967).

150. Bangham, A. D., M. M. Standish, and N. Miller, *Nature,* **208**, 1295 (1965). Bangham, A. D., M. M. Standish, and G. Weissmann, *J. Mol. Biol.,* **13**, 253 (1965)

151. Bangham, A. D., M. M. Standish, and G. Weissmann, *J. Mol. Biol.,* **13**, 253 (1965).

152. Bangham, A. D., J. De Gier, and G. D. Greville, *Chem. Phys. Lipids,* **1**, 225 (1967).

153. Stoeckenius, W., *J. Biochem. Biophys. Cytol.*, **5**, 491 (1959).
154. Reeves, J. P., and R. M. Dowben, *J. Cell. Physiol.*, **73**, 49 (1969).
155. Razin, S., H. J. Morowitz, and T. M. Terry, *Proc. Natl. Acad. Sci. (U.S.)*, **54**, 219 (1965). Also see Chapter 9.
156. McConnel, D. G., A. Tzagoloff, D. H. MacLennan, and D. E. Green, *J. Biol. Chem.*, **241**, 2373 (1966). Also see Chapter 7.
157. Danielli, J. F., *J. Theoret. Biol.*, **12**, 439 (1966).
158. Ohki S., and N. Fukuda, *J. Theoret. Biol.*, **15**, 362 (1967).
159. Good, R. J., *J. Colloid. Interf. Sci.*, **31**, 540 (1969).
160. Bahadur, K., *Synthesis of Jeewanu the Protocell*, Ram Narain Lal Beni Prasad, Allahabad (India), 1966.
161. Kenyon, D. H., and G. Steinman, *Biochemical Predestination*, McGraw-Hill Book Co., New York (1969).
162. Coster, H. G. L., and Simon R., *Biochim. Biophys. Acta.*, **163**, 234 (1968).
163. Tien, H. T., *J. Phys. Chem.*, **72**, 2723 (1968).
164. Taylor, J., and D. A. Haydon, *Disc. Faraday Soc.*, **42**, 51 (1966).
165. Verwey, E. J. W., and J. Th. G. Overbeek with the collaboration of K. Van Nes, *Theory of the Stability of Lyophobic Colloids: The Interaction of Sol Particles Having an Electric Double Layer*, Elsevier Publishing Co., Amsterdam (1948).
166. Bell, G. M., and S. Z. Levine, *Phys. Chemie.*, **231**, 289 (1966).
167. Haydon, D. A., *J. Am. Oil Chemist's Soc.*, **45**, 230 (1968).
168. Parsegian, V. A., *Trans. Faraday Soc.*, **62**, 848 (1966).
169. Lifson, S., and A. Katchalsky, *J. Polymer Sci.*, **13**, 43 (1954).
170. Small, D., and M. Bourges, *J. Mol. Crystals*, **1**, 541 (1966).
171. Parsegian, V. A., *J. Theoret Biol.*, **15**, 70 (1967).
172. Parsegian, V. A., *Science*, **156**, 939 (1967).
173. Pethica, B. A., *Protoplasma*, **63**, 181 (1967).
174. Vaidyanathan, V. S., and N. S. Goel, *J. Theoret. Biol.*, **21**, 331 (1968).
175. DeVries, A. J., *Rec. Trav. Chim.*, **77**, 383 (1958).
176. Parsegian, V. A., and B. W. Ninham, *Nature*, **224**, 1197 (1969); *Biophys. J.*, **10**, 664 (1970). Ninham, B. W., and V. A. Parsegian, *J. Chem. Phys.*, **52**, 4578 (1970); *Biophys. J.*, **10**, 646 (1970).
177. Thompson, T. E., *Protoplasma*, **63**, 194 (1967). Danielli, J. F., in *Formation and Fate of Cell Organelles*, p. 239 (1967).
178. Ochs, A. L., R. M. Burton, Twelfth Annual Meeting of Biophysical Society (Abstr.) p. A-27 (1968). Mann, J. A., H. R. Hohl, and M. J. Sborov, *Nature*, **222**, 471 (1969).
179. Curtis, A. S. G., *The Cell Surface: Its Molecular Role in Morphogenesis*, Logas Press and Academic Press, London (1967). Trinkaus, J. P., *Cells into Organs: The Forces that Shape the Embryo*, Prentice-Hall Inc., Englewood-Cliffs, N.J. (1969). Also see Barton, P. G., *J. Biol. Chem.*, **243**, 3884 (1968).
180. Bockris, J. O'M., and A. K. N. Reddy, *Modern Electrochemistry*, Vols. I and II, Plenum Press, New York (1970).
181. Moran, A., and A. Ilani, *Chem. Phys. Lipids*, **4**, 169 (1970).
182. Dervichian, D. G., in *Surface Phenomena in Chemistry and Biology*, Eds.

J. F. Danielli, K. G. A. Pankhurst, and A. C. Riddiford, Pergamon Press, New York, (1958). De Bernard L., *Bull. Soc. Chim. Biol.*, **40**, 161 (1958).

182a. Casley-Smith, J. R., and A. J. Day, *Quart. J. Exp. Physiol.*, **51**, 1 (1966). Also see p. 129 in Curtis, Ref. 179.

183. Curtis, A. S. G., *Exptl. Cell Res.*, **8**, S 107 (1961).

184. Joly, M., *J. Colloid Sci.*, **11**, 519 (1956).

185. Liberman, E. A., and V. A. Nenashev, *Biofizika*, **13**, 193 (1968).

186. Yamauchi, A., A. Matsubara, H. Kimizuka, and L. G. Abood, *Biochim. Biophys. Acta*, **150**, 181 (1968). Schulman, J. H., M. Rosoff, and H. Erbring, *Protoplasma*, **63**, 165 (1967).

187. Karnovesky, M. L., *Physiol. Rev.*, **42**, 143 (1962). Lilian, J. E., *Current Topics in Develop. Biol.*, **4**, 169 (1969).

188. De Duve, C., *Protoplasma*, **63**, 95 (1967).

189. Tien, H. T., *J. Gen. Physiol.*, **52** (Suppl. II), 125s (1968). Tien, H. T., and A. L. Diana, *Chem. Phys. Lipids*, **2** 55 (1968). Goldup, A., S. Ohki, and J. F. Danielli, *Progress in Surface Science*, **3**, 193 (1970).

190. Stevens, J. D., *Nature*, **214**, 199 (1967).

191. Koenig, F. A., *Handbook of Experimental Phys.*, Vol. 12 (2), p. 376, Leipzig (1933).

192. Huang, C., and T. E. Thompson, *J. Mol. Biol.*, **13**, 183 (1965).

193. Tien, H. T., *J. Theoret. Biol.*, **16**, 97 (1967).

194. Cherry, R. J., and D. Chapman, *J. Mol. Biol.*, **40**, 19 (1969).

195. Huang, C., and T. E. Thompson, *J. Mol. Biol.*, **16**, 576 (1966).

196. Tien, H. T., *J. Mol. Biol.*, **16**, 577 (1966).

197. Simon, R., *Biochim. Biophys. Acta.*, **203**, 209 (1970).

198. Cole, K. S., and R. H. Cole, *J. Chem. Phys.*, **10**, 99 (1942).

199. Rosen, D., and A. M. Sutton, *Biochim. Biophys. Acta.*, **163**, 226 (1968).

200. Ohki, S., *J. Theoret. Biol.*, **19**, 97 (1968).

201. Lauger, P., W. Lesslauer, E. Marti, and J. Richter, *Biochim. Biophys. Acta.*, **135**, 20 (1967).

202. Everitt, C. T., and D. A. Haydon, *J. Theoret. Biol.*, **18**, 371 (1968). For capacitance of vesicles see Schwan, H. P., S. Takashima, V. K. Miyamoto, and W. Stoeckenius, *Biophys. J.*, **10**, 1102 (1970).

203. Haydon, D. A., in *Membrane Models Formation and Biological Membranes*, Ed L. Bolis, North-Holland Publ. Co., Amsterdam, (1967).

204. Ohki, S., *Biophys. J.*, **9**, 1195 (1969).

205. White, S. H., *Biochim. Biophys. Acta.*, **196**, 354 (1970).

206. Hanai, T., D. A. Haydon, and J. Taylor, *J. Theoret. Biol.*, **9**, 278 (1965).

207. Pauly, H., in *Physikalishe Grundlagen der Medizin Abhandlungen aus der Biophysik*, Heft 7, VEB Georg Thieme Verlag, Leipzig (1967).

208. Tayler, R. E., *J. Cell. Comp. Physiol.*, **66** (suppl. 2), 21 (1965).

209. Falk, G., and P. Fatt, *Proc. Roy. Soc. B*, **160**, 69 (1964).

210. Huang, C., L. Wheeldon, and T. C. Thompson, *J. Mol. Biol.*, **8**, 148 (1964).

211. Andreoli, T. E., M. Tieffenberg, and D. C. Tosteson, *J. Gen. Physiol.*, **50**, 2527 (1967).

212. Thompson, T. E., in *Cellular Membranes in Development*, Ed. M. Locke, Academic Press, New York, p. 83 (1964).

213. Drost-Hansen, W., and A. Thorhang, *Nature,* **215**, 506 (1967).
214. Bean, R. C., *Symposium on Molecular Basis of Membrane Structure and Function,* Durham, N.C., Aug. 1968.
215. Coster, H. G. L., and R. Simons, *Biochim. Biophys. Acta.,* **203**, 17 (1970).
216. Seufert, W. D., G. Beauchesne, and M. Belanger, *Biochim. Biophys. Acta.,* **211**, 358 (1970).
217. Ohki, S., and A. Goldup, *Nature,* **217**, 458 (1968). Ohki, S., *J. Colloid Interface Sci.,* **30**, 413 (1969).
218. Lesslauer, W., J. Richter, and P. Lauger, *Nature,* **213**, 1224 (1967).
219. Sutton, A. M., and D. Rosen, *Nature,* **219**, 153 (1968); its retraction in *Nature,* **225**, 274 (1970).
220. Gutman, F., *Nature,* **219**, 1359 (1968).
221. Tosteson, D. C., *Fed. Proc.,* **27**, 1269 (1968).
222. Onsager, L., *J. Chem. Phys.,* **2**, 599 (1934).
223. Neumcke, D. Walz, and P. Lauger, *Biophys. J.,* **10**, 172 (1970). Kauffman, J. W., and C. A. Mead, *Biophys. J.,* **10**, 1084 (1970).
224. Bruner, L. J., *Biophys. J.,* **7**, 627 (1967).
225. Bruner, L. J., *Biophys. J.,* **5**, 867, 887 (1965).
226. Bruner, L. J., *Biophys. J.,* **7**, 947 (1967).
227. Lewis, T. J., *Advancement of Science,* p. 501, March (1964).
228. Pethica, B. A., *Protoplasma,* **63**, 147 (1967).
229. Barfort, P., E. R. Arquilla, and P. O. Vogelhut, *Science,* **160**, 1119 (1968).
230. Merriam, R. W., *J. Biophys. Biochem. Cytol.,* **6**, 353 (1959).
231. Korohoda, W., J. A. Forrester, K. G. Moreman, and E. J. Ambrose, *Nature,* **217**, 615 (1968).
232. Loewenstein, W. R., *Ann. N.Y. Acad. Sci.,* **137**, 441 (1966).
233. Gingell, D., *J. Theoret. Biol.,* **19**, 340 (1968).
234. Brandt, P. W., and A. R. Freeman, *Science,* **155**, 582 (1967).
235. Ranck, J. B., *Science,* **144**, 187 (1964).
236. See Refs. 181, 187, and 233.
237. Weiss, L., *The Cell Periphery, Metastasis and other Contact Phenomena,* North-Holland Publishing Co., Amsterdam (1967).
238. Levine, Y. K., A. I. Bailey, and M. H. F. Wilkins, *Nature,* **220**, 577 (1968).
239. Ohki, S., *J. Theoret. Biol.,* **23**, 158 (1969); **26**, 277 (1970).
240. Jacobs, M. H., *Diffusion Processes,* Springer-Verlag, New York (1967).
241. Fick, A., *Pogg. Ann.,* **94**, 59 (1855).
242. Spandau, H., *Ber Deutsch. Chem Ges.,* **74**, 1028 (1941).
243. Stein, W. D., *The Movement of Molecules Across Cell Membranes,* Academic Press, New York (1967).
244. Kadem, O., and A. Katchalsky, *J. Gen. Physiol.,* **45**, 143 (1961); *Biochim. Biophys. Acta,* **27**, 229 (1958). Staverman, A. J., *Rec. Trav. Chim.,* **70**, 344 (1951).
245. Dainty, J., and B. Z. Ginzburg, *J. Theoret. Biol.,* **5**, 256 (1963).
246. Solomon, A. K., *Scientific American,* **203**, 146 (Dec., 1960). Goldstein, D. A., and A. K. Solomon, *J. Gen. Physiol.,* **44**, 1 (1961).
247. Koefoed-Jonson, V., and H. H. Ussing, *Acta. Physiol. Scand.,* **28**, 60 (1953).

248. Pappenheimer, J. R., E. M. Renkin, and L. M. Borrero, *Am. J. Physiol.*, **167**, 13 (1951).
249. Durbin, R. P., H. Frank, and A. K. Solomon, *J. Gen. Physiol.*, **39**, 535 (1956).
250. Paganelli, C. V., and A. K. Solomon, *J. Gen. Physiol.*, **41**, 259 (1957).
251. Villegas, R., T. C. Barton, and A. K. Solomon, *J. Gen. Physiol.*, **42**, 355 (1958).
252. Dick, D. A. T., *Physical Basis of Circulation, Transport, and Regular Exchange*, Proceedings Conference, Denver, p. 217 (1966).
253. Prescott, D. M., and E. Zeuthen, *Acta. Physiol. Scand.*, **28**, 77 (1953).
254. Rich, G. T., R. I. Sha'afi, T. C. Barton, and A. K. Solomon, *J. Gen. Physiol.*, **50**, 2391 (1967).
255. Schatzberg, P., *J. Polymer Sci. C*, **10**, 87 (1965).
256. Blank, M., *J. Physical Chem.*, **68**, 2793 (1964).
257. Essig, A., *J. Theoret. Biol.*, **13**, 63 (1966).
258. Wang, J. H., C. V. Robinson, and I. S. Edelman, *J. Amer. Chem. Soc.*, **75**, 466 (1953).
259. Nakayama, F. S., and R. D. Jackson, *J. Phys. Chem.*, **67**, 932 (1963).
260. McCall, D. W., and D. C. Douglass, *J. Phys. Chem.*, **69**, 2001 (1965).
261. Hodgkin, A. L., and R. D. Keynes, *J. Physiol. (London)*, **128**, 61 (1955).
262. Harris, E. J., *Transport and Accumulation in Biological Systems*, Academic Press, Inc., New York (1960).
263. Dainty, J., *Advances in Botanical Res.*, **I**, 279 (1963).
264. Robinson, R. A., and R. H. Stokes, *Electrolyte Solutions*, Butterworths Scientific Publications, Lond (1959).
265. Kubin, M., and P. Spacek, *Coll. Czech. Chem. Commun*, **32**, 2733 (1967).
266. Tuwiner, S. B., *Diffusion and Membrane Technology*, Van Nostrand Reinhold, New York (1962). Also see Wedner, H. J., and J. M. Diamond, *J. Membrane Biol.*, **1**, 92 (1969).
267. Sha'afi, R. I., G. T. Rich, V. W. Sidel, W. Bossert, and A. K. Solomon, *J. Gen. Physiol.*, **50**, 1377 (1967).
268. Kedem, O., and A. Katchalsky, *Biochim. Biophys. Acta.*, **27**, 229 (1958). Vreeman, H. J., *Kon Ned. Acad. Wetensch. Proc., Ser B*, **69**, 542, 555, 564 (1966).
269. Redwood, W. R., and D. A. Haydon, *J. Theoret. Biol.*, **22**, 1 (1969). C. T. Everitt, W. R. Redwood, and D. A. Haydon, *J. Theoret. Biol.*, **22**, 20 (1969). For data on permeability of phospholipid vesicles see Reeves, J. P., and R. M. Dowben, *J. Membrane Biol.*, **3**, 123 (1970).
270. Cass, A., and A. Finkelstein, *J. Gen. Physiol.*, **50**, 1765 (1967). See also 343.
271. For effect of lipid composition on biological membranes see McElhanev. R. N., J. De Gier, and L. L. M. Van Deenen. *Biochim. Biophys. Acta.*, **219** 245 (1970); Williams, D. D., and W. S. Platener, *J. Amer. Physiol.*, **212**, 167 (1967); Anderson, O. R., *Amer. Biol. Teacher*, **32**, 154 (1970). For model systems see Moiseeva, L. N., and A. G. Pasynskii, *Biokhimiya*, **31**, 1159 (1966).

272. Shiratori, A., H. Mizuno, and H. Arita, *J. Tokyo Univ. Fish.*, **55**, 21, 29, 33, 37, 41, 45, 55, 63 (1968).
273. Archer, R. J., and V. K. LeMer, *J. Phys. Chem.*, **59**, 200 (1955).
274. Dainty, J., and C. R. House, *J. Physiol.*, **185**, 172 (1966).
275. Gutknecht, J., *Biochim. Biophys. Acta.*, **163**, 20 (1968). However, see Zimmermann, U., and E. Steudle, *Z. Naturforsch. B*, **25**, 500 (1970).
276. Jung, C. Y., and F. M. Snell, *Fed. Proc.*, **27**, 286 (Abstr. No. 395) (1968).
277. Bean, R. C., W. C. Shepherd, and H. Chan, *J. Gen. Physiol.*, **52**, 495 (1968).
278. De Gier, J., J. G. Mandersloot, and L. L. M. Van Deenen, *Biochim. Biophys. Acta,* **150**, 666 (1968).
279. Wood, R. E., F. P. Wirth, and H. E. Morgan, *Biochim. Biophys. Acta.*, **163**, 171 (1968).
280. De Gier, J., L. L. M. Van Deenen, and K. G. Van Senden, *Experientia*, **22**, 20 (1966).
281. Lippe, C., *Nature*, **218**, 197 (1968).
282. Holz, R., and A. Finkelstein, *J. Gen. Physiol.*, **56**, 125 (1970).
283. Andreoli, T. E., V. W. Dennis, and A. M. Weigl, *J. Gen. Physiol.*, **53**, 133 (1969).
284. Lippe, C., *J. Mol. Biol.*, **39**, 669 (1969).
285. Overton, E., *Pflugers Arch.*, **92**, 115, 346 (1902); *Handbuch der Physiologie des Menschen*, **2**, 744–898 (1907).
286. Overton, E., *Studien Uber die Narkose*, Jena- Fischer (1912).
287. Maddy, A. H., C. Huang, and T. E. Thompson, *Fed. Proc.*, **25**, 933 (1966).
288. Maddy, A. H., *Biochim. Biophys. Acta.*, **88**, 448 (1964).
289. Collander, R., *Physiologia Plantarum*, **2**, 300 (1949).
290. Wortivaara, V., and R. Collander, in *Protoplasmologia*, Vol. II, C. 8d p. 1, (1960).
291. Wright, E. M., and J. M. Diamond, *Proc. Roy. Soc. B*, **172**, 227 (1969).
292. Diamond, J. M., and E. M. Wright, *Proc. Roy. Soc. B*, **172**, 272 (1969). For data on permeability of small non-electrolytes in liposomes see B. E. Cohen and A. D. Bangham, *Nature*, **236**, 173 (1972).
293. Diamond, J. M., and E. M. Wright, *Proc. Roy. Soc. B*, **172**, 203 (1969).
294. Pfeffer, W., *Osmotische Untersuchungen*, Leipzig (1877).
295. Ruhland, W., *Jb. wiss. Bot.*, **51**, 376 (1912).
296. Ruhland, W., *Jb. wiss. Bot.*, **51**, 391 (1914).
297. Jacobs, M. H., H. N. Glassman, and A. K. Parpart, *J. Cell. Comp. Physiol.*, **7**, 197 (1935).
298. Hober, R., and S. L. Orskov, *Pflugers Arch.*, **231**, 599 (1933).
299. Kozawa, S., *Biochem. Z.*, **60**, 231 (1914).
300. Diamond, J. M., and E. M. Wright, *Ann. Rev. Physiol.*, **31**, 581 (1969).
301. Hays, R. M., and N. Franki, *J. Membrane Biol.*, **2**, 263 (1970). Also see Fong, C. T. O., L. Silver, and G. Lankau, *Proc. Natl. Acad. Sci. (U.S.)*, **67**, 1221 (1970); Civan, M. M., *J. Theoret. Biol.*, **27**, 387 (1970).
302. Glasstone, S., K. J. Laidler, and H. Eyring, *The Theory of Rate Processes*, McGraw-Hill, New York (1941).

303. Zwolinski, B. J., H. Eyring, and C. E. Reese, *J. Physical and Colloid Chemistry*, **53**, 1426 (1949).
304. Hopfer, U., A. L. Lehninger, and W. J. Lennerz, *J. Membrane Biol.*, **2**, 41 (1970).
305. Teorell, T., *Progr. Biochem. Biophys. Chem.*, **3**, 305 (1953).
306. Fast, P. G., *Science*, **155**, 1680 (1967).
307. E. Deutsch, E., E. Gerlach, and K. Moser (Eds.), *Metabolism and Membrane Permeability of Erythrocytes and Thrombocytes*, Georg Thieme Verlag, Stuttgart (1968).
308. Passow, H., *Progress in Biophys. Biochem.*, **19** (II), 423 (1968). Latis, G. G. in *Cellular Membranes in Development*, Academic Press, Inc., New York, p. 299 (1964).
309. Charnock, J. S., and L. J. Opit. *Biol. Basis Medicine*, **1**, 69 (1968). Michaelis, A. S., *Progr. Separ. Purif.*, **1**, 143 (1968). Friedenberg, R. M. and T. B. Day, *Curr. Modern Biol.*, **3**, 122 (1969). Lakshminarayanaiah, K., *Chem. Rev.*, **65**, 491 (1965). Simon, R., and A. Taloni, *Biochim. Biophys. Acta.*, **211**, 448 (1970). Castleden, J. A., and R. Fleming, *Biochim. Biophys. Acta.*, **211**, 478 (1970).
310. Nomura, M., *Proc. Natl. Acad. Sci. (U.S.)*, **52**, 1514 (1964).
311. Trembley, G. Y., M. J. Daniels, and M. Schaechter, *J. Mol. Biol.*, **40**, 65 (1969).
312. Parsegian, P., *Nature*, **221**, 844 (1969).
313. See Refs. 66 and 910.
314. Ciani, S., G. Eisenman, and G. Szabo, *J. Membrane Biol.*, **1**, 1 (1969).
315. Eisenman, G., S. Ciani, and G. Szabo, *J. Membrane Biol.*, **1**, 294 (1969). Also Higuchi, W. I., A. H. Ghanem, and A. B. Bikhazi, *Fed. Proc.*, **29**, 1327 (1970).
316. Szabo, G., G. Eisenman, and S. Ciani, *J. Membrane Biol.*, **1**, 346 (1969).
316a. McLaughlin, S. G. A., G. Szabo, G. Eisenman, and S. M. Ciani, *Proc. Natl. Acad. Sci.*, **67**, 1268 (1970).
317. Conti, F., and G. Eisenman, *Biophys. J.*, **6**, 227 (1966).
318. Sandblom, J. P., *Biophys. J.*, **7**, 243 (1967).
319. Eisenman, G., J. P. Sandblom, and J. L. Walker, Jr., *Science*, **155**, 965 (1967).
320. Schogle, R., *Strofftransport durch Membranes*, Steinkopft, Dermstadt, Germany (1964).
321. Conway, B. E., in *Physical Chemistry: An Advanced Treatise*, Vol. IXA, *Electrochemistry*, Ed. H. Eyring, Academic Press, New York, (1970). Also Hertz, H. G., *Angew. Chem. (Intern. Ed.)*, **9**, 124 (1970).
322. Moelwyn-Hughes, E. A., *Physical Chemistry*, Pergamon Press, (1957).
323. Eisenman, G., and F. Conti, *J. Gen. Physiol.*, **48**, 55s (1965).
324. Born, M., *Z. Physik*, **1**, 45 (1920).
325. Korosy, F., *Kem. Kozlem.*, **27**, 89 (1967).
326. Diamond, J. M., and E. M. Wright, *Ann. Rev. Physiol.*, **31**, 581 (1969).
326a. Lindley, B. D., *J. Theoret. Biol.*, **17**, 213 (1967).
327. Diebler, H., M. Eigen, G. Ilgenfritz, G. Maass, and R. Winkler, *Pure Appl. Chem.*, **20**, 93 (1969).

328. Lea, E. J. A., and P. C. Croghan, *J. Membrane Biol.*, **1**, 225 (1969).

329. Finkelstein, A., *Biochim. Biophys. Acta.*, **205**, 1 (1970).

329a. Liberman, E. A., and V. P. Topaly, *Biochim. Biophys. Acta.*, **163**, 125 (1968).

330. Hopfer, U., A. L. Lehninger, and T. E. Thompson, *Proc. Natl. Acad. Sci. (U.S.)*, **59**, 484 (1968).

331. Hopfer, U., A. L. Lehninger, and W. I. Lennarz, *J. Membrane Biol.*, **3**, 142 (1970).

332. Rosenberg, B., and G. L. Gendrasiak, *Chem. Phys. Lipids*, **2**, 47 (1968). Rosenberg, B., B. B. Bhowmik, *Chem. Phys. Lipids*, **3**, 109 (1969). Bhowmik, B. B., and B. Rosenberg, to be published.

333. Skulachev, V. P., A. A. Sharaf, L. S. Yagujzinsky, A. A. Jasaitis, E. A. Liberman, and V. P. Topali, *Currents in Modern Biol.*, **2**, 98 (1968).

334. Skulachev, V. P., A. A. Sharaf, and E. A. Liberman, *Nature*, **216**, 718 (1967). Liberman, Y. A., Y. N. Mokhova, V. P. Skulachev, and V. P. Topaly, *Biofizika* (in Russian), **13**, 226 (1968).

335. LeBlanc, O. H., *Biochim. Biophys. Acta.*, **193**, 350 (1969).

336. Liberman, E. A., V. P. Topaly, and L. M. Tsofina, *Biofizika* (in Russian), **15**, 69 (1970).

337. Noguchi, S., and S. Koga, *J. Gen. Appl. Microbiol.*, **15**, 41 (1969).

338. Liberman, E. A., and V. P. Topaly, *Biofizika* (in Russian), **14**, 452 (1969).

339. Liberman, E. A., V. P. Topaly, and A. Y. Silberstein, *Biochim. Biophys. Acta.*, **196**, 221 (1970).

340. Lauger, P., J. Richter, and W. Lesslauer, *Ber. Bunges. Phys. Chem.*, **71**, 906 (1967).

341. Liberman, Y. A., V. P. Topaly, L. M. Tsofina, and A. M. Shkrob, *Biofizica* (in Russian), **14**, 55 (1969).

342. Pashaev, P. A., and L. M. Tsofina, *Biofizika* (in Russian), **13**, 360 (1968).

343. Finkelstein, A., and A. Cass, *J. Gen. Physiol*, **52**, 145s (1968).

344. Pant, H. C., and B. Rosenberg, *Chem. Phys. Lipids*, **4**, 203 (1970). Jendrasiak, G. L., *Chem. Phys. Lipids*, **3**, 98 (1969). Bhowmik, B. B., G. L. Gendrasiak, and B. Rosenberg, *Nature*, **215**, 842 (1967). Rosenberg, B., and G. L. Jendrasiak, *Chem. Phys. Lipids*, **2**, 47 (1968).

345. Foster, R., *Organic Charge-Transfer Complexes*, Academic Press, New York (1969).

346. Jendrasiak, G. L., *Chem. Phys. Lipids*, **4**, 85 (1970).

347. Sharp, J. H., in *Physical Chemistry: An Advanced Treatise*, Vol. X, *Solid State*, Ed. W. Jost, Academic Press, New York (1970).

348. Kallmann, H., and M. Pope, *J. Chem. Phys.*, **30**, 585 (1959); **32**, 300 (1960); *Nature*, **185**, 753 (1960); **186**, 31 (1960).

349. Soma, M., *J. Amer. Chem. Soc.*, **92**, 3289 (1970). Soma, M., and A. Yamagishi, *Biochim. Biophys. Acta.*, **205**, 183 (1970).

350. Kay, R. E., and H. Chen, *Radiation Research*, **10**, 177 (1969).

351. Mauzerall, D., and A. Finkelstein, *Nature*, **224**, 690 (1969).

352. Jain, M. K., F. P. White, A. L. Strickholm, and E. H. Cordes, *Nature*, **227**, 705 (1970).

353. Jendrasiak, G. L., and R. Hays, *Nature*, **225**, 278 (1970).

354. Szent-Gyorgyi, A., *Proc. Natl. Acad. Sci. (U.S.)*, **58**, 2012 (1967). Tien, H. T., and S. P. Verma, *Nature*, **227**, 1232 (1970); Bockris, J. O'M., *Nature*, **224**, 775 (1969).

355. Rosenberg, B., and E. Postow, *Ann. N.Y. Acad. Sci.*, **158**, 161 (1969).

356. Gruenstein, E., and J. Wynn, *J. Theoret. Biol.*, **26**, 343 (1970).

357. Haynes, D. H., and B. C. Pressman, *Fed. Proc.*, **29**, 886 (1970). Pressman, B. C., E. J. Harris, W. S. Jagger, and J. H. Johnson, *Proc. Natl. Acad. Sci. (U.S.)*, **58**, 1949 (1967). Pressman, B. C., *Proc. Natl. Acad. Sci. (U.S.)*, **53**, 1076 (1965).

358. Graven, S. N., H. A. Lardy, D. Johnson, and A. Rutter, *Biochemistry*, **5**, 1729 (1966). Graven, S. N., H. A. Lardy, and A. Rutter, *Biochemistry*, **5**, 1735 (1966). Graven, S. N., H. A. Lardy, and S. Estrada-O, *Biochemistry*, **6**, 365 (1967).

359. Mueller, P., and D. O. Rudin, in *Current Topics in Bioenergetics*, **3**, 157 (1969).

360. Shemyakin, M. M., Yu A. Ovchinnikov, V. T. Ivanov, V. K. Antonov, E. I. Vinogradova, A. M. Shkrob, G. G. Malenkov, A. V. Evstratov, I. A. Laine, E. I. Melnik, and I. D. Ryabova, *J. Membrane Biol.*, **1**, 402 (1969).

361. Tosteson, D. C., *Fed. Proc.*, **27**, 1269 (1968). Tosteson, D. C., T. E. Andreoli, M. Tiffenberg, and P. Cook, *J. Gen. Physiol.*, **51**, 373s (1968). Tosteson, D. C., P. Cook, T. Andreoli, and M. Tiffenberg, *J. Gen. Physiol.*, **50**, 2513 (1967).

362. Henderson, P. J. F., J. D. McGivan, and J. B. Chappell, *Biochem. J.*, **111**, 521 (1969).

363. Harold, F. M., and J. R. Baarda, *J. Bacteriol.*, **95**, 816 (1968).

364. Shemyakin, M. M., E. I. Vinogradova, M. Yu. Feiginia, N. A. Aldanova, N. F. Loginova, I. D. Ryabova, and I. A. Pavlenko, *Experientia*, **21**, 548 (1965).

365. Stefanac, Z., and W. Simon, *Microchem. J.*, **12**, 125 (1967). Shemyakin, M. M., Yu A. Ovchinnikov, V. T. Ivanov, A. A. Kiriyushkin, G. L. Zhdanov, and I. D. Ryabova, *Experientia*, **19**, 566 (1963).

366. Shemyakin, M. M., V. K. Antonov, L. D. Bergelson, V. T. Ivanov, G. G. Ivanov, G. G. Malenkov, Yu A. Ovchinnikov, and A. M. Shkrob, in *The Molecular Basis of Membrane Function*, Ed. D. C. Tosteson, Prentice-Hall, Inc., Englewood, N.J., p. 173 (1969).

367. Mueller, P., and D. O. Rudin, *Biochim. Biophys. Res. Comm.*, **26**, 398 (1967).

368. Johnson, S. M., and A. D. Bangham, *Biochim. Biophys. Acta.*, **193**, 82 (1969). Saha, J., D. Papahadjopoulos, and C. E. Wenner, *Biochim. Biophys. Acta.*, **196**, 10 (1970). DeGier, J., C. W. M. Heast, J. G. Mandersloot, and L. L. M. Van Deenen, *Biochim. Biophys. Acta.*, **211**, 373 (1970).

368a. *Stability Constants*, Spec. Publ. No. 17, Ed. A. Martell, and L. G. Sillen, Chemical Society, London (1964).

368b. Izatt, R. M., J. H. Rytting, D. P. Nelson, and B. L. Haymore, *Science*, **164**, 443 (1969).

368c. Rais, J., and M. Krys, *J. Inorg. Nuclear Chem.*, **31**, 2903 (1969).

369. Ashton, R., and L. K. Steinrauf, *J. Mol. Biol.*, **49**, 547 (1970). Pioda, L. A. R., and W. Simon, *Chimia*, **23**, 72 (1969). Wipf, H. K., W. Pache, P. Jordan, H. Zahner, W. Keller-Schierlein, and W. Simon, *Biochem. Biophys. Res. Comm.*, **36**, 387 (1969).

370. Wipf, H. K., and W. Simon, *Biochem. Biophys. Res. Comm.*, **34**, 707 (1969).

371. Shemyakin, M. M., L. A. Shchukina, E. I. Vinogradova, G. A. Ravdel, and Yu A. Ovchinnikov, *Experientia*, **22**, 535 (1966).

372. Shemyakin, M. M., Yu A. Ovchinnikov, V. T. Ivanov, and A. V. Evstratov, *Nature*, **213**, 412 (1967).

373. Pauling, L., *The Nature of Chemical Bond*, Cornell University Press, Ithaca, N.Y. (1960).

373a. Wieland, T., H. Faulstisch, W. Burgermeister, W. Otting, W. Moehle, M. M. Shemyakin, Yu A. Ovchinnikov, V. T. Ivanov, and G. G. Malenkov, *FEBS Letters*, **9**, 89 (1970).

374. Stefanac, Z., and W. Simon, *Chimia*, **20**, 436 (1966).

375. Wipf, H. K., L. A. R. Pioda, Z. Stefanac, and W. Simon, *Helv. Chim. Acta.*, **51**, 377 (1968).

376. Lev, A. A., and E. P. Buzhinskii, *Tsitologiya*, **9**, 102 (1967).

377. Pioda, L. A. R., H. A. Wachter, R. E. Dohner, and W. Simon, *Helv. Chim. Acta.*, **50**, 1373 (1967).

378. Pederson, C. J., *J. Amer. Chem. Soc.*, **89**, 7017 (1968). Izatt, R. M., J. H. Rytting, D. P. Nelson, B. L. Haymore, and J. J. Christensen, *Science*, **164**, 443 (1969). Eisenman, G., S. M. Ciani, and G. Szabo, *Fed. Proc.*, **27**, 1289 (1968).

379. Sarges, R., and B. Witkop, *J. Amer. Chem. Soc.*, **87**, 2011 (1965).

380. Podelski, T., and J. P. Changeux, *Nature*, **221**, 541 (1969).

381. Schwyzer, V. R., *Experientia*, **26**, 577 (1970).

382. Hall, M. J., *Biochim. Biophys. Res. Comm.*, **38**, 590 (1969). Bevan, K., J. S. Davies, J. Salmon, M. J. Hall, C. H. Hassall, R. B. Morton, D. A. S. Phillips, Y. Ogihara, and W. A. Thomas, *Experientia*, **26**, 122 (1970).

383. Ivanov, V. T., I. A. Laine, N. D. Abdulaev, L. B. Senyavina, E. M. Popov, Yu A. Ovchinnikov, and M. M. Shemyakin, *Biochim. Biophys. Res. Comm.*, **34**, 803 (1969). Shemyakin, M. M. et. al., *ibid.*, **29**, 834 (1967).

384. Ohnishi, M., and D. W. Urry, *Science*, **168**, 1091 (1970).

385. Pioda, L. A. R., H. K. Wipf, and W. Simon, *Chimia*, **22**, 189 (1968). W. E. Morf, and W. Simon, *Helv. Chim. Acta*, **54**, 2683 (1971).

386. Haynes, D. H., A. Kowalsky, and B. C. Pressman, *J. Biol. Chem.*, **244**, 502 (1969).

387. Prestegard, J. H., and S. I. Chan, *Biochemistry*, **8**, 3921 (1969). For data on ^{23}Na-complexation by ionophores see D. H. Haynes, B. C. Pressman, and A. Kowalsky, *Biochemistry*, **10**, 852 (1971).

388. Prestegard, J. H., and S. I. Chan, *J. Amer. Chem. Soc.*, **92**, 4440 (1970).

389. Ovchinnikov, Yu A., V. T. Ivanov, A. V. Evstratov, V. F. Bystrov, N. D. Abdullaev, E. M. Popov, G. M. Lipkind, S. F. Arkhipova, E. S. Efremov,

and M. M. Shemyakin, *Biochim. Biophys. Res. Comm.*, **37**, 668 (1969).

390. Shemyakin, M. M., Yu A. Ovchinnikov, V. T. Ivanov, and I. D. Ryaboda, *Experientia*, **23**, 326 (1967).

391. Ovchinnikov, Yu A., V. T. Ivanov, B. F. Bystrov, A. I. Miroshnikov, E. N. Shepel, N. D. Abdullaev, E. S. Efremov, and L. B. Senyavina, *Biochem. Biophys. Res. Comm.*, **39**, 217 (1970).

392. Waki, M., and N. Izumiya, *Tetrahedron Letters*, 3083 (1968).

393. Sigel, H., and D. B. McCormick, *Accounts Chem. Res.*, **3**, 201 (1970). M. Szwarc, *Accounts Chem. Res.*, **2**, 87 (1969).

394. Botre, C., W. Dorst, M. Marchetti, and A. Memoli, *Biochim. Biophys. Acta.*, **193**, 333 (1969). Yamauchi, A., A. Matsubara, H. Kimizuka, and L. G. Abood, *Biochim. Biophys. Acta.*, **150**, 181 (1968). N. Lakshminarayanaiah, and P. C. Bianchi, *Curr. Mod. Biol.*, **2**, 75 (1968).

395. Bobler, M., J. D. Dunitz, and J. Krajewski, *J. Mol. Biol.*, **42**, 603 (1969).

396. Kilbourn, B. T., J. D. Dunitz, L. A. R. Pioda, and W. Simon, *J. Mol. Biol.*, **30**, 559 (1967).

397. Pinkerton, M., L. K. Steinrauf, and (in part) P. Dawkins, *Biochem. Biophys. Res. Comm.*, **35**, 512 (1969).

398. Steinrauf, L. K., M. Pinkerton, and J. W. Chamberlin, *Biochim. Biophys. Res. Comm.*, **33**, 29 (1968).

399. Pinkerton, M., and L. K. Steinrauf, *J. Mol. Biol.*, **49**, 533 (1970).

400. Paul, I. C., S. M. Johnson, J. Herrin, and S. J. Liu, *Acta. Cryst.*, (Abstr.), **A25**, S196 (1969).

401. Stein, W. D., *Nature*, **218**, 570 (1968).

402. Colacicco, G., E. E. Gordan, and G. Berchenko, *Biophys. J.* (Abstr.), **A-22** (1968).

403. Lauger, P., and G. Stark, *Biochim. Biophys. Acta.*, **211**, 458 (1970).

404. Markin, V. S., *Mol. Biol.*, **3**, 610 (1969). Markin, V. S., V. F. Pastushenko, L. J. Kristalik, E. A. Liberman, and V. P. Topaly, *Biofizika* (in Russian), **14**, 256, 462 (1969). Liberman, E. A., L. A. Pronevich, and V. P. Topaly, *Biofizika*, **15**, 612 (1970).

405. Fassina, G., and P. Dorigo, *Adv. Exp. Med. Biol.*, **4**, 117 (1969).

406. Stillman, I. M., D. L. Gilbert, and M. Robbins, *Biochim. Biophys. Acta.*, **203**, 338 (1970).

407. Ermishkin, L. N., E. A. Liberman, and V. V. Smolyaninov, *Biofizika*, **13**, 205 (1968). Yermishkin, L. N., Ye A. Liberman, and V. V. Smolyaninov, *Biofizika*, **13**, 248 (1968).

408. Lucy, J. A., *J. Theoret. Biol.*, **7**, 360 (1964).

409. Kinsky, S. C., *J. Bacteriol.*, **82**, 889 (1961); *Ann. Rev. Pharmacol.*, **10**, 119 (1970). Lampen, J. O., *Symp. Soc. Gen. Microbiol.*, **16**, 111 (1966).

410. Bean, R. C., W. C. Shepherd, H. Chen, and J. Eichner, *J. Gen. Physiol.*, **53**, 741 (1969). Also Ref. 1318.

411. Ehrenstein, H. Lecar, and R. Nossal, *J. Gen. Physiol.*, **55**, 119 (1970).

412. Hladky, S. B., and D. A. Haydon, *Nature*, **225**, 451 (1970).

413. Redwood, W. R., H. Muldner, and T. E. Thompson, *Proc. Natl. Acad. Sci. (U.S.)*, **64**, 989 (1969).

414. Mueller, P., *Symp. Molecular Basis of Membrane Function*, Durham, N.C., Aug. 1968. Also see Ref. 359.
415. Goodall, M. C., *Nature*, 225, 1258 (1970).
416. Goodall, M. C., *Biochim. Biophys. Acta.*, 203, 28 (1970).
417. *Ibid.*, 219, 28 (1970) and unpublished data.
418. Jain, M. K., and E. H. Cordes, unpublished data.
419. Damel, R. A., F. J. L. Crombag, L. L. M. Van Deenen, and S. C. Kinsky, *Biochim. Biophys. Acta.*, 150, 1 (1968).
420. Weissmann, G., and G. Sessa, *J. Biol. Chem.*, 242, 616 (1967).
421. Weissmann, G., R. Hirshhorn, M. Pras, G. Sessa, and V. A. H. Bevans, *Biochem. Pharmacol.*, 16, 1057 (1967).
422. Ceder, O., and R. Ryhage, *Acta. Chem. Scand.*, 18, 558 (1964). Weber, M. M., and S. C. Kinsky, *J. Bacteriol.*, 89, 306 (1965). Gottlieb, D., H. E. Carter, J. H. Sloneker, L. C. Wu, and E. Gaudy, *Phytopathology*, 51, 321 (1961). Schlosser, E., and D. Gottlieb, *Z. Naturforsch*, 21B, 74 (1966).
423. Damel, R. A., L. L. M. Van Deenen, and S. C. Kinsky, *J. Biol. Chem.*, 240, 2749 (1965).
424. Ghosh, B. K., and A. N. Chatterjee, *Ann. Biochem. Exptl. Med.*, 23, 173, 309 (1963). Kinsky, S. C., *Proc. Natl. Acad. Sci. (U.S.)*, 48, 1049 (1962).
425. Feingold, D. S., *Biochem. Biophys. Res. Comm.*, 19, 261 (1965).
426. Van Etten, J. L., and D. Gottlieb, *J. Gen. Microbiol.*, 46, 377 (1967).
427. Rinehart, K. L., Jr., V. F. German, W. P. Tucker, and D. Gottlieb, *Justus Liebigs Ann. Chem.*, 668, 77 (1963).
428. Kinsky, S. C., G. R. Gronau, and M. M. Weber, *Mol. Pharmacol.*, 1, 190 (1965).
429. Andreoli, T. E., V. W. Dennis, and A. M. Weigl, *J. Gen. Physiol.*, 53, 133 (1969).
430. Cass, A., A. Finkelstein, and V. Krespi, *J. Gen. Physiol.*, 56, 100 (1970).
431. Dennis, V. W., N. W. Stead, and T. E. Andreoli, *J. Gen. Physiol.*, 55, 375 (1970).
432. Kinsky, S. C., S. A. Luse, D. Zopf, L. L. M. Van Deenen, and J. Haxby, *Biochim. Biophys. Acta.*, 135, 844 (1967).
433. Del Castillo, J., A. Rodriguez, and C. A. Romero, *Ann. N.Y. Acad. Sci.*, 144, 803 (1967).
434. Del Castillo, J., A. Rodriguez, C. A. Romero, and V. Sanchez, *Science*, 153, 185 (1966).
435. Barfort, P., E. R. Arquilla, and P. O. Vogelhut, *Science*, 160, 1119 (1968).
436. Hubbel, W. L., and H. M. McConnell, *Proc. Natl. Acad. Sci. (U.S.)*, 63, 16 (1969).
437. Ruttenberg, M. A., T. P. King, and L. C. Craig, *Biochemistry*, 5, 2857 (1966).
438. Laiken, S., M. Printz, and L. C. Craig, *J. Biol. Chem.*, 244, 4454 (1969).
439. Chapman, D., R. J. Cherry, E. G. Finer, H. Hauser, M. C. Phillips, G. G. Shipley, and A. I. McMullen, *Nature*, 224, 692 (1969). Hauser, H., E. G. Finer, and D. Chapman, *J. Mol. Biol.*, 53, 419 (1970). McMullen, A. I., *Biochem. J.*, 119, 10P (1970). Patkau, A., and W. S. Chalack, *Biochim.*

Biophys. Acta, **255**, 161 (1972). For recent data on discrete conductance by alamethicin see L. G. Gordon and D. A. Haydon, *Biochim. Biophys. Acta,* **255**, 1014 (1972). See also Ref. 1317

440. Glazer, V. M., S. A. Silaeva, and S. V. Shestakov, *Biokhimia* (in Russian), **31**, 1135 (1966). Stachiewicz, E., and J. H. Quastel, *Can. J. Biochem. Physiol.,* **41**, 397 (1963). Gordan, H. W., and C. P. Scheffner, *Proc. Natl. Acad. Sci. (U.S.),* **60**, 1201 (1968). Silman, H. I., and A. Karlin, *Proc. Natl. Acad. Sci. (U.S.),* **61**, 674 (1968). Nikiforova, A. A., I. M. Matveeva, and I. M. Treshin, *Antibiotiki (Moscow),* **15**, 31 (1970). Betina, V., H. Barathova, P. Nemec, and Z. Barath, *J. Antibiot. (Tokyo),* **22**, 129 (1969). Wong, R. G., T. H. Adams, P. A. Roberts, and A. W. Norman, *Biochim. Biophys. Acta.,* **219**, 61 (1970).

441. Kinsky, S. C., S. A. Luse, and L. L. M. Van Deenen, *Fed. Proc.,* **25**, 1503 (1966).

442. Seeman, P., and J. Weinstein, *Biochem. Pharm.,* **15**, 1737 (1966).

443. Inglot, A. D., and E. Wolna, *Biochem. Pharm.,* **17**, 269 (1968).

444. Seeman, O., W. O. Kwant, T. Sauks, and W. Argent, *Biochim. Biophys. Acta.,* **183**, 490 (1969). Seeman, P., W. O. Kwant, and T. Sauks, *Biochim. Biophys. Acta.,* **183**, 499 (1969).

445. Seeman, P., and W. O. Kwant, *Biochim. Biophys. Acta.,* **183**, 512 (1969).

446. Seeman, P., *Biochim. Biophys. Acta.,* **183**, 520 (1969).

447. Kwant, W. O., and P. Seeman, *Biochim. Biophys. Acta.,* **183**, 530 (1969). See also 1230.

448. Meyer, K. H., *Trans. Faraday Soc.,* **33**, 1062 (1937).

449. Pauling, L., *Science,* **134**, 15 (1961).

450. Miller, S. L., *Proc. Natl. Acad. Sci.,* **47**, 1515 (1961).

451. Bangham, A. D., *Protoplasma,* **63**, 183 (1967).

452. Papahadjopoulos, D., *Biochim. Biophys. Acta.,* **211**, 467 (1970).

453. Sessa, G., and G. Weissmann, *Biochim. Biophys. Acta.,* **150**, 173 (1968).

454. Katchalsky, A., *Biochim. Biophys. Acta.,* **33**, 120 (1959).

455. Maldrum, B. S., *Parmacol. Rev.,* **17**, 393 (1965).

456. Ponder, E., *J. Exptl. Biol.,* **16**, 38 (1939).

457. Van Zutphen, H., and L. L. M. Van Deenen, *Chem. Phys. Lipids,* **1**, 389 (1967). Also see Henrickson, K. P., and R. C. Henrickson, *Experientia,* **26**, 842 (1970).

458a. Gutknecht, J., and D. C. Tosteson, *J. Gen. Physiol.,* **55**, 359 (1970).

458b. Cherry, R. J., G. H. Good, and D. Chapman, *Biochim. Biophys. Acta.,* **211**, 409 (1970).

459. Seufert, W. D., *Nature,* **207**, 174 (1965).

460. Ohki, S., *Biochim. Biophys. Acta.,* **219**, 18 (1970). Ritchie, J. M., and P. Greengard, *Ann. Rev. Pharmacol.,* **6**, 405 (1966).

461. Voss, J. G., *J. Gen. Microbiol.,* **48**, 391 (1967).

462. Gilby, A. R., and A. V. Few, *J. Gen. Microbiol.,* **23**, 19 (1960).

463. Nash, T., A. C. Allison, and J. S. Harrington, *Nature,* **210**, 259 (1966).

464. Schulman, J. H., B. A. Pethica, A. V. Few, and M. R. J. Salton, *Progr. Biophys. Chem.,* **5**, 41 (1955). Also see M. R. J. Salton, *J. Gen. Physiol.,* **52**, 227s (1968).

465. Kondo, T., and M. Tomizawa, *J. Colloid, Interface Sci.*, **21**, 224 (1966).

466. Eibl, B. H. J., and F. Reman unpublished results quoted in L. L. M. Van Deenen in Ref. 366.

467. Rideal, E., and F. H. Taylor, *Proc. Roy. Soc. B*, **148**, 450 (1958); the Figure is from Rideal, E., and F. H. Taylor, *Proc. Roy. Soc. B*, **146**, 225 (1956–57).

468. Green, H., and B. Goldberg, *Ann. N.Y. Acad. Sci.*, **87**, 352 (1960). Also see Ellory, J. C., and E. M. Tucker, *Nature*, **222**, 477 (1969).

469. Ting, T. P., and R. E. Zirklc, *J. Cell. Comp. Physiol.*, **16**, 189 (1940). Sheppard, C. W., and M. Stewart, *J. Cell. Comp. Physiol.*, **39**. Maroney, S. P. R., *J. Gen. Physiol.*, **44**, 469 (1961). Cook, J. S., *J. Cell. Comp. Physiol.*, **47**, 55 (1956).

470. Inoue, K., and S. C. Kinsky, *Biochemistry*, **9**, 4767 (1970). Haxby, J. A., C. B. Kinsky, and S. C. Kinsky, *Proc. Natl. Acad. Sci. (U.S.)*, **61**, 300 (1968).

471. Berg, H. C., J. M. Diamond, and P. S. Murfey, *Science*, **150**, 64 (1965).

472. Metcalfe, J. C., and A. S. V. Burgen, *Nature*, **220**, 587 (1968).

473. Wagoner, A. S., O. H. Griffith, and C. R. Christensen, *Proc. Natl. Acad. Sci. (U.S.)*, **57**, 1198 (1967). Hubbell, W. L., and H. M. McConnell, *ibid.*, **64**, 20 (1969).

474. Rickenberg, H. W., G. N. Cohen, G. Buttin, and J. Manod, *Ann. L'Institut Pasteur*, **91**, 829 (1956). Cohen, G. N., and J. Manod., *Bacteriol. Rev.*, **21**, 169 (1957).

475. Wilbrandt, W., and T. Rosenberg, *Parmacol. Rev.*, **13**, 109 (1961). Wilbrandt, W., *Protoplasma*, **63**, 299 (1967).

476. Kepes, A., and G. N. Cohen, in *The Bacteria*, Vol. IV, p. 179, Academic Press, New York, (1962). Kepes, A., in *The Cellular Functions of Membrane Transport*, Ed. J. E. Hoffman, Prentice-Hall, Englewood Cliffs, N. J. (1964).

477. Christensen, H., *Biological Transport*, W. A. Benjamin Inc., New York (1962).

478. Whittam, R., *Transport and Diffusion in Red Blood Cells*, Edward Arnold, Publishers, Ltd., London, (1964).

479. Koch, A. L., *Biochim. Biophys. Acta.*, **79**, 177 (1964).

480. Wilbrandt, W., *Experientia*, **25**, 673 (1969).

481. Cohen, G., *The Regulation of Cell Metabolism*, Holt, Rinehart and Winston, Inc., New York (1967). Also see Lacko, L., M. Burger, L. Hejmova, and J. Rejnkova, in *Membrane Transport and Metabolism*, Eds. Kleinzeller, A., and A. Kotyk, Czechoslav Acad of Science Publishing House, and Academic Press, p. 399, (1961). Orlowski, M., and A. Meister, *Proc. Natl. Acad. Sci. (U.S.)*, **67**, 1248 (1970).

482. Shyamala, M. B., *J. Sci. Indust. Res.*, **28**, 118 (1969).

483. Stein, W. D., *Brit. Med. Bull.*, **24**, 146 (1968).

484. Berlin, R. D., *Science*, **168**, 1539 (1970).

485. Mitchell, P., *Advances in Enzymology*, **29**, (1967).

486. Lusk, J. E., and E. P. Kennedy, *J. Biol. Chem.*, **244**, 1653 (1969). Silver, S., *Proc. Natl. Acad. Sci. (U.S.)*, **62**, 764 (1969).

487. Pardee, A. B., *Science,* **156,** 1627 (1967). Pardee, A. B., and L. S. Prestidge, *Proc. Natl. Acad. Sci. (U.S.),* **55,** 189 (1966). Pardee, A. B., L. S. Prestidge, M. B. Whipple, and J. Dreyfuss, *J. Biol. Chem.,* **241,** 3962 (1966).

488. Fox, C. F., and E. P. Kennedy, *Proc. Natl. Acad. Sci.,* **54,** 891 (1965).

489. Winkler, H. H., *Biochim. Biophys. Acta.,* **117,** 231 (1966).

490. Dietz, G., and L. A. Heppel, *J. Biol. Chem.,* **246,** 2881, 2885, 2891 (1971).

491. Schleif, R., *J. Mol. Biol.,* **46,** 185 (1969).

492. Wiesmeyer, H., and M. Cohn, *Biochim. Biophys. Acta.,* **39,** 440 (1960).

493. Carter, J. R., C. F. Fox, and E. P. Kennedy, *Proc. Natl. Acad. Sci. (U.S.),* **60,** 725 (1968).

494. Stoeber, F., *Compt. Rend.,* **244C,** 1091 (1957).

495. Anraku, Y., *J. Biol. Chem.,* **243,** 3116, 3123, 3128 (1968).

496. Kaback, H. R., and E. R. Stadtman, *Proc. Natl. Acad. Sci. (U.S.),* **55,** 920 (1966). See also Ref. 1319

497. Kaback, H. R., and E. R. Stadtman, *J. Biol. Chem.,* **243,** 1390 (1968).

498. Nakane, P. K., G. E. Nicoalds, and D. L. Oxender, *Science,* **161,** 182 (1968).

499. Wilson, O. H., and J. T. Holden, *J. Biol. Chem.,* **244,** 2743 (1969).

500. Weiner, J. H., E. A. Berger, M. N. Hamilton, and L. A. Heppel, *Fed. Proc.,* **29,** 341 (abstr.) (1970). Weiner, J. H., C. E. Furlong, and L. A. Heppel, *Arch. Biochem. Biophys.,* **142,** 715 (1971).

501. Winter, C. G., and H. N. Christensen, *J. Biol. Chem.,* **240,** 3594 (1965).

502. Widdas, W. F., in *Carbohydrate Metabolism and its Disorders,* Vol. I, p. 1, Eds. Dickens, F., P. J. Randle, and W. J. Whelen, Academic Press, New York (1968).

503. Miller, D. M., in *Red Cell Membrane: Structure and Function,* p. 240, Eds. Jamieson, G. A., and T. J. Greenwalt, J. B. Lippincott Co., Philadelphia, (1969).

504. Hawkins, R. A., and R. D. Berlin, *Biochim. Biophys. Acta.,* **173,** 324 (1969), and Ref. therein.

505. Askari, A., *J. Gen. Physiol.,* **49,** 1147 (1966).

506. Arme, C., and C. P. Read, *Biol. Bull.,* **135,** 80 (1968).

507. Chappell, L. H., C. Arme, and C. P. Read, *Biol. Bull.,* **136,** 313 (1969).

508. Read, C. P;, A. H. Rothman, and J. E. Simmons, Jr., *Ann. N.Y. Acad. Sci.,* **113,** 154 (1963).

509. Segal, S., and I. Smith, *Proc. Natl. Acad. Sci. (U.S.),* **63,** 926 (1969). Scriver, C. R., and O. H. Wilson, *Science,* **155,** 1428 (1967).

510. Shannon, J. A., *Physiol. Rev.,* **19,** 63 (1939). Wesson, L. G., *Physiology of the Human Kidney,* Grune and Stratton, New York (1969).

511. Silverman, M., M. A. Agnon, and F. P. Chinard, *Amer. J. Physiol.,* **218,** 743 (1970).

512. Hillman, R. E., and L. E. Rosenberg, *J. Biol. Chem.,* **244,** 4494 (1969). Segal, S., and I. Smith, *Biochim. Biophys. Res. Comm.,* **35,** 771 (1969).

513. Neame, K. D., *Progress in Brain Research,* **29,** 185 (1968).

514. Quastel, J. H., *Neuroscience Res.,* **3,** 1 (1970).

515. Bourke, R. S., *Exp. Brain Res.*, **8**, 219 (1969).
516. Itokawa, Y., and J. R. Cooper, *Biochem. Parmacol.*, **18**, 545 (1969). Potter, L. T., in *The Interaction of Drugs and Subcellular Components in Animal Cells*, p. 293, Ed. P. N. Campbell, and J. & A. Churchill Ltd., London (1968).
517. Chaudry, I. H., and M. K. Gould, *Biochim. Biophys. Acta.*, **196**, 320 (1970).
518. Korenman, S. G., and B. R. Rao, *Proc. Natl. Acad. Sci. (U.S.)*, **61**, 1028 (1968).
519. Rogers, T. O., and H. C. Lichstein, *J. Bacteriol.*, **100**, 565 (1969).
520. Cirillo, V. P., in *Membrane Transport and Metabolism*, Eds. A. Kleinzeller, and A. Kotyk, p. 343, Academic Press (1960). Cirillo, V. P., P. O. Wilkins, and J. Anton, *J. Bacteriol.*, **86**, 1259 (1963). Divies, C., and J. N. Morfauz, *Ann. Technol. Agr.*, **17**, 355 (1968).
521. LeFevre, P. G., and R. I. Davies, *J. Gen. Physiol.*, **34**, 515 (1951).
522. LeFevre, P. G., and J. K. Marchall, *Amer. J. Physiol.*, **194**, 333 (1958).
523. LeFevre, P. G., *Pharmacol. Rev.*, **13**, 39 (1961).
524. Wilbrandt, W., *J. Cell Physiol.*, **47**, 137 (1956).
525. Lacko, L., and M. Burger, *Nature*, **191**, 881 (1961).
526. Sen, A. K., and W. F. Widdas, *J. Physiol.*, **160**, 392 (1962).
527. Shimmin, E. R. A., and W. D. Stein, *Biochim. Biophys. Acta.*, **211**, 308 (1970).
528. Metthews, R. H., C. A. Lestie, and P. G. Scholefield, *Biochim. Biophys. Acta.*, **203**, 457 (1970).
529. Schwencke, N. M., and J. Schwencke, *Biochim. Biophys. Acta.*, **173**, 313 (1969). Pall, M. L., *Biochim. Biophys. Acta.*, **211**, 139, 513 (1970). Scarborough, G. A., *J. Biol. Chem.*, **245**, 1694 (1970). Yabu, K., *Biochim. Biophys. Acta.*, **135**, 181 (1967). Hopfer, M., *J. Membrane Biol.*, **3**, 73 (1970).
530. Gryder, R. M., and E. Adams, *J. Bacteriol.*, **101**, 948 (1970). Kay, W. W., and A. F. Gronlund, *Biochim. Biophys. Acta.*, **193**, 444 (1969).
531. Widdas, W. F., *J. Physiol. (London)*, **118**, 23 (1952).
532. Morgan, H. E., D. M. Regen, and C. R. Park, *J. Biol. Chem.*, **239**, 369 (1964).
533. Winkler, H. H., and T. H. Wilson, *J. Biol. Chem.*, **241**, 2200 (1966).
534. Rosenberg, T., and W. Wilbrandt, *J. Gen. Physiol.*, **41**, 289 (1957). Wilbrandt, W., S. Frei, and C. Becker, unpublished results quoted in Ref. 480.
535. Park, C. R., R. L. Post, C. F. Kalman, J. H. Wright, Jr., L. H. Johnson, and H. E. Morgan, *Ciba Foundation Symp. on Endocrinology*, **9**, 240 (1956).
536. Alvarado, F., T. Ramon, L. Mateu, and J. W. L. Robinson, *FEBS Letters*, **8**, 153 (1970).
537. Reuter, H., and N. Seitz, *J. Physiol. (London)*, **195**, 451 (1968).
538. Blaustein, M. P., and A. L. Hodgkin, *J. Physiol. (London)*, **198**, 46P (1968).
539. Britton, H. G., *Nature*, **225**, 746 (1970).
540. Levine, M., D. L. Oxender, and W. D. Stein, *Biochim. Biophys. Acta.*, **109**, 151 (1965).

541. Bolis, L., P. Luly, B. A. Pethica, and W. Wilbrandt, *J. Membrane Biol.*, **3**, 83 (1970).

542. Van Den Bosch, J. R. Williamson, and P. R. Vagelos, *Nature*, **228**, 338 (1970).

543. Pardee, A. B., *Science*, **162**, 632 (1968).

544. Kotyk, A., in *Membrane: Structure and Function*, Eds. J. R. Villanueva, and F. Ponz, Academic Press, New York (1970), p. 99.

545. Heppel, L. A., *J. Gen. Physiol.*, **54**, 95s (1969). For osmotic shock release of lipids see Gale, E. F., and J. M. Llewellin, *Biochim. Biophys. Acta.*, **222**, 546 (1970).

546. Taylor, A. N., and R. H. Wasserman, *Arch. Biochem. Biophys.*, **119**, 536 (1967).

547. Jones, T. H. D., and E. P. Kennedy, *Fed. Proc.*, **27**, 644 (1968); *J. Biol. Chem.*, **244**, 5981 (1969).

548. Haskovec, C., and A. Kotyk, *Europ. J. Biochem.*, **9**, 343 (1969). Cirillo, V. P., *J. Bacteriol.*, **95**, 603 (1968).

549. Kotyk, A., *Folia microbiol.*, *Praha*, **12**, 121 (1967). Azam, F., and A. Kotyk, *FEBS Letters*, **2**, 333 (1969).

550. Bobinksy, H., and W. D. Stein, *Nature*, **211**, 1366 (1966). Bonsall, R. B., and S. Hunt, *Nature*, **211**, 1368 (1966). Langdon, R. G., and H. R. Sloan, *Proc. Natl. Acad. Sci. (U.S.)*, **57**, 401 (1967).

551. Wasserman, R. H., R. A. Corrandino, and A. N. Taylor, *J. Biol. Chem.*, **243**, 3978 (1968).

552. Rosen, B. P., and F. D. Vashington, *Fed. Proc.*, **29**, 342 (abstr.) (1970).

553. Medvezky, N., and H. Rosenberg, *Biochim. Biophys. Acta.*, **211**, 158 (1970).

554. Wilson, G., S. P. Rose, and C. F. Fox, *Biochem. Biophys. Res. Comm.*, **38**, 617 (1970).

555. Smith, P. F., *Lipids*, **4**, 331 (1969).

556. Hsia, D. Y., *Inborn Errors of Metabolism*, Part I, Year-Book Medical Publishers, Inc., Chicago (1967). Halvorson, S., O. Hygstedf, R. Jagenburg, and O. Sjaastad, *J. Clin. Invest.*, **48**, 1552 (1969). Hess, C. E., and D. N. Moher, *Proc. Soc. Exp. Biol. Med.*, **128**, 1042 (1968). Wasserman, R. H., and F. A. Kallfelz, *Biological Calcification: Cellular and Molecular Aspects*, Ed. H. Schraer, Appleton-Century-Crofts, New York (1970), p. 313. Lin, E. C. C., *Ann. Rev. Genetics*, **4**, 225 (1970). Wilson, T. H., M. Kusch, and E. R. Kashket, *Biochem. Biophys. Res. Comm.*, **40**, 1409 (1970). Scriver, C. R., *Advances in Human Genetics*, **1**, 211 (1970). Baerlocher, K. E., C. R. Scriver, and F. Mohyuddin, *Proc. Natl. Acad. Sci. (U.S.)*, **65**, 1009 (1970).

557. Frye, B. E., *Hormonal Control in Vertebrates*, The Macmillan Co., New York (1967).

558. Hendler, R. W., *Protein Biochemistry and Membrane Biochemistry*, John Wiley, New York (1968). Segal, S., *Nature*, **203**, 17 (1964). Wool, I. G., *Fed. Proc.*, **24**, 1060 (1965); Wool, I. G., W. S. Stirewatt, K. Kurihara, R. B. Low, and P. Bailey, *Recent Progress in Hormone Res.*, **24**, 139

(1968). Wisniewski, K., *Postepy Biochem.*, 16, 33 (1970). Also see Ref. 1125.

559. Randle, P. J., P. B. Garland, C. N. Hales, E. A. Newsholme, R. M. Denton, and C. I. Pogson, *Recent Progress in Hormone Research,* 22, 1 (1966).

560. Randle, P. J., S. J. H. Ashcroft, and J. R. Gill, in *Carbohydrate Metabolism and its Disorders,* Vol. I, p. 427, Eds. F. Dickens, P. J. Randle, and W. J. Whelan, Academic Press, N.Y. (1968).

561. Levine, R., *Fed. Proc.,* 24, 1071 (1965). Crofford, O. B., and A. E. Renold, *J. Biol. Chem.,* 240, 14 (1965).

562. Denton, R. M., R. E. Yorke, and P. J. Randle, *Biochem. J.,* 100, 407 (1966).

563. Weis, L. S., and H. T. Narahara, *J. Biol. Chem.,* 244, 3084 (1969).

564. Holloszy, J. O., and H. T. Narahara, *J. Biol. Chem.,* 240, 3493 (1965). Morgan, H. E., J. R. Neely, R. E. Wood, C. Liebecq, H. Liebermeister, and C. R. Park, *Fed. Proc.,* 24, 1040 (1965).

565. Riggs, T. R., in *Actions of Hormones on Molecular Processes,* p. 1, Eds. G. Litwack, and D. Kritchevsky, John Wiley, New York (1964). Sharp, G. W. G., and A. Leaf, *Recent Progress in Hormone Research,* 22, 1 (1966). E. Schoffeniels, *Cellular Aspects of Membrane Permeability,* Pergamon Press, New York (1967). Clegg, P. C., and A. C. Clegg, *Hormones, Cells and Organisms,* Stanford University Press, Stanford, Calif. (1969). Riggs, T. R., Biochemical Action of Hormones, 1, 157 (1970).

566. Maddaiah, V. T., *J. Theoret. Biol.,* 25, 495 (1969).

567. LeFevre, P. G., *Am. J. Physiol.,* 203, 286 (1962).

568. Winter, C. G., and H. N. Christensen, *J. Biol. Chem.,* 239, 872 (1964).

569. Mawe, R. C., and H. G. Hempling, *J. Cell. Physiol.,* 66, 95 (1965).

570. Levine, M., D. L. Oxender, and W. D. Stein, *Biochim. Biophys. Acta.,* 109, 151 (1965).

571. Miller, D. M., *Biophys. J.,* 8, 1339 (1968).

572. Miller, D. M., *Biophys. J.,* 5, 417 (1965).

573. Miller, D. M., *Biophys. J.,* 8, 1329 (1968).

574. Levine, M., and S. Levine, *J. Theoret. Biol.,* 24, 85 (1969).

575. Vidaver, G. A., *J. Theoret. Biol.,* 10, 301 (1966).

576. Patlak, C. S., *Bull. Math. Biophys.,* 19, 209 (1957).

577. Hill, T. L., and O. Kedem, *J. Theoret. Biol.,* 10, 399 (1966).

578. Silverman, M., and C. A. Goresky, *Biophys. J.,* 5, 487 (1965).

579. LeFevre, P. G., K. I. Habich, H. S. Hess, and M. R. Hudson, *Science,* 143, 955 (1964). Mawdsley, D. A., and W. F. Widdas, *J. Physiol.,* 189, 75P (1967). Moore, T. J., and B. Schlowsky, *Chem. Phys. Lipids,* 3, 273 (1969).

579a. Ames, G. F., and J. Lever, *Proc. Natl. Acad. Sci. (U.S.),* 66, 1096 (1970). Reid, K. G., N. M. Utech, and J. T. Holden, *J. Biol. Chem.,* 245, 5261 (1970).

580. Scholtz, S. G., and P. F. Curran, *Physiol. Rev.,* 50, 637 (1970). Also Crane, R. K., in *Carbohydrate Metabolism and its Disorders,* Vol. I, p. 25,

Eds. F. Dickens, P. J. Randle, and W. J. Whelan, Academic Press, New York (1968).

581. Csaky, T. Z., *Fed. Proc.*, **22**, 3 (1963). Crane, R. K., G. Forstner, and A. Eichholz, *Biochim. Biophys. Acta.*, **109**, 467 (1965).

582. Goldner, A. M., S. G. Schultz, and P. F. Curran, *J. Gen. Physiol.*, **53**, 362 (1969).

583. Curran, P. F., S. G. Schultz, R. A. Chez, and R. E. Fuisz, *J. Gen. Physiol.*, **50**, 1261 (1967).

584. Chez, R. A., R. R. Palmer, S. G. Schultz, and P. F. Curran, *J. Gen. Physiol.*, **50**, 2357 (1969).

585. Kleinzeller, A., J. Koliska, and I. Benes, *Biochem. J.*, **104**, 852 (1967).

586. Kolinska, J., and G. Semenza, *Biochim. Biophys. Acta.*, **146**, 181 (1967).

587. Vidaver, G. A., *Biochemistry*, **3**, 795 (1964).

588. Read, C. P., *Biol. Bull.*, **133**, 630 (1967).

589. Jacqez, J. A., and J. H. Sherman, *Biochim. Biophys. Acta.*, **109**, 128 (1965).

590. Limmich, G. A., *Biochemistry*, **9**, 3669 (1970).

591. Jaquez, J. A., and J. A. Schaffer, *Biochim. Biophys. Acta.*, **193**, 368 (1969).

592. Stevenson, J., *Biochem. J.*, **99**, 257 (1966).

593. Schultz, S. C., and R. Zalusky, *J. Gen. Physiol.*, **47**, 567 (1964).

594. Wright, E. M., *J. Physiol.*, **185**, 486 (1966).

595. Vidaver, G. A., L. F. Romain, and F. Haurowitz, *Arch. Biochim. Biophys.*, **107**, 82 (1964).

596. Crane, R. K., in *Intracellular Transport*, Symp. International Society for Cell Biology, Vol. V, p. 71, Ed. K. B. Warren, Academic Press (1966).

597. Rose, R. C., and S. G. Schultz, *Biochim. Biophys. Acta.*, **211**, 376 (1970); *J. Gen. Physiol.*, **57**, 639 (1971).

598. Barry, R. J. C., D. H. Smyth, and E. M. Wright, *J. Physiol.*, **181**, 410 (1965). Barry, R. J. C., J. Eggenton, and D. H. Smyth, *J. Physiol.*, **204**, 299 (1969).

599. White, J. F., and W. McD. Armstrong, *Biophys. Soc. Ann. Meetings*, Baltimore (1970) (Abstr., p. 36a).

600. Kitner, W. D., and Wilson, T. H., *J. Cell. Biol.*, **25** (2), 19 (1965).

601. Semenza, G., in *Protides of the Biological Fluids*, Vol. 15, p. 201, Elsevier Publishing Co., Amsterdam (1967). Semenza, G., in *Membranes: Structure and Function*, p. 117, Eds., J. R. Villaneueva, and F. Ponz, Academic Press, New York (1970). Gitzelmann, R., Th. Bachi, H. Binz, J. Lindenmann, and G. Semenza, *Biochim. Biophys. Acta.*, **196**, 20 (1970). Also see Ref. 586.

602. Srivastava, L. M., P. Shakespeare, and G. Hubscher, *Biochem. J.*, **109**, 35 (1968). Shakespeare, P., L. M. Srivastava, and G. Hubscher, *Biochem. J.*, **111**, 63 (1969). Mayer, R. J., P. Shakespeare, and G. Hubscher, *Biochem. J.*, **116**, 43 (1970).

603. Simoni, R. D., M. F. Smith, and S. Roseman, *Biochim. Biophys. Res. Comm.*, **31**, 804 (1968).

604. Hangstenberg, W., J. B. Egan, and M. L. Morse, *J. Biol. Chem.*, **243**, 1881 (1968).
605. Laue, P., and R. E. MacDonald, *Biochim. Biophys. Acta.*, **165**, 410 (1968). Egan, J. B., and M. L. Morse, *Biochim. Biophys. Acta.*, **97**, 310 (1965); **109**, 172 (1965); **112**, 63 (1966).
606. Kashket, E. R., and T. H. Wilson, *Bacteriol. Proc. Abstr.*, p. 49 (1969). Rose, S. P., and C. F. Fox, *Fed. Proc.*, (Abstr.) **28**, 1145 (1969).
607. Silbert, D. F., F. Ruch, and P. R. Vagelos, *J. Bacteriol.*, **95**, 1658 (1968).
608. Hansen, T. E., and R. L. Anderson, *Proc. Natl. Acad. Sci. (U.S.)*, **61**, 269 (1968).
609. Simoni, R. D., M. Levinthal, F. D. Kundig, W. Kundig, B. Anderson, P. E. Hartman, and S. Roseman, *Proc. Natl. Acad. Sci. (U.S.)*, **58**, 1963 (1967).
610. Kaback, H. R., *Ann. Rev. Biochem.*, **39**, 561 (1970); *Current Topics in Membranes and Transport*, **1**, 35 (1970).
611. Phibbs, Jr., P. V., and R. G. Eagon, *Arch. Biochim. Biophys.*, **138**, 470 (1970).
612. Winkler, H. T., and T. H. Wilson, *J. Biol. Chem.*, **241**, 2200 (1966).
613. Kundig, W., F. D. Kundig, B. Anderson, and S. Roseman, *J. Biol. Chem.*, **241**, 3243 (1966).
614. Tanaka, S., and E. C. C. Lin, *Proc. Natl. Acad. Sci. (U.S.)*, **57**, 913 (1967).
615. Tanaka, S., D. G. Fraenkel, and E. C. C. Lin, *Biochim. Biophys. Res. Comm.*, **27**, 63 (1967).
616. Hengstenberg, W., J. B. Egan, and M. L. Morse, *Proc. Natl. Acad. Sci.*, **58**, 274 (1967).
617. Roseman, S., *J. Gen. Physiol.*, **5**, 139s (1969). Gachelin, G., and A. Kepes, in *Membrane: Structure and Function*, p. 87, Eds. J. R. Villanueva, and F. Ponz, Academic Press (1970).
618. Kaback, H. R., *Proc. Natl. Acad. Sci. (U.S.)*, **63**, 724 (1969).
618a. Milner, L. S., and H. R. Kaback, *Proc. Natl. Acad. Sci. (U.S.)*, **65**, 683 (1970).
619. Pavlasava, E., and F. M. Harold, *J. Bacteriol.*, **98**, 198 (1969). Harold, F. M., and J. R. Baarda, *J. Bacteriol.*, **94**, 53 (1967); *J. Bacteriol.*, **95**, 816 (1968). Herold, F. M., J. R. Baarda, and E. Pavlasova, *J. Bacteriol.*, **101**, 152 (1970). West, I. C., *Biochem. Biophys. Res. Comm.*, **41**, 655 (1970). Also see Spear, D. G., J. K. Barr, and C. E. Barr, *J. Gen. Physiol.*, **54**, 397 (1969). Gershanovich, V. N., and D. N. Mandzhgaladze, *Biokhimiya*, **35**, 657 (1970).
620. Winkler, H. H., and T. H. Wilson, *J. Biol. Chem.*, **241**, 2200 (1966). Scarborough, G. A., M. K. Rumley, and E. P. Kennedy, *Proc. Natl. Acad. Sci. (U.S.)*, **60**, 951 (1968).
621. Wong, P. T. S., E. R. Kashket, and T. H. Wilson, *Proc. Natl. Acad. Sci. (U.S.)*, **65**, 63 (1970). Schachter, D., and A. J. Mindlin, *J. Biol. Chem.*, **244**, 1808 (1969).
622. Mackenzie, S., and C. R. Scriver, *Biochim. Biophys. Acta.*, **196**, 110 (1970). Scarborough, G. A., *J. Biol. Chem.*, **245**, 1694 (1970).
623. Koch, A. L., *J. Mol. Biol.*, **59**, 447 (1971).

624. Ring, K., W. Gross, and E. Heinz, *Arch. Biochim. Biophys.*, **137**, 243 (1970).
625. Gross, W., K. Ring, and E. Heinz, *Arch. Biochim. Biophys.*, **137**, 253 (1970).
626. Rotman, B., and J. Radojkovic, *J. Biol. Chem.*, **239**, 3153 (1964).
627. Barnes, E. M., Jr., and H. R. Kaback, *Proc. Natl. Acad. Sci. (U.S.)*, **66**, 1190 (1970).
628. Kaback, H. R., and L. S. Milner, *Proc. Natl. Acad. Sci. (U.S.)*, **66**, 1008 (1970).
629. Smith, F. A., *Biochim. Biophys. Acta.*, **126**, 94 (1966).
630. Rosenberg, H., N. Medveczky, and J. M. La Nauze, *Biochim. Biophys. Acta.*, **193**, 159 (1969).
631. Cerbon, J., *J. Bacteriol.*, **97**, 658 (1969); **102**, 97 (1970).
632. Macrobbie, E. A. C., *Biochim. Biophys. Acta.*, **94**, 64 (1965).
633. Luttge, U., C. K. Pallaghy, and C. B. Osmond, *J. Membrane Biol.*, **1**, 17 (1970).
634. Snoswell, A. M., *Biochemistry*, **5**, 1660 (1966).
635. Henin, S., A. Bianchi, and C. Lippe, *Experientia*, **25**, 701 (1969).
636. White, T. D., and P. Keen, *Biochim. Biophys. Acta.*, **196**, 285 (1970).
637. Kinter, W. B., and A. L. Cline, *Amer. J. Physiol.*, **201**, 309 (1961).
638. Cunningham, D. D., and A. B. Pardee, *Proc. Natl. Acad. Sci. (U.S.)*, **64**, 1049 (1969).
639. Berlin, R. D., *Science*, **163**, 1194 (1969).
640. Peterson, R. N., and A. L. Koch, *Biochim. Biophys. Acta.*, **126**, 129 (1966).
641. Maretzki, A., and M. Thom, *Biochemistry*, **9**, 2731 (1970).
642. Wise, W. C., and J. W. Archdeacon, *J. Gen. Physiol.*, **53**, 487 (1969).
643. Klotz, I. M., *Energy Changes in Biochemical Reactions*, Academic Press, New York (1967). Also see Ref. 719
644. Boyd, D. B., and W. N. Lipscomb, *J. Theoret. Biol.*, **25**, 403 (1969). Wald, G., in *Horizons in Biochemistry*, p. 127, Eds. M. Kasha, and B. Pullman, Academic Press, New York (1962). Banks, B. E. C., *Chemistry in Britain*, **5**, 514 (1969). Banks, B. E. C., and C. A. Vernon, *J. Theoret. Biol.*, **29**, 301 (1970). Also see comments by L. Pauling, *Chemistry in Britain*, **6**, 468 (1970). Huxley, A. E., *Chemistry in Britain*, **6**, 477 (1970). Wilkie, D., *Chemistry in Britain*, **6**, 472 (1970).
645. Shen, L. C., and D. E. Atkinson, *Biochem. J.*, **245**, 5974 (1970). D. L. Purich, and H. J. Fromm, *J. Biol. Chem.*, **247**, 249 (1972).
646. Crane, E. E., and R. E. Davies, *Biochem. J.*, **49**, 169 (1951).
647. Ussing, H. H., *Protoplasma*, **63**, 292 (1967).
648. Kushmerick, M. J., and R. E. Davies, *Proc. Roy. Soc. B*, **174**, 315 (1969).
649. O'Kelley, J. C., G. E. Becker, and A. Nason, *Biochim. Biophys. Acta.*, **205**, 409 (1970). O'Kelley, J. C., and A. Nason, *Biochim. Biophys. Acta.*, **205**, 426 (1970). Sinclair, P. R., and D. C. White, *J. Bacteriol.*, **101**, 365 (1970).
650. Sojka, G. A., A. Baccarini, and H. Gest, *Science*, **166**, 113 (1969).
651. Bowen, E. J. (Ed.), *Luminescence in Chemistry*, D. Van Nostrand Co.,

Ltd., London (1968). Seliger, H. H., and W. D. McElroy, *Light: Physical and Biological Action*, Academic Press, New York (1965). McElroy, W. D., H. H. Seliger, and E. H. White, *Photochemistry and Photobiology*, **10**, 153 (1969). Karkhanis, Y. D., and M. J. Cormier, *Biochemistry*, **10**, 317 (1971).

652. Shimomura, O., and F. H. Johnson, *Biochemistry*, **8**, 3991 (1969).

653. Jardetzky, O., *Biochim. Biophys. Acta.*, **79**, 631 (1964).

654. Lewis, P. R., *Biochem. J.*, **52**, 330 (1952).

655. Canessa-Fischer, M., F. Zambrano, and M. Rojas, *J. Gen. Physiol.*, **51**, 162s (1968).

656. Hodgkin, A. L., and R. D. Keynes, *J. Physiol. (London)*, **128**, 28 (1955).

657. Garrahan, P. J., and I. M. Glynn, *J. Physiol. (London)*, **192**, 159 (1967).

658. *Ibid.*, **192**, 175 (1967).

659. *Ibid.*, **192**, 189 (1967).

660. *Ibid.*, **192**, 217 (1967).

661. *Ibid.*, **192**, 237 (1967).

662. Baker, P. F., *Biochim. Biophys. Acta.*, **88**, 458 (1964). R. A. Sjodin (p. 96) in Ref. 1307.

663. Baker, P. F., *Brit. Med. Bull.*, **24**, 179 (1968).

664. Hoffman, J. F., and F. M. Kregnow, *Ann. N.Y. Acad. Sci.*, **137**, 566 (1966).

665. Glynn, I. M., and V. L. Lew, *J. Physiol. (London)*, **207**, 393 (1970). Reid, R. A., *Biochem. J.*, **116**, 12P (1970). Lant, A. F., and R. Whittam, *J. Physiol. (London)*, **197**, 66P (1968). Lew, V. L., I. M. Glynn, and J. C. Ellory, *Nature*, **225**, 865 (1970). Lant, A. F., R. N. Priestland, and R. Whittam, *J. Physiol. (London)*, **207**, 291 (1970).

666. Martin, K., and T. I. Shaw, *J. Physiol. (London)*, **208**, 171 (1970).

667. Curran, P. F., and M. Cereijido, *J. Gen. Physiol.*, **48**, 1011 (1965). Ussing, H. H., and K. Zerahn, *Acta. Physiol. Scand.*, **23**, 110 (1951). Ussing, H., *Quart. Rev. Biophys.*, **1**, 365 (1969). Keynes, R. D., *Quart. Rev. Biophys.*, **2**, 177 (1969). Also see Refs. 1242–1245.

668. Koefoed-Johnson, V., and H. H. Ussing, *Acta. Physiol. Scand.*, **42**, 298 (1958).

669. Ussing, H. H., in *The Molecular Basis of Membrane Function*, p. 577, Ed. D. C. Tosteson, Prentice-Hall Inc., Englewood-Cliffs, N.J. (1969). Cereijido, M., and C. A. Rotunno, *Introduction to the Study of Biological Membranes*, p. 163, Gordon & Beach Science Publishers, New York, (1970).

670. Snell, F. M., and T. K. Chowdhury, in *Intracellular Transport*, Intern. Symp., Soc. Cell Biology, Vol. 5, p. 141, Academic Press, New York, (1966). Bentley, P. J., *J. Physiol.*, **195**, 317 (1968). Vanatta, J. C., and L. A. Bryant, *Proc. Soc. Exp. Biol. Med.*, **133**, 385 (1970). See also Ref. 1245.

671. Kernan, R. P., *Nature*, **193**, 986 (1962). Adrian, R. H., and C. L. Slayman, *J. Physiol. (London)*, **184**, 970 (1966). Cross, S. B., R. D. Kaynes, and R. Rybova, *J. Physiol. (London)*, **181**, 865 (1965).

672. Tamani, T., and S. Kagiyama, *Circ. Res.*, **22**, 423 (1968). Page, E., and S. R. Strom, *J. Gen. Physiol.*, **48**, 957 (1965).

673. Nicholls, J. G., and D. A. Baylor, *Science*, **162**, 279 (1968). Marmor,

M. F., *Nature,* **226,** 1252 (1970). Kerkut, G. A., and R. C. Thomas, *Comp. Biochem. Physiol.,* **14,** 167 (1965). Thomas, R. C., *J. Physiol.,* **201,** 495 (1969). Carpenter, D. O., and B. O. Alving, *J. Gen. Physiol.,* **52,** 1 (1968).

674. Nakajima, S., and K. Takahashi, *J. Physiol. (London),* **187,** 105 (1966).

675. Rang, H. P., and J. M. Ritchie, *J. Physiol. (London),* **196,** 183 (1968). Also see Mullins, L. J., and F. J. Brinley, *J. Gen. Physiol.,* **53,** 704 (1969). See also Ref. 1243.

676. Taylor, G. S., D. M. Paton, and E. E. Daniel, *J. Gen. Physiol.,* **56,** 360 (1970).

677a. Moriarty, C. M., and A. R. Terepka, *Arch. Biochem. Biophys.,* **135,** 160 (1969).

677b. Gonzalez, C. F., *Biochim. Biophys. Acta.,* **193,** 146 (1969).

677c. Holland, M. G., and C. C. Gipson, *Investigative Ophthalmology,* **9,** 20 (1970). For active transport across plant cells see, N. Higinbotham, J. S. Graves, and R. F. Davis, *J. Membrane Biol.,* **3,** 210 (1970).

678. Skou, J. C., *Physiol. Rev.,* **45,** 596 (1965). Skou, J. C., in *Molecular Basis of Membrane Function,* Ed. D. C. Tosteson, Prentice-Hall, Englewood-Cliffs, N.J., (1969).

679. Jorgensen, P. L., and J. C. Skou, *Biochim. Biophys. Res. Comm.,* **37,** 39 (1969); *Biochim. Biophys. Acta.,* **233,** 366 (1971).

680. Nagano, K., T. Kanazawa, N. Mizuno, Y. Tashima, T. Nakao, and M. Nakao, *Biochim. Biophys. Res. Comm.,* **19,** 759 (1965).

681. Kahlenberg, A., P. R. Galsworthy, and L. E. Hokin, *Arch. Biochem. Biophys.,* **126,** 331 (1968).

682. Schoner, W., R. Beusch, and R. Kramer, *Eur. J. Biochem.,* **7,** 102 (1968).

683. Tashmukhamedov, B. A., and N. A. Niyazmetova, *Uzb. Biol. Zh.,* **13,** 76 (1969). Skou, J. C., and C. Hilberg, *Biochim. Biophys. Acta.,* **185,** 198 (1969).

684. Ruoho, A. E., L. E. Hokin, R. J. Hemingway, and S. M. Kupchan, *Science,* **159,** 1354 (1968).

685. Fenster, L. J., and J. H. Copenhaver, Jr., *Biochim. Biophys. Acta.,* **137,** 406 (1967). Hegyvary, C., and R. L. Post, in *The Molecular Basis of Membrane Function,* p. 519, Ed. D. C. Tosteson, Prentice-Hall, Inc., Englewood-Cliffs, N.J., (1969). Wheeler, K. P., R. Whittam, *J. Physiol. (London),* **207,** 303 (1970). Also see Faria, R. N., A. L. Goldemberg, and R. E. Trucco, *Arch. Biochem. Biophys.,* **139,** 38 (1970). However see K. D. Wheeler, *Biochem. J.,* **125,** 71P (1971).

686. Tanaka, R., and T. Sakamoto, *Biochim. Biophys. Acta.,* **193,** 384 (1969).

687. Jain, M. K., A. Strickholm, and E. H. Cordes, *Nature,* **222,** 871 (1969). M. Goodall, personal communication.

688. Bader, H., R. L. Post, and G. H. Bond, *Biochim. Biophys. Acta.,* **150,** 41 (1968).

689. Askari, A., and D. Koval, *Biochem. Biophys. Res. Comm.,* **32,** 227 (1968). Also see Whittam, R., and K. P. Wheeler, *Ann. Rev. Physiol.,* **32,** 21 (1970).

690. Dunham, P. B., and J. F. Hoffman, *Proc. Natl. Acad. Sci. (U.S.)*, **66**, 936 (1970). Chignell, C. F., and E. Titus, *Proc. Natl. Acad. Sci. (U.S.)*, **64**, 324 (1969).

691. Suelter, C. H., *Science*, **168**, 789 (1970).

692. Tobin, T., S. P. Banerjee, and A. K. Sen, *Nature*, **225**, 745 (1970). Tobin, T., and A. K. Sen, *Biochim. Biophys. Acta*, **198**, 120 (1970). Robinson, J. D., *Arch. Biochem. Biophys.*, **139**, 17 (1970). Sen, A. K., T. Tobin, and R. L. Post, *J. Biol. Chem.*, **244**, 6596 (1969). Bader, H., A. B. Wilkes, and D. H. Jean, *Biochim. Biophys. Acta*, **198**, 583 (1970).

693. Lindenmayer, G. E., and A. Schwartz, *Arch. Biochem. Biophys.*, **140**, 371 (1970). Allen, J. C., G. E. Lindenmayer, and A. Schwartz, *Arch. Biochem. Biophys.*, **141**, 322 (1970). Barnett, R. E., *Biochemistry*, **9**, 4644 (1970).

694. Nagai, K., G. E. Lindenmayer, and A. Schwartz, *Arch. Biochem. Biophys.*, **139**, 252 (1970). Lishko, V. K., M. K. Malysheva, and N. M. Polyakova, *Doklady. Akad. Nauk SSSR*, **187**, 1191 (1969).

695. Post, R. L., S. Kume, T. Tobin, B. Orcutt, and A. K. Sen, *J. Gen. Physiol.*, **54**, 306s (1969).

696. Askari, A., and S. N. Rao, *Biochim. Biophys. Res. Comm.*, **36**, 631 (1969).

697. Whittham, R., in *The Neurosciences*, p. 313, Eds. G. C. Quarton, T. Melnechuk, and F. O. Schmitt, The Rockefeller University Press, New York (1967).

698. Jardetzky, O., *Nature*, **211**, 969 (1966).

699. Opit, L. J., and J. S. Charnock, *Nautre*, **208**, 471 (1965).

700. Lowe, A. G., *Nature*, **219**, 934 (1968).

701. Middleton, H. W., *Arch. Biochem. Biophys.*, **136**, 280 (1970).

702. Hill, T. L., *Proc. Natl. Acad. Sci. (U.S.)*, **65**, 409 (1970).

703. Triggle, D. J., *Progress in Surface Science*, **3**, 273 (1970). F. M. Kregenow, *J. Gen. Physiol.*, **58**, 372, 396 (1971).

704. Whitlock, R. T., and H. O. Wheeler, *J. Clin. Invest.*, **43**, 2249 (1964).

705. Patlack, C. S., D. A. Goldstein, and J. F. Hoffman, *J. Theoret. Biol.*, **5**, 426 (1963).

706. Rehm, W. S., C. F. Butler, S. G. Spangler, and S. S. Sanders, *J. Theoret. Biol.*, **27**, 433 (1970).

707. Whittam, R., *J. Physiol.*, **131**, 542 (1956).

708. Coppen, A. J., *Int. Psychiat. Clin.*, **6**, 53 (1969). Randrup, A., and I. Munkvad, *Brit. J. Psychiat.*, **112**, 173 (1966). Ismail-Beigi, F., and I. S. Edelman, *Proc. Natl. Acad. Sci. (U.S.)*, **67**, 1071 (1970). Mangos, J. A., and N. R. McSherry, *Science*, **158**, 135 (1967). Elithorn, A., P. K. Bridges, M. C. Lobban, and B. E. Tredre, *Brit. Med. J.*, **2**, 1620 (1966). Simon, K. A., S. L. Bontig, and N. M. Hawkins, *Exptl. Eye Res.*, **1**, 253 (1962). Kupfer, S., and J. D. Kosovsky, *J. Clin. Invest.*, **44**, 1132 (1965). Gonda, O., and J. H. Quastel, *Biochem. J.*, **84**, 394 (1962). Smith, E. K. M., and P. D. Samuel, *Clin. Sci.*, **38**, 49 (1970).

709. Watten, R. H., F. M. Morgan, Y. N. Songkhla, B. Vanikiati, and R. A. Phillips, *J. Clin. Invest.*, **38**, 1879 (1959). Phillips, R. A., A. H. G. Love, E. M. Mitchell, and E. M. Neptune, Jr., *Nature*, **206**, 1367 (1964).

710. Chignell, C. F., *Biochem. Pharmacol.*, **17**, 1207 (1968). Pickford, G. E., R. W. Griffith, J. Torretti, F. Hendlez, and F. H. Epstein, *Nature*, **228**, 378 (1970). Karlsson, K. A., B. S. Samuelsson, and G. O. Steen, *J. Membrane Biol.*, **5**, 169 (1970). Also see Horton, C. R., W. Q. Cole, and H. Bader, *Biochem. Biophys. Res. Comm.*, **40**, 505 (1970).

711a. Karnovsky, M. L., *Physiol. Rev.*, **42**, 143 (1962).

711b. Chapman-Andresen, C., *Compt. Rend. Trav. Lab. Carlsberg*, **33**, 73 (1963).

711c. Jacques, P. J., in *Lysosomes: Biology and Pathology*, p. 395, Ed. J. T. Dingle, North-Holland Publ. Co., Amsterdam (1969).

712. Bennett, H. S., *J. Biophys. Biochem. Cytol.*, **2** (Suppl.), 99 (1956).

713. Palade, G. E., and R. R. Bruns, *J. Cell. Biol.*, **37**, 633 (1968).

714. Brandt, P. W., and A. R. Freeman, *Science*, **155**, 582 (1967). Cohn, Z. A., and E. Parks, *J. Exptl. Med.*, **125**, 213 (1967). Najjar, V. A., and K. Nishioka, *Nature*, **228**, 672 (1970).

715. Schultz, J. (Ed.), *Biochemistry of the Phagocytic Process*, North-Holland Publishing Co., Amsterdam (1970).

716. Terepka, A. R., M. E. Stewart, and N. Merkel, *Exptl. Cell Res.*, **58**, 107 (1969).

717. Mahler, H. R., and E. H. Cordes, *Biological Chemistry*, Harper & Row, New York (1971) 2nd edition.

718. Lehninger, A. L., E. Carafoli, and C. S. Rossi, *Advances in Enzymology*, **29**, 259 (1967). Green, D. E., and D. H. MacLennan, *Bioscience*, **19**, 213 (1969).

719. Lehninger, A. L., *The Mitochondrion: Molecular Basis of Structure and Function*, W. A. Benjamin, New York (1964). Lehninger, A. L., *Bioenergetics: The Molecular Basis of Biological Energy Transformations*, W. A. Benjamin, New York (1965).

720. Parsons, D., G. Williams, W. Thompson, W. Wilson, and B. Chance, in *Proc. Symp. Mitochondrial Structure and Function*, Bari, Italy, 1967.

721. Ernster, L., and B. Kuylenstierna, in *Membranes of Mitochondria and Chloroplasts*, Ed. E. Racker, p. 172, Van Nostrand Reinhold Co., New York (1970).

722. Mitchell, P., and J. Moyle, *Nature*, **208**, 147 (1965); in *Biochemistry of Mitochondria*, Ed. E. C. Slater, Z. Kaninga, and L. Wojtczak, Academic Press, London (1967). Also see Robertson, R. N., *Protons, Electrons, Phosphorylation and Active Transport*, Cambridge University Press, Cambridge (1968). Grinius, L. L., A. A. Jasaitis, Yu P. Kadziauskas, E. A. Liberman, V. P. Skulachev, V. P. Topali, L. M. Tsofina, and M. A. Vladimirova, *Biochim. Biophys. Acta.*, **216**, 1 (1970). Bakeeva, L. E., L. L. Grinius, A. A. Jasaitis, V. V. Kuliene, D. O. Levitsky, E. A. Liberman, I. I. Severina, and V. P. Skulachev, *Biochim. Biophys. Acta.*, **216**, 13 (1970). Isaev, P. I., E. A. Liberman, V. D. Samuilov, V. P. Skulachev, and L. M. Tsofina, *Biochim. Biophys. Acta.*, **216**, 22 (1970). Liberman, E. A., and V. P. Skulachev, *Biochim. Biophys. Acta.*, **216**, 30 (1970).

723. Stoffel, W., and H. G. Schiefer, *Hoppe Seyler's Z. Physiol.*, **349**, 1017 (1968). Smoly, J. M., B. Kuylenstierna, and L. Ernster, *Proc. Natl. Acad.*

Sci. (U.S.), **66**, 125 (1970). Tyler, D. D., *Biochem. J.*, **116**, 30P (1970). Ernster, L., and B. Kuylenstierna, in *Mitochondria: Structure and Function*, p. 5, Eds. L. Ernster, and Z. Drahota, Academic Press, New York (1969). Bachmann, E., G. Lenaz, J. F. Pardue, N. Orme-Johnson, and D. E. Green, *Arch. Biochem. Biophys.*, **121**, 73 (1967). Levy, M., R. Toury, M. T. Sauner, and J. Andre, in *Mitochondira: Structure and Function*, p. 33.

724. Chance, B., and G. R. Williams, *Advances in Enzymol.*, **17**, 65 (1956).

725. Chance, B., M. Erecinska, and C. P. Lee, *Proc. Natl. Acad. Sci. (U.S.)*, **66**, 928 (1970). Montal, M., B. Chance, and C. Lee, *J. Membrane Biol.*, **2**, 201 (1970). Chance, B., A. Azzi, I. Y. Lee, C. P. Lee, and L. Mela, in *Mitochondria: Structure and Function*, p. 233 (see Ref. 723). Lee, C. P., L. Ernster, and B. Chance, *Eur. J. Biochem.*, **8**, 153 (1969).

725a. Lenaz, G., A. M. Sechi, G. P. Castelli, and L. Masotti, *Arch. Biochem. Biophys.*, **141**, 79 (1970). Lenaz, G., A. M. Sechi, L. Masotti, and G. P. Castelli, *Arch. Biochem. Biophys.*, **141**, 89 (1970).

726. Yamashita, S., and E. Racker, *J. Biol. Chem.*, **243**, 2446 (1968).

727. Klingenberg, M., *Ber. Bunsen. Ges. Physik. Chem.*, **68**, 747 (1964).

728. Pressman, B. C., in *Mitochondria: Structure and Function*, (see Ref. 723). Skulachev, V. P., A. A. Jasaitis, V. V. Navickite, L. S. Vaguzhinsky, E. A. Liberman, V. P. Topali, and L. M. Zofina, in *Mitochondria: Structure and Function*, p. 275 (see Ref. 723). Caswell, A. H., *J. Membrane Biol.*, **1**, 53 (1969). Weinbach, E. C., and J. Garbus, *Nature*, **221**, 1016 (1969).

729. Pressman, B. C., and J. K. Park, *Biochim. Biophys. Res. Comm.*, **11**, 182 (1963).

730. Büchel, K. H., and F. Korte, *Angew. Chem. (Intern. Ed.)*, **4**, 788, 789 (1965). See Ting, H. P., D. F. Wilson, and B. Chance, *Arch. Biochem. Biophys.*, **141**, 141 (1970). Also see Ref. 334 and 1249.

731. Wilson, D. F., and A. Azzi, *Arch. Biochem. Biophys.*, **126**, 724 (1968). Kaplay, M., C. K. R. Kurup, K. W. Lam, and D. R. Sanadi, *Biochemistry*, **9**, 3599 (1970).

732. Singer, T. P., *Biological Oxidations*, Interscience Publishers, New York (1968).

733. Glynn, I. M., *Nature*, **216**, 1318 (1967).

734. Lardy, H. A., S. N. Graven, and S. Estrada-O, *Fed. Proc.*, **26**, 1355 (1967). See B. C. Pressman, in Ref. 721.

735. Kagawa, Y., and E. Racker, *J. Biol. Chem.*, **241**, 2461 (1966).

736. Kagawa, Y., and E. Racker, *J. Biol. Chem.*, **241**, 2467 (1966).

737. *Ibid.*, **241**, 2475 (1966).

738. Greville, G. D., *Curr. Topics in Bioenergetics*, **3**, 1 (1969).

739. Slater, E. C., *Nature*, **172**, 975 (1953).

740. Lehninger, A. L., E. Carafoli, and C. S. Rossi, *Adv. in Enzymol.*, **29**, 259 (1967).

741. Wang, J. H., *Science*, **167**, 25 (1970). Tu, S., and J. H. Wang, *Biochemistry*, **9**, 4505 (1970). Also see Korman, E. F., and J. McLick, *Proc. Natl. Acad. Sci.*, **67**, 1130 (1970).

742. Painter, A. A., and F. E. Hunter, Jr., *Science*, **170**, 552 (1970).

743. Mitchell, P., *Biological Rev., Cambridge Phil. Soc.*, **41**, 445 (1966); *Chemiosmotic Coupling and Energy Transduction*, Glynn Res. Ltd., Bodmin, Cornwall, England (1968). *Theor. Exp. Biophys.*, **2**, 159 (1969). Also see Bakeeva, L. E., L. L. Grinius, A. Jasaitis, V. Kuliene, D. O. Levitskii, E. A. Liberman, I. I. Severina, and V. P. Skulachev, *Biochim. Biophys. Acta.*, **216**, 13 (1970).

744. Cohn, M., *J. Biol. Chem.*, **201**, 735 (1953).

745. Mitchell, R. A., R. D. Hill, and P. D. Boyer, *J. Biol. Chem.*, **242**, 1793 (1967).

746. Wojtczak, L., K. Bogucka, M. G. Sarzala, and H. Zaluska, p. 79, and Hittelman, K. J., B. Cannon, and O. Lindberg, p. 153, in *Mitochondria: Structure and Function*, (cf. Ref. 723).

747. Jagendorf, A. T., and E. Uribe, in *Energy Conservation by the Photosynthetic Apparatus*, p. 215, *Brookhaven Symp. Biol.*, No. 19 (1967).

748. Hall, D. O., and J. M. Palmer, *Nature*, **221**, 717 (1969). For a theoretical discussion see Caplan, S. R., and A. Essig, *Proc. Natl. Acad. Sci. (U.S.)*, **64**, 211 (1969).

749. Chance, B., *Proc. Natl. Acad. Sci. (U.S.)*, **67**, 560 (1970). Azzi, A., B. Chance, G. K. Radda, and C. P. Lee, *Proc. Natl. Acad. Sci.*, **62**, 612 (1969).

750. Penniston, J. T., R. A. Harris, J. Asai, and D. E. Green, *Proc. Natl. Acad. Sci. (U.S.)*, **59**, 624 (1968).

751. *Ibid.*, **59**, 830 (1968).

752. Green, D. E., *Proc. Natl. Acad. Sci. (U.S.)*, **67**, 544 (1970).

753. Young, J. H., G. A. Blondin, G. Vanderkooi, and D E. Green, *Proc. Natl. Acad. Sci. (U.S.)*, **67**, 550 (1970). D. E. Green and S. Ji, *ibid.*, **69**, 727 (1972).

754. Chance, B., D. F. Wilson, P. L. Dutton, and M. Erecinska, *Proc. Natl. Acad. Sci. (U.S.)*, **66**, 1175 (1970).

755a. Duysens, L. N. M., *Progr. Biophys. Mol. Biol.*, **14**, 1 (1964). Walker, D. A., and A. R. Crofts, *Ann. Rev. Biochem.*, **39**, 389 (1970).

755b. Kok, B., in *The Physiology of Plant Growth and Development*, p. 335, Ed. M. B. Wilkins, McGraw-Hill, New York (1969). See also Ref. 1252.

756. Arnon, D. I., H. Y. Tsujimoto, and B. D. McSwain, *Proc. Natl. Acad. Sci. (U.S.)*, **51**, 1274 (1964).

757. McCarty, R. E., and E. Racker, *Brookhaven Symp. in Biology*, No. 19 (1966). McCarty, R., and E. Racker, *J. Biol. Chem.*, **243**, 129 (1968).

758. Muhlethaler, K., *Biochemistry of Cloroplasts*, Vol. 1, 49 (1966).

759. Arnold, W., and H. K. Sherwood, *Proc. Natl. Acad. Sci. (U.S.)*, **43**, 105 (1957).

760. Calvin, M., *J. Theoret. Biol.*, **1**, 258 (1961).

761. Hesketh, T. R., *Nature*, **224**, 1026 (1969). Tien, H. T., *J. Phys. Chem.*, **72**, 4512 (1968). Also see Ref. 132, and Ullrich, H. M., and H. Kuhn, *Z. Naturforsch.* **B24**, 1342 (1969).

762. Alamuti, N., and P. Lauger, *Biochim. Biophys. Acta.*, **211**, 362 (1970).

763. Cherry, R. J., *Quart. Rev.*, **22**, 160 (1968).

764. Vinnikov, Y. A., in *The Structure and Function of Nervous Tissue*, p. 265, Ed., G. H. Bourne, Academic Press, New York (1969). Also see Catton, W. T., *Physiol. Rev.*, **50**, 297 (1970).

765. Davis, H., *Physiol. Rev.*, **41**, 391 (1961).

766. Mellon, D., Jr., *The Physiology of Sense Organs*, W. H. Freeman & Co., San Francisco (1968).

767. Duncan, C. J., *The Molecular Properties and Evolution of Excitable Cells*, Pergamon Press, London (1967).

768. Patton, H. D., in *Neurophysiology*, Eds. T. C. Ruch, H. D. Patton, J. W. Woodbury, and A. L. Towe, W. B. Saunders Co., Philadelphia (1965), p. 95.

769. Bonting, S. L., *Current Topics in Bioenergetics*, **3**, 351 (1969). Morton, R. A., and G. A. J. Pitt, *Adv. in Enzymology*, **32**, 97 (1969). Arden, G. B., *Progr. in Biophys. Mol. Biol.*, **19** (2), 373 (1969). See also Ref. 1303.

770. Wald, G., P. K. Brown, and I. R. Gibbons, in *Biological Receptor Mechanisms*, Academic Press, New York (1962). Also see Young, R. W., *Scientific American*, **223** (Oct.), 81 (1970). Tomita, T., *Quart. Rev. Biophys.*, **3**, 179 (1970).

771. Tollins, G., *Current Topics in Bioenergetics*, **3**, 417 (1969).

772. Heller, J., *Biochemistry*, **7**, 2906, 2914 (1968).

773. Ostroy, S. E., F. Erhardt, and E. W. Abrahamson, *Biochim. Biophys. Acta.*, **112**, 265 (1966). Also Poincelot, R. P., P. G. Millar, R. L. Kimbel, and E. W. Abrahamson, *Nature*, **221**, 256 (1969). Figure 7–15 is modified from R. Hubbard and A. Kroff, *Ann. N.Y. Acad. Sci.*, **81**, 442 (1959).

774. Abrahamson, E. W., and S. E. Ostroy, *Progr. Biophys. Mol. Biol.*, **17**, 179 (1967).

775. Hagins, W. A., R. D. Penn, and S. Yoshikami, *Biophys. J.*, **10**, 380 (1970).

776. Wald, G., *Nature*, **219**, 800 (1968).

777. Morton, R. A., and G. A. J. Pitt, *Advances in Enzymology*, **32**, 97 (1969).

778. Daeman, F. J. M., and S. L. Bonting, *Biochim. Biophys. Acta.*, **163**, 212 (1968).

779. Biedler, L. M., *Progr. Biophys. Mol. Biol.*, **12**, 107 (1962); *NRP Bull.*, in press (1971).

780. Amoore, J. E., *Molecular Basis of Odor*, Charles C. Thomas, Sprinfgield, Ill. (1970).

781. Beets, M. G. J., *Pharmacol. Rev.*, **22**, 1 (1970).

782. Dastoli, F. R., D. V. Lopiekes, and S. Price, *Biochemistry*, **7**, 1160 (1968) and earlier papers. Hiji, Y., N. Kobayashi, and M. Sato, *Kumamoto Med. J.*, **21**, 137 (1968).

783. Rosenburg, B., T. N. Misra, and R. Switzer, *Nature*, **217**, 423 (1968).

784. Scalzi, H. A., *Z. Zellforsch. Mikroskop*, **80**, 413 (1967).

785. Jahn, T. L., and E. C. Rovee, *Physiol. Rev.*, **49**, 793 (1969).

786. Allen, R. D., in *Symp. of the Society for Experimental Biology*, XXII, 151, (1968), and other reports.

787. Brokaw, C. J., and B. Benedict, *Arch. Biochem. Biophys.*, **125**, 770 (1968).

788. Curtis, A. S. G., *The Cell Surface: Its Molecular Role in Morphogenesis*, Academic Press, London (1967).

789. Ambrose, E. J., *Endeavour,* **24** (91), 27 (1965).
790. Allen, R. D., and N. Kamiya, *Primitive Motile Systems in Cell Biology,* Academic Press, New York (1964).
791. Bingley, M. S., and C. M. Thompson, *J. Theoret. Biol.,* **2**, 16 (1962).
792. Brandt, P. W., and A. R. Freeman, *Science,* **155**, 582 (1967).
793. Bennett, M. V. L., *Ann. Rev. Physiol.,* **32**, 471 (1970). Lissmann, H. W., in *Bioelectrogenesis,* p. 223, Elsevier Publishing Co., Amsterdam (1961).
794. Adelman, W. J., and R. E. Taylor, *Biophys. J.,* **4**, 451 (1964).
795. Cole, K. S., R. Guttman, and F. Bezanilla, *Proc. Natl. Acad. Sci. (U.S.),* **65**, 884 (1970).
796. Moore, J. W., and K. S. Cole, in *Physical Techniques in Biological Research,* Vol. 6, p. 263, Ed. W. L. Nastuk, Academic Press, New York (1963); also p. 143 in Ref. 1307.
797. Palti, Y., and W. J. Adelman, Jr., *J. Membrane Biol.,* **1**, 431 (1969). Fishman, H. M., *Biophys. J.,* **10**, 799 (1970).
798. Baker, P. F., A. L. Hodgkin, and T. I. Shaw, *J. Physiol.,* **164**, 330 (1962).
799. Tasaki, I., I. Singer, and A. Watanabe, *J. Gen. Physiol.,* **50**, 989 (1967) and earlier papers. See also D. L. Gilbert (p. 264) in Ref. 1307.
800. Chandler, W. K., and H. Meves, *J. Physiol. (London),* **180**, 788 (1965).
801. Cole, K. S., *Membranes, Ions and Impulses,* University of California Press, Los Angeles (1968).
802. Grundfest, H., in *Advances in Comparative Physiology and Biochemistry,* Vol. II, p. 1 (1966). Van Harreveld, A., *Progress in Neurology and Psychiatry,* **23**, 21 (1968). Ungar, G., *Excitation,* C. C. Thomas, Springfield, Ill. (1963).
803. Hodgkin, A. L., *The Conduction of the Nervous Impulse,* C. C. Thomas, Springfield, Ill. (1964).
804. Katz, B., *Nerve, Muscle and Synapse,* McGraw-Hill, New York (1966). See also Ref. 1255.
805. Plonsey, R., *Bioelectric Phenomena,* McGraw-Hill, New York (1969).
806. Tasaki, I., and Y. Kobatake, in *Nerve Excitation,* Charles C. Thomas, Springfield, Ill. (1967). Also see Tasaki, I., and A. Watanabe, *Fed. Proc.,* **27**, 703 (1968). Tasaki, I., L. Lerman, and A. Watanabe, *Amer. J. Physiol.,* **216**, 130 (1969).
807. Schoffeniels, E., *Arch. Internat. Physiol. Biochim.,* **68**, 1 (1960).
808. Saddler, H. D. W., *J. Gen. Physiol.,* **55**, 802 (1970). Marmor, M. F., and A. L. F. Gorman, *Science,* **167**, 65 (1970). Izquierdo, J. A., and I. Izquierdo, *Ann. Rev. Pharmacol.,* **7**, 125 (1967). Vieth, J. B., *Elektromedizin,* **15**, 99 (1970). Carpenter, D. O., *Comp. Biochem. Physiol.,* **35**, 371 (1970).
809. Strickholm, A., personal communication.
810. Spyropoulos, C. S., *Fed. Proc.,* **27**, 1252 (1968). Spyropoulos, C. S., and T. Teorell, *Proc. Natl. Acad. Sci. (U.S.),* **60**, 118 (1968).
811. Barry, P. H., and J. M. Diamond, *J. Membrane Biol.,* **3**, 93 (1970). Williams, J. A., *J. Theoret. Biol.,* **28**, 287 (1970). Gorman, A. L. F., and M. F. Marmor, *J. Physiol. (London),* **210**, 897 (1970). Findlay, G. P., A. B.

Hope, and E. J. Williams, *Austr. J. Biol. Sci.*, **22**, 1163 (1969). Also see 808 and Slesak, E., *Wiad. Bot.*, **14**, 37 (1970).

811a. Sabah, N. H., and K. N. Leibovic, *Biophys. J.*, **9**, 1206 (1969). Wauro, A., F. Conti, F. Dodge, and R. Schor, *J. Gen. Physiol.*, **55**, 497 (1970).

812. Blinks, L. R., *J. Gen. Physiol.*, **13**, 793 (1930); **19**, 633 (1936).

813. Candia, O. A., *Biophys. J.*, **10**, 323 (1970).

814. Mueller, P., *J. Gen. Physiol.*, **42**, 137 (1958).

815. Bennett, M. V. L., and H. Grundfest, *J. Gen. Physiol.*, **50**, 141 (1966).

816. Julian, F. J., J. W. Moore, and D. E. Goldman, *J. Gen. Physiol.*, **45**, 1217 (1962).

817. Moore, J. W., *Nature*, **183**, 265 (1959).

818. Shashoua, V. E., in *Molecular Basis of Membrane Function*, p. 147, Ed. D. C. Tosteson, Prentice-Hall, Inc., Englewood-Cliffs (1969).

819. Franck, U. F., *Ber. der Bunsenges.*, **67**, 657 (1963).

820. Caswell, A. H., *J. Membrane Biol.*, **1**, 53 (1969).

821. Ridley, B. K., *Proc. Phys. Soc.*, **82**, 954 (1963). Gunn, J. B., *I.B.M. Journal of Research and Development*, **8** (2), 14 (1964). Bowers, R., *Sci. Amer.*, **215** (Aug.), 22 (1966).

822. Cole, K. S., and J. W. Moore, *Biophys. J.*, **1**, 1 (1960).

823. Julian, F. J., J. W. Moore, and D. E. Goldman, *J. Gen. Physiol.*, **45**, 1217 (1962).

824. Coster, H. G. L., *Biophys. J.*, **5**, 669 (1965).

825. Mueller, P., and D. O. Rudin, *J. Theoret. Biol.*, **4**, 268 (1963).

826. Verveen, H. E., and A. A. Derksen, *Proc. IEEE*, **56**, 906 (1968); *Science*, **151**, 1388 (1966); *Acta. Physiol. Pharmacol. Neer*, **15**, 353 (1969). Poussart, D. J. M., *Proc. Natl. Acad. Sci. (U.S.)*, **64**, 95 (1969); *Biophys. J.*, **11**, 211 (1971). See also Ref. 1266.

827. Hille, B., *J. Gen. Physiol.*, **51**, 199 (1968). Also see Katz, B., and R. Miledi, *Nature*, **226**, 962 (1970).

828. Walz, D., E. Bamberg, and P. Lauger, *Biophys. J.*, **9**, 1150 (1969). Neumcke, B., and P. Lauger, *Biophys. J.*, **9**, 1160 (1969). Agin, D., *Biophys. J.*, **9**, 209 (1969). *Proc. Natl. Acad. Sci. (U.S.)*, **57**, 1232 (1967). Hamel, B. B., and I. Zimmerman, *Biophys. J.*, **10**, 1029 (1970). Bothorel, P., and C. Lussan, *C. R. Acad. Sci. (Paris)*, Ser D, **266**, 2492 (1968). See also Ref. 1256.

829. Segal, J. R., *Amer. J. Physiol.*, **215**, 467 (1968).

830. Hodgkin, A. L., and A. F. Huxley, *J. Physiol.*, **116**, 497 (1952).

831. *Ibid.*, **117**, 500 (1952).

832. Hodgkin, A. L., and B. Katz, *J. Physiol. (London)*, **109**, 240 (1949).

833. Fitzhugh, R., *J. Gen. Physiol.*, **49**, 989 (1966). R. Guttman with the Technical assistance of R. Barnhill, *J. Gen. Physiol.*, **49**, 1007 (1966). Guttman, R., and R. Barnhill, *J. Gen. Physiol.*, **55**, 104 (1970).

834. Abbott, B. C., J. V. Howarth, and J. M. Ritchie, *J. Physiol.*, **178**, 368 (1965).

835. Howarth, J. V., R. D. Keynes, and J. M. Ritchie, *J. Physiol. (London)*, **194**, 745 (1968).

836. Brinley, F. J., and L. J. Mullins, *J. Neurophysiol.*, **28**, 526 (1966). Rojas, E., F. Bezanilla, and R. E. Taylor, *Nature*, **225**, 747 (1970); *J. Physiol. (London)*, **207**, 151 (1970).

837. Gilbert, D. L., and G. Ehrenstein, *Biophys. J.*, **6**, 553 (1966). Also see Rojas, E., R. E. Taylor, I. Atwater, and F. Bezanilla, *J. Gen. Physiol.*, **54**, 532 (1970).

838. Armstrong, C. M., *Nature*, **219**, 1262 (1968); *J. Gen. Physiol.*, **54**, 553 (1969).

839. Ashley, C. C., *J. Physiol. (London)*, 32P (1969). Fishman, H., Biophysical Society Annual Meeting (14th), Baltimore, (Feb. 1970), and personal communication.

840. Moore, J. W., T. Narahashi, and W. Ulbricht, *J. Physiol. (London)*, **172**, 163 (1964).

841. Binstock, L., and H. Lecar, *J. Gen. Physiol.*, **53**, 342 (1969).

842. Singer, I., and I. Tasaki, in *Biological Membranes: Physical Fact and Function*, p. 347, Ed. D. Chapman, Academic Press (1968). Also see Ref. 806.

843. Adelman, W. J., and J. P. Senft, *J. Gen. Physiol.*, **51**, 102s (1968).

844. Hafeman, D. R., *Chemical Aspects of the Effects of Calcium on Nerve Membrane Properties*, Ph.D. Thesis, California Institute of Technology, Pasadena (1967). Manery, J. F., *Fed. Proc.*, **25**, 1804 (1966).

845. Frankenhaeuser, B., and A. F. Huxley, *J. Physiol. (London)*, **137**, 218 (1957).

846. Gilbert, D. L., and G. Ehrenstein, *Biophys. J.*, **9**, 447 (1969).

847. Cox, F. H., H. L. Fernandez, and B. H. Smith, *Biophys. J.*, **6**, 675 (1966).

848. Rojas, E., *Proc. Natl. Acad. Sci. (U.S.)*, **53**, 306 (1965).

849. Huneeus-Cox and Fernandez, unpublished data quoted in *The Neurosciences: A Study Program*, p. 265, Rockefeller University Press, New York (1967).

850. Rudenburg, F. H., and J. M. Tobias, *J. Cell. Comp. Physiol.*, **55**, 149 (1960).

851. Narahashi, T., J. W. Moore, and W. R. Scott, *J. Gen. Physiol.*, **47**, 965 (1964). Moore, J. W., and T. Narahashi, *Fed. Proc.*, **26**, 1655 (1967). Narahashi, T., J. W. Moore, and D. T. Frazier, *J. Pharmacol. Exp. Therp.*, **169**, 224 (1969). Also see Phillis, J. W., A. K. Tebecis, and D. H. York, *Nature*, **217;** 271 (1968).

852. Cuervo, L. A., and W. J. Adelman, Jr., *J. Gen. Physiol.*, **55**, 309 (1970).

853. Hille, B., *Nature*, **210**, 1220 (1966).

854. Narahashi, T., H. G. Haas, and E. F. Therrien, *Science*, **157**, 1441 (1967). For a review on tetrodotoxin and sexitoxin see C. Y. Kao, *Pharmacol. Rev.*, **18**, 997 (1966).

855. Narahashi, T., M. Yamada, and D. T. Frazier, *Nature*, **223**, 748 (1969).

856. Frank, G. B., *Fed. Proc.*, **27**, 132 (1968).

857. Narahashi, T., and J. W. Moore, *J. Gen. Physiol.*, **51**, 93s (1968).

858. Biro, G., and K. Gabor, *Acta Biochim. Biophys. Acad. Sci. Hung.*, **4**, 313 (1969).

859. Narahashi, T., J. W. Moore, and R. N. Poston, *J. Neurobiol.*, **1**, 3 (1969).

Narahashi, T., D. T. Frazier, and M. Yamada, *J. Pharmacol. Expt. Therap.*, **171**, 32 (1970). Frazier, D. T., T. Narahashi, and M. Yamada, *J. Pharm. Expt. Therap.*, **171**, 45 (1970).

860. Schultz, O. E., U. P. Hansen, and K. Vanselow, *Z. Naturforsch B*, **24**, 785 (1969).

861. Schauf, C., and D. Agin, *Nature*, **221**, 768 (1969).

862. Frazier, D. T., T. Narahashi, and J. W. Moore, *Science*, **163**, 820 (1968).

863. Narahashi, T., and H. G. Haas, *Science*, **157**, 1438 (1967); *J. Gen. Physiol.*, **51**, 177 (1968).

864. Holan, G. *Nature*, **221**, 1025 (1969).

865. Ulbricht, W., *J. Cell. Comp. Physiol.*, **66** (Suppl. 2), 91 (1965); *Reviews of Physiology*, **61**, 18 (1969), Springer-Verlag, Berlin.

866. Narahashi, T., J. W. Moore, and B. I. Shapiro, *Science*, **163**, 680 (1969).

867. Zlotkin, E., and A. S. Shulov, *Toxicon.*, **7**, 217 (1969). Koppenhofer, E., and H. Schmidt, *Experientia*, **24**, 41 (1968).

868. Narahashi, T., and N. C. Anderson, *Toxicol. Appl. Pharmacol.*, **10**, 529 (1967).

869. Shrager, P. G., A. Strickholm, and R. I. Macey, *J. Cell Physiol.*, **74**, 91 (1969). Also see Hutter, O. F., *Nature*, **224**, 1215 (1969).

870. Shrager, P. G., R. I. Macey, and A. Strickholm, *J. Cell Physiol.*, **74**, 77 (1969).

871. Armstrong, C. M., *Nature*, **211**, 322 (1966); **219**, 1262 (1968); *J. Gen. Physiol.*, **50**, 491 (1966). See also Ref. 1261.

872. Woodin, A. M., and A. A. Wieneke, *Nature*, **227**, 460 (1970). Dettbarn, W. D., and E. Bartels, *Biochem. Pharmacol.*, **17**, 1833 (1968). Podelski, T., J. C. Meunier, and J. P. Changeux, *Proc. Natl. Acad. Sci.*, **63**, 1239 (1969).

873. Cooke, I. M., J. M. Diamond, A. D. Grinnell, S. Hagiwara, and H. Sakata, *Proc. Natl. Acad. Sci. (U.S.)*, **60**, 470 (1968).

874. Pooler, J., *Biophys. J.*, **8**, 1009 (1968).

875. Kem, W. R., K. Murayama, and T. Narahashi, *Fed. Proc.*, **29** (Abstr.), 795 (1970).

876. Rojas, E., and I. Atwater, *Nature*, **215**, 850 (1967).

877. Armstrong, C. M., *J. Gen. Physiol.*, **54**, 553 (1969).

878. Bryant, S. H., and J. M. Tobias, *J. Cell. Comp. Physiol.*, **46**, 71 (1955).

879. Cohen, L. B., R. D. Keynes, and B. Hille, *Nature*, **218**, 438 (1968).

880. Tasaki, I., A. Watanabe, R. Sandlin, and L. Carnay, *Proc. Natl. Acad. Sci. (U.S.)*, **61**, 883 (1968). Tasaki, I., L. Carnay, and A. Watanabe, *Proc. Natl. Acad. Sci. (U.S.)*, **64**, 1362 (1969). Kasai, M., J. P. Changeux, and L. Monnerie, *Biochim. Biophys. Rec. Comm.*, **36**, 420 (1969).

881. Girardier, L., *J. Gen. Physiol.*, **47**, 189 (1963).

882. Papano, A. J., and N. Sperelakis, *Amer. J. Physiol.*, **217**, 615 (1969).

883. Noble, D., and R. W. Tsien, *J. Physiol. (London)*, **200**, 205, 233 (1969).

884. Belton, P., and H. Grundfest, *Amer. J. Physiol.*, **203**, 588 (1962).

885. Gaffey, C. T., and L. J. Mullins, *J. Physiol.*, **144**, 505 (1958).

886. Kishimoto, V., *Jap. J. Physiol.*, **14**, 515 (1964). Williams, E. J., and J. Bradley, *Biochim. Biophys. Acta.*, **150**, 626 (1968).

887. Del Castillo, J., and T. Morales, *J. Gen. Physiol.*, **50**, 603 (1967).
888. Koketsu, K., and S. Nishi, *Nature*, **217**, 468 (1968).
889. Gerasimov, V. D., *Fed. Proc.*, **24**, T371 (1965).
890. Chamberlain, S. G., and G. A. Kerkut, *Com. Biochem. Physiol.*, **28**, 787 (1969).
891. Wareham, A. C., C. J. Duncan, and K. Bowler, *Nature*, **217**, 970 (1968).
892. Strickholm, A., *Nature*, **212**, 835 (1966). For electrophysiological data on frog skin see Fishman, H. M., and R. I. Macey, *Biophys. J.*, **9**, 127, 140 (1969).
893. Umrath, K., *Handbook der Pflanzenphysiol.*, **17** (2), 542 (1962).
894. Motokizawa, F., J. P. Reuben, and H. Grundfest, *J. Gen. Physiol.*, **54**, 437 (1969).
895. Bonhoeffer, K. F., *J. Gen. Physiol.*, **32**, 69 (1948). Also see Ref. 924.
896. Franck, U. F., *Progress in Biophys. Biphysical Chem.*, **6**, 172 (1956).
897. Grundfest, H., *Fed. Proc.*, **26**, 1613 (1967).
898. Fulpius, B., and F. Baumann, *J. Gen. Physiol.*, **53**, 541 (1969).
899. Bishop, G. H., *Physiol. Rev.*, **36**, 376 (1956). Also see Nelson, P., W. Ruffner, and M. Nirenberg, *Proc. Natl. Acad. Sci. (U.S.)*, **64**, 1004 (1969).
900. Baumann, G., *Nature*, **223**, 316 (1969).
901. Shashoua, V., *Nature*, **215**, 846 (1967).
902. Teorell, T., *J. Gen. Physiol.*, **42**, 831 (1959).
903. *Ibid.*, **42**, 847 (1959).
904. See Ref. 66 and 910.
905. Bernstein, J., *Arch. Ges. Physiol.*, **92**, 521 (1902).
906. Hodgkin, A. L., A. F. Huxley, and B. Katz, *J. Physiol.*, **116**, 424 (1952).
907. Hodgkin, A. L., and A. F. Huxley, *J. Physiol.*, **116**, 449 (1952).
908. *Ibid.*, **116**, 473 (1952).
909. Boyle, P. J., and E. J. Conway, *J. Physiol.*, **100**, 1 (1940).
910. Goldman, D. E., *J. Gen. Physiol.*, **27**, 37 (1943). Zelman, D. A., *J. Theoret. Biol.*, **18**, 396 (1968). Also A. D. MacGillivray, and D. Hare, *J. Theoret. Biol.*, **25**, 113 (1969).
911. Eyring, H., H. B. Eyring, and J. W. Woodbury, *Proc. Natl. Acad. Sci. (U.S.)*, **58**, 462 (1967). Also see Fitzhugh, R., *J. Cell. Comp. Physiol.*, **66**, (Suppl. 2), 111 (1965); also in *Biological Engineering*, Ed. H. P. Schwan, McGraw-Hill Book Co., Inc., New York, p. 1 (1969). J. Evans, N. Shenk, *Biophys. J.*, **10**, 1090 (1970). Chizmadzhev, Y. A., and V. S. Markin, *Priroda (Moscow)*, **6**, 18 (1970).
912. Mullins, L. J., *J. Gen. Physiol.*, **52**, 550, 555 (1968); Narahashi, T., and J. W. Moore, *ibid.* p. 553.
913. Segal, J. R., *J. Theoret. Biol.*, **24**, 159 (1969).
914. Johnson, F. H., *The Kinetic Basis of Molecular Biology*, J. Wiley, New York (1954).
914a. Offner, F. F., *J. Gen. Physiol.*, **56**, 272 (1970).
915. Minorsky, N., and E. Leimanis, *Dynamics and Non-linear Mechanics*, J. Wiley, New York (1958).
916. Kobatake, Y., and H. Fujita, *J. Chem. Phys.*, **40**, 2213 (1964).

917. *Ibid.,* **40**, 2219 (1964). Also Kobatake, Y., M. Yuasa, and H. Fujita, *J. Phys. Chem.,* **72**, 1752 (1968).

918. Bearman, R., and J. I. Kirkwood, *J. Chem. Phys.,* **28**, 136 (1958).

919. Aranow, R. H., *Proc. Natl. Acad. Sci. (U.S.),* **50**, 1066 (1963).

920. Aranow, R. H., *Protoplasma,* **63**, 206 (1967).

921. Vaidyanathan, V. S., and H. M. Phillips, *J. Theoret. Biol.,* **10**, 460 (1966).

922. Vaidyanathan, V. S., *J. Theoret. Biol.,* **13**, 18 (1966).

923. Wei, L. Y., *IEEE Spectrum,* **3**, 123 (Sept. 1966); **4**, 192 (March 1967).

924. Wei, L. Y., and R. H. Neuman, *Biophys. J.,* **10**, 818 (1970).

925. Johnson, R. N., and G. R. Hanna, *J. Theoret. Biol.,* **22**, 401 (1969).

926. Katachalsky, A., in *The Neurosciences: A Study Program,* p. 326, The Rockefeller University Press, New York (1967).

927. Steinberg, I. Z., A. Oplatka, and A. Katchalsky, *Nature,* **210**, 568 (1966).

928. Kalvin, M., H. H. Wang, G. Entine, D. Gill, P. Ferruti, M. A. Harpold, and M. P. Klein, *Proc. Natl. Acad. Sci. (U.S.),* **63**, 1 (1969). Hubbell, W. L., and H. M. McConnell, *Proc. Natl. Acad. Sci. (U.S.),* **61**, 12 (1968). Also see Peracchia, C., and J. D. Robertson, *J. Cell Biol.,* (Abstr.), **39**, 103a (1968).

929. Monod, J., J. P. Changeux, and F. Jacob, *J. Mol. Biol.,* **6**, 306 (1963). Monod, J., J. Wyman, and J. P. Changeux, *J. Mol. Biol.,* **12**, 88 (1965).

930. Wyman, J., in *Symmetry and Functions of Biological systems at the Macromolecular Level,* p. 267, Ed. A. Engstrom and B. Strandberg, John Wiley, Interscience Division, New York (1969).

931. Nachmansohn, D., in *New Perspectives in Biology,* p. 176, Ed. M. Sela, Elsevier, Amsterdam (1964); *Proc. Natl. Acad. Sci. (U.S.),* **61**, 1034 (1968); *Ann. N. Y. Acad. Sci.,* **137**, 877 (1966). Also see Belleau, B., and J. L. Lavoie, *Can. J. Biochem.,* **46**, 1397 (1968). D. Nachmansohn, *Proc. Natl. Acad. Sci.* (U.S.), **68**, 3170 (1971).

932. Noble, R. W., *J. Mol. Biol.,* **39**, 479 (1969). Also Changeux, J. P., J. Thiery, Y. Tung, C. Kittel, *Proc. Natl. Acad. Sci. (U.S.),* **57**, 335 (1967).

933. Blumenthal, R., J. P. Changeux, and R. Lefever, *J. Membrane Biol.,* **2**, 351 (1970).

934. Karlin, A., *J. Theoret. Biol.,* **16**, 306 (1967).

935. Hill, T. L., *Proc. Natl. Acad. Sci. (U.S.),* **58**, 111 (1967); also see Ref. 1267.

936. Gordon, R., *J. Chem. Phys.,* **49**, 570 (1968). Also see D. E. Weiss, *Nature,* **223**, 634 (1969).

937. Eccles, J. C., *The Physiology of Synapses,* Springer-Verlag, Berlin (1964). For recent developments see Bloom, F. E., L. L. Iversen, and F. O. Schmitt, *NRP Bulletin,* **8**, 325 (1970).

938. Weiss, D. E., *Autr. J. Sci.,* in press.

939. Adam, G., *Z. Naturforsch. B.,* **23**, 181 (1968). Also see Ref. 1264.

940. Wien, M., *Ann. Physik (IV),* **83**, 327 (1927).

941. Wien, M., *Physik. Z.,* **32**, 545 (1931).

942. Onsager, L., *J. Chem. Phys.,* **2**, 599 (1934).

943. Patterson, A., and F. E. Bailey, *J. Polymer Sci.,* **33**, 225 (1958). Hilbers, C. W., and C. MacLean, *Mol. Phys.,* **16**, 275 (1969); **17**, 433, 517 (1969). Fukuda, E., *Biorheology,* **5**, 199 (1968). Schwan, H. P., and L. D. Sher, *J. Electrochem. Soc. (Japan),* **116**, 22c (1969).

944. Bass, L., and W. J. Moore, in *Structural Chemistry and Molecular Biology*, p. 356, Ed. Rich, A., and N. Davidson, W. H. Freeman & Co., San Francisco (1968).

945. Stephens, W. G. S., *Nature*, **224**, 547 (1969).

946. Jain, M. K., R. H. L. Marks, and E. H. Cordes, *Proc. Natl. Acad. Sci. (U.S.)*, **67**, 799 (1970).

946a. Tsien, R. W., D. Noble, *J. Membrane Biol.*, **1**, 248 (1969).

947. Engstrom, A., and B. Strandberg (Eds.), *Symmetry and Function of Biological Systems at the Macromolecular Level*, John Wiley, New York (1969).

948. Sundaralingam, M., *Nature*, **217**, 35 (1968).

949. Pethica, B. A., *J. Gen. Microbiol.*, **18**, 473 (1958). Also in *Surface Activity and the Microbial Cell*, p. 85, Academic Press, N.Y. (1966).

950. Privalov, P. L., *Biofizika*, **13**, 163 (1968). Berendsen, H. J. C., *Theoret. Exptl. Biophys.*, **1**, 1 (1967).

951. Schulz, R. D., and S. K. Asumna, *Recent Progr. Surface Science*, **3**, 291 (1970). Also see Ref. 1269 and 1322.

952. Vandenheuvel, F. A., *J. Amer. Oil Chemist's Soc.*, **40**, 455 (1963); *Ann. N.Y. Acad. Sci.*, **122**, 57 (1965); *J. Amer. Oil Chemist's Soc.*, **43**, 258 (1966).

953. Husson, F. R., and V. Luzzati, *Adv. Biol. Med. Phys.*, **11**, 87 (1967). Sheard, B., *Nature*, **223**, 1057 (1969). Veksli, Z., N. J. Salsbury, and D. Chapman, *Biochim. Biophys. Acta.*, **183**, 434 (1969). Castleden, J. A., *J. Pharm. Sci.*, **58**, 149 (1969). Stewart, G. T., in *Molecular Crystals and Liquid Crystals*, **7**, 75 (1969). Levine, Y. K., and M. H. F. Wilkins, *Nature New Biol.*, **230**, 69 (1971).

954. Finean, J. B., S. Knutton, A. R. Limbrick, and R. Coleman, *Mol. Cryst. Liquid Cryst.*, **7**, 347 (1969). Finean, J. B., *Quart. Rev. Biophys.*, **2**, 123 (1969). Jenkinson, T. J., V. B. Xamat, and D. Chapman, *Biochim. Biophys. Acta.*, **183**, 427 (1969). Also see Ref. 37, and Coleman, R., and J. B. Finean, *Comparative Biochemistry*, **23**, 99 (1968). Blaurock, A. E., *J. Mol. Biol.*, **56**, 35 (1971). Wilkins, M. H. F., A. E. Blaurock, and D. M. Engelman, *Nature New Biol.*, **230**, 72 (1971). Kirschner, D. A., and D. L. D. Casper, *ibid.*, **231**, 46 (1971).

955. Butler, K. W., H. Dugas, I. C. P. Smith, and H. Schneider, *Biochem. Biophys. Res. Comm.*, **40**, 770 (1970). Chapman, D., V. B. Kamat, J. de Gier, and S. A. Penkett, *J. Mol. Biol.*, **31**, 101 (1968). Chapman, D., and S. A. Penkett, *Nature*, **211**, 1304 (1966).

956. Sears, B., *Ph.D. Thesis*, Indiana University, Bloomington (1971). E. H. Cordes et al. Unpublished data. Also see Ref. 1276 and D. Doddrell, and A. A. Allerhand, *J. Amer. Chem. Soc.*, **93**, 1558 (1971).

957. Howell, J. I., J. A. Lucy, R. C. Pirola, and I. A. D. Bouchier, *Biochim. Biophys. Acta*, **210**, 1 (1970). Also see Ref. 76.

958. Eldjarn, L., in *Molecular Basis of Some Aspects of Mental Acitivty*, Ed. O. Walaas, Academic Press, New York (1967).

959. Melchiro, D. L., H. J. Morowitz, J. M. Sturtevant, and T. Y. Tsong,

Biochim. Biophys. Acta., **219,** 114 (1970). Cerbon, J., *Biochim. Biophys. Acta.,* **211,** 389 (1970). Steim, J. M., M. E. Tourtellottem, J. C. Reinert, R. N. McElhaney, and R. L. Rader, *Proc. Natl. Acad. Sci. (U.S.),* **63,** 104 (1969). Reinert, J. C., and J. M. Steim, *Science,* **168,** 1580 (1970). Steim, J. M., in *Liquid Crystals and Ordered Fluids,* p. 1, Eds. J. F. Johnson and R. F. Porter, Plenum Press, New York (1969).

960. Starr, P. R., and L. W. Parks, *J. Cell. Comp. Physiol.,* **59,** 107 (1962). Haskin, R. H., *Science,* **150,** 1615 (1965). House, H. D., D. F. Riordan, and J. S. Barlow, *Can. J. Zool.,* **36,** 629 (1958). Deuticke, B., and W. Gruber, *Biochim. Biophys. Acta.,* **211,** 369 (1970). Yamanaka, W., and R. Ostwald, *J. Nutr.,* **95,** 381 (1968). McElhaney, R. N., J. de Gier, and L. L. M. Van Deenen, *Biochim. Biophys. Acta.,* **219,** 245 (1970). Gunasekaran, M., P. K. Raju, and S. D. Lyda, *Phytopathology,* **60,** 1027 (1970). Cullen, J., M. C. Phillips, G. G. Shipley, *Biochem. J.,* **125,** 733 (1971).

961. Cronan, J. E., Jr., T. K. Ray, and P. R. Vagelos, *Proc. Natl. Acad. Sci. (U.S.),* **65,** 737 (1970). Sherr, S. I., *Bacteriol. Proc. Abstr.,* P24 (1969). Holden, J. T., O. Hild, Y. Wong-Leung, and G. Rouser, *Biochem. Biophys. Res. Comm.,* **40,** 123 (1970). Hayashi, M., T. Muramatsu, I. Hara, *Biochim. Biophys. Acta,* **255,** 98 (1972).

962. Meyer, F., and K. Bloch, *J. Biol. Chem.,* **238,** 2654 (1963).

963. Reman, F. C., and L. L. M. Van Deenan, *Biochim. Biophys. Acta.,* **137,** 592 (1967).

964. Rouser, G., G. J. Nelson, S. Fleischer, and G. Simon, in *Biological Membranes: Physical Fact and Function,* p. 5, Ed., D. Chapman, Academic Press, New York (1968). Also see Ref. 77.

965. Lovelock, J. E., *Biochem. J.,* **60,** 692 (1955). Tucker, A. N., and D. C. White, *J. Bacteriol.,* **102,** 498, 508 (1970). Hoogeween, J. Th., R. Juliano, J. Coleman, and A. Rothstein, *J. Membrane Biol.,* **3,** 156 (1970).

966. Heppel, L. A., *J. Gen. Physiol.,* **54,** 95s (1969).

967. Marchesi, S. L., E. Steers, V. T. Marchesi, and T. W. Tillack, *Biochemistry,* **9,** 50 (1970). Lewin, S., *J. Theoret. Biol.,* **23,** 279 (1969).

968. Scheraga, H. A., G. Nemethy, and I. Z. Steinberg, *J. Biol. Chem.,* **237,** 2506 (1962). Tracey, M. V., *Proc. Roy. Soc. B,* **171,** 59 (1968).

969. Tria, E., and A. M. Scanu, *Structural and Functional Aspects of Lipoproteins in Living Systems,* Academic Press, New York (1969).

970. Schmitt, F. O., *Neuroscience Res. Program, Bull.,* **7,** 281 (1969).

971. Manson, L. A., (Ed.), *Biological Properties of Mammalian Surface Membrane,* Wistar Inst. Press, Philadelphia, Pa. (1967).

972. Maddy, A. H., *Intern. Rev. Cytology,* **20,** 1 (1966).

973. Green, D. E., and A. Tzagoloff, *J. Lipid Res.,* **7,** 587 (1966). Also see Ref. 703.

974. Chapman, D. (Ed.), *Biological Membranes: Physical Facts and Functions,* Academic Press, New York (1968). Cereijido, M., and C. A. Rotunno, *Introduction to the Study of Biological Membranes,* Gordon and Breach Science Publishers, New York (1970).

975. Giorgio, L., *Ital. J. Biochem.,* **17,** 129 (1968).

976. Stoeckenius, W., and D. M. Engelman, *J. Cell. Biol.*, **42**, 613 (1969).

977. Malhotra, S. K., and A. Van Harreveld, *Biol. Basis Med.*, **1**, 3 (1968).

978. Oncley, J. L., and N. R. Harvie, *Proc. Natl. Acad. Sci. (U.S.)*, **64**, 1107 (1969). Hammes, G. G., and S. E. Schullery, *Biochemistry*, **9**, 2555 (1970). Green, D. H., and M. R. J. Salton, *Biochim. Biophys. Acta.*, **211**, 139 (1970). Trump, B. F., S. M. Duttera, W. L. Byrne, and A. U. Arstila, *Proc. Natl. Acad. Sci. (U.S.)*, **66**, 433 (1970).

979. Cordes, E. H., and C. Gitler, *Progr. Bioorganic Chem.*, in press (1972).

980. Day, C. E., and R. S. Levy, *J. Theoret. Biol.*, **22**, 541 (1969). Triggle, D. J., *J. Theoret. Biol.*, **25**, 499 (1969). Also see Ref. 1282 and 1283.

981. Strom, D. R., D. E. Koshland, Jr., *Proc. Natl. Acad. Sci.*, **66**, 445 (1970). However see Ref. 1284.

982. Roelofsen, B., *Some Studies on the Extractability of Lipids and the ATPase activity of the Erythrocyte Membrane*, Ph.D. Thesis, University of Amsterdam, Amsterdam (1968).

983. Mizutani, H., and H. Mizutani, *Nature*, **204**, 781 (1964). Gordon, A. S., D. F. H. Wallach, and J. H. Straus, *Biochim. Biophys. Acta.*, **183**, 405 (1969). Lenard, J., and S. J. Singer, *Science*, **159**, 738 (1968).

983a. Gel'man, N. S., *Sovre. Biol. (Russ.)*, **68**, 3 (1969). Reynolds, J. A., and C. Tanford, *Proc. Natl. Acad. Sci. (U.S.)*, **66**, 1002 (1970). Emmelot, P., and H. V. Dias, *Biochim. Biophys. Acta.*, **203**, 172 (1970).

984. Fleischer, S., G. Brierley, H. Klouwen, and D. B. Slautterback, *J. Biol. Chem.*, **237**, 3264 (1962).

985. Dallner, G., P. Siekevitz, and G. E. Palade, *J. Cell. Biol.*, **30**, 97 (1966). Jones, P. D., and S. J. Wakil, *J. Biol. Chem.*, **242**, 5267 (1967).

986. Sekuzu, I., P. Jurtshuk, Jr., and D. E. Green, *J. Biol. Chem.*, **238**, 975 (1963).

987. Brierley, G. P., A. J. Merola, and S. Fleischer, *Biochim. Biophys. Acta.*, **64**, 218 (1962). Crane, F. L., *Biochemistry*, **1**, 510 (1962). Yamashita, S., and E. Racker, *J. Biol. Chem.*, **244**, 1220 (1969).

988. Machinist, J. M., and T. P. Singer, *J. Biol. Chem.*, **240**, 3182 (1965).

989. Brierley, G. P., and A. J. Merola, *Biochim. Biophys. Acta.*, **64**, 205 (1962). McConnell, D. G., A. Tzagoloff, D. H. MacLennan, D. E. Green, *J. Biol. Chem.*, **241**, 2373 (1966). Awasthi, Y. C., T. F. Chuang, T. W. Keenan, and F. L. Crane, *Biochim. Biophys. Res. Comm.*, **39**, 822 (1970).

990. Racker, E., *J. Gen. Physiol.*, **54**, 38s (1969).

991. Bulos, B., E. Racker, *J. Biol. Chem.*, **243**, 3901 (1968).

992. Duttera, S. M., W. L. Byrne, and M. C. Ganoza, *J. Biol. Chem.*, **243**, 2216 (1968).

993. Soodsma, J. F., and R. C. Nordlei, *Biochim. Biophys. Acta.*, **191**, 636 (1969). Stetten, M. R., S. Malamed, and M. Federman, *Biochim. Biophys. Acta.*, **193**, 260 (1969).

994. Martonoshi, A., J. Donley, and R. A. Halpin, *J. Biol. Chem.*, **243**, 61 (1968). G. Meissnner and S. Fleischer, *Biochim. Biophys. Acta*, **255**, 19 (1972).

995. Widnell, C. C., and J. C. Unkeless, *Proc. Natl. Acad. Sci. (U.S.)*, **61**, 1050 (1968).

996. Saiki, T., T. Suzuki, Y. Takagi, T. Narasaki, G. Tamura, and K. Arima, *Agr. Biol. Chem.*, **33**, 1101 (1969).
997. Sawada, H., M. Takeshita, Y. Sugita, and Y. Yoneyama, *Biochim. Biophys. Acta.*, **178**, 145 (1969).
998. Jones, P. D., P. W. Holloway, R. O. Peluffo, and S. J. Wakil, *J. Biol. Chem.*, **244**, 744 (1969).
999. Raju, P. K., and R. Reiser, *Lipids*, **5**, 487 (1970). See also Ref. 1286.
1000. Wisnieski, B. J., A. D. Keith, and M. R. Resnick, *J. Bacteriol.*, **101**, 160 (1970).
1001. Thomas, P. J., and J. H. Law, *J. Biol. Chem.*, **241**, 5013 (1966).
1002. Smith, P. F., *J. Bacteriol.*, **99**, 480 (1969).
1003. Lennarz, W. J., and B. Talamo, *J. Biol. Chem.*, **241**, 2707 (1966).
1004. Fox, C. F., *Proc. Natl. Acad. Sci. (U.S.)*, **63**, 850 (1969). Schairer, H. U., and P. Overath, *J. Mol. Biol.*, **44**, 209 (1969). Wilson, G., S. P. Rose, and C. F. Fox, *Biochem. Biophys. Res. Comm.*, **38**, 617 (1970).
1005. Hradec, J., and Z. Dusek, *J. Biochem.*, **115**, 873 (1969).
1006. Maddy, A. H., *Biochim. Biophys. Acta.*, **117**, 193 (1966).
1007. Bremer, J., and K. R. Norum, *J. Biol. Chem.*, **242**, 1749 (1967). Graham, A. B., and G. C. Wood, *Biochem. Biophys. Res. Comm.*, **37**, 567 (1969). Imai, Y., and R. Sato, *Biochim. Biophys. Acta.*, **42**, 164 (1960). Canal, J., and M. L. Girard, *Bull. Soc. Chim. Biol.*, **50**, 1523 (1968). Also see Refs. 1172, 1173.
1008. Barton, P. G., and D. J. Hanahan, *Biochim. Biophys. Acta.*, **187**, 319 (1969).
1009. Overath, P., H. U. Schairer, and W. Stoffel, *Proc. Natl. Acad. Sci. (U.S.)*, **67**, 606 (1970).
1010. White, D. C., and A. N. Tucker, *J. Bacteriol.*, **97**, 199 (1969).
1011. Friedman, R. M., and I. Pastan, *J. Mol. Biol.*, **40**, 107 (1969).
1012. Green, D. E., N. F. Haard, G. Lenaz, and H. I. Silman, *Proc. Natl. Acad. Sci. (U.S.)*, **60**, 277 (1968). Also see Ref. 1020.
1013. Das, M. L., *Biochemistry*, **4**, 859 (1965).
1014. Wang, J. H., *Proc. Natl. Acad. Sci. (U.S.)*, **67**, 916 (1970).
1015. Hatch, F. T., and A. L. Bruce, *Nature*, **218**, 1166 (1968).
1016. Morgan, T. E., and D. J. Hanahan, *Biochemistry*, **5**, 1050 (1966). Also see Eylar, E. H., *Proc. Natl. Acad. Sci. (U.S.)*, **67**, 1425 (1970). D. F. H. Wallach, *Biochim. Biophys. Acta*, **265**, 61 (1972). However see H. J. Morowitz and T. M. Terry, *Biochim. Biophys. Acta*, **183**, 276 (1969).
1017. Lenard, J., and S. J. Singer, *Proc. Natl. Acad. Sci. (U.S.)*, **56**, 1828 (1966).
1018. Wallach, D. F. H., *J. Gen. Physiol.*, **54**, 3s (1969). Wallach, D. F. H., and P. H. Zahler. *Proc. Natl. Acad. Sci. (U.S.)*, **56**, 1552 (1966).
1019. Urry, D. W., and T. H. Ji, *Arch. Biochem. Biophys.*, **128**, 802 (1968). Schmeider, A. S., M. T. Schneider, and K. Rosenheck, *Proc. Natl. Acad. Sci. (U.S.)*, **66**, 793 (1970). D. W. Urry, *Biochim. Biophys. Acta*, **265**, 115 (1972). W. L. Zahler, D. Puett and S. Fleischer, *Biochim. Biophys. Acta*, **255**, 365 (1972).

1020. See Refs. 1012 and 1048. Woodward, D. D., and K. D. Munkers, *Proc. Sci. (U.S.)*, **55**, 872 (1966).

1021. McClare, C. W. F., *Nature,* **216**, 766 (1967). See also Ref. 1306.

1022. Robinson, J. D., *Nature,* **212**, 199 (1966).

1023. Glick, M. C., and L. Warren, *Proc. Natl. Acad. Sci. (U.S.)*, **63**, 563 (1969). Hoober, J. K., P. Siekevitz, and G. E. Palade, *J. Biol. Chem.,* **244**, 2621 (1969).

1024. Peterson, J. A., and H. Rubin, *Exptl. Cell Res.,* **60**, 383 (1970). Lepetina, E. G., G. G. Lunt, and E. DeTobertis, *J. Neurobiol.,* **1**, 295 (1970).

1025. Eder, H. A., *Amer. J. Med.,* **23**, 269 (1957).

1026. Kunkel, H. G., and A. G. Bearn, *Proc. Soc. Exptl. Biol. Med.,* **86**, 887 (1954).

1027. Porte, D., Jr., and R. J. Havel, *J. Lipid Res.,* **2**, 357 (1961).

1028. Nichols, A. V., and E. L. Coggiola, *J. Lipid Res.,* **7**, 215 (1966).

1029. Roheim, P. S., L. Miller, and H. A. Eder, *J. Biol. Chem.,* **240**, 2994 (1965).

1030. Sakagami, T., O. Minari, and T. Orii, *Biochim. Biophys. Acta.,* **98**, 111 (1965).

1031. Trewhell, M. A., and F. D. Frederick, *Lipids,* **4**, 304 (1969).

1032. Wirtz, K. W. A., and D. B. Zilversmit, *Biochim. Biophys. Acta.,* **193**, 105 (1969); Zilversmit, D. B., *J. Biol. Chem.,* **246**, 2645 (1971).

1033. Branton, D., *Ann. Rev. Plant Physiol.,* **20**, 209 (1969). Meyer, H. W., and H. Winkelmann, *Protoplasma,* **68**, 253 (1969). D. Branton, *Trans. Royal Soc.* (Lond.) B, **261**, 133 (1971). Muhlethaler, K., *Intern. Rev. Cytol.,* **31**, 1 (1971). V. Speth, D. F. H. Wallach, E. Weidekamm, and H. Knufermann, *Biochim. Biophys. Acta,* **255**, 386 (1972).

1034. Deamer, D. W., R. Leonard, A. Tardieu, and D. Branton, *Biochim. Biophys. Acta.,* **219**, 47 (1970). Staeheltin, L. A., *J. Ultrastr. Res.,* **22**, 326 (1968).

1035. Cunningham, W. P., and F. L. Crane, *Exptl. Cell Res.,* **44**, 31 (1966). Sjostrand, F. S., *Nature,* **199**, 1262 (1963). Remsen, C. C., S. W. Watson, and A. W. Bernheimer, *Biochem. Biophys. Res. Comm.,* **40**, 1297 (1970). Also see Ref. 1176.

1036. Green, D. E., and J. F. Perdue, *Proc. Natl. Acad. Sci. (U.S.)*, **55**, 1295 (1966).

1037. Gent, W. L. G., N. A. Gregson, D. B. Gammack, and J. H. Raper, *Nature,* **204**, 553 (1964).

1038. Razin, S., H. J. Morowitz, and T. M. Terry, *Proc. Natl. Acad. Sci. (U.S.),* **54**, 219 (1965). For similar results on B. subtilis see Joyeux, Y., and H. Jouin, *C. R. Acad. Sci., Ser. D,* **271**, 434 (1970).

1039. Engelman, D. M., and H. J. Morowitz, *Biochim. Biophys. Acta.,* **150**, 376, 385 (1968). Razin, S., Z. Neeman, and I. Ohad, *Biochim. Biophys. Acta.,* **193**, 277 (1969). Tillack, T. W., R. Carter, and S. Razin, *Biochim. Biophys. Acta.,* **219**, 123 (1970). Also see Ref. 1177.

1040. Rothfield, L., and M. Perlman-Kothencz, *J. Mol. Biol.,* **44**, 477 (1969). Butler, T. F., G. L. Smith, and E. A. Grula, *Can. J. Microbiol.,* **13**, 1471 (1967).

1041. Kuehn, G. D., B. A. McFadden, R. A. Johanson, J. M. Hill, and L. K. Shumway, *Proc. Natl. Acad. Sci. (U.S.)*, **62**, 407 (1969).

1042. Blaurock, A. E., and M. H. F. Wilkins, *Nature*, **223**, 906 (1969). Blasie, J. K., C. R. Worthington, and M. M. Dewey, *J. Mol. Biol.*, **39**, 407 (1969). Blasie, J. K., and C. R. Worthington, *J. Mol. Biol.*, **39**, 417 (1969).

1043. Yanagida, M., and H. Noda, *J. Biochem. (Tokyo)*, **65**, 709, 721 (1969).

1044. Rogers, S. W., H. E. Gilleland, Jr., and R. G. Eagon, *Can. J. Microbiol.*, **15**, 743 (1969).

1045. Vergara, J., W. Longley, and J. D. Robertson, *J. Mol. Biol.*, **46**, 593 (1970). For similar results on rat urinary bladder see Warren, R. C., and R. M. Hicks, *Nature*, **227**, 280 (1970).

1046. DiCarlo, V., *Nature*, **213**, 833 (1967).

1047. Gemne, G., in Ref. 947, p. 305.

1048. Braun, P. E., and N. S. Radin, *Biochemistry*, **8**, 4310 (1969). Patterson, R. H., and W. J. Lennarz, *Bichem. Biophys. Res. Comm.*, **40**, 408 (1970). See also Ref. 1041 and R. S. Criddle and J. Willemot in Ref. 969.

1049. R. T. Swank, G. T. Sheir, in *Autonomy and Biogenesis of Mitochondria and Chloroplasts*, N. K. Boardman, A. W. Linnane, and R. M. Smillie, (eds.). North-Holland Publishing Co., Amsterdam, p. 152 (1971). Kiehn, E. D., and J. J. Holland, *Proc. Natl. Acad. Sci. (U.S.)*, **61**, 1370 (1968). Demus, H., and E. Mehl, *Biochim. Biophys. Acta.*, **203**, 291 (1970). Also see Ref. 1189.

1050. Senior, A. E., and D. H. MacLennan, *J. Biol. Chem.*, **245**, 5086 (1970). Bont, W. S., P. Emmelot, and H. V. Dias, *Biochim. Biophys. Acta.*, **173**, 389 (1969). Schnaitman, C. A., *Proc. Natl. Acad. Sci. (U.S.)*, **63**, 412 (1969).

1051. Laico, M. T., E. I. Ruoslahti, D. S. Papermaster, and W. J. Dreyer, *Proc. Natl. Acad. Sci. (U.S.)*, **67**, 120 (1970). However see Ref. 1187.

1052. Blair, J. E., G. Lenaz, and N. F. Haard, *Arch. Biochem. Biophys.*, **126**, 753 (1968). Gwynne, J. T., and C. Tanford, *J. Biol. Chem.*, **245**, 3269 (1970).

1053. Casper, D. L. D., *Protoplasma*, **63**, 197 (1967). Rosen, R., *J. Theoret. Biol.*, **28**, 415 (1970) and references therein. Wyman, J., *J. Mol. Biol.*, **39**, 533 (1969). Vanderkooi, G., and D. E. Green, *Proc. Natl. Acad. Sci. (U.S.)*, **66**, 615 (1970). See p. 235–312 in Ref. 947. Also see Rothfield, L., and A. Finkelstein, *Ann. Rev. Biochem.*, **37**, 463 (1968). Schofeniels, E., *Arch. Int. Physiol. Biochem.*, **78**, 205 (1970). Green, N. M., *Biochem. J.*, **122**, 37P (1971). Also see Ref. 37k.

1054. Steahelin, L. A., and M. C. Probine, *Adv. Bot. Res.*, **3**, 1 (1970).

1055. Hendler, R. W., *Physiol. Rev.*, **51**, 66 (1971).

1056. Cook, G. M. W., *Ann. Rev. Plant Physiol.*, **22**, 97 (1971).

1057. Abramson, M. B., and R. Katzman, *Adv. Exptl. Med. Biol.*, **7**, 85 (1970).

1058. Tinoco, J., and D. J. McIntosh, *Chem. Phys. Lipids*, **4**, 72 (1970).

1059. Henson, A. F., R. B. Leslie, L. Rayner, and N. Sanders, *Chem. Phys. Lipids*, **4**, 345 (1970).

1060. Kamel, A. M., A. Felmeister, and N. D. Weiner, *J. Lipid Res.*, **12**, 155 (1971). Joos, P., *Chem. Phys. Lipids*, **4**, 162 (1970).

1061. Huang, C.-H., and J. P. Charlton, *J. Biol. Chem.*, **246**, 2555 (1971).

1062. Miyamoto, V. K., and W. Stoeckenius, *J. Membrane Biol.*, **4**, 252 (1971).

1063. Seufert, W. D., *Biophysik*, **7**, 60 (1970).

1064. Tien, H. T. In *The Chemistry of Biosurfaces*, M. Hair (ed.), Mercel Dekker, Inc., New York, p. 266 (1971).

1065. Rand, R. P., *J. Gen. Physiol.*, **52**, 173s (1968).

1066. Liberman, E. A., and V. A. Nenashev, *Biofizika*, **15**, 1014 (1970).

1067. Simons, R., and P. E. Ciddor, *Biochim. Biophys. Acta*, **233**, 296 (1971).

1068. White, S. H., *Biophys. J.*, **10**, 1127 (1970).

1069. Kauffman, J. W., and C. A. Mead, *Biophys. J.*, **10**, 1084 (1970).

1070. Parisi, M., and E. Rivas, *Biochim. Biophys. Acta*, **233**, 469 (1971).

1071. Ververgaert, P. H. J. Th., and P. F. Elbers, *J. Mol. Biol.*, **58**, 431 (1971).

1072. Reeves, J. P., and R. M. Dowben, *J. Membrane Biol.*, **3**, 123 (1970).

1073. Fenichel, I. R., and S. B. Horowitz, *J. Physical Chem.*, **74**, 2966 (1970).

1074. DeGier, J., J. G. Mandersloot, J. V. Hupkes, R. N. McElhaney, and W. P. van Breek, *Biochim. Biophys. Acta*, **233**, 610 (1971). Haest, C. W. M., J. DeGier, J. A. F. Op Den Kemp, P. Bartels, L. L. M. Van Deenen, *Biochim. Biophys. Acta*, **255**, 720 (1972). See also Ref. 1291 for the effect of cholesterol derivatives on passive permeability.

1075. Klein, R. A., M. J. Moore, and M. W. Smith, *J. Physiol.*, **210**, 33P (1970).

1076. Andreoli, T. E., and S. L. Troutman, *J. Gen. Physiol.*, **57**, 464 (1971).

1077. Andreoli, T. E., J. A. Schafer, and S. L. Troutman, *J. Gen. Physiol.*, **57**, 479 (1971). A. K. Solomon, C. M. Gary-Bobo, *Biochim. Biophys. Acta*, **255**, 1019 (1972).

1078. Lieb, W. R., and W. D. Stein, *Nature*, **224**, 240 (1969).

1079. Trauble, H., *J. Membrane Biol.*, **4**, 193 (1971). See also Ref. 1215.

1080. Chen, Li-Fu, D. B. Lund, and T. Richardson, *Biochim. Biophys. Acta*, **225**, 89 (1971). See also Ref. 1192.

1081. Higuchi, W. I., A. H. Ghanem, and A. B. Bikhazi, *Fed. Proc.*, **29**, 1327 (1970).

1082. Moore, T. J., and B. Schlowsky, *Chem. Phys. Lipids*, **3**, 273 (1969). Moore, T. J., in *Surface Chemistry of Biological Systems*, p. 295, Ed. M. Blank, Plenum Press, New York (1970).

1083. Karan, D. M., R. I. Macey, and R. E. L. Farmer, *Biophys. Soc. Abstr.*, 281a (1971).

1084. Jung, C. Y., *J. Membrane Biol.*, **5**, 200 (1971).

1085. Cussler, E. L., D. F. Evans, and M. A. Matesich, *Science*, **172**, 377 (1971).

1086. LeBlanc, O. H., Jr., *J. Membrane Biol.*, **4**, 227 (1971).

1087. Noguchi, S., and Koga, *J. Gen. Appl. Microbiol.*, **16**, 489 (1970).

1088. Cope, F. W., *Adv. Biol. Medicinal Physics*, **13**, 1 (1970).

1089. Pant, H. C., and B. Rosenberg, *Chem. Phys. Lipids*, **6**, 39 (1971).

1090. Bouslavskii, L. I., A. V. Lebedev, and F. I. Bogolepova, *Dokl. Akad. Nauk SSSR*, **195**, 475 (1970).

1091. Ovchinnikov, Yu. A., V. T. Ivanov, and I. I. Michaleva, *Tetrahedron Letters.*, 159 (1971). Roeske, R. W., S. Isaac, L. K. Steinrauf, and T. King, *Fed. Proc.* (Abstr.), **30**, 1282 (1971).

1092. Lutz, W. K., H. K. Wipf, and W. Simon, *Helv. Chim. Acta.*, **53**, 1741 (1970).

1093. Estrada-O., S., C. Gomez-Lojero, *Biochemistry*, **10**, 1598 (1971).

1094. Hamilton, W. A., *Biochem. J.*, **118**, 46P (1970). Harold, F. M., *Adv. Microbiol. Physiol.*, **4**, 45 (1970).

1095. Wipf, H. K., A. Olivier, and S. Eilhelm, *Helv. Chim. Acta.*, **53**, 1605 (1970). Andreoli, T. E., and D. C. Tosteson, *J. Gen. Physiol.*, **57**, 526 (1971).

1096. Ivanov, V. T., A. I. Miroshnikov, N. D. Abdullaev, L. B. Senyavina, S. F. Arkhipova, N. N. Uvarova, K. Kh. Khalilulina, V. F. Bystrov, and Yu. A. Ovchinnikov, *Biochim. Biophys. Res. Comm.*, **42**, 654 (1971).

1097. Haynes, D. H., B. C. Pressman, and A. Kowalsky, *Biochemistry*, **10**, 852 (1971).

1098. Blondin, G. A., and R. M. Hull, *Biophys. Soc. Abstr.*, 120a (1971).

1099. Calissano, P., and A. D. Bangham, *Biochem. Biophys. Res. Comm.*, **43**, 504 (1971). See also Ref. 1297.

1100. Parisi, M., E. Rivas, and E. DeRobertis, *Science*, **172**, 56 (1971). (It may be noted that the transient conductance change has time constant of the order of 30 sec, and may be due to bulk diffusion.) Vasquez, C. M., M. Parisi, and E. DeRobertis, *J. Membrane Biol.*, **6**, 353 (1971).

1101. Margoliash, E., G. H. Barlow, and V. Byers, *Nature*, **228**, 723 (1970). Schejter, A., and Margalit R., *FEBS Letters*, **10**, 179 (1970).

1102. Osterhout, M. J. V., *J. Gen. Physiol.*, **23**, 429 (1940); **27**, 91 (1944).

1103. Laiken, S. L., M. P. Printz, and L. C. Craig, *Biochem. Biophys. Res. Comm.*, **43**, 595 (1971).

1104. Urry, D. W., *Proc. Natl. Acad. Sci.*, **68**, 672 (1971).

1105. Kemp, G., T. Dougherty, K. Jacobson, and C. E. Wenner, *Biophys. Soc. Abstr.*, 311a (1971).

1106. Goodall, M. C., *Biochim. Biophys. Acta*, **219**, 471 (1970).

1107. Roth, S., and P. Seeman, *Nature New Biol.*, **231**, 284 (1971).

1108. Seeman, P., S. Roth, and H. Schneider, *Biochim. Biophys. Acta*, **225**, 171 (1971).

1109. Watkins, J. C., *J. Theoret. Biol.*, **9**, 37 (1965).

1110. Sweet, C., and J. E. Zull, *Biochim. Biophys. Acta*, **173**, 94 (1969).

1111. Kimelberg, H. K., and D. Papahadjopoulos, *J. Biol. Chem.*, **246**, 1142 (1971).

1112. Ter-Minassian-Saraga, L., and J. Weitzerbin-Falszpan, *Biochem. Biophys. Res. Comm.*, **41**, 1231 (1970).

1113. Inoue, K., T. Kataoka, and S. C. Kinsky, *Biochemistry*, **10**, 2574 (1971).

1114. Tsukagoshi, N., G. Tamura, and K. Arima, *Biochim. Biophys. Acta.*, **196**, 211 (1970).

1115. Kotyk, A., and K. Jenacek, *Cell Membrane Transport: Principles and Techniques*, Plenum Press, New York (1970).

1116. Kepes, A., *Current Topics in Membranes and Transport*, **1**, 101 (1970). Kepes, A., *J. Membrane Biol.*, **4**, 87 (1971).

1117. Krupka, R. M., *Biochemistry*, **10**, 1143, 1148 (1971).

1118. Baker, P. F. in *Calcium and Cellular Function*, p. 96, Ed. A. W. Cuthbert, McMillan & Co. Ltd., 1970.

1119. Schairer, H. U., and P. Overath, *J. Mol. Biol.*, **44**, 209 (1969).

1120. Mindich, L., *Proc. Natl. Acad. Sci.*, **68**, 420 (1971). Kahlenberg, A. and B. Banjo, *J. Biol. Chem.*, **247**, 1156 (1972).

1121. Kirschmann, C., I. Ten-Ami, I. Smorodinsky, and A. DeVries, *Biochim. Biophys. Acta*, **233**, 644 (1971).

1122. Gale, E. F., and J. M. Llewellin, *Biochim. Biophys. Acta*, **222**, 546 (1970).

1123. Poncova, M., and A. Kotyk, *Curr. Mod. Biol.*, **1**, 189 (1967).

1124. Boos, W., and A. S. Gordon, *J. Biol. Chem.*, **246**, 621 (1971).

1125. Manchester, K. L., *Biochemical Action of Hormones*, **1**, 267 (1970).

1126. Faust, R. G., M. J. Burns, and D. W. Misch, *Biochim. Biophys. Acta.*, **219**, 507 (1970).

1127. Kundig, W., and S. Roseman, *J. Biol. Chem.*, **246**, 1393, 1407 (1971).

1128. Hechtman, P., and C. R. Scriver, *Biochim. Biophys. Acta*, **219**, 428 (1970). Wilson, T. H., and M. Kusch, *Biochim. Biophys. Acta*, **255**, 786 (1972).

1129. Wong, J. T., A. Pincock, and P. M. Bronskill, *Biochim. Biophys. Acta*, **233**, 176 (1971).

1130. Thomas, E. L., and H. N. Christensen, *J. Biol. Chem.*, **246**, 1682 (1971).

1131. Ussing, H. H., *Physiol. Veg.*, **9**, 1 (1971). Herrera, F. C., *Membranes and Transport*, **3**, 1 (1971).

1132. Skou, J. C., *Adv. Exp. Medicine Biol.*, **14**, 175 (1971).

1133. Klingenberg, M., and E. Pfaff, in *Regulation of Metabolic Processes in Mitochondria*, p. 180, Eds. J. M. Tager et al., Adriatica Editrice, Bari, Italy, 1966. Waino, W. W., *The Mammalian Mitochondrial Respiratory Chain*, Academic Press, New York (1970).

1134. MacLennan, D. H., *Current Topics in Membranes and Transport*, **1**, 177 (1970). Crane, F. L., J. W. Stiles, K. S. Prezbindowski, F. J. Ruzicka, and F. F. Sun, in *BBA Library*, **11**, 21 (1968).

1135. Whittingham, C. P., *Progress in Biophysics*, **21**, 127 (1970). Kreutz, W., *Adv. Bot. Res.*, **3**, 53 (1970).

1136. Nultsch, W., in *Photobiology of Microorganisms*, p. 213, Ed. P. Halldel, Wiley-Interscience, 1970.

1137. Bitensky, M. W., R. E. Gorman, and W. H. Miller, *Proc. Natl. Acad. Sci.*, **68**, 561 (1971).

1138. Brown, P. K., *Biophys. Soc. Abstr.*, 248a (1971).

1139. Scalzi, H. A., *Z. Zellforsch. Microskop.*, **80**, 413 (1967). Farbman, A. I.,

1139. in *Symp. on Foods: The Chemistry and Physiology of Flavors*, Ed. H. W. Schultz, The Avi Publ. Co., Inc., Westport, Conn., 1967.

1140. Mackey, M. C., *Biophys. J.*, **11**, 75, 91 (1971). Woodbury, J. W., *Adv. Chem. Phys.*, **21**, 601 (1971). Schlogl, R., *Quart. Rev. Biophys.*, **2**, 305 (1969).

1141. Lorento de No, R., *Proc. Natl. Acad. Sci.*, **68**, 192 (1971).

1142. Taylor, R. E., and E. Rojas, *Biophys. Soc. Abstr.*, 56a (1971).

1143. Thornhang, A., *Biochim. Biophys. Acta,* **225**, 151 (1971).

1144. Rojas, E., and C. Armstrong, *Nature,* **229**, 177 (1971).

1145. Narahashi, T., T. Deguchi, and E. X. Albuquerque, *Nature New Biology,* **229**, 221 (1971). Albuquerque, E. X., J. W. Daly, and B. Witkop, *Science,* **172**, 995 (1971). Also see Ref. 1260.

1146. Keynes, R. D., J. M. Ritchie, and E. Rojas, *J. Physiol.,* **213**, 235 (1971).

1147. Barry, P. H., *J. Membrane Biol.,* **3**, 313, 335 (1970).

1148. Tasaki, I., *Advances in Biological and Medical Physics,* **13**, 307 (1970). Tasaki, I., A. Watanabe, and M. Hallett, *Proc. Natl. Acad. Sci.,* **68**, 938 (1971).

1149. Pant, H. C., and B. Rosenberg, *Biochim. Biophys. Acta,* **225**, 379 (1971).

1150. Botre, C., W. Dorst, M. Marchetti, A. Memoli, and E. M. Scarpelli, *Biochim. Biophys. Acta,* **219**, 283 (1970).

1151. Hout, R. C., *Biophys. J.,* **11**, 110 (1971).

1152. Hille, B., *Progr. in Biophys.,* **21**, 1 (1970).

1153. Hille, B., *Biophys. Soc. Abstr.,* 54a (1971); *Proc. Natl. Acad. Sci.,* **68**, 280 (1971).

1154. McLaughlin, S. G. A., G. Szabo, and G. Eisenman, *Biophys. Soc. Abstr.,* 146a (1971).

1155. Phillips, M. C., V. B. Kamat, and D. Chapman, *Chem. Phys. Lipids,* **4**, 409 (1970). Haque, R., I. J. Tinsley, D. Schmedding, *J. Biol. Chem.,* **247**, 157 (1972). Horwitz, A. F., W. J. Horslay, M. P. Klein, *Proc. Natl. Acad. Sci.* (U.S.), **69**, 590 (1972). See also Ref. 1273.

1156. Hsia, J. C., H. Schneider, and I. C. P. Smith, *Chem. Phys. Lipids,* **4**, 238 (1970); *Biochim. Biophys. Acta,* **202**, 399 (1970). Tourtellotte, M. E., D. Branton, and A. Keith, *Proc. Natl. Acad. Sci.,* **66**, 909 (1970). See also Ref. 1274.

1157. Kruckdorfer, K. R., P. A. Edwards, and C. Green, *Eur. J. Biochem.,* **4**, 506 (1968).

1158. McFarland, B. G., and H. M. McConnell, *Proc. Natl. Acad. Sci.,* **68**, 1274 (1971).

1159. McConnell, H. M., and R. D. Kornberg, *Biochemistry,* **10**, 1111 (1971). See also Ref. 1272.

1160. Raison, J. K., J. M. Lyons, R. J. Mehlhorn, and A. D. Keith, *J. Biol. Chem.,* **246**, 4036 (1971).

1161. Abramson, M. B., in *Surface Chemistry of Biological Systems,* p. 37, Ed. M. Blank, Plenum Press, New York, 1970.

1162. Abramson, M. B., *J. Coll. Interface Sci.,* **34**, 571 (1970).

1163. Chapman, D., and J. Urbina, *FEBS Letters,* **12**, 169 (1971). R. M. Williams, D. Chapman, *Progr. Chemistry of Fats and Other Lipids,* XI (part i), p. 1 (1971). See also Ref. 1271.

1164. Kamat, V. B., D. Chapman, R. F. A. Zwaal, and L. L. M. Van Deenen, *Chem. Phys. Lipids,* **4**, 323 (1970). See also Ref. 1019.

1165. Butler, K. W., H. Dugas, I. C. P. Smith, and H. Schneider, *Biochem. Biophys. Res. Comm.,* **40**, 770 (1970).

1166. Lewin, S., *Biochem. J.,* **122**, 20P (1971); *Nature New Biol.,* **231**, 80 (1971).

1167. Van Deenen, L. L. M., *BBA Library*, **11**, 72 (1968). Joyce, G. H., R. K. Hammond, and D. C. White, *J. Bacteriol.*, **104**, 323 (1970).

1168. Lesslauer, W., F. C. Wissler, and D. F. Parsons, *Biochim. Biophys. Acta.*, **203**, 199 (1970). Curtis, P. J., *ibid.*, **183**, 239 (1969). Maddy, A. H., *ibid.*, **88**, 448 (1964). Curtis, P. J., *Biochim. Biophys. Acta*, **255**, 833 (1972). H. Sandermann, Jr., and J. L. Strominger, *Proc. Natl. Acad. Sci.* (U.S.), **68**, 2441 (1971). G. Lenaz, G. Parenti-Castelli, A. M. Sechi, L. Masotti, *Arch. Biochem. Biophys.*, **148**, 391 (1972).

1169. Lovrein, R., *J. Amer. Chem. Soc.*, **85**, 3677 (1963). Helenius, A., and K. Simons, *Biochemistry*, **10**, 2542 (1971). S. F. Estrugo, V. Larraga, M. A. Corrales, C. Duch, E. Munoz, *Biochim. Biophys. Acta*, **255**, 960 (1972). P. Swanljung, *Anal. Biochem.*, **43**, 382 (1971). Also see Ref. 1280.

1170. Daniels, M. J., *Biochem. J.*, **122**, 197 (1971). Also see Refs. 1232 and 1281.

1171. Wetlaufer, D. B., and R. Lovrein, *J. Biol. Chem.*, **239**, 596 (1964).

1172. Cunningham, C. C., and L. P. Hager, *J. Biol. Chem.*, **246**, 1583 (1971).

1173. Oreland, L., and T. Olivecrona, *Arch. Biochem. Biophys.*, **142**, 710 (1971). Chaplin, M. D., and G. J. Mannering, *Mol. Pharmacol.*, **6**, 631 (1970). Wilson, H. L., *Fed. Proc.*, (Abstr.) **30**, 1126 (1971). Zorn, M., and S. Futterman, *J. Biol. Chem.*, **246**, 881 (1971). Jones, P. D., and S. J. Wakil, *J. Biol. Chem.*, **242**, 5267 (1967). DeRobertis, E., S. Fiszer, and E. F. Soto, *Science*, **158**, 928 (1967). Cunningham, C. C., and P. H. Lowell, *J. Biol. Chem.*, **246**, 1575 (1971). Ficus, W. G., and W. C. Schneider, *Fed. Proc.*, **24**, 476 (1965). Schulze, H. U., and H. Staudinger, *Hoppe-Seyler's Z. Physiol. Chem.*, **352**, 309 (1971). Lenaz, G., A. Castelli, G. P. Littarru, E. Bertoli, and K. Folkers, *Arch. Biochem. Biophys.*, **142**, 407 (1971). Richardson, S. H., H. O. Hultin, and S. Fleischer, *Arch. Biochem. Biophys.*, **105**, 254 (1964).

1174. Rothstein, A., *Current Topics in Membranes and Transport*, **1**, 135 (1970).

1175. Saunder, M. T., and M. Levy, *J. Lipid Res.*, **12**, 71 (1971).

1176. Malhotra, S. K., *J. Ultrastruct. Res.*, **15**, 14 (1966). Sjostrand, F. S., *ibid.*, **9**, 340 (1963).

1177. Engelman, D. M., J. C. Metcalfe, and S. M. Metcalfe, *Brit. J. Pharmacol.*, **41**, 382P (1971).

1178. Robertson, J. D., *J. Cell. Biol.*, **19**, 201 (1963).

1179. Kaplan, D. M., and R. S. Criddle, *Physiol. Rev.*, **51**, 249 (1971).

1180. Masoro, E. J., and B. P. Yu, *Biochem. Biophys. Res. Comm.*, **34**, 686 (1969).

1181. Fairbanks, G., T. L. Steck, and D. F. H. Wallach, *Biochemistry*, **10**, 2606 (1971).

1182. Williams, D. J., and B. R. Rabin, *Nature*, **232**, 102 (1971).

1183. Oppenheimer, S. B., and T. H. Humphreys, *Nature*, **232**, 125 (1971), and references therein.

1184. Shinitzky, M., A.-C. Dianoux, C. Gitler, and G. Weber, *Biochemistry*, **10**, 2106 (1971). Also see Refs. 1215 and 1275.

1185. Seeling, J., *J. Amer. Chem. Soc.*, **92**, 3881 (1970).
1186. Chan, S. I., G. W. Feigenson, and C. H. A. Seiter, *Nature*, **231**, 111 (1971).
1187. Cole, R. M., T. J. Popkin, B. Prescott, R. M. Chanock, and S. Razin, *Biochim. Biophys. Acta*, **233**, 76 (1971). Zahler, P., and E. R. Weibel, *Biochim. Biophys. Acta.*, **219**, 320 (1970).
1188. Hober, R., *Physical Chemistry of Cells and Tissues*, Blakiston, Philadelphia, Pa., 1945.
1189. Lenard, J., *Biochemistry*, **9**, 1119, 5037 (1970).
1190. Tomita, T., in *Smooth Muscle*, Eds. E. Bulbring, A. F. Brading, A. W. Jones, and T. Tomita, The Williams and Wilkins Co., (1970), p. 197.
1191. Hinkle, P., *Biochem. Biophys. Res. Comm.*, **41**, 1375 (1970), and unpublished data.
1192. Bikhazi, A. B., and W. I. Higuchi, *Biochim. Biophys. Acta*, **233**, 676 (1971).
1193. Kuhn, H., and D. Mobius, *Angew. Chem.*, **10**, 620 (1971).
1194. Richter, G. W., D. G. Scarpelli, and N. Kaufmann, Eds., *Cell Membranes: Biological and Pathological Aspects*, The Williams and Wilkins Co., Baltimore (1971).
1195. Rouser, G., A. Yamamota, and G. Kritchevsky, *Arch. Intern. Med.*, **127**, 1105 (1971).
1196. Brown, G. H., J. W. Doane, and V. D. Neff, *A Review of the Structure and Physical Properties of Liquid Crystals*, CRC Press, Cleveland, (1971).
1197. Fox, K. K., *Farad. Soc. Trans.*, **67**, 2802 (1971).
1198. Majer, H., *Agents and Actions*, 2, 33 (1971).
1199. Szubinska, B., *J. Cell Biol.*, **49**, 747 (1971).
1200. Finer, E. G., A. G. Flook, and H. Hauser, *FEBS Letters*, **18**, 331 (1971).
1201. Tien, H. Ti, in *The Chemistry of Biosurfaces*, Ed. M. L. Hair, Marcel Dekker Inc., New York (1971), p. 233.
1202. Wobschall, D., *J. Coll. Int. Sci.*, **36**, 385 (1971).
1203. Bair, R. E., in *Adhesives in Biological Systems*, Ed., R. S. Manly, Academic Press, New York (1970), p. 15.
1204. Roseman, S., *Chem. Phys. Lipids*, **5**, 270 (1970).
1205. Badzhinyan, S. A., S. A. Kovalev, and L. M. Chailakhyan, *Biol. Zh. Arm.*, **24**, 98 (1971).
1206. Fettiplace, R., D. M. Andrews, and D. A. Haydon, *J. Mem. Biol.*, **5**, 277 (1971).
1207. Crossley, J., *R.I.C. Reviews*, **4**, 69 (1971).
1208. Rastogi, R. P., P. C. Shukla, and B. Yadava, *Biochim. Biophys. Acta*, **249**, 454 (1971).
1209. Naftalin, R. J., *Biochim. Biophys. Acta*, **233**, 635 (1971).
1210. Gary-Bobo, C. M., and A. K. Solomon, *J. Gen. Physiol.*, **57**, 610 (1971).
1211. Kroes, J., and R. Ostwald, *Biochim. Biophys. Acta*, **249**, 647 (1971).
1212. Leo, A., C. Hansch, and D. Elkins, *Chem. Rev.*, **71**, 525 (1971).
1213. Smulders, A. P., and E. M. Wright, *J. Membrane Biol.*, **5**, 297 (1971).
1214. Graziani, Y., and A. Livne, *Biochem. Biophys. Res. Comm.*, **45**, 321 (1971).
1215. Trauble, H., *Naturwiss.*, **58**, 277 (1971).

1216. Scatchard, G., *J. Phys. Chem.*, **68**, 1056 (1964).

1217. Papahadjopoulos, D., *Biochim. Biophys. Acta*, **241**, 254 (1971).

1218. Hope, A. B., *Ion Transport and Membranes*, Butterworths, London (1971).

1219. Hinton, J. F., and E. S. Amis, *Chem Rev.*, **71**, 627 (1971).

1220. Botre, C., S. Borghi, M. Morchetti, A. Memoli, and L. Bolis, *Farmaco, Ed. Sci.*, **25**, 939 (1970).

1221. Ketterer, B., B. Neumcke, and P. Lauger, *J. Membrane Biol.*, **5**, 225 (1971).

1222. Kay, R. E., and R. C. Bean, *Advances in Biological and Medical Physics*, **13**, 235 (1970).

1223. Wenner, C. E., and T. J. Dougherty, *Progress in Surface and Membrane Science*, **4**, 351 (1971).

1224. Feinstein, M. B., and H. Felsenfeld, *Proc. Natl. Acad. Sci.*, **68**, 2037 (1971). Haynes, D. H., *Biochim. Biophys. Acta*, **255**, 406 (1972).

1225. Steinrauf, L. K., E. W. Czerwinski, and M. Pinkerton, *Biochem. Biophys. Res. Comm.*, **45**, 1279 (1971).

1226. Stark, G., and R. Benz, *J. Membrane Biol.*, **5**, 133 (1971). Stark. G., B. Ketterer, R. Benz, and P. Lauger, *Biophys. J.*, **11**, 981 (1971). Haynes, D. H., *FEBS Letters*, **20**, 221 (1972).

1227. Poole. D. T., T. C. Butler, and M. E. Williams, *J. Membrane Biol.*, **5**, 261 (1971).

1228 McMullen, A. I., D. I. Marlborough, and P. M. Bayley, *FEBS Letters*, **16**, 278 (1971).

1229. Urry, D. W., M. C. Goodall, J. D. Glickson, and D. F. Mayers, *Proc. Natl. Acad. Sci.*, **68**, 1907 (1971).

1230. Roth, S., and P. Seeman, *Nature New Biol.*, **231**, 284 (1971). Seeman, P., and S. Roth, *Biochim. Biophys. Acta*, **255**, 171 (1972). H. Machleidt, S. Roth, P. Seeman, *ibid.*, 178. S. Roth and P. Seeman, *ibid.*, 190. S. Roth, P. Seeman, S. B. A. Akerman, M. Chau-Wong, *ibid.*, 199. S. Roth and P. Seeman, *ibid.*, 207.

1231. Inoue, K., T. Kataoka, and S. C. Kinsky, *Biochem.*, **10**, 2574 (1971).

1232. Jain, M. K., *Current Topics in Membrane and Transport*, **4**, (1973) in press.

1233. Lin, E. C. C., *Ann. Rev. Genetics*, **4**, 225 (1970).

1234. Barash, H., and Y. S. Halpern, *Biochem. Biophys. Res. Comm.*, **45**, 681 (1971).

1235. Kuzuya, H., K. Bromwell, and G. Guroff, *J. Biol. Chem.*, **246**, 6371 (1971).

1236. Miller, D. M., *Biophys. J.*, **11**, 915 (1971).

1237. Thompson, J., and R. A. McLeod, *J. Biol. Chem.*, **246**, 4066 (1971).

1238. Thomas, E. L., T.-C. Shao, and H. N. Christensen, *J. Biol. Chem.*, **246**, 1677 (1971). Koser, B. H., and H. N. Christensen, *Biochim. Biophys. Acta*, **241**, 9 (1971).

1239. Anderson, B., N. Weigel, W. Kundig, and S. Roseman, *J. Biol. Chem.*, **246**, 7023 (1971).

1240. Barnes, E. M., Jr., and H. R. Kaback, *J. Biol. Chem.*, **246**, 5518 (1971); Kaback, H. R., and E. M. Barnes, Jr., **246**, 5523 (1971); Konings, W. N., E. M. Barnes, Jr., and H. R. Kaback, *J. Biol. Chem.*, **246**, 5857 (1971). Kerwar, G. K., A. S. Gordon, H. R. Kaback, *J. Biol. Chem.*, **247**, 291 (1972). S. A. Short, D. C. White, H. R. Kaback, *J. Biol. Chem.*, **247**, 298 (1972).

1241. Gunn, R. B., and D. C. Tosteson, *J. Gen. Physiol.*, **57**, 593 (1971).

1242. MacRobbie, E. A. C., *Quart. Rev. Biophys.*, **3**, 251 (1970).

1243. Ritchie, J. M., *Current Topics in Bioenergetics*, **4**, 327 (1971).

1244. Koketsu, K., *Advances in Biophys.*, **2**, 77 (1971).

1245. Herrera, F., *Membranes and Transport*, Ed. E. E. Bittar, Vol. 3 (1971), p. 1.

1246. Saddler, H. D. W., *J. Membrane Biol.*, **5**, 250 (1971).

1247. Yonath, J., and M. M. Civan, *J. Membrane Biol.*, **5**, 366 (1971).

1248. Skou, J. C., *Current Topics in Bioenergetics*, **4**, 357 (1971).

1249. Stockdale, M., and M. J. Selwyn, *Eur. J. Biochem.*, **21**, 565 (1971).

1250. Henderson, P. J., *Ann. Rev. Microbiol.*, **25**, 393 (1971).

1251. Bennun, A., *Nature New Biol.*, **233**, 5 (1971).

1252. Gibbs, M., (Ed), *Structure and Function of Chloroplasts*, Springer-Verlag, Berlin (1971).

1253. Wolken, J. J., *Invertebrate Photoreceptors: Comparative Analysis*, Academic Press, New York (1971).

1254. Jain, M. K., F. P. White, A. Strickholm, E. Williams, and E. H. Cordes, *J. Membrane Biol.*, **9**, in press (1972).

1255. Aidley, D. J., *The Physiology of Excitable Cells*, Cambridge University Press, London (1971).

1256. Fishman, S. N., B. I. Chodorov, and M. V. Volkenstein, *Biochim. Biophys. Acta*, **225**, 1 (1971).

1257. Hille, B., *J. Gen. Physiol.*, **58**, 599 (1971).

1258. Guttman, R., and L. Hachmeister, *J. Gen. Physiol.*, **58**, 304 (1971).

1259. McLaughlin, S. G. A., G. Szabo, and L. Eisenman, *J. Gen. Physiol.*, **58**, 667 (1971).

1260. Albuquerque, E. X., M. Sasa, B. P. Avner, and J. W. Daly, *Nature New Biol.*, **234**, 93 (1971). B. Witkop, *Experientia*, **27**, 1123 (1971).

1261. Armstrong, C. M., *J. Gen. Biol.*, **58**, 413 (1971).

1262. Scuka, M., *Life Sci.*, **10**, 355 (1971).

1263. Hess, B., and A. Boiteux, *Ann. Rev. Biochem.*, **40**, 237 (1971).

1264. Adam, G., *J. Membrane Biol.*, **3**, 291 (1970).

1265. Schmidt, P. P., *Biochim. Biophys. Acta.*, **233**, 765 (1971).

1266. Lecar, H., and R. Nossal, *Biophys. J.*, **11**, 1048, 1068 (1971).

1267. Hill, T. L., and Yi-der Chen, *Proc. Natl. Acad. Sci.*, **68**, 1711, 2488 (1971).

1268. Ehrenstein, G., and H. M. Fishman, *Nature New Biol.*, **233**, 16 (1971).

1269. Vasilescu, V., and D. Margineanu, *Rev. Roum. Physiol.*, **8**, 217 (1971). Walter, W. V., and R. G. Hays, *Biochim. Biophys. Acta*, **249**, 528 (1971).

1270. Vandenheuvel, F. A., *Adv. in Lipid Res.*, **9**, 161 (1971).

1271. Esfahani, M., A. R. Limbrick, S. Knutton, T. Oka, and S. J. Wakil, *Proc. Natl. Acad. Sci.*, **68**, 3180 (1971).

1272. Kornberg, R. D., and M. H. McConnell, *Proc. Natl. Acad. Sci.*, **68**, 2564 (1971).

1273. Lecar, H., G. Ehrenstein, and I. Stillman, *Biophys. J.*, **11**, 140 (1971). Dea, P., S. I. Chan, F. J. Dea, *Science*, **175**, 206 (1972).

1274. Schraier-Muccillo, S., H. Dugas, H. Schneider, and I. P. S. Smith, unpublished data. Also see D. A. Cadenhead and S. S. Katti, *Biochim. Biophys. Acta*, **241**, 709 (1971). I. P. S. Smith, *Chimia*, **25**, 349 (1971). E. Oldfield, K. M. Keough, D. Chapman, *FEBS Letters*, **20**, 344 (1972).

1275. Lussan, C., and J. F. Faucon, *FEBS Letters*, **19**, 186 (1971). Radda, G. K., *Biochem. J.*, **122**, 385 (1971). Fortes, P. A. G., and J. F. Hoffman, *J. Membrane Biol.*, **5**, 154 (1971).

1276. Metcalfe, J. C., M. J. M. Birdsall, J. Feeney, A. G. Lee, Y. K. Levine, and P. Partington, *Nature*, **233**, 199 (1971).

1277. Bretscher, M. S., *J. Mol. Biol.*, **58**, 775 (1971). Bothorel, P., E. Dulos, and J. P. Desmazes, *Comptes Rend. Acad. Sci.*, *Ser D*, **272**, 1683 (1971).

1278. Raison, J. K., and J. M. Lyons, *Proc. Natl. Acad. Sci.*, **68**, 2092 (1971). Karlsson, K.-A., B. E. Samuelsson, and G. O. Steen, *J. Membrane Biol.*, **5**, 169 (1971). Fulco, A. J., *J. Biol. Chem.*, **245**, 2985 (1970). Raison, J. K., J. M. Lyons, R. J. Mehlhorn, and A. D. Keith, *J. Biol. Chem.*, **246**, 4036 (1971).

1279. Pascaud, M., *Cah. Nutrit. Diet.*, **5**, 71 (1970). Overath, P., F. F. Hill, and I. Lamnek-Hirsch, *Nature New Biol.*, **234**, 264 (1971). Getz, G. S., *Advances in Lipid Res.*, **8**, 175 (1970).

1280. Auborn, J. J., E. M. Eyring, and G. L. Choules, *Proc. Natl. Acad. Sci.*, **68**, 1996 (1971).

1281. Marchesi, V. T., and E. P. Andrews, *Science*, **174**, 1247 (1971). Brunette, D. M., and J. E. Till, *J. Membrane Biol.*, **5**, 215 (1971). Beug, H., G. Gerisch, and E. Muller, *Science*, **173**, 742 (1971). Also see Ellar, D. J., E. Munoz, and M. R. J. Salton, *Biochim. Biophys. Acta*, **225**, 140 (1971).

1282. Giannoni, G., F. J. Padden, and R. J. Roe, *Biophys. J.*, **11**, 1018 (1971). Dudek, G., *Postepy Biochemii*, **17**, 333 (1971). Fidge, N. H., *Autr. J. Exptl. Biol. Med. Sci.*, **49**, 153 (1971). Porcellati, G., and F. Di Jeso, *Membrane-Bound Enzymes*, Plenum Press, New York (1971).

1283. Vessey, D. A., and D. Zakin, *J. Biol. Chem.*, **246**, 4649 (1971).

1284. Bruice, T. C., A. Brown, and D. O. Harris, *Proc. Natl. Acad. Sci.*, **68**, 658 (1971).

1285. Goth, A., H. R. Adams, and M. Knochuizen, *Science*, **173**, 1034 (1971).

1286. Coon, M. J., A. P. Autor, and H. W. Strohel, *Chemico-Biological Interactions*, **3**, 248 (1971).

1287. Harrison, S. C., D. L. D. Casper, R. D. Camerini-Otero, and R. M. Franklin, *Nature New Biol.*, **229**, 197 (1971).

1288. Romeo, D., A. Girard, and L. Rothfield, *J. Mol. Biol.*, **53**, 475 (1970); Romeo, D., A. Hinkley, and L. Rothfield, *J. Mol. Biol.*, **53**, 491 (1970).

1289. Singer, S. J., and G. L. Nicolson, *Amer. J. Pathol.*, **65**, 427 (1971); *Science*, **175**, 720 (1972).

1290. Kobamoto, N., and H. Ti Tien, *Biochim. Biophys. Acta,* **241,** 129 (1971).
1291. Demel, R. A., K. R. Bruckdorfer, L. L. M. Van Deenen, *Biochim. Biophys. Acta,* **255,** 321 (1972). B. De Kruyff, R. A. Demel, L. L. M. Van Deenen, *Biochim. Biophys. Acta,* **255,** 331 (1972).
1292. Fromter, E., J. Diamond, *Nature New Biol.,* **235,** 9 (1972).
1293. Ohnishi, M., M. -C. Fedarko, J. D. Baldesch-Wieler, L. F. Johnson, *Biochem. Biophys. Res. Comm.,* **46,** 312 (1972). Bystrov, V. F., V. T. Ivanov, S. A. Kozmin, I. I. Mikhaleva, K. Kh. Kalilulina, Yu. A. Ovchin-nikov, E. I. Fedin, P. V. Petrovskii, *FEBS Letters,* **21,** 34 (1972).
1294. Krasne, S., G. Eisenman, G. Szabo, *Science,* **174,** 412 (1971). Also M. K. Jain, unpublished data.
1295. Tosteson, D. C. and coworkers, unpublished data (1971).
1296. Patkau, A., and W. S. Chalack, *Canad. J. Biochem.,* in press (1972).
1297a. Goodall, M. C., G. Sachs, *Nature New Biol.,* in press (1972). M. K. Jain, unpublished data.
1297b. Jain, M. K., and E. H. Cordes, in preparation (1972).
1298. Boos, W., A. S. Gordon, R. E. Hall, H. D. Price, *J. Biol. Chem.,* **247,** 917 (1972).
1299. Ochoa, E., S. Fiszer de Plazas, E. De Robertis, *J. Pharm. Pharmacol.,* **24,** 75 (1972).
1300. Lieb, W. R., W. D. Stein, *Nature New Biol.,* **230,** 108 (1971). S. J. D. Karlish, W. R. Lieb, D. Ram, W. D. Stein, *Biochim. Biophys. Acta,* **255,** 126 (1972).
1301. Kagawa, Y., E. Racker, *J. Biol. Chem.,* **246,** 5477 (1971).
1302. Racker, E., A. Kandrach, *J. Biol. Chem.,* **246,** 7069 (1971). Hinckle, P. C., J. J. Kim, E. Racker, *J. Biol. Chem.,* **247,** 1338 (1972).
1303. Heitzman, H., *Nature New Biol.,* **235,** 114 (1972).
1304. Brown, P. K., *Nature New Biol.,* **236,** 35 (1972).
1305. Cone, R. A., *Nature New Biol.,* **236,** 39 (1972).
1306. Oesterhelt, D., W. Stoeckenius, *Nature New Biol.,* **233,** 149 (1971). A. E. Blaurock and W. Stoeckenius, *ibid.,* 152
1307. Adelman, W. J., (ed.). *Biophysics and Physiology of Excitable Membranes,* Van Nostrand Reinhold Co., New York (1971).
1308. Sherebrin, M. H., *Nature New Biol.,* **235,** 122 (1972).
1309. Kung, C., and R. Ecker, *Proc. Natl. Acad. Sci.* (U.S.), **69,** 93 (1972).
1310. Denburg, J. L., M. E. Eldefrawi, R. D. O'Brien, *Proc. Natl. Acad. Sci.* (U.S.), **69,** 177 (1972).
1311. Chan, S. I., C. H. A. Seiter, G. W. Feigenson, *Biochem. Biophys. Res. Comm.,* **46,** 1488 (1972).
1312. Livne, A., P. J. C. Kuiper, N. Mayerstein, *Biochim. Biophys. Acta,* **255,** 745 (1972).
1313. Miller, E. K., R. M. C. Dawson, *Biochem. J.,* **126,** 823 (1972).
1314. Mavis, R. D., and P. R. Vagelos, *J. Biol. Chem.,* **247,** 652 (1972).
1315. Novak, R. F., T. J. Swift, *Proc. Natl. Acad. Sci.* (U.S.), **69,** 640 (1972). I. J. Tinsley, R. Haque, D. Schmedding, *Science,* **174,** 145 (1971).
1316. Green, J. R., P. A. Edwards, C. Green, *Biochem. J.,* **125,** 41P (1971).

1317. Cherry, R. J., D. Chapman, and D. E. Graham, *J. Membrane Biol.*, **7**, 325 (1972).

1318. Bean, R. C., *J. Membrane Biol.*, **7**, 15 (1972).

1319. Carter, J. R., J. Avruch, and D. B. Martin, *J. Biol. Chem.*, **247**, 2682 (1972).

1320. Hafeman, D. R., *Biochim. Biophys. Acta*, **266**, 548 (1972).

1321. Razin, S., *Biochim. Biophys. Acta*, **265**, 241 (1972).

1322. Drost-Hansen, W., in *The Chemistry of Cell Interface*, Part B., H. D. Brown, ed., p. 2, Academic Press, 1971.

Index